Protein–Ligand Interactions

METHODS IN MOLECULAR BIOLOGY™

John M. Walker, SERIES EDITOR

316. **Bioinformatics and Drug Discovery**, edited by *Richard S. Larson, 2005*
315. **Mast Cells:** *Methods and Protocols*, edited by *Guha Krishnaswamy and David S. Chi, 2005*
314. **DNA Repair Protocols:** *Mammalian Systems, Second Edition*, edited by *Daryl S. Henderson, 2005*
313. **Yeast Protocols:** *Second Edition*, edited by *Wei Xiao, 2005*
312. **Calcium Signaling Protocols:** *Second Edition*, edited by *David G. Lambert, 2005*
311. **Pharmacogenomics:** *Methods and Applications*, edited by *Federico Innocenti, 2005*
310. **Chemical Genomics:** *Reviews and Protocols*, edited by *Edward D. Zanders, 2005*
309. **RNA Silencing:** *Methods and Protocols*, edited by *Gordon Carmichael, 2005*
308. **Therapeutic Proteins:** *Methods and Protocols*, edited by *C. Mark Smales and David C. James, 2005*
307. **Phosphodiesterase Methods and Protocols**, edited by *Claire Lugnier, 2005*
306. **Receptor Binding Techniques:** *Second Edition*, edited by *Anthony P. Davenport, 2005*
305. **Protein–Ligand Interactions:** *Methods and Applications*, edited by *G. Ulrich Nienhaus, 2005*
304. **Human Retrovirus Protocols:** *Virology and Molecular Biology*, edited by *Tuofu Zhu, 2005*
303. **NanoBiotechnology Protocols**, edited by *Sandra J. Rosenthal and David W. Wright, 2005*
302. **Handbook of ELISPOT:** *Methods and Protocols*, edited by *Alexander E. Kalyuzhny, 2005*
301. **Ubiquitin–Proteasome Protocols**, edited by *Cam Patterson and Douglas M. Cyr, 2005*
300. **Protein Nanotechnology:** *Protocols, Instrumentation, and Applications*, edited by *Tuan Vo-Dinh, 2005*
299. **Amyloid Proteins:** *Methods and Protocols*, edited by *Einar M. Sigurdsson, 2005*
298. **Peptide Synthesis and Application**, edited by *John Howl, 2005*
297. **Forensic DNA Typing Protocols**, edited by *Angel Carracedo, 2005*
296. **Cell Cycle Control:** *Mechanisms and Protocols*, edited by *Tim Humphrey and Gavin Brooks, 2005*
295. **Immunochemical Protocols**, *Third Edition*, edited by *Robert Burns, 2005*
294. **Cell Migration:** *Developmental Methods and Protocols*, edited by *Jun-Lin Guan, 2005*
293. **Laser Capture Microdissection:** *Methods and Protocols*, edited by *Graeme I. Murray and Stephanie Curran, 2005*
292. **DNA Viruses:** *Methods and Protocols*, edited by *Paul M. Lieberman, 2005*
291. **Molecular Toxicology Protocols**, edited by *Phouthone Keohavong and Stephen G. Grant, 2005*
290. **Basic Cell Culture Protocols**, *Third Edition*, edited by *Cheryl D. Helgason and Cindy L. Miller, 2005*
289. **Epidermal Cells**, *Methods and Applications*, edited by *Kursad Turksen, 2005*
288. **Oligonucleotide Synthesis**, *Methods and Applications*, edited by *Piet Herdewijn, 2005*
287. **Epigenetics Protocols**, edited by *Trygve O. Tollefsbol, 2004*
286. **Transgenic Plants:** *Methods and Protocols*, edited by *Leandro Peña, 2005*
285. **Cell Cycle Control and Dysregulation Protocols:** *Cyclins, Cyclin-Dependent Kinases, and Other Factors*, edited by *Antonio Giordano and Gaetano Romano, 2004*
284. **Signal Transduction Protocols,** *Second Edition*, edited by *Robert C. Dickson and Michael D. Mendenhall, 2004*
283. **Bioconjugation Protocols,** edited by *Christof M. Niemeyer, 2004*
282. **Apoptosis Methods and Protocols,** edited by *Hugh J. M. Brady, 2004*
281. **Checkpoint Controls and Cancer, Volume 2:** *Activation and Regulation Protocols*, edited by *Axel H. Schönthal, 2004*
280. **Checkpoint Controls and Cancer, Volume 1:** *Reviews and Model Systems*, edited by *Axel H. Schönthal, 2004*
279. **Nitric Oxide Protocols,** *Second Edition*, edited by *Aviv Hassid, 2004*
278. **Protein NMR Techniques,** *Second Edition*, edited by *A. Kristina Downing, 2004*
277. **Trinucleotide Repeat Protocols,** edited by *Yoshinori Kohwi, 2004*
276. **Capillary Electrophoresis of Proteins and Peptides,** edited by *Mark A. Strege and Avinash L. Lagu, 2004*
275. **Chemoinformatics,** edited by *Jürgen Bajorath, 2004*
274. **Photosynthesis Research Protocols,** edited by *Robert Carpentier, 2004*
273. **Platelets and Megakaryocytes, Volume 2:** *Perspectives and Techniques*, edited by *Jonathan M. Gibbins and Martyn P. Mahaut-Smith, 2004*
272. **Platelets and Megakaryocytes, Volume 1:** *Functional Assays*, edited by *Jonathan M. Gibbins and Martyn P. Mahaut-Smith, 2004*
271. **B Cell Protocols,** edited by *Hua Gu and Klaus Rajewsky, 2004*
270. **Parasite Genomics Protocols,** edited by *Sara E. Melville, 2004*
269. **Vaccina Virus and Poxvirology:** *Methods and Protocols*,edited by *Stuart N. Isaacs, 2004*
268. **Public Health Microbiology:** *Methods and Protocols*, edited by *John F. T. Spencer and Alicia L. Ragout de Spencer, 2004*

METHODS IN MOLECULAR BIOLOGY™

Protein–Ligand Interactions

Methods and Applications

Edited by

G. Ulrich Nienhaus

Department of Biophysics, University of Ulm
Ulm, Germany

HUMANA PRESS TOTOWA, NEW JERSEY

Softcover reprint of the hardcover 1st edition 2005
999 Riverview Drive, Suite 208
Totowa, New Jersey 07512

www.humanapress.com

This publication is printed on acid-free paper. ∞
ANSI Z39.48-1984 (American Standards Institute) Permanence of Paper for Prined Library Materials.

Production Editor: Nicole E. Furia

Cover design by Patricia F. Cleary

Cover Illustration: (Foreground) Figure 1 from Chapter 18, "High-Throughput Screening of Interactions Between G Protein-Coupled Receptors and Ligands Using Confocal Optics Microscopy," by Lenka Zemanová, Andreas Schenk, Martin J. Valler, G. Ulrich Nienhaus, and Ralf Heilker; (Background) Figure 3B from Chapter 9, "Combined Use of XAFS and Crystallography for Studying Protein–Ligand Interactions in Metalloproteins," by Richard W. Strange and S. Samar Hasnain.

For additional copies, pricing for bulk purchases, and/or information about other Humana titles, contact Humana at the above address or at any of the following numbers: Tel.: 973-256-1699; Fax: 973-256-8341; E-mail: orders@humanapr.com; or visit our Website: www.humanapress.com

10 9 8 7 6 5 4 3 2 1

ISBN 1-59259-912-5 (eBook)

Library of Congress Cataloging in Publication Data
Protein-ligand interactions : methods and applications / edited by G. Ulrich Nienhaus.
p. ; cm. -- (Methods in molecular biology, ISSN 1064-3745 ; 305)
Includes bibliographical references and index.
ISBN 978-1-61737-525-5 ISBN 978-1-59259-912-7 (eBook)
1. Protein binding--Laboratory manuals. I. Nienhaus, G. Ulrich (Gerd Ulrich) II. Methods in molecular biology (Clifton, N.J.) ; 305.
[DNLM: 1. Ligands--Laboratory Manuals. 2. Proteins--chemistry--Laboratory Manuals. 3. Chemistry, Analytical--methods--Laboratory Manuals. 4. Protein Binding--Laboratory Manuals. QU 25 P96665 2005]
QP551.P69599 2004
572'.6--dc22

2004042387

Preface

The genomes of several organisms have been sequenced in recent years, and the efficient exploration of interactions among tens of thousands of gene products has moved to center stage in our quest for a detailed understanding of life at the molecular level. Molecular recognition and binding of ligands (atoms, ions, molecules) by proteins with high sensitivity and selectivity is of central importance to essentially all biomolecular processes. Therefore, a thorough understanding of protein–ligand interactions is of key importance for the basic and applied sciences. Techniques to study protein–ligand interactions have been established and refined for many years. They continue to be improved by the development of new reagents, protocols, and instrumentation. A variety of powerful experimental and theoretical tools have become available in recent years, and novel techniques are continually being introduced to meet new demands.

Protein–Ligand Interactions: Methods and Applications features a collection of methods for studying the interaction between proteins and ligands, including biochemical/bulk techniques, structure analysis, spectroscopy, single-molecule studies, and theoretical/computational tools. The presen volume aims to provide the researcher with technical background information that will enable him or her to develop strategies for characterizing protein–ligand interactions in the most effective way. Life scientists in both academia and industry will find hands-on information regarding both established and novel approaches for the study of protein–ligand interactions. We have attempted to present a broad selection of widely applicable techniques. We hope that *Protein–Ligand Interactions: Methods and Applications* will provide a good starting point from which to embark on other, more specialized techniques.

I wish to thank all contributing authors for their hard work and considerable patience. I greatly appreciate the high quality of their presentations that made compiling this volume a particularly pleasurable experience.

Gerd Ulrich Nienhaus

Contents

Contributors

PHILIP A. ANFINRUD • *Laboratory of Chemical Physics, National Institute of Diabetes and Digestive and Kidney Diseases, National Institutes of Health, Bethesda, MD*

ABEL BAERGA-ORTIZ • *Department of Chemistry and Biochemistry, University of California San Diego, La Jolla, CA*

UTE CURTH • *Department of Biophysical Chemistry, Medizinische Hochschule, Hannover, Germany*

ROBERT H. EIBL • *Department of Physiology and Biophysics, University of Miami School of Medicine, Miami, FL*

ARNOLD M. FALICK • *Department of Chemistry and Biochemistry, University of California San Diego, La Jolla, CA*

JOEL M. FRIEDMAN • *Department of Physiology and Biophysics, Albert Einstein College of Medicine, Bronx, NY*

KLAUS GERWERT • *Department of Biophysics, Ruhr-University Bochum, Bochum, Germany*

FRANK GESELLCHEN • *Department of Biochemistry, University of Kassel, Kassel, Germany*

LUIS GRACIA • *Department of Physiology and Biophysics, Weill Medical College, Cornell University, New York, NY*

HELMUT GRUBMÜLLER • *Theoretical and Computational Biophysics Department, Max-Planck-Institute for Biophysical Chemistry, Göttingen, Germany*

MARK S. HARGROVE • *Department of Biochemistry, Biophysics and Molecular Biology, Iowa State University, Ames, IA*

S. SAMAR HASNAIN • *Molecular Biophysics Group, College of Biology and Medicine, CCLRC Daresbury Laboratory, Daresbury, Warrington, Cheshire, UK*

SERGIO A. HASSAN • *Center for Molecular Modeling, Division of Computational Bioscience, Center for Information Technology, National Institutes of Health, Bethesda, MD*

THEODORE L. HAZLETT • *Laboratory for Fluorescence Dynamics, Department of Physics, University of Illinois at Urbana-Champaign, Urbana, IL*

RALF HEILKER • *Department of Integrated Lead Discovery, Boehringer Ingelheim Pharma GmbH & Co KG, Biberach, Germany*

FRIEDRICH W. HERBERG • *Department of Biochemistry, University of Kassel, Kassel, Germany*

HYOTCHERL IHEE • *Department of Chemistry and School of Molecular Science, KAIST, Daejeon, South Korea*

DAVID M. JAMESON • *Department of Cell and Molecular Biology, John A. Burns School of Medicine, University of Hawaii, Honolulu, HI*

Andreas Janshoff • *Department of Physical Chemistry, Johannes-Gutenberg-University, Mainz, Germany*
Herbert P. Jennissen • *Department of Physiological Chemistry, University of Duisburg-Essen, Germany*
Elizabeth A. Komives • *Department of Chemistry and Biochemistry, University of California San Diego, La Jolla, CA*
Carsten Kötting • *Department of Biophysics, Ruhr-University Bochum, Bochum, Germany*
Edwin A. Lewis • *Department of Chemistry and Biochemistry, Northern Arizona University, Flagstaff, AZ*
Manho Lim • *Department of Chemistry, Pusan National University, Busan, South Korea*
H. Peter Lu • *William R. Wiley Environmental Molecular Sciences Laboratory, Fundamental Science Division, Pacific Northwest National Laboratory, Richland, WA*
Jeffrey G. Mandell • *Department of Chemistry and Biochemistry, University of California San Diego, La Jolla, CA*
Rachel Marrington • *Department of Chemistry, University of Warwick, Coventry, UK*
Till Maurer • *Department of Lead Discovery, Boehringer Ingelheim Pharma GmbH & Co KG, Biberach, Germany*
Gabor Mocz • *Biotechnology Program, Pacific Biomedical Research Center, University of Hawaii, Honolulu, HI*
Vincent T. Moy • *Department of Physiology and Biophysics, University of Miami School of Medicine, Miami, FL*
Kenneth P. Murphy • *Department of Biochemistry, University of Iowa, Iowa City, IA*
G. Ulrich Nienhaus • *Department of Biophysics, University of Ulm, Ulm, Germany*
Karin Nienhaus • *Department of Biophysics, University of Ulm, Ulm, Germany*
Reinhard Pahl • *Consortium for Advanced Radiation Sources, The University of Chicago, Chicago, IL*
Alison Rodger • *Department of Chemistry, University of Warwick, Coventry, UK*
David Roper • *Department of Chemistry, University of Warwick, Coventry, UK*
Carme Rovira • *Centre de Recerca en Química Teòrica, Parc Científic de Barcelona, Barcelona, Spain*
Qiaoqiao Ruan • *Core R&D Biotechnology, Abbott Diagnostic Division, Abbott Laboratories, Abbott Park, IL*
Uri Samuni • *Department of Physiology and Biophysics, Albert Einstein College of Medicine, Bronx, NY*
Andreas Schenk • *Tecan Austria GmbH, Grödig, Austria*

ILME SCHLICHTING • *Department of Biomolecular Mechanisms, Max Planck Institute for Medical Research, Heidelberg, Germany*
MARIUS SCHMIDT • *Department of Physics, Technical University of Munich, Garching, Germany*
VUKICA ŠRAJER • *Consortium for Advanced Radiation Sources and Department of Biochemistry and Molecular Biology, The University of Chicago, Chicago, IL*
PETER J. STEINBACH • *Center for Molecular Modeling, Division of Computational Bioscience, Center for Information Technology, National Institutes of Health, Bethesda, MD*
CLAUDIA STEINEM • *Institute of Analytical Chemistry, Chemo- & Biosensors, University of Regensburg, Regensburg, Germany*
RICHARD W. STRANGE • *Molecular Biophysics Group, College of Biology and Medicine, CCLRC Daresbury Laboratory, Daresbury, Warrington, Cheshire, UK*
SERGEY Y. TETIN • *Core R&D Biotechnology, Abbott Diagnostic Division, Abbott Laboratories, Abbott Park, IL*
CLAUS URBANKE • *Department of Biophysical Chemistry, Medizinische Hochschule, Hannover, Germany*
MARTIN J. VALLER • *Department of Integrated Lead Discovery, Boehringer Ingelheim Pharma GmbH & Co KG, Biberach, Germany*
GEETHA VASUDEVAN • *Scientific Computing, Medarex Inc., Sunnyvale, CA*
STUART WINDSOR • *Biotechnology Group, National Physical Laboratory, Teddington, Middlesex, UK*
GREGOR WITTE • *Department of Biophysical Chemistry, Medizinische Hochschule, Hannover, Germany*
LENKA ZEMANOVÁ • *Department of Biophysics, University of Ulm, Ulm, Germany*
BASTIAN ZIMMERMANN • *Biaffin GmbH & Co KG, Kassel, Germany*

1

Isothermal Titration Calorimetry

Edwin A. Lewis and Kenneth P. Murphy

Summary

Isothermal titration calorimetry is an ideal technique for measuring biological binding interactions. It does not rely on the presence of chromophores or fluorophores, nor does it require an enzymatic assay. Because the technique relies only on the detection of a heat effect upon binding, it can be used to measure the binding constant, *K*, the enthalpy of binding, ΔH° and the stoichiometry, or number of binding sites, n. This chapter describes instrumentation, experimental design, and the theoretical underpinnings necessary to run and analyze a calorimetric binding experiment.

Key Words: Binding; thermodynamics; proton linkage; enthalpy; heat capacity; data analysis.

1. Introduction

Titration calorimetry was first described as a method for the simultaneous determination of *K* and ΔH about 40 yr ago by Christensen and Izatt ***(1,2)***. The method was originally applied to a variety of weak acid-base equilibria and to metal ion complexation reactions ***(3–5)***. These systems could be studied with the calorimetric instrumentation available at the time that was limited to the determination of *K* values less than about 10^4 to 10^5 M^{-1} ***(6)***. The determination of larger association constants requires more dilute solutions and the calorimeters of that day were simply not sensitive enough.

Beaudette and Langerman published one of the first calorimetric binding studies of a biological system using a small volume TRONAC titration calorimeter ***(7)***. Their data for the titration of an enzyme, bovine liver glutamate dehydrogenase (GDH), with an inhibitor, adenosine diphosphate (ADP), are shown in **Fig. 1.** In 1979, Langerman and Biltonen published a description of microcalorimeters for biological chemistry, including a discussion of available

From: *Methods in Molecular Biology, vol. 305: Protein–Ligand Interactions: Methods and Applications*
Edited by: G. U. Nienhaus © Humana Press Inc., Totowa, NJ

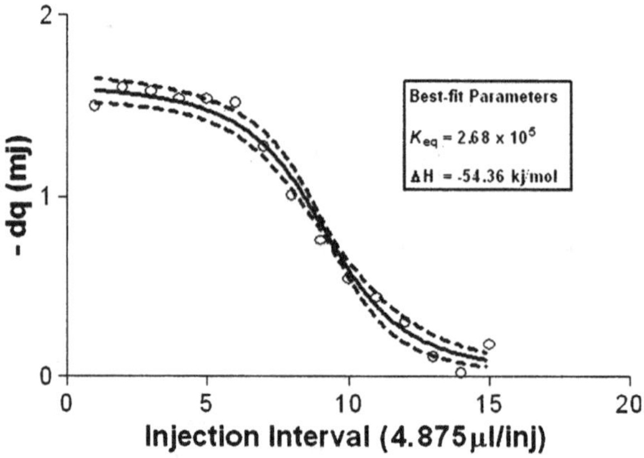

Fig. 1. Titration of 2.00 mL of 0.1340 m*M* GDH with 6.17 m*M* ADP at pH 7.6 and 25ºC. (Data taken from **ref. *8***.)

instrumentation, applications, experimental design, and data analysis and interpretation ***(8,9)***. This was really the beginning of the use of titration calorimetry to study biological equilibria. It took another 10 yr before the first commercially available titration calorimeter specifically designed for the study of biological systems became available from MicroCal ***(10)***.

Isothermal titration calorimetry (ITC) is now routinely used to directly characterize the thermodynamics of biopolymer binding interactions ***(11–13)***. This is largely a result of improvements in the ITC instrumentation and data analysis software. Modern instruments, like the MicroCal and Calorimetry Sciences Corporation ITCs, make it possible to measure heat effects as small as 0.4 µJ (0.1 µcal) allowing the determination of binding constants, *K*'s, as large as 10^8 to 10^9 M^{-1}.

In order to take full advantage of the powerful ITC technique, the user must be able to design the optimum experiment, understand the nonlinear fitting process, and appreciate the uncertainties in the fitting parameters *K*, ΔH, and n. ITC experiment design and data analysis have been the subject of numerous papers ***(14–17)***. This chapter reviews the planning of optimal ITC experiments, guides the reader through a sample experiment, the titration of RNase A with 2'-cytidine monophosphate (2'-CMP), and reviews theory underlying the nonlinear fitting of ITC data and the interpretation of ITC results.

2. Instrument Description

A schematic diagram of an isothermal titration calorimeter is shown in **Fig. 2.** The essential components of the ITC instrument are: (a) a matched pair of sample and reference cells contained within a thermostatted environment,

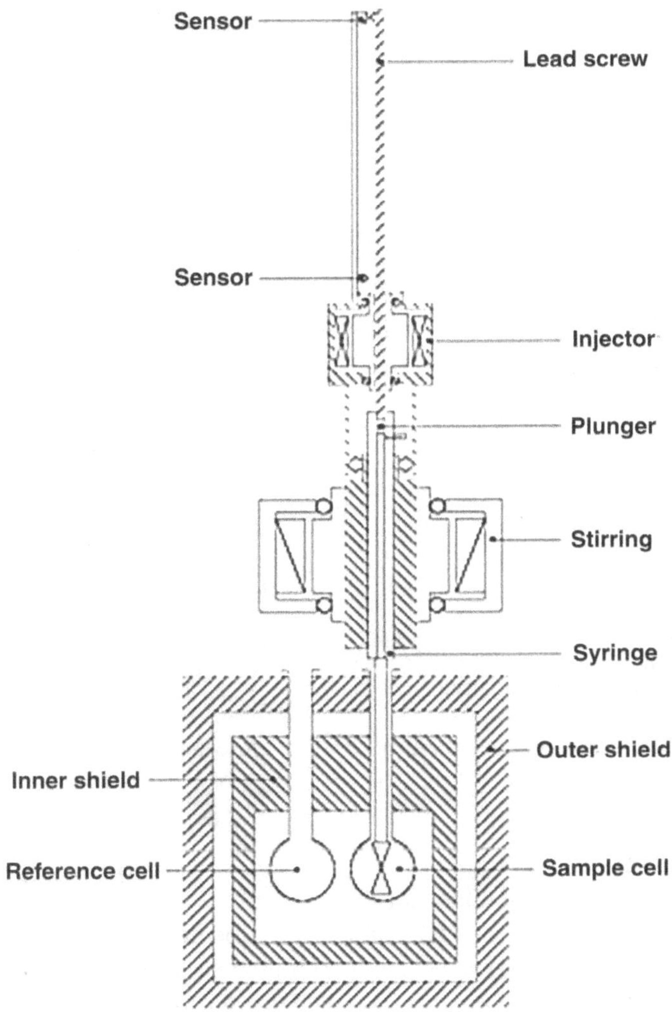

Fig. 2. Diagram of the MicroCal VP-ITC measuring unit (taken from the MicroCal website, http://www.microcalorimetry.com/).

(b) a stepper-motor-driven syringe for injecting titrant (ligand solution) into the sample cell, (c) a stirrer for keeping the contents of the sample cell homogeneous, and (d) a means for compensating (and measuring) the heat flow to the sample cell so that it is maintained at the same temperature as the reference cell. In modern ITC instruments, the cell volumes are nominally 1.5 mL, the temperature of the thermostat can be set from about 5 to 80ºC, the injected volume can range from about 1 to 20 μL, and heats as small as 0.4 μJ (0.1 μcal) can be measured.

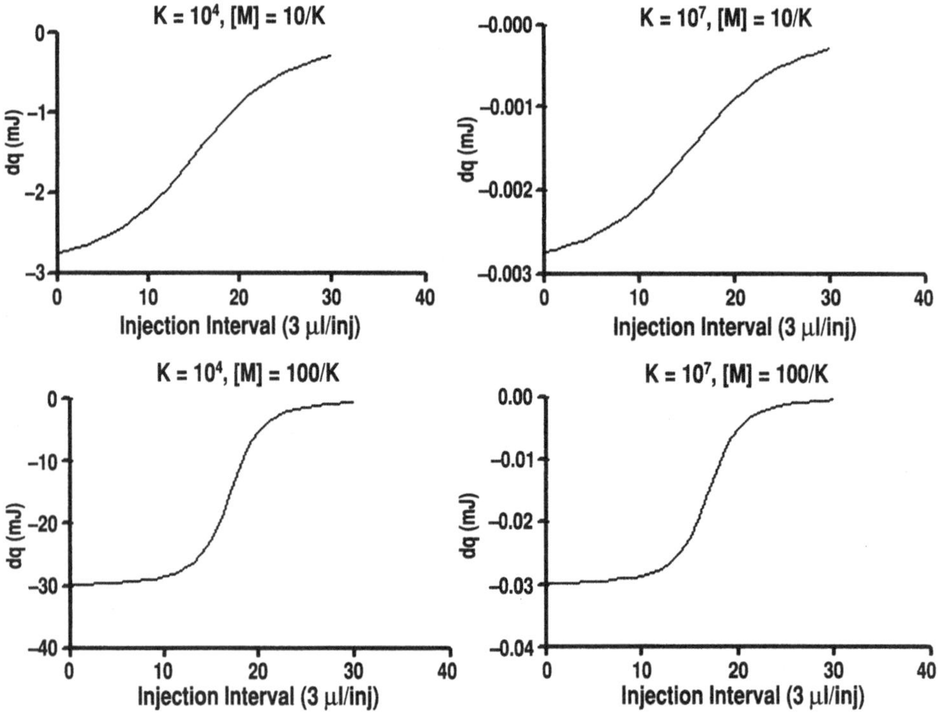

Fig. 3. Relationship between thermogram curvature and the c parameter, ([M] × K), for two different values of c, 10 and 100.

The ITC signal is dependent on the concentrations of the macromolecule, [M], and the ligand, [L]; the cell volume; the injected volume; and the values of *K*, ΔH, and n (or a larger set of parameters for binding models more complicated than the n independent sites model). In order to obtain an estimate for *K*, the ITC experiment must yield a curved thermogram. Furthermore, of course, the ITC experiment must also be done under conditions that produce detectable amounts of heat for each titrant addition. These points are illustrated graphically in **Fig. 3.** The upper panels in **Fig. 3** show the curvature that would be observed in two experiments for systems with different binding constants, 10^4 and 10^7 M^{-1}, if the concentration of the macromolecule was chosen to be (10/*K*), i.e., [M] = 1 m*M* for data shown in the upper left panel, and [M] = 1 µ*M* for the data shown in the upper right panel. The lower two panels in **Fig. 3** show thermograms for the same two systems with the exception that the macromolecule concentration was increased to be (100/*K*). The first point that can be made from the data in **Fig. 3** is that the curvature is the same as long as the product of macromolecule concentration and *K* is held constant. It has been widely reported that the c parameter, ([M] × *K*), must be between 1 and 1000 in order to produce a thermogram with the curvature required for the simultaneous determination of *K* and ΔH ***(10)***. The authors of this chapter believe that the best experiments will be done with the c parameter having a value between 10 and 100.

At first glance, each of the simulated thermograms in **Fig. 3** would seem to be representative of an experiment in which *K* and ΔH could be accurately determined. On closer inspection, it is apparent that two of the experiments shown are less than optimal. The experiment shown in the upper right panel would yield heats that are too small to be determined accurately, even for the first injections where the largest heats would only be about 2.7 mJ (0.65 mcal), whereas the experiment shown in the lower left panel yields heats that are too large, approx 7000 mcal for the first injections, and the experiment would require excessive amounts of reagents. Clearly, simulations are important in optimizing the ITC experiment and in achieving a balance between detectable heats and curvature in the thermogram.

3. Methods

There are seven steps to running the ITC experiment. These are: 1) planning the experiment (simulations), 2) preparing the L and M solutions, 3) collecting the raw ITC data, 4) collecting the blank (L solution dilution), 5) correcting the raw ITC data, 6) nonlinear regression of the corrected titration data to provide estimates of the thermodynamic parameter values, and 7) interpretation of the model data. Each step will be discussed herein.

In our discussion of running the ITC experiment, we will use the binding of cytidine-2'-monophosphate, 2'-CMP, to bovine ribonuclease A, RNase, as a test system ***(10,18,19)***. These chemicals are available from Sigma Aldrich (St. Louis, MO) in suitable purity and have been widely used as a test system by ITC manufacturers. The approximate thermodynamic parameters for the 2'-CMP/RNase system are $K \approx 6 \times 10^4\ M^{-1}$ and $\Delta H \approx -45$ kJ/mol with a stoichiometry of 1 at 25 ºC ***(19)***. An alternative test system is the binding of Ba^{+2} ion by the cyclic poly ether, 18-crown-6 ***(19,20)***.

3.1. Planning the Experiment

The first step in running the ITC experiment is to determine the concentrations for the macromolecule and ligand solutions. If the objective of the ITC experiment is only to determine the binding enthalpy change, ΔH, then the only consideration is that the concentration of the ligand will be large enough that an accurately measurable heat effect, ≥ 40 µJ (10 µcal), will be observed and that the macromolecule concentration will be in excess. In the case of our test system, the binding of 2'-CMP to RNase, these conditions would be met

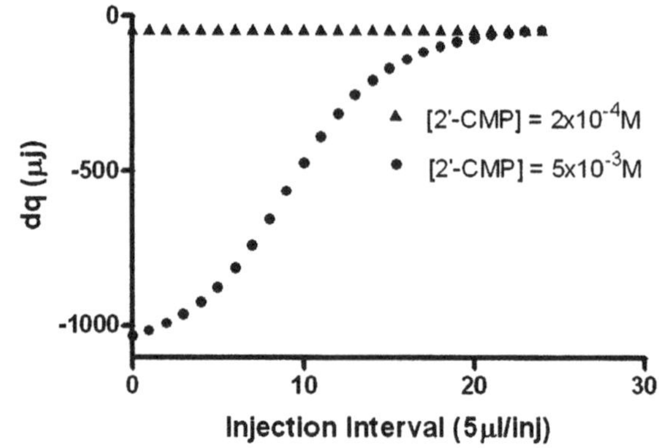

Fig. 4. Simulated experiments for the titration of a 1.7×10^{-4} *M* solution of RNase with two different titrant solutions. (2'-CMP at a concentration of either 2×10^{-4} or 5×10^{-3} *M*.)

with [2'-CMP] = 2×10^{-4} *M*, and [RNase] = 1.7×10^{-4} *M*. With an injection volume, v_{inj}, of 5 µL, the heat per injection would be given by **Eq. 1** below and there would be no curvature in the thermogram:

$$q_{inj} = \Delta H \times [L] \times v_{inj} = -45 \text{ kJ/mol} \times \left(2 \times 10^{-4}\right) \text{M} \times \left(5 \times 10^{-6}\right) 1 \approx -45 \ \mu\text{J} \left(-11 \ \mu\text{cal}\right) \quad (1)$$

If the concentration of the L (2'-CMP) is increased to 5×10^{-3} *M*, the thermogram would show curvature similar to that shown in the lower panels of **Fig. 3** (c = 10) and an endpoint would be reached after approx 20 (5 µL) injections. The integrated heat values for the first injections would now be more than –1000 µJ. Increasing the concentration of RNase to 1.7×10^{-3} *M* (c = 100) and the ligand concentration to 5×10^{-2} *M* would yield a thermogram showing the same curvature as that shown in the upper panels of **Fig. 3**. In this last case, the heat observed in the early injections would be too large, more than –10,000 µJ. **Fig. 4** shows simulated ITC data for experiments done under the first set of conditions where only ΔH would be determined and under the second set of set of conditions where both *K* and ΔH would be determined.

3.2. Solution Preparation and Handling

The final results of the ITC experiment depend on exact knowledge of the titrate and titrant solution concentrations, so it is imperative that the concentrations be determined as accurately as possible. Perhaps the ITC solutions can be made by volumetric dilution of stock solutions that were made up by weight. Whenever possible the concentrations should be verified by another analytical procedure (e.g., absorbance, kinetic activity, other analysis, etc.). As will be noted later, it is especially important that the L concentration be known precisely, as errors in this value will affect the determination of both K and ΔH.

It also is extremely important that the two solutions be matched with regard to composition, e.g., pH, buffer, salt concentration, etc. If the two solutions are not perfectly matched, there may be heat of mixing (or dilution) signals that overwhelm the heat signals for the binding reaction. It is typical that the solution of the macromolecule is dialyzed against a large volume of the buffer. The artifact heats of mixing can be minimized by using the dialysate from preparation of the macromolecule solution as the *solvent* for preparation of the ligand solution.

3.3. Correcting the Raw ITC Data

Obviously, the dialysis/dialysate approach will virtually eliminate the mixing or dilution effects for all solute species in common between the macromolecule and ligand solutions. The exception is that the heat of dilution for the ligand itself must be measured in a blank experiment. In this blank experiment, the ligand solution is titrated into buffer in the sample cell. The heat of dilution of the macromolecule should also be measured in a second blank experiment. This is done by simply injecting buffer from the syringe into the macromolecule solution in the sample cell. Usually the heat of dilution of the macromolecule measured in this way is negligible. To be completely rigorous, a third blank experiment should also be done. This buffer into buffer experiment may be thought of as an instrument blank. The equation to correct the heat data for dilution effects is:

$$Q_{\text{corr}} = Q_{\text{meas}} - Q_{\text{dil, macromolecule}} + Q_{\text{instrument blank}} \qquad (2)$$

The blank corrections are for the same injection volumes as used in the collection of the actual titration data. In the case of the 2'-CMP/RNase titration experiment shown in **Fig. 5**, the only significant correction is for the dilution of the titrant (the results of the 2'-CMP dilution blank experiment are also shown in **Fig. 5**).

Another complicating reaction encountered in many biological binding experiments results from the release (or uptake) of protons as binding occurs. The released protons are taken up by the buffer conjugate base, and there are contributions to the heat both from binding protons to buffer and from the heat of removing protons from the macromolecule ***(17)***. The treatment of this complicating reaction requires knowledge of the number of protons released (or taken up) and the heat of ionization of the buffer. The measured enthalpy is given by:

$$\Delta H_{\text{meas}} = \Delta H_0 - \Delta H_{\text{ion}} \times n_p \qquad (3)$$

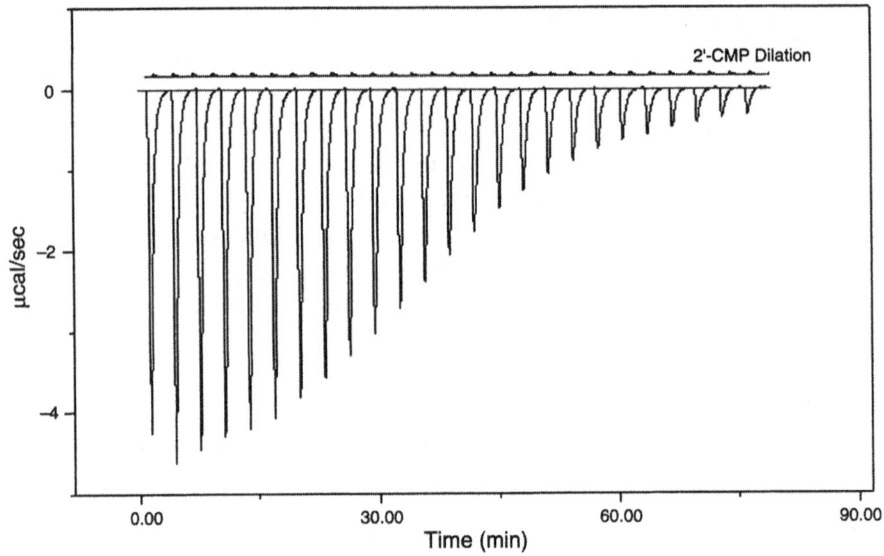

Fig. 5. The raw ITC data (power vs time) are shown for two titrations. The larger heat pulses are from the titration of 1.4 mL of a 1.55×10^{-4} *M* RNase solution with a 3.19×10^{-3} *M* solution of 2'-CMP. The smaller heat pulses are for the dilution of 5 mL of the 2'-CMP titrant into 1.4 mL of the acetate buffer. The power is given in units of mcal/s (where 1 mcal/s = 4.184 mJ/s = 4.184 mW).

where ΔH_0 is the enthalpy of binding in the absence of a heat from protons binding to the buffer, ΔH_{ion} is the heat of proton ionization for the buffer and n_p is the number of protons released on binding 1 mole of L. The value of n_p is determined from titrations done in at least two buffers with different heats of ionization. It should be emphasized that ΔH_0 includes the heat of protons being released from the protein upon binding and, as such, does not represent the *intrinsic* heat of the protein L interaction ***(17)***. Instead, it simply removes the contribution of the buffer. This phenomenon also provides an approach to manipulating the heat signal for a reaction that is accompanied by proton release. By simply using a buffer with a large heat of ionization, the heat signal can be enhanced. Alternatively, the use of a buffer with a small ΔH_{ion} (≈ 0) could be used to minimize the *artifact signal*.

Finally, because the generation of bubbles in the sample (or reference) solutions during an ITC experiment will generate spurious heat signals, the solutions should be degassed prior to filling the cell and injection syringe. The ITC manufacturers provide vacuum degassing accessories for this purpose. Precautions need to be taken to avoid boiling the solutions and changing the concentrations. Also, the ITC manufacturers supply cell loading syringes and instructions on cell filling that should be followed to avoid the problem of introducing bubbles.

3.4. Example ITC Experiment

The example ITC experiment described is for the binding of 2'-CMP to ribonuclease A (RNase), the same experiment that was simulated in **Fig. 4**. The 2'-CMP (cat. no. C-7137) and RNase (cat. no. R-5500) were purchased from SigmaAldrich and used without further purification. The RNase solution was prepared by dissolving a weighed amount of the RNase in acetate buffer and then dialyzing for 16 h at 4ºC against 4 L of acetate buffer using 3500 MWCO Spectrofluor dialysis tubing. The 2'-CMP solution was prepared by dissolving a weighed amount of the 2'-CMP in the acetate buffer dialysate. The final concentrations for both the RNase and 2'-CMP were determined spectrophotometrically as described by Wiseman et al. ***(10)***. The buffer was 0.2 *M* in sodium acetate, 0.2 *M* in sodium chloride and was adjusted to pH 5.5.

The ITC experiment was run at 25ºC and was set to deliver 25 (5 μL) injections at 300-s intervals. The raw ITC data are shown in **Fig. 5**. Data are shown for one titration experiment in which the 2'-CMP solution was added to the RNase solution in the cell and one titrant dilution experiment in which the 2'-CMP solution was added to buffer (dialysate) in the cell. The dilution of the 2'-CMP titrant is slightly endothermic and contributes less than + 0.6 μcal to the total heat observed for the addition of 5 μL of CMP in the RNase titration. The 2'-CMP titrant dilution represents less than 0.5% of the heat signal observed for the initial titrant additions. The dilution experiment in which buffer was added to the RNase solution in the cell is not shown since the heat of dilution of the RNase was even less significant than the 2'-CMP dilution under the conditions of these experiments.

The dilution-corrected and integrated heat data are shown in **Fig. 6**. The integrated heat data were fit with a one site-binding model using the Origin-7™ software provided with the MicroCal VP-ITC. The *best-fit* parameters resulting from the nonlinear regression fit of these data are also shown in **Fig. 6** along with the fitted curve. The *K* and ΔH values determined in this experiment are consistent with those reported by Horn et al. ***(21)***, but significantly different from those reported by Wiseman et al. ***(10)***. The differences are the result of the experiments being done at different temperatures, different salt concentrations, and perhaps at slightly different pH. A more detailed discussion of the nonlinear regression fitting and data interpretation follows.

3.5. Analysis of Calorimetric Data

In the basic ITC experiment, one is seeking to determine three parameters: the stoichiometry (i.e., the number of binding sites), n; the binding constant, *K*;

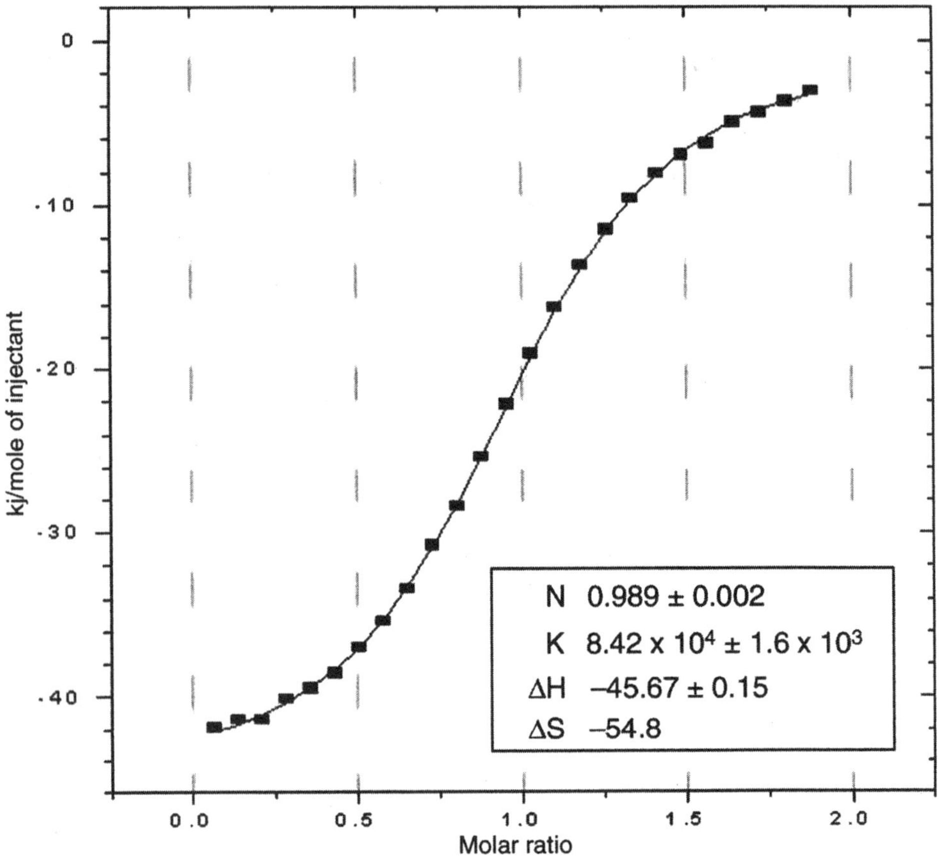

Fig. 6. Nonlinear regression fit of the RNase/2'-CMP ITC data shown in **Fig. 5** to a 1 binding site model. This analysis of the ITC experiment yields the *best-fit* values of *K*, ΔH, and the stoichiometry, n, shown in the box within the figure.

and the change in enthalpy upon binding, ΔH. The basic approach discussed here can also be modified to deal with more complicated systems, such as binding to multiple binding sites with different affinities. Fitting procedures are implemented in software provided by the instrument manufacturers, but can also be applied using commercially available software.

The integrated heat for each injection in an ITC experiment is the difference in heat content before and after the injection. Under the usual conditions of an ITC experiment, the cell of the calorimeter is filled with the macromolecule solution and the injection of ligand solution results in the ejection of an equivalent volume. The heat content is given relative to a reference state that is defined as the unligated macromolecule. The heat content is just the product of the excess enthalpy per mole of macromolecule, <ΔH>, the total concentration of macromolecule (both bound and free), $[M]_{tot}$, and the volume. Thus, the heat for injection number *i* is given as:

$$Q_i = \langle \Delta H \rangle_i [M]_{tot,i} V_{cell} - \langle \Delta H \rangle_{i-1} [M]_{tot,i-1} (V_{cell} - V_{inj}) \tag{4}$$

In the case of one-to-one binding, the excess enthalpy is given as:

$$\langle \Delta H \rangle = \Delta H \frac{K[L]}{1 + K[L]} \tag{5}$$

where ΔH is the change in enthalpy for binding one mole of ligand to one mole of macromolecule, *K* is the binding constant, and [L] is the concentration of unbound ligand. Experimentally, the total concentration of ligand (i.e., the sum of both bound and free) is known, but the concentration of unbound ligand is not and must be calculated from known quantities.

In setting up the experiment, the concentration of macromolecule initially in the cell of the calorimeter and the concentration of titrant initially in the syringe are both known quantities (typically with uncertainties on the order of 5 to 10%). Because an equal volume of the solution in the calorimeter cell is ejected from the cell with each injection of titrant, the concentration of macromolecule decreases at injection number *i* as:

$$[M]_{tot,i} = [M]_{tot,0} D^i \tag{6}$$

where $[M]_{tot,i}$ is the total concentration of macromolecule in the cell after injection step *i*, $[M]_{tot,0}$ is the total concentration of macromolecule in the cell before any injections are made, and D is a dilution factor defined as:

$$D = 1 - \frac{V_{inj}}{V_{cell}} \tag{7}$$

Here, V_{inj} is the injection volume and V_{cell} is the volume of the calorimetric cell. The total concentration of the titrant increases with each injection step as:

$$[L]_{tot,i} = [L]_{tot,0} (1 - D^i) \tag{8}$$

where $[L]_{tot,i}$ is the total concentration of titrant solution after injection step *i* and $[L]_{tot,0}$ is the total concentration of titrant in the injection syringe.

Using **Eqs. 6–8**, the total concentration of macromolecule and ligand at any step can be determined. The total concentration of macromolecule is equal to the sum of the bound and the free and is given as (the subscript *i* is omitted for clarity):

$$[M]_{tot} = [M] + [ML] = [M](1 + K[L]) \tag{9}$$

Likewise, the total concentration of ligand is given as:

$$[L]_{tot} = [L] + [ML] = [L](1 + K[M]) \quad (10)$$

Eq. 9 can be rearranged to give the concentration of free [M] as:

$$[M] = \frac{[M]_{tot}}{1 + K[L]} \quad (11)$$

which can be substituted into **Eq. 10** and rearranged to give the following quadratic expression:

$$K[L]^2 + (1 + K([M]_{tot} - [L]_{tot}))[L] - [L]_{tot} = 0 \quad (12)$$

Consequently, the concentration of free ligand, [L], can be determined after any injection step from the known total concentrations of M and L using the quadratic equation:

$$[L] = \frac{-[1 + K([M]_{tot} - [L]_{tot})] + \sqrt{[1 + K([M]_{tot} - [L]_{tot})]^2 + 4K[L]_{tot}}}{2K} \quad (13)$$

It should be noted that the positive root is used as the negative root can result in a physically meaningless negative concentration of L.

The free concentration of ligand from **Eq. 13** can be substituted into **Eq. 5** at each injection step. This result can then be substituted into **Eq. 4** in order to give the expected heat for each injection. The fitting programs provided by instrument manufacturers use the above equations to find the values of *K* and ΔH that give a result in the best agreement between the calculated and experimental values of q at each injection.

In addition to determining *K* and ΔH, ITC data are also fit for the best value of the stoichiometry, n, which does not appear in the above equations. Even for a system that is known to have a single ligand binding site, the value of n can differ from unity. There are several possibilities for nonunitary values of n in a one-to-one binding system: error in determining macromolecule concentration; the presence of macromolecule that is damaged or otherwise unable to bind ligand; and error in ligand concentration.

The effect of either of the first two possibilities is the same. The total concentration of active binding sites is equal to the product of n and $[M]_{tot}$. In fitting the experimental data, **Eq. 6** is then written as:

$$[M]_{tot,i} = n[M]_{tot,0} D^i \quad (14)$$

In practice, **Eq. 14** is used in place of **Eq. 6** and the data are fit for n, in addition to *K* and ΔH.

If the reason for a nonunitary value of n is, in reality, error in the determination of the concentration of binding-competent macromolecule, then the value of *K* and ΔH determined in the fit of the data accurately reflect the true values that would be determined if the true concentration were known *(22)*. This is because the product of n and $[M]_{tot}$ yields the concentration of binding sites. Thus, the value of n determined in this way serves as an excellent measure of the quality of macromolecule preparation and can be used to assess whether changes in storage or of a purification protocol result in changes in the macromolecule.

On the other hand, if the reason for a nonunitary value of n is error in the determination of the ligand concentration, the fitted values of *K* and ΔH will *not* reflect the true values. The fitting procedures assume that the ligand concentration is accurately known and that **Eq. 13** is appropriate. It is thus critically important that the ligand concentration is determined accurately. In practice, if the fitted value of n is within approx 5% of unity it can be assumed that the values of K and ΔH are accurately determined.

4. Conclusions

The conclusions that can be drawn from the discussions of the ITC experiment, the nonlinear fitting of ITC data, and data interpretation are as follows:

- It is important in planning the ITC experiment, that reasonable concentrations be chosen for the macromolecule and the ligand. This is most easily done by simulating the thermogram with reasonable guesses for K and ΔH (although the guess for ΔH is less critical).
- The linear parameters ΔH and n will be better determined than the nonlinear parameter *K*.
- The best results will be obtained at $10/K \leq [M] \leq 100/K$, and $[L] \approx 20$ to $50 \times [M]$, subject to solubility and heat signal considerations.
- The best results will be obtained when the initial integrated heat(s) are larger than 40 μJ (10 μcal).
- Titrant and titrate concentrations must be accurately known. (Nonintegral values for n are often the result of concentration errors. Errors in titrate concentration contribute directly to a similar systematic error in n. Errors in titrant concentration or titrant delivery contribute directly to errors in *K* and ΔH.)

References

1. Hansen, L. D., Christensen, J. J., and Izatt, R. M. (1965) Entropy titration. A calorimetric method for the determination of ΔG° (k), ΔH° and ΔS°. *Chem. Comm.* 36–38.
2. Christensen, J. J., Izatt, R. M., Hansen, L. D., and Partridge, J. M. (1966) Entropy titration. A calorimetric method for the determination of ΔG, ΔH and ΔS from a single thermometric titration. *J. Phys. Chem.* **70,** 2003–2010.

3. Christensen, J. J., Wrathall, D. P., Oscarson, J. L., and Izatt, R. M. (1968) Theoretical evaluation of entropy titration method for calorimetric determination of equilibrium constants in aqueous solution. *Anal. Chem.* **40,** 1713–1717.
4. Christensen, J. J., Izatt, R. M., and Eatough, D. (1965) Thermodynamics of metal cyanide coordination. V. Log K, ΔH°, and ΔS° values for the Hg^{2+}-cn-system. *Inorg. Chem.* **4,** 1278–1280.
5. Eatough, D. (1970) Calorimetric determination of equilibrium constants for very stable metal-ligand complexes. *Anal. Chem.* **42,** 635–639.
6. Eatough, D. J., Lewis, E. A., and Hansen, L. D. (1985) Determination of ΔH and K_{eq} values. In *Analytical Solution Calorimetry* (Grime, K., ed.). John Wiley & Sons, New York, NY, pp. 137–161.
7. Beaudette, N. V. and Langerman, N. (1978) An improved method for obtaining thermal titration curves using micromolar quantities of protein. *Anal. Biochem.* **90,** 693–704.
8. Langerman, N. and Biltonen, R. L. (1979) Microcalorimeters for biological chemistry: Applications, instrumentation and experimental design. *Methods Enzymol.* **61,** 261–286.
9. Biltonen, R. L. and Langerman, N. (1979) Microcalorimetry for biological chemistry: Experimental design, data analysis, and interpretation. *Methods Enzymol.* **61,** 287–318.
10. Wiseman, T., Williston, S., Brandts, J. F., and Lin, L.-N. (1989) Rapid measurement of binding constants and heats of binding using a new titration calorimeter. *Anal. Biochem.* **179,** 131–137.
11. Freire, E., Mayorga, O. L., and Straume, M. (1990) Isothermal titration calorimetry. *Anal. Chem.* **62,** 950A–959A.
12. Doyle, M. L. (1997) Characterization of binding interactions by isothermal titration calorimetry. *Curr. Opin. Biotechnol.* **8,** 31–35.
13. Holdgate, G. A. (2001) Making cool drugs hot: Isothermal titration calorimetry as a tool to study binding energetics. *Biotechniques* **31,** 164–166, 168, 170 passim.
14. Bundle, D. R. and Sigurskjold, B. W. (1994) Determination of accurate thermodynamics of binding by titration microcalorimetry. *Methods Enzymol.* **247,** 288–305.
15. Fisher, H. F. and Singh, N. (1995) Calorimetric methods for interpreting protein-ligand interactions. *Methods Enzymol.* **259,** 194–221.
16. Indyk, L. and Fisher, H. F. (1998) Theoretical aspects of isothermal titration calorimetry. *Methods Enzymol.* **295,** 350–364.
17. Baker, B. M. and Murphy, K. P. (1996) Evaluation of linked protonation effects in protein binding reactions using isothermal titration calorimetry. *Biophys. J.* **71,** 2049–2055.
18. Straume, M. and Freire, E. (1992) Two-dimensional differential scanning calorimetry: Simultaneous resolution of intrinsic protein structural energetics and ligand binding interactions by global linkage analysis. *Anal. Biochem.* **203,** 259–268.
19. Horn, J. R., Russell, D. M., Lewis, E. A., and Murphy, K. P. (2001) Van't hoff and calorimetric enthalpies from isothermal titration calorimetry: Are there significant discrepancies? *Biochemistry* **40,** 1774–1778.

20. Briggner, L.-E. and Wadsö, I. (1991) Test and calibration processes for microcalorimeters, with special reference to heat conduction instruments used with aqueous systems. *J. Biochem. Biophys. Methods* **22,** 101–118.
21. Horn, J. R., Brandts, J. F., and Murphy, K. P. (2002) Van't hoff and calorimetric enthalpies ii: Effects of linked equilibria. *Biochemistry* **41,** 7501–7507.
22. Murphy, K. P., Freire, E., and Paterson, Y. (1995) Configurational effects in antibody-antigen interactions determined by micro-calorimetry. *Proteins* **21,** 83–90.

2

Direct Optical Detection of Protein–Ligand Interactions

Frank Gesellchen, Bastian Zimmermann, and Friedrich W. Herberg

Summary

Direct optical detection provides an excellent means to investigate interactions of molecules in biological systems. The dynamic equilibria inherent to these systems can be described in greater detail by recording the kinetics of a biomolecular interaction. Optical biosensors allow direct detection of interaction patterns without the need for labeling. An overview covering several commercially available biosensors is given, with a focus on instruments based on surface plasmon resonance (SPR) and reflectometric interference spectroscopy (RIFS). Potential assay formats and experimental design, appropriate controls, and calibration procedures, especially when handling low molecular weight substances, are discussed. The single steps of an interaction analysis combined with practical tips for evaluation, data processing, and interpretation of kinetic data are described in detail. In a practical example, a step-by-step procedure for the analysis of a low molecular weight compound interaction with serum protein, determined on a commercial SPR sensor, is presented.

Key Words: Optical biosensors; surface plasmon resonance; reflectometric interference spectroscopy; biomolecular interaction analysis; kinetics; low molecular weight ligands.

1. Introduction

A functional description of biomolecules has to extend beyond a solely statical description of the protein content within a cell, a cellular compartment, or a tissue. A detailed kinetic description of the interaction patterns has to be added, because these molecules are involved in a dynamic equilibrium. Several methods employing different physical principles have been adopted to determine the binding of one biomolecule to another or to monitor complex formation. This can be done by combining a detector, very often an optical mass detector,

From: *Methods in Molecular Biology, vol. 305: Protein–Ligand Interactions: Methods and Applications*
Edited by: G. U. Nienhaus © Humana Press Inc., Totowa, NJ

with a microfluidics system and monitoring the interaction of a component immobilized to a solid phase, in the following termed **ligand**, and an **analyte** in flow phase (*see* **Fig. 1**). Biomolecular interaction analysis (BIA) describes highly accurately relevant binding events between compounds under physiological conditions, thereby providing detailed kinetic data of the association and dissociation of binding partners. Being a part of functional genomics, BIA is also implemented in drug development and quantifies the interactions between small molecules, proteins (e.g., receptors, enzymes, antibodies), peptides, nucleotides, carbohydrates, and other biomolecules. In combination with systematic molecular and cellular analyses of proteins, BIA assigns function to arbitrary listings of gene products.

A typical interaction analysis is based on three steps: 1) coupling of the ligand, 2) interaction analysis, and 3) regeneration.

1. The ligand has to be immobilized in an appropriate manner maintaining the biological activity as well as providing a stable binding to the solid support (*see* **Subheadings 1.3.1.** and **3.1.** for details).
2. Once the ligand surface displays a stable baseline, the analyte is added using a well controlled flow system. This allows monitoring association and dissociation phases separately and plotting them in form of a *sensorgram* (*see* **Fig. 2**). Using serial dilutions of analyte, the association rate constant (k_a) and dissociation rate constant (k_d) are extracted with an appropriate software applying pseudo-first order kinetics. With the known concentration of the analyte, apparent equilibrium binding constants (K_D or K_A) can be calculated. Besides the kinetic constants, EC_{50} values for competitors can be determined by solution competition or surface competition as described later. Special care has to be taken to subtract nonspecific or unspecific binding events. In a multichannel system, this can be performed by subtracting the binding on a reference surface. A reference surface either lacks the specific ligand or, preferably, is modified with an appropriate negative control.
3. After the interaction has taken place, bound analyte has to be removed completely from the ligand surface to perform another interaction analysis. However, except in the case of transient interactions, baseline level is seldom reached in an acceptable time frame. Therefore, a procedure referred to as regeneration has to be performed, removing bound analyte with appropriate agents without destroying the biological activity of the immobilized ligand (*see* **Fig. 2**). Appropriate conditions should be optimized for individual interaction partners, common methods include for example treatment with glycine at acidic pH in the case of antibody mediated interactions. Optimally, a biospecific regeneration procedure can be used, as shown in **Fig. 2** for the interaction between the catalytic (C-) and the regulatory (R-) subunit of cAMP dependent protein kinase (PKA). Dissociation of the regulatory subunit from the catalytic subunit is initiated by cAMP, therefore a C-subunit surface can be regenerated by an injection of this physiological effector. An overview of possible regeneration conditions for differently immo-

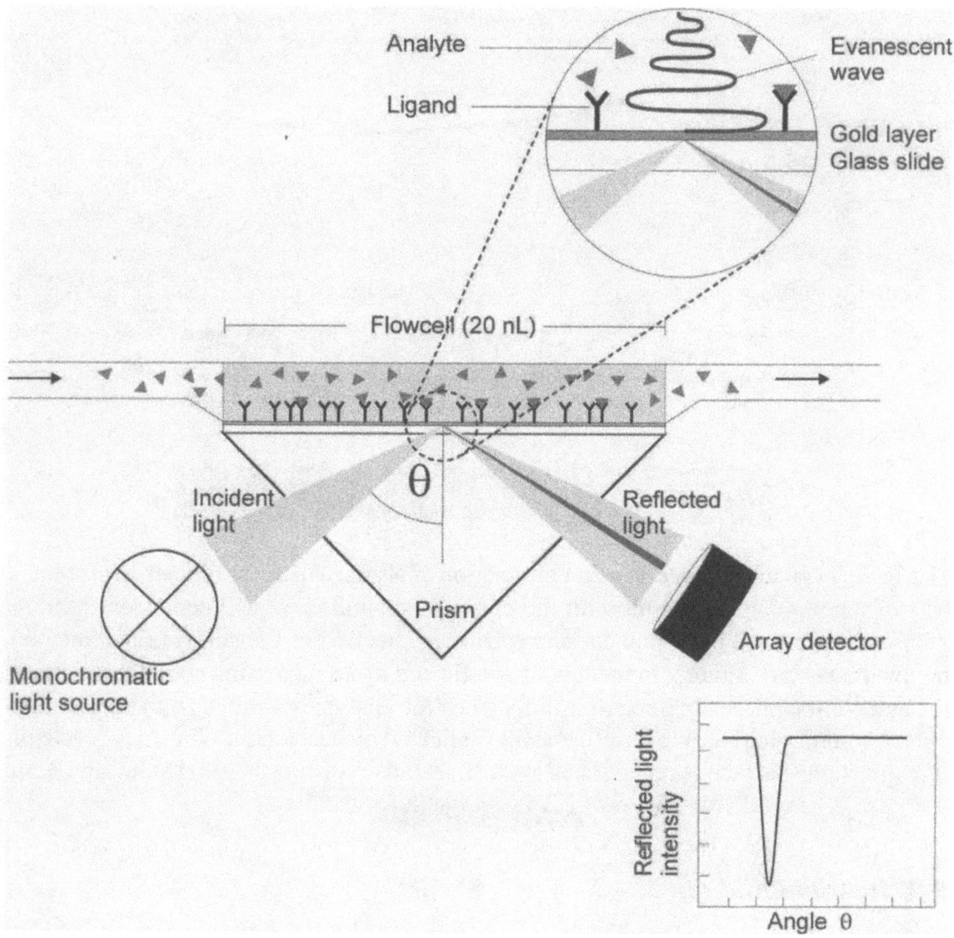

Fig. 1. Detection principle of a SPR-biosensor (*see* **Subheading 1.1.1.** for details). Monochromatic light focused in a wedge on a metal-coated interface between two media with a high refractive index (glass) and a low refractive index (solution), respectively, is totally internally reflected. At a specific angle (the so-called resonance angle θ) the incident light is coupled into the plasmons of the metal layer that results in emission of an evanescent wave into the lower refractive index medium (*see* inset). The ensuing drop in light intensity appears as a shadow in the reflected light wedge, which is monitored on the position-sensitive diode array detector. The resonance angle is dependent on the refractive index of the solution close to the surface layer and hence on the amount of analyte bound to the immobilized ligand (*see* text for details).

bilized ligands is given by Herberg and Zimmermann *(1)*. Biacore AB is compiling a database of immobilization and regeneration procedures on their website (http://www.biacore.com) that should prove very helpful to the user.

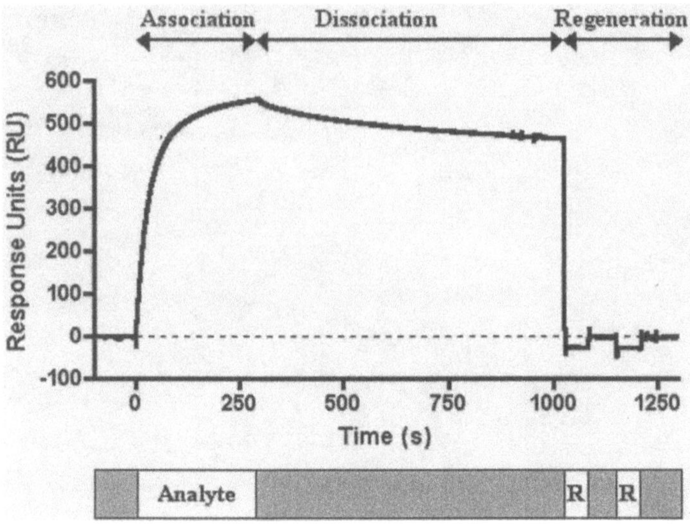

Fig. 2. Typical sensorgram of an interaction analysis. The figure shows the interaction of the regulatory subunit with the catalytic subunit of cAMP dependent protein kinase immobilized by amine coupling. During injection of the analyte (as indicated below the *x*-axis), binding to the ligand is reflected by an increasing signal throughout the association phase. Dissociation starts when the analyte is omitted from the running buffer. In order to return to baseline level (indicated by dashed line) for a new cycle of injections, the surface is regenerated with short pulses of regeneration solution (R, in this case 0.1 m*M* cAMP, 2.5 mM EDTA, *see* text for details).

1.1. Instrumentation

Optical detection principles for monitoring of biomolecular interactions have been implemented into various commercial biosensors. Most optical biosensors consist of three main components: a detector based on different physical/optical principles, a sample delivery system (microfluidics), and the sensor surface where one of the interaction partners is immobilized either covalently or noncovalently. In the following, surface plasmon resonance (SPR)-based detectors—with special emphasis on the Biacore technology—will be discussed in more detail. Additionally, reflectometric interference spectroscopy (RIfS) is presented as a very promising technology in biomolecular interaction analysis, followed by a brief introduction of commercially available biosensors.

1.1.1. Surface Plasmon Resonance-Detectors

Traditionally, SPR has been used to determine binding constants for macromolecular interactions, owing to the fact that SPR sensors can be utilized as optical mass detectors. SPR occurs when light illuminates thin conducting films (gold in the case of Biacore instruments) under specific conditions. The resonance is a result of the interaction between electromagnetic vectors in the incident light and free electron clouds, called plasmons, in the conductor. SPR arises as a result of a resonant coupling between the incident light energy and the surface plasmons in the conducting film at a specific angle of incident light. Absorption of the light energy results in the emission of an evanescent wave into the low refractive index medium, which causes a characteristic drop in the reflected light intensity at that specific angle (*see* **Fig. 1**).

The resonance angle θ is sensitive to a number of factors, including the wavelength of the incident light, the nature and thickness of the conducting film and the temperature. Most importantly for this technology, the angle depends on the refractive index of the medium opposite of the incident light. When all other factors are kept constant, the resonance angle is a direct measure of the refractive index of the medium. Only the angle at which SPR occurs is altered and detected with the diode array detector; the intensity of the shadow in the reflected light is unchanged. Binding events cause changes in the refractive index at the surface layer, which are detected as changes in the SPR signal. In general, the refractive index change for a given change in mass concentration at the surface layer of a sensor chip is practically the same for all proteins and peptides *(2)*, thereby providing a sensitive real-time mass detector. However, for glycoproteins, lipids, and nucleic acids this refractive index change is slightly different. Suitable calibration procedures using standard substances still allow determination of correct values for binding, plotted as response units (RU). For a general purpose sensor chip, the CM5 chip (Biacore AB, Sweden), 1000 RU correspond to 1 ng protein/.mm^2 sensor surface *(2)* and generate a shift of 0.1° in the resonance angle θ.

The technology has been implemented into several instruments already available or under development from Biacore AB. For basic research, Biacore X (two flow cells), Biacore 2000, and 3000 (four flow cells), and for higher throughput, Biacore TAS (eight flow cells), were developed, and an array instrument for high throughput is under construction. The Biacore S51 instrument with lower sample consumption and higher resolution is aimed specifically at drug screens.

1.1.2. Reflectometric Interference Spectroscopy

Reflectometric interference spectroscopy (RIfS) is a BIA technology that has—although already in use for several years—only recently been implemented into a commercially available biosensor, the BIAffinity instrument just introduced by Analytik Jena AG (Jena, Germany). So far, however, almost all research data based on RIfS have been collected using custom built detectors in academic research laboratories as described below.

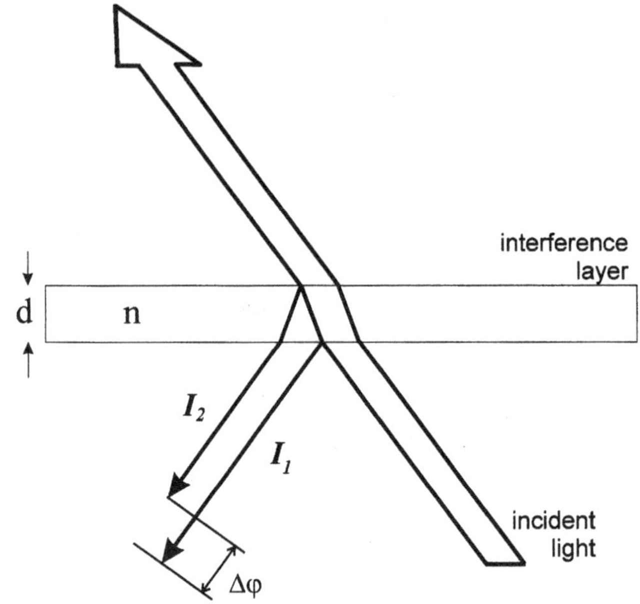

Fig. 3. Detection principle of RIfS. Incident white light is partially reflected at each surface of an interference layer with a refractive index n and a physical thickness d. Reflected light beams of intensities I_1 and I_2 travel different optical path lengths, resulting in a phase difference $\Delta\varphi$, which produces a distinct interference pattern of alternating minima and maxima. Binding of biological material at the surface increases the optical thickness ($n \times d$) of the interference layer, leading to a corresponding shift in the interference pattern. (Adapted from **ref. *3***.)

RIfS exploits an optical phenomenon occurring at thin transparent films: a light beam passing a weakly reflecting thin film will be reflected in part at each of the interfaces (*see* **Fig. 3**). As the reflected light beams travel different optical pathlengths, a phase difference is introduced, resulting in a distinct interference pattern of alternating minima and maxima. This phenomenon is dependent on the angle of incidence and the wavelength of the incident light as well as the physical thickness of the film and its refractive index. Binding of biological material to the surface causes a change in the optical thickness of the film. The increase in optical thickness results in a shift of the reflectance pattern which can be monitored with high resolution in real-time with a simultaneous diode array spectrophotometer.

For biomolecular interaction analysis, glass chips with an interference layer of 500 nm SiO_2 deposited on top of 10 nm Ta_2O_5 for reflection enhancement are used. To reduce nonspecific binding the interference layer must be modified. This is done by silanization with aminobutyl-dimethylmethoxysilane and covalent coupling of hydrophilic polymers (dextran, polyethylenglycol) *(3)*. Binding curves can be recorded using a simple setup consisting of a 20 W tungsten halogen lamp and a diode array spectrophotometer connected by a 2:1 fiber-optic coupler.

The general applicability of RIfS for monitoring interactions between biomolecules and low molecular weight ligands has been demonstrated by several applications including the characterization of the streptavidin-biotin interaction *(4)*, antibodies binding to a hapten *(3)*, as well as the interaction between DNA and DNA-binding compounds *(5)*. The technique is also well suited for parallelization, and has been used for high-throughput screens of thrombin inhibitors *(6)*, for screening of a combinatorial triazine library with different antibodies *(7)*, and for epitope mapping *(8)*.

The applications described above suggest sensitivity and detection limits for RIfS in the same order of magnitude as for other commonly employed optical biosensors *(9)*. The RIfS technology provides a simple and robust alternative to the Biacore SPR sensors. While not quite reaching the high sensitivity of the Biacore instruments, it is nevertheless capable of generating reproducible interaction data over a high dynamic range despite the lack of a sophisticated microfluidics system *(9)*. An advantage of RIfS is that the detection principle itself is not temperature sensitive, and thus does not require the expensive temperature control systems crucial for SPR-based sensors. Finally, the capability of parallelization and high-throughput screenings make RIfS an attractive technology in the field of optical biosensors.

1.1.3. Other Optical-Based Technologies

Aside from the Biacore instruments, the SPR technology has also been implemented into other commercially available instruments, such as the Instrument of Biomolecular Interaction Sensing *(10)* (IBIS, Windsor Scientific, Slough, UK), which utilizes a cuvet based setup, or the Spreeta device from Texas Instruments (Dallas, TX), a miniaturized portable SPR platform *(11)*. Another related physical principle used to monitor biomolecular interactions is the resonant mirror *(12)* that is implemented in the IAsys system (ThermoFinnigan, San Jose, CA). Aside from the differences in the detection principle, the instrument uses a cuvette based sample delivery system.

Another application of SPR technology takes advantage of the evanescent waves generated in fiber-optics waveguides when the propagated light beam is totally internally reflected at the wall of the fiber. A fiber-optic sensor specifically aimed at quantifiying protein and small molecule interactions, the LunaScan device, has been patented by Luna Analytics (Blacksburg, VA). This biosensor uses long period gratings (LPG) inside an optical fiber to scatter light from the optical fiber at a predetermined wavelength that is dependent on the grating period. The light scatter in turn is dependent on the refractive index of the fiber and its surrounding environment. Upon adsorption of target analytes to the surface coating, the resulting refractive index change causes a spectral shift in the wavelength of the scattered light that is proportional to the mass of analyte bound.

The FLEX CHIP Kinetic Analysis System by HTS Biosystems (Hopkinton, MA) employs grating coupled SPR, where the incident light is coupled into the surface plasmon via an optical grating. An instrument based on the same principle, the Applied Biosystems 8500 Affinity Chip Analyzer is also aimed at high throughput analyses. According to the manufacturer, the instrument is capable of measuring interactions with K_D from the pM to the μM range.

A novel biosensor developed by SRU Biosystems (Woburn, MA) utilizes a colorimetric resonant diffractive grating surface as surface binding platform ***(13)***. The grating is designed in such a way that, when illuminated with white light, it reflects only light of a single wavelength. Attachment of molecules to the surface shifts the reflected light wavelength due to the change of the optical path of light coupled into the grating. This method is capable of resolving changes of 0.1 nm thickness of protein binding on the surface and is well suited for miniaturization and parallelization ***(13)***. For other recent developments in the field of optical biosensors the reader is referred to a comprehensive review by Baird and Myszka ***(14)***.

1.2. Applications

1.2.1. Basic Considerations

Based on sales and on the amount of scientific literature published, SPR-based devices, for example the Biacore instruments, are the most commonly used commercially available biosensors ***(15)***. Typical applications include analysis of protein–protein or protein–DNA interactions, characterization of antibodies (epitope mapping), elucidation of the influence of post-translational modifications on interaction kinetics, but also the investigation of macromolecular complexes up to supramolecular compounds like viruses, microorganisms, or entire cells. On the opposite end of the scope stands BIA of low molecular weight compounds with proteins, which is of special interest in drug research ***(16)***. For hit validation and optimization of lead compounds a detailed kinetic characterization of pharmaceutical substances is required. A potential limitation of SPR sensors lies in the detection principle: a mass change on the sensor surface is transduced into a proportional optical signal, i.e., a small increase in mass on the surface results in an accordingly small signal. Therefore, it appears favorable to immobilize the low molecular weight ligand on the sensor surface and use the larger interaction partner as the analyte in the soluble phase. Immobilization of small ligands, however, often requires their previous derivatization and care must be taken to determine the effect of the modification on ligand functionality. Another caveat of this approach is that high-density surfaces of ligand are prone to mass transfer limitations (*see* **Note 2**), whereas low density surfaces—which are suitable for kinetic analyses—are difficult to adjust with small ligands, because the immobilization process cannot be observed easily as a result of the low response. Another problem is a reduction in degrees of freedom inherent to the immobilization process that can severely affect the rate constants. On the other hand, immobilization procedures for large molecules (i.e., proteins) are well established ***(1)*** and recent advances in instrumentation (SPR based biosensors like the Biacore 3000 and S51), in the analysis software, and suitable calibration procedures allow direct optical detection of small molecule binding.

1.2.2. Ligand Interaction in the Indirect Assay Format

Alternatively, binding of low molecular weight substances can be assessed by solution or surface competition assays (*see* **Fig. 4**) comparable to the procedures already in use for immunoassays such as radioimmunoassay (RIA) and enzyme-linked immunosorbent assay (ELISA) ***(17)***. In solution competition experiments, the competitor molecule interferes with binding of the analyte to the immobilized ligand, whereas in surface competition the molecule of interest competes with the analyte for the same binding site (*see* **Fig. 4**).

Shortly after the introduction of the first SPR biosensor by Biacore in 1990, Karlsson ***(18)*** described a competitive kinetics approach for characterization of low molecular weight ligands, using the binding of HIV p24-derived peptides (competitor 1) vs the intact antigen (competitor 2) to a monoclonal antibody as a model system. This assay format continues to be a valuable tool, as it has been used to determine levels of small metabolites like morphine-3-glucuronide, the main metabolite of heroine and morphine ***(19)***, or deoxynivalenol, a highly toxic fungal metabolite, that may contaminate food and animal feed ***(20)***.

1.2.3. Ligand Interaction in the Direct Assay Format

However, in recent years researchers have increasingly employed the direct binding assay for BIA analysis of low molecular weight ligands ***(15)***. The validity of interaction data acquired with SPR has been ascertained by comparison with stopped-flow fluorescence and isothermal titration calorimetry measurements ***(21,22)***. Compared to other approaches, SPR biosensors have the advantage of providing a relatively robust readout, a simple assay format, and low sample consumption. The binding event can be monitored directly without the need for labeling of one or both of the interaction partners. Further-

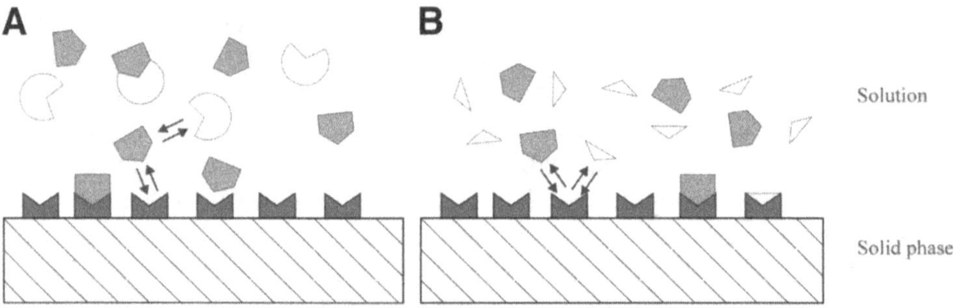

Fig. 4. Competition experiments. Schematic setup of competition experiments. By varying the concentration of the competitor, EC_{50} values can be deduced. **(A)** *Solution competition.* Soluble receptor molecules (white) bind to the analyte in solution (light gray), thereby competing with analyte binding to the ligand immobilized to the surface (dark gray). Only binding of free analyte to the ligand is detected. **(B)** *Surface competition.* A competitor molecule (white) competes with the analyte (light gray) for the same binding site on the ligand (dark gray). For use of this assay with biosensors based on SPR, a significant difference in molecular weight between analyte and competitor is required.

more, higher information content is given, because this technology allows the user to measure association and dissociation rate constants separately. One can determine kinetic as well as equilibrium binding data in a single experiment. Additionally, even thermodynamic data of the respective binding event can be extracted once a series of experiments is performed at different temperatures *(23)*. Moreover, the whole process can be automated and used for screening of compounds in a medium throughput format.

The interactions of low molecular weight substances with target molecules have been investigated by several groups using Biacore technology. Among many similar studies the usefulness of SPR-based biosensors for drug screens has been demonstrated by describing binding kinetics and affinities of 58 different inhibitors to HIV-1 proteinase *(24)*, and by a screen of 170 compounds for binding to immobilized thrombin *(25)*.

On the very extreme end of the spectrum Gestwicki and co-workers *(26)* were able to detect binding of maltose to maltose-binding protein (MBP) and calcium ions to tissue transglutaminase (tTG), respectively, using a Biacore instrument. The resulting SPR signal could not be attributed to the actual binding event, because it was either negative (in the case of maltose-MBP interaction) or too high to be explained by binding of the small analyte (in the case of Ca^{2+} binding to tTG). According to the authors, a possible explanation could be a conformational change of the receptor protein upon analyte binding, which in turn may induce refractive index changes close to the matrix.

Aside from protein interaction studies, the binding of small molecules to nucleic acids is an important issue in the biomedical field. The coupling of DNA/RNA to the sensor surface is simplified by a previous biotinylation of the nucleic acid followed by immobilization on a streptavidin coated sensor chip (*see* **Note 1**).

Using biotinylated DNA-hairpin oligomers the mode of action and sequence specificity of the DNA binding antitumor antibiotic AT2433-B1 was successfully identified with SPR analyses complementing DNase footprinting experiments *(27)*. This approach has been used in several similar studies with antitumor drugs binding to DNA. Accordingly, the same strategy is also applicable to RNA, as exemplified by several studies of therapeutics binding to RNA molecules *(28–30)*.

For interaction screens in the direct assay format it is crucial that the immobilized ligand retains its biological activity during the entire set of experiments, which should be checked routinely by injection of a reference analyte. Another issue generally relevant for the analysis of small analytes is unspecific binding either to the chip matrix or to the immobilized protein. The extent of such unspecific binding should be assessed and corrected for by injection over appropriate reference surfaces. Likewise, the influence of solvents such as DMSO can be adressed by calibration on a reference surface. A detailed description for the characterization of the binding behavior of a low molecular weight compound (here: warfarin) with immobilized serum albumins from different species is given under **Subheading 3.5.** of this article.

1.3. Interaction Analysis

An essential feature of direct optical detection systems is the immobilization of the ligand molecule in order to detect the binding of a soluble analyte. Most optical biosensors are based on glass and/or metal surfaces that have to be derivatized to generate a biocompatible environment. Carboxymethylated dextrans with a low degree of crosslinking have been proven to be excellent for ligand coupling as they allow to generate high surface densities and display low unspecific binding. At the same time, a hydrophilic matrix for the biological interaction close to the sensor surface is provided.

1.3.1. Immobilization

An accurate kinetic analysis can only be performed, if the ligand molecule is immobilized in a biologically active conformation. Steric hindrance caused by the immobilization strategy can be a serious problem, either prohibiting, reducing, or modulating the respective binding. Therefore, a suitable experi-

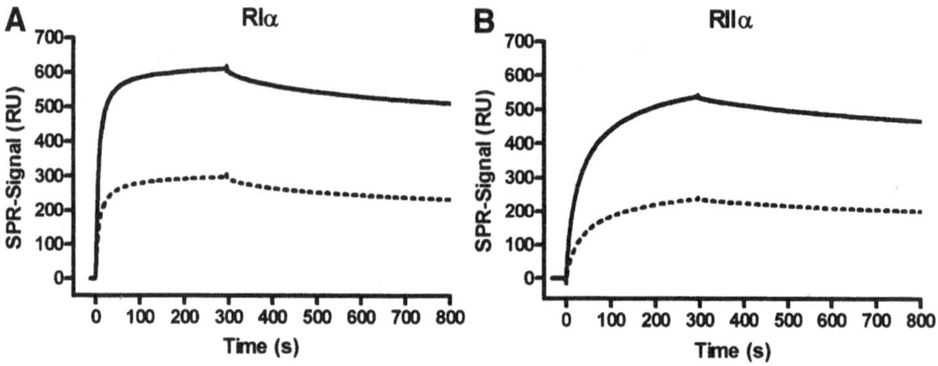

Fig. 5. Influence of MgATP on immobilization of the C-subunit of PKA. Recombinant C-subunit was immobilized on a CM-dextran surface via primary amines. 730 RU and 870 RU of C-subunit in 10 m*M* acetate buffer pH 6.0 were coupled, in the absence and presence of MgATP (1 m*M* ATP, 2 m*M* $MgCl_2$), respectively. A reference surface was treated accordingly without injection of a ligand. After immobilization, 400 n*M* recombinant RIα (**A**) or RIIα subunits (**B**) were injected in running buffer (20 m*M* MOPS pH 7, 150 m*M* NaCl, 2 m*M* $MgCl_2$, 1 m*M* ATP, 0.005% Tween-20) and the association monitored for 5 min before the dissociation phase was started by injection of running buffer. No unspecific binding was monitored on the reference surface. After 10 min of dissociation, the surface was regenerated to baseline level by injection of 0.1 m*M* cAMP, 2.5 m*M* EDTA (not shown). The binding stoichiometry to both inhibitors was increased by a factor of two when adding MgATP (solid lines) during the immobilization compared to C-subunit immobilized without MgATP (dashed lines).

mental setup has to be generated to check for biological activity of the immobilized ligand.

This is exemplified when looking at interaction partners of the catalytic subunit of cAMP dependent protein kinase. If the catalytic subunit is immobilized to the sensor surface using primary amines, coupling can also occur via Lys72, a residue that is essential for the correct conformation of the catalytic site, which in turn is a prerequisite for efficient binding of physiological kinase inhibitors. To avoid coupling via Lys72, MgATP is added during the immobilization procedure to occupy the active site cleft of this protein kinase. When this cosubstrate of the enzyme is bound, the critical lysine residue is protected from modification. If Lys72 is not protected during immobilization, binding of physiological inhibitors is negatively influenced, reflected in a decreased binding stoichiometry. **Figure 5** shows the interaction of two different physiological inhibitors of the catalytic subunit, the regulatory subunit type I and type II.

Although those inhibitors occupy distinctly different surface areas of the catalytic subunit, both require interaction with the active site for high-affinity binding ***(31)***. The binding stoichiometry in the presence of the protective cofactor MgATP is increased by approx 100% (from 33 to 66% in the case of the type I regulatory subunit and from 33 to 62% in the case of RII subunit, assuming binding in a 1:1 molar ratio). Interestingly, the binding pattern, as indicated by the shape of the curves, does not differ significantly between catalytic subunit immobilized in the presence and absence of MgATP, respectively, suggesting that only catalytic subunit immobilized in an active conformation participates in the binding. This is also reflected in the apparent association and dissociation rate constants calculated from serial dilutions of the regulatory subunit (data not shown).

1.3.2. Detection of Small Molecule Ligand Interactions

A wide range of biological interactions can be analyzed using SPR biosensors. For reasons discussed earlier, direct optical detection of low molecular weight compounds is a challenge. As an example, in the early phase of the drug development process the determination of adsorption, distribution, metabolism, and excretion (ADME) parameters has become very important ***(16)***. In this context, the interaction of low molecular weight compounds to serum proteins such as human serum albumin or alpha acid glycoprotein is investigated. High-affinity binding to serum proteins significantly changes the physiologically active concentration of a compound and thus reduces its bioavailability, but also prolongs its duration of action because of a slower release from the bound state. Although a wide range of methods such as equilibrium dialysis, ultracentrifugation, spectroscopic, and chromatographic approaches are available to monitor binding of small analytes to serum proteins, direct optical detection methods have advantages as a result of low sample consumption, high accuracy, and reproducibility and do not require labeling. Furthermore, the potential for parallelization and automation makes this technique suitable for screening assays with increased throughput. The experiment depicted in **Figs. 6** and **7**, shows the interaction of warfarin, a low molecular weight compound (molecular weight 308.3 Da), with serum albumins of human (HSA), bovine (BSA), and murine (MSA) origin, examined in parallel on a Biacore instrument. This allows a direct comparison of serum albumin binding levels and a prediction of bioavailibility in different biological systems such as cell culture, transgenic animals, or in humans. Interestingly, significantly different binding patterns could be observed for the three mammalian serum albumins investigated, yielding K_D-values of 3.6 µ*M*, 7.8 µ*M*, and 49.8 µ*M* for HSA, BSA, and MSA, respectively (*see* **Fig. 7**). The binding data for HSA are in excellent agreement with results provided by Rich et al. ***(32)***.

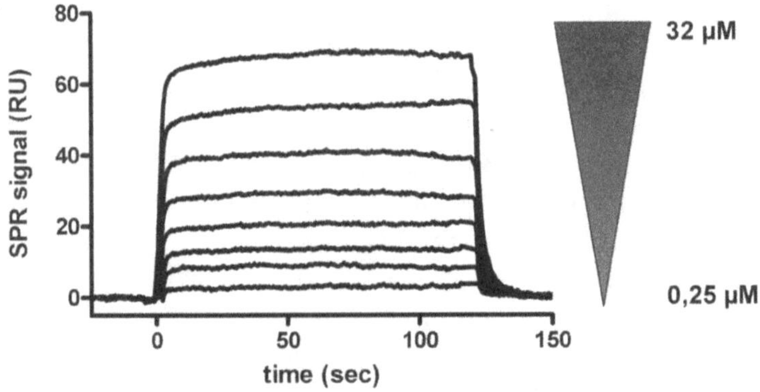

Fig. 6. Reference subtracted binding curves of warfarin to human serum albumin (HSA, 20000 RU immobilized). Warfarin in running buffer containing 3% DMSO was injected in concentrations ranging from 250 n*M* to 32 µ*M*. A representative set of data is presented.

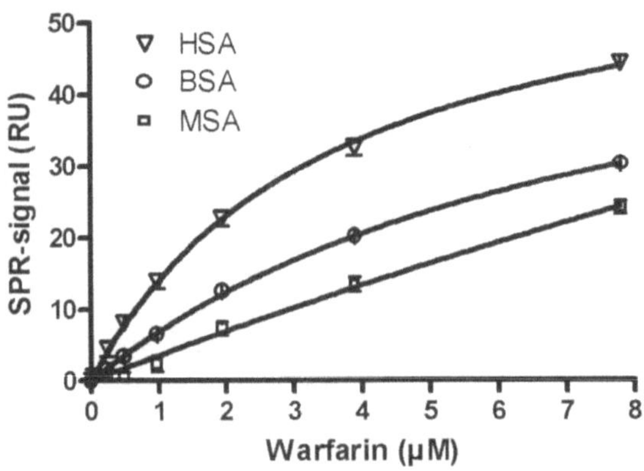

Fig. 7. Equilibrium binding analysis of warfarin to three different serum albumins of HSA, bovine BSA, and MSA origin. Here, the reference subtracted and solvent corrected steady-state binding levels are plotted against the respective warfarin concentrations. K_D-values for each data set are extracted by nonlinear regression plots of the saturation curves.

In principle, serum protein binding assays using SPR biosensors are relatively simple to perform and can be used in routine analysis. However, some experimental details are crucial for a successful realization of the experiment and will be described in the following. Serum albumin proteins are coupled to high-surface densities using standard amine coupling (*see* **Subheading 3.5.1.**)

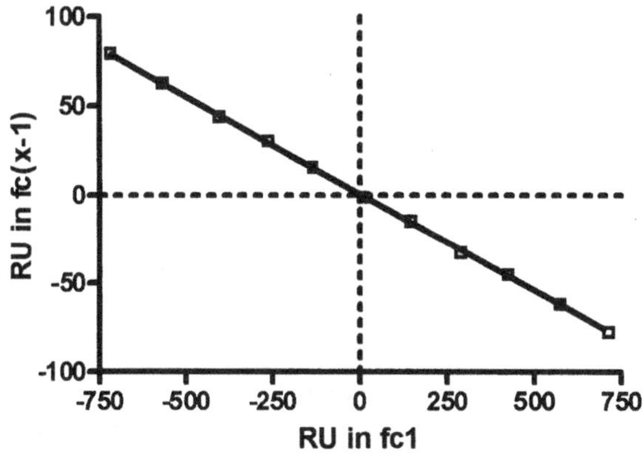

Fig. 8. DMSO calibration curve for concentration series of solvent in running buffer. DMSO in running buffer, ranging in concentrations 0.5% above and below (in 0.1% increments) the DMSO concentration in running buffer, was injected over the specific (fcx) as well as the reference surface (fc1). The signal on the reference cell (fc1, *x*-axis) is plotted against the reference subtracted signal on the specific surface (fc[x-1], *y*-axis). Correction factors are calculated by inserting the signal on the reference surface (RU in fc1) as x-value into the equation obtained from the linear regression of the calibration curve. The resulting *y*-value (RU in fc[x-1]) is the corresponding correction factor. For further details *see* **Subheading 3.5.3.**

to obtain an acceptable signal-to-noise ratio. Common problems related to high-surface densities like mass transfer limitations (*see* **Note 2**) do not apply to the kinetics of small molecule analytes because small molecules have favorable diffusion properties. However, as a result of their limited solubility in aequous solutions, most low molecular weight compounds have to be dissolved in buffers containing organic solvents like dimethyl sulfoxide (DMSO). DMSO itself has a high refractive index and small differences in concentrations induce large increases in signal that have to be subtracted. Furthermore, the refractive index changes might be slightly different for the reference and the specific surfaces demanding a sophisticated calibration procedure (*see* **Subheading 3.5.2.**). The resulting calibration curve enables the calculation of correction factors for the specific samples (*see* **Fig. 8**). Once the sample is diluted, consider to match the DMSO concentration exactly to the running buffer because the high refractive index changes induced by varying DMSO concentrations will increase the correction factors and make the assay less sensitive. It is essential to check if the compound is still soluble after dilution. At this step it is also very important to avoid evaporation because this will significantly change the buffer composition and the concentration of the compound.

Most low molecular weight compound interactions with serum albumins display fast on- and off-rates, hence it is sufficient to inject each sample for 1 min and report the equilibrium binding response in the middle of the injection phase. A sufficient number of blank injections should be performed between samples in order to prevent carry-over. As a result of the fast off-rates a regeneration of the serum albumin surface is usually not necessary. The reference subtracted and solvent corrected equilibrium binding responses are then plotted against the compound concentration, and the K_D-values for each data set can be calculated by nonlinear regression (*see* **Subheading 3.5.** for details). Still, interpretation of the data is not always trivial because HSA is known to have several binding sites for small ligands with varying affinities. Therefore, two or more binding sites for one ligand might be observed and reflected in the binding kinetics. For data evaluation it is important to select an adequate range of concentrations because ligand binding to multiple sites causes large variances in the calculated K_D-values depending on the selected concentration range.

1.3.3. Interpretation of Kinetic Data

Several software packages based on linear or nonlinear analysis are available to analyze high-resolution kinetic data (*see* **Subheading 3.4.**). However, library screens—either of expression libraries or chemically generated libraries—produce a vast number of interaction data that can be difficult to interpret with available software. Software tools have to be developed to classify and visualize bulk kinetic data automatically and to perform basic kinetic evaluations. KineticXpert by Microdiscovery GmbH (Berlin, Germany) is a software tool under development performing bulk analyses and evaluation of interaction data and facilitating data management (http://www.microdiscovery.com).

Interpretation of kinetic data can be helpful in elucidating more complex biological mechanisms. Yaqub et al. *(33)* investigated the domain interaction of C-terminal src kinase (Csk), a member of the src kinase family using surface plasmon resonance. After immobilization of the Csk kinase domain using amine coupling, the immobilized protein was phosphorylated by another kinase (PKA) on the chip, and thereby modulated in its binding behavior. This is reflected in a change in the binding kinetics of an isolated SH3 domain run over the chip before and after phosphorylation *(33)*. Evaluation of the interaction patterns after phosphorylation demonstrated that the kinetics obtained were not compatible with a 1:1 binding model. Careful examination of several different binding models (*see* **Fig. 9** and legend) led to the conclusion that the immobilized Csk kinase domain had not been phosphorylated completely, resulting in a heterogeneous ligand surface. When the soluble SH3 domain was injected, a slow phase for the unphosphorylated species superimposed a fast phase for the modified, phosphorylated Csk *(33)*. Comparing several models

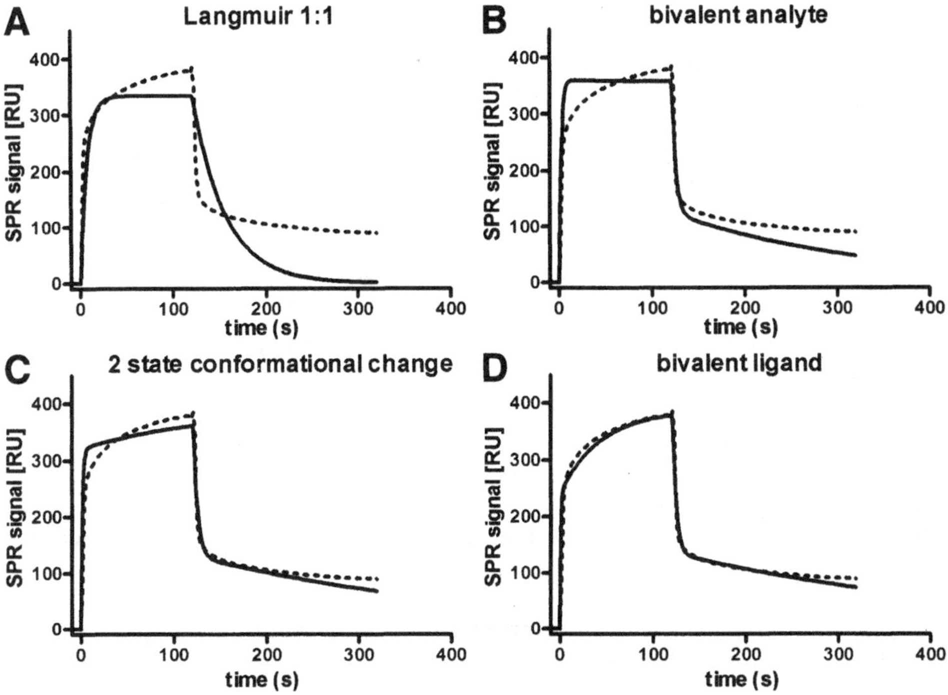

Fig. 9. Interpretation of biological binding data using different kinetic models. 34 μ*M* of an isolated SH3 domain were injected over a surface with immobilized Csk kinase domain which had been on-chip phosphorylated. Panels **A–D** show the respective sensorgram (dashed line) superimposed with fits from different interaction models (solid lines), generated with the BIAevaluation software (v3.2, Biacore AB). Neither a 1:1 binding model **(A)**, a bivalent analyte model **(B)** or a two-state conformational change **(C)** yield adequate fitting results. Only the bivalent ligand model **(D)** gives a near-perfect fit with the interaction data, indicating that as a result of incomplete on-chip phosphorylation two distinctly different kinetics were superimposed.

of interaction, the bivalent ligand model yielded by far the best fit with the experimental data (*see* **Fig. 9**, panel D). These data demonstrate that biological function can be maintained and modulated even if a molecule is immobilized to a solid phase. Furthermore, these data show that appropriate models describing the biomolecular interaction of choice can generate significant information on the nature of the biological interaction.

2. Materials

The chemicals listed are intended for use with the Biacore system, however, they can easily be adopted for other biosensor devices where one interaction partner has to be immobilized.

2.1. Coupling

Basically, two different covalent coupling chemistries are commonly used: coupling via primary amines using NHS/EDC or coupling via thiols, either by surface or ligand thiol coupling.

2.1.1. Amine Coupling

When performing an *amine coupling* a ligand with a primary amine function (for example a free N-terminus or a lysine residue) is needed.

1. Appropriate immobilization buffer with low ionic strength ranging from pH 3.0 to 6.0, this buffer should contain no primary amines (i.e., do not use Tris-buffers, instead prepare acetate, phosphate or MES (2-(*N*-Morpholinoethanesulfonic acid)) buffer in the intended pH range).
2. 100 m*M* *N*-hydroxysuccinimide (NHS).
3. 400 m*M* *N*-ethyl-*N*'-(dimethylaminopropyl)-carbodiimide (EDC).
4. 1 *M* Ethanolamine hydrochloride, pH 8.5.
5. Ligand solution: 1–100 µg/mL ligand in an appropriate immobilization buffer.

2.1.2. Thiol Coupling

Thiol coupling provides an alternative to amine coupling, and is recommended for ligands where amine coupling cannot be used or yields unsatisfactory results, e.g., for acidic proteins or peptides. Generally, thiol coupling is performed in two different approaches: coupling via intrinsic thiol groups of the ligand (e.g., cysteines), or coupling via thiol groups introduced into carboxyl or amino groups of the ligand (e.g., engineered cysteine residues).

2.1.2.1. Intrinsic Ligand Thiol Coupling

1. Ligand solution: 10–200 µg/mL ligand in an appropriate immobilization buffer.
2. 80 m*M* 2-(2-pyridinyldithio)-ethaneamine hydrochloride (PDEA) in 0.1 *M* borate buffer pH 8.5, freshly prepared.
3. 50 m*M* L-cysteine, 1 *M* NaCl in 0.1 *M* formiate buffer, pH 4.3 (cysteine/NaCl), freshly prepared.

2.1.2.2. Surface Thiol Coupling

1. Ligand solution: 1 mg/mL in 0.1 *M* MES buffer, pH 5.0.
2. Fast desalting column (NAP10 column [Amersham Biosciences] or equivalent).
3. 40 m*M* Cystamine dihydrochloride in 0.1 *M* borate buffer, pH 8.5.
4. 0.1 *M* DTT (dithiothreitol) or DTE (1,4-dithioerythritol) in 0.1 *M* borate buffer, pH 8.5.
5. 20 m*M* PDEA, 1 *M* NaCl in 0.1 *M* sodium formate buffer, pH 4.3 (PDEA / NaCl), freshly prepared.

2.1.3. Noncovalent Coupling

Noncovalent coupling is performed using fusion tags, by employing biotinylated components, or by generating lipid-containing sensor surfaces. Fusion proteins are captured via site-specific antibodies against the fusion tag, for example, anti GST or anti poly-His antibodies. For poly-His fusion proteins a patented Ni-NTA sensor chip can be used (*see* **Note 1**) and the following materials are needed:

1. Running buffer: 10 m*M* HEPES (*N*-[2-Hydroxyethyl]piperazine-*N'*-[2-ethanesulfonic acid]), pH 7.4, 150 m*M* NaCl, 50 μ*M* EDTA (ethylenediaminetetraacetic acid), 0.005% Tween-20, filtered and degassed.
2. Nickel solution: 500 μ*M* $NiCl_2$ in running buffer.
3. Ligand solution (do not use additional EDTA or bivalent metal ions in the buffer; nonspecific binding can be prevented by varying ionic strength and pH; addition of 10–20 m*M* imidazole can be advantageous).
4. Regeneration solution: 10 m*M* HEPES, pH 8.3, 150 m*M* NaCl, 350 m*M* EDTA, 0.005% Tween-20.
5. Dispensor buffer: 10 m*M* HEPES, pH 7.4, 150 m*M* NaCl, 3 m*M* EDTA, 0.005% Tween-20.

For the different surfaces an appropriate regeneration solution has to be chosen. An overview of Biacore compatible solutions (e.g., urea, guanidinium hydrochloride, SDS, NaOH) is given under *(1)*.

3. Methods

The following methods are described in detail for Biacore systems, however, they are in principle easily transferable to other biosensor devices. A detailed procedure for the most commonly used immobilization strategy, coupling via primary amines, is given below, for other coupling chemistries refer to *(1)*.

3.1. Immobilization—Step-By-Step Procedure for Coupling Via Primary Amines

1. Let the sensor chip reach ambient temperature before insertion into the sensor device.
2. Equilibrate the system with running buffer (Run *Prime* procedure on the Biacore instrument).
3. Start a sensorgram and wait until a stable baseline is reached (preferably at a high flow rate between 50 and 100 μL/min). For CM5 chips, perform an injection of 10–20 m*M* NaOH for 20–30 s.
4. Switch to the flow cell where the immobilization should take place (if this has not been chosen at the start of the sensorgram). The flow rate should now be set to 5 μL/min. Again, make sure that the baseline is stable.
5. Perform pre-concentration runs, i.e., perform short injections of ligand at different pH and different concentrations to ensure that electrostatic attraction of protein to the dextran matrix yields a sufficient amount for subsequent immobilization. If this is not the case or if excessive nonspecific binding is observed (i.e., signal does not return to baseline levels after switching to running buffer), try a different immobilization buffer/different pH. It is advisable to change the pH in small increments, i.e., 0.1 pH units.

6. Mix the thawed solutions of NHS and EDC in a 1:1 ratio (50 µL each).
7. Inject 40 µL of the mixture (corresponds to 8 min) to activate the CM surface. Select the command *Extraclean* to wash the integrated flow cartridge (IFC) and the needle.
8. Inject the interaction partner until the desired immobilization level is reached. Several injections can be performed.
9. Inject 40 µL of 1 *M* ethanolamine hydrochloride, pH 8.5 (corresponds to 8 min) to deactivate excess reactive groups and to remove noncovalently bound material from the surface. Again perform *Extraclean.*
10. Wash the surface(s) with a washing solution, i.e., a regeneration solution that is tolerated by the ligand and monitor if the baseline is stable.
11. The sensor chip can either be used directly or stored at suitable conditions (*see* **Note 3**). Sometimes it is recommended to run a sensor chip overnight in buffer to assure a stable baseline for the following interaction analysis.

3.2. Kinetic Experiment

1. Insert the sensor chip with immobilized ligand into the Biacore instrument. Make sure that neither side of the chip contains salt deposits or storage solution (you can carefully rinse the chip surfaces with deionized water and soak excess water off the sensor surfaces with a precision wipe placed to one edge on the surface; do not touch the center of the surface containing the immobilized ligand!).
2. Equilibrate the chip in running buffer using the *Prime* procedure.
3. Choose desired flow path.
4. Inject an appropriate dilution of analyte. For kinetic analyses use the *Kinject* command that consumes more analyte, but monitors the dissociation phase without any disturbing peaks caused by needle movements. Note that shifts and bulk effects may occur at the beginning and the end of the injection if the buffer composition of the analyte solution differs from the running buffer. The refractive index of solutions changes dramatically, if even small additional amounts of glycerol, sucrose, detergents or other buffer components are added. This can be overcome by subtracting sensorgrams recorded on reference surfaces. Still, it is highly recommended to match running buffer and analyte buffer as closely as possible, for example, by the use of buffer exchange columns like PD10 or NAP5 (Amersham Biosciences) during analyte preparation.
5. After interaction analysis an appropriate regeneration of the sensor chip surfaces has to be developed to disrupt the analyte-ligand interaction without damaging the biological function of immobilized ligand. For example, a 30 s injection of 10 m*M* glycine pH 2.2 is suitable for immobilized antibodies. If, in case of antibodies, baseline level is not reached, try longer injections, lower the pH carefully in 0.1 pH steps down to pH 1.9 or alternatively use 0.05% sodium dodecyl sulfate (SDS, note that use of SDS may result in a drifting baseline and that you have to wash the surface for a longer time with running buffer or water).
6. Perform a second injection with the same analyte solution at identical conditions to control for stability of the immobilized ligand during the regeneration procedure. No loss of binding activity should be detected.

7. Once these technical details have been established, a series of experiments with several cycles are started. Biacore systems offer the possibility to write methods for automation; additonally a wizard function is available.

3.3. Data Processing

Raw data need to be processed before the sensorgrams are evaluated. Besides the raw data, on-line referencing in advanced Biacore systems provides data where a control surface is already subtracted. With the Biaevaluation software (Biacore AB) rate and equilibrium binding constants can be calculated. The following steps have to be performed before data evaluation (*see* **Subheading 3.4.** and **ref.** ***34***).

1. Zeroing:
   - *y*-axis: zero the response just prior to the start of the association phase.
   - *x*-axis: align the starting points of each injection.
2. Reference subtraction:
   Correct for refractive index changes and nonspecific binding by subtraction of the reference cell. If the binding curves contain bulk shifts data may be difficult to fit. A software routine is available to detect and subtract bulk refractive index changes, however, you should always verify those data manipulations yourself (*see* **Subheading 3.2., step 4**).
3. Overlay:
   All curves of one data set, i.e., a series of concentrations of one analyte, should be overlayed (after **steps 1** and **2** have been performed). In some instances it might be appropriate to overlay data derived from different ligand surfaces and subtract an additional reference surface (for example, *see* **Fig. 5**).

3.4. Evaluation of Kinetic Data

Pre-processed data are now ready to be evaluated. Several kinetic modules are available in the Biaevaluation software. **References** ***34*** and ***35*** discuss potential models for data evaluation. Biaevaluation supports three ways of data evaluation. A global fit module allows for fitting of an entire set of association and dissociation curves with one set of rate constants which improves the robustness of the fitting procedure. Separate fitting of the association and dissociation phase, respectively, is another option. Furthermore, transient kinetics that are often observed with the binding of small molecules are fitted with equilibrium binding analysis using the equation $Y = B_{max}X/ (K_D + X)$. As an example *see* **Figs. 6** and **7**.

A 1:1 Langmuir fit model should be applied as a first try (*see* **Note 4**). However, it is important to consider the biological system first when deciding on the fit model. More complex models of interaction are available. As the complexity of those models increases, the ability to fit the equations to given experimental data will improve automatically! This is simply because there are more degrees of freedom if an increasing number of parameters is applied to generate a close fit. Therefore, assumptions about the mechanism of interaction should be decided on before applying a more complex model. Complex systems are extremely difficult to interpret, and even sophisticated evaluation software cannot substitute for careful experimental design (*see* **Fig. 9** and **Subheading 1.3.3.**).

A global fit where one set of rate constants is used for the approximation to the association and dissociation phase should be performed in order to test the reaction model of choice. *See* **Note 6** for a discussion of pseudo-first order binding kinetics.

The following protocol describes a global fit analysis using BIAevaluation (v3 and higher). For details regarding different models refer to **Note 4**.

1. Open overlayed plot of the processed data from one data set (*see* **Subheading 3.3.**).
2. Choose *Fit kinetics simultaneous* k_a/k_d (= global fit).
3. Select the injection start and end points as well as the area for the association and dissociation phase. This is simplified with the option *split view* (*see* **Note 5**).
4. Enter the concentration of analyte for each curve, choose the appropriate model and press *Fit* (for selecting a model *see* **Note 4**).

3.5. Binding of Warfarin to Serum Proteins: A Practical Approach

In the following the procedures described in **Subheadings 3.1.** to **3.4.** are exemplified in the analysis of the interaction of the low molecular weight compound warfarin, a coumarin derivative, with HSA immobilized to a CM5 chip by amine coupling.

3.5.1. Human Serum Albumin Immobilization

1. Equilibrate a new CM5 sensor chip to room temperature while still enclosed in the nitrogen atmosphere, dock and prime the sensor chip with running buffer (e.g., PBS without DMSO for immobilization).
2. Prepare a solution of 50 µg/mL HSA (essentially fatty acid and globulin free) in 10 m*M* sodium acetate buffer (pH 5.2).
3. Before activation clean the CM5 sensor surfaces with at least two short pulses (20 s) of 20 m*M* NaOH and wait until the baseline is stable.
4. Activate the CM5 surface by injecting a freshly prepared 1:1 mixture of NHS (100 m*M*) and EDC (400 m*M*) for 8 min at a flow rate of 5 µL/min.
5. Inject the HSA solution until a surface density of at least 10.000 RU is reached.
6. Deactivate the HSA surface by injecting 1 *M* ethanolamine (pH 8.5) for 8 min at a flow rate of 5 µL/min.
7. Run the chip overnight in running buffer containing DMSO to achieve a stable baseline.

3.5.2. Dimethyl Sulfoxide Calibration

1. Use freshly prepared, filtered, and degassed running buffer with a well adjusted concentration of organic solvent (DMSO).

2. Prime the IFC and the sensor surfaces at least three times with the appropriate running buffer containing a clearly defined percentage of DMSO.
3. Run a normalize procedure as described for Biacore instruments in order to minimize differences in the refractive indices of the different sensor surfaces followed by another prime procedure.
4. Prepare concentration series of running buffer with different percentages of DMSO, e.g., between 0.5% below and above the DMSO concentration of the running buffer in 0.1% steps.
5. It is very important to avoid evaporation from sample tubes since this will significantly change the buffer composition, in particular the percentage of DMSO.
6. Inject the DMSO concentration series for 1 min each with a flow rate of 30 µL/min using the *Kinject* command and set report points in the middle of the injection phase.
7. Run each calibration curve in duplicate.
8. Plot the original SPR signal from the reference surface (flowcell 1) in RU vs the reference subtracted SPR signal from the HSA surface (flowcell x-1, *see* **Fig. 8**).
9. Perform a linear regression for the DMSO calibration curve from which the correction factors for the samples can be calculated.

3.5.3. Interaction Analysis of Low-Molecular-Weight Ligands

1. Dissolve the low-molecular-weight compound completely in 100% DMSO at room temperature.
2. Centrifuge the solution at 16,000*g* in a tabletop centrifuge for 10 min to remove undissolved constituents.
3. Prepare dilution series of the compound in running buffer and try to match the DMSO concentration exactly to the running buffer. Remember that high refractive index changes induced by varying DMSO concentrations increase the correction factors and thus make the assay less sensitive.
4. Check the solubility of the compound in running buffer at different concentrations. If the compound is not dissolved completely at certain concentrations, centrifuge as described above to remove undissolved constituents but keep in mind that the concentration of compound might be changed significantly by this procedure. Again, it is very important to avoid evaporation.
5. Inject the compound concentration series for 1 min each with a flow rate of 30 µL/min using the *Kinject* command and set report points in the middle of the injection phase to determine the respective signal.
6. Run at least independent duplicates for each concentration series of compound.
7. Consider to perform a sufficient number of blank injections (running buffer without compound) between the samples in order to prevent sample carry over.
8. Since most small ligand interactions display transient kinetics, a regeneration procedure is usually not necessary to remove bound ligand. However, if the solubility of the compound in running buffer is limited, unspecific binding might occur on the serum protein surfaces. In those cases a regeneration procedure using detergents or mild basic solutions like 10 m*M* NaOH is usually effective and will not disturb the biological activity of the immobilized protein.

9. Calculate correction factors for each sample by inserting the SPR signal on the reference flowcell (fc1) in the calibration curve as described in the legend to **Fig. 8**. Subtract the respective correction factors from the reference subtracted signal (fc x-1).
10. Plot the corrected SPR signal against the concentration of the compound and perform a nonlinear regression analysis to yield the K_D-value.

4. Notes

1. Sensor surfaces: the following overview is based on the sensor surfaces produced and marketed by Biacore AB, Uppsala, Sweden. Similar surfaces are also employed in other commercially available or custom-built biosensors.
   Carboxymethylated (CM-) dextran surfaces: the most widely used sensor surface in BIA. It facilitates coupling of biomolecules via primary amine, sulfhydryl, aldehyde, or carboxyl groups. The dextran matrix provides a hydrophilic environment for the biological interaction to take place. These surfaces are available with dextran polymers of different length and different degrees of carboxymethylation. The shorter the dextrans the lower the overall immobilization capacity of the surface which can be helpful in reducing steric effects, while carboxymethylation affects the charge density of the dextran matrix, which can reduce non-specific binding of positively charged molecules.
   Carboxylated surfaces: this surface is devoid of any further modifications besides carboxylation, but supports the same immobilization chemistries as the CM-dextran surfaces. Because of the lack of dextran polymers this surface is more hydrophobic and has a lower immobilization capacity that may be helpful to reduce steric effects when working with high molecular weight ligands.
   Streptavidin (SA)-surfaces: Surfaces with pre-immobilized streptavidin molecules allow efficient capturing of biotinylated ligand molecules, ranging from small molecules, DNA, peptides, and proteins to vesicles containing biotinylated lipids. Unlike covalent coupling, the capturing results in an oriented immobilization of the ligand. Electrostatic preconcentration of the ligand on the surface is not necessary.
   NTA-surfaces: CM-dextran surfaces derivatized with nitrilotriacetic acid (NTA) for capturing of recombinant proteins with a poly-His tag. The matrix is first loaded with Ni^{2+}, then the poly-His tagged ligand is immobilized to the Ni^{2+}-NTA complex via free coordination sites. After binding analysis, the surface can be regenerated with an injection of EDTA.
   Hydrophobic surfaces: Flat hydrophobic surfaces allow lipid vesicles to adsorb directly to the surface, thereby forming a lipid monolayer with the hydrophilic head groups directed towards the soluble phase. Alternatively, dextran surfaces are available modified with lipophilic compounds that permit the immobilization of intact bilayers together with integral membrane proteins.
2. Mass transfer limitations: Mass transfer limitation is a phenomenon that occurs when the association and dissociation rate constants are faster than the diffusion rate of the analyte from the laminar flow zone to the relatively undisturbed sur-

face layer. This leads to a depletion of analyte close to the matrix during the association phase, meaning that binding is no longer interaction-controlled but diffusion-limited. The same holds true for dissociation, as the analyte is not transported away fast enough. A related problem is referred to as rebinding since the analyte might rather bind to the ligand than diffuse into the laminar flow zone. This problem may be overcome by injecting soluble ligand during the dissociation phase. Mass transfer limitations slow down both the association and the dissociation rate. It should be noted that these effects are most pronounced with high molecular weight analytes, because of their low diffusion coefficient.

To test for mass transfer effects, perform interaction analyses at different flow rates and with different surface densities. If kinetics look different the interaction may be prone to mass transfer limitations. In addition, a plot of ln(dR/dt) vs time will display a straight line in mass transfer-limited reactions.

Consequently, mass transfer limitations can be overcome by:

- reduction of the surface density (decreasing available ligand sites on the surface).
- increasing the flow rate (increasing rate of transfer of the analyte to the surface).

3. Storage conditions: Once a covalent immobilization has been performed, the sensor chip with bound ligand may be taken out of the Biacore instrument and put into a 50 mL screw cap tube filled with approx 35 mL buffer (the sensor surface should be covered). It is not recommended to immerse the chip completely, otherwise the buffer may be contaminated when the chip is taken out and/or the labeling may come off further contaminating the buffer.
4. 1:1 (Langmuir) binding:

$$A + L \underset{k_d}{\overset{k_a}{\rightleftharpoons}} AL$$

The Langmuir model displays the simplest situation of an interaction between an analyte (A) and an immobilized ligand (L). It is equivalent to the Langmuir isotherm for adsorption to a surface. The Langmuir isotherm was developed by Irving Langmuir in 1916 to describe the dependence of the surface coverage of an adsorbed gas on the pressure of the gas above the surface at a fixed temperature (***36,37***). The equilibrium that exists between gas adsorbed on a surface and molecules in the gas phase is dynamic, i.e., the equilibrium represents a state in which the rate of adsorption of molecules onto the surface is exactly counterbalanced by the rate of desorption of molecules back into the gas phase. Therefore, it should be possible to derive an isotherm for the adsorption process simply by considering and equating the rates for these two processes. These considerations are also applied to the SPR detection system.

The 1:1 Langmuir module also allows for deviation in the raw data. Sometimes the baseline shows a slight drift that is largely eliminated by the use of a reference cell. However, in analysis with low surface binding capacity (R_{max} levels

100 RU or less) a model including linear drift may be appropriate (1:1 binding with drifting baseline).

A third 1:1 binding model considering mass transfer limitations is also included in the Biaevaluation software. Thus, kinetic data are produced even though the interaction analysed is mass transfer-limited. Yet, it is recommended to perform the experiment in a way to avoid mass transfer limitations as described in **Note 2** and **ref. *1***.

Alternative models for more complex interaction patterns are available such as bivalent analyte, heterogeneous analyte (competing reactions), heterogeneous ligand (parallel reactions), and two-state reaction (conformational change). Refer to the Biaevaluation (v3.0 or later) manual for details.

It is recommended to use the global fit module for data evaluation. However, for some data sets it is necessary to perform a separate k_a/k_d determination, e.g., if one of the phases is obscured by bulk shifts or if different conditions apply during association and dissociation phase. Biaevaluation also includes a module to fit the association and dissociation phases separately. Additionally, a general fit module including 4-parameter equation, linear fit, solution affinity and steady state affinity is available. More models may be imported into the software.

5. Evaluation with *split view*: When evaluating a curve set it is important to know which area should be selected for implementing the fit. The Biaevaluation software offers a *split view* function where the plot window is split into two panels with the original curves in the top panel and derivative functions in the bottom panel. Depending on the part of the sensorgram which should be analyzed the user has the option to choose between several mathematical transformations: for the *dissociation phase*: $\ln(dR_0/R_t)$ vs time (termed $\ln(Y_0/Y)$ in Biaevaluation); for the *association phase* $\ln(dR/dt)$ vs time (termed ln(abs(dY/dX)) in Biaevaluation). This helps to judge whether the model and the parts of the sensorgram selected are appropriate for data evaluation. The functions $\ln(dR/dt)$ and $\ln(dR_0/R_t)$ are linear for 1:1 interactions, constant for mass transfer-limited interactions and curved for more complex systems. It is easier to judge curves in *split view* when the overlay function is turned off. Do not forget to perform the overlay again before proceeding to the next step in the evaluation procedure, i.e., a global fit analysis.
6. The binding of an analyte to a ligand under constant flow is regarded as a pseudo-first order reaction, since the concentration of the analyte is constant in the flow cell. This is not absolutely true, especially with a cuvet system ***(38,39)***; the depletion of analyte may have a significant effect on the analyte concentration. The same might also be true for flow systems; as a result of mass transfer limitations the concentration of analyte might be reduced close to the dextran matrix, where interaction with the immobilized ligand takes place ***(40)*** (*see* **Note 2**). This inherent problem may produce the same kind of deviations from pseudo-first order binding processes. Therefore, global fitting may potentially result in conclusions as doubtful as those derived from conventional linear analysis of data ***(41)***.

Acknowledgments

The authors would like to thank Claudia Hahnefeld and Oliver Diekmann for valuable input. This work was supported by the Bundesministerium für Bildung und Forschung (BMBF, 031U102F) and the European Union (EU CRAFT QLK2-CT-2002-72419).

References

1. Herberg, F. W. and Zimmermann, B. (1999) Analysis of protein kinase interactions using biomolecular interaction analysis. In *Protein Phosphorylation-A Practical Approach* (Hardie, D. G., ed.), Vol. 2, pp. 335–371. Oxford University Press, Oxford.
2. Stenberg, E., Persson, B., Roos, H. and Urbaniczky, C. (1991) Quantitative determination of surface concentration of protein with surface plasmon resonance using radiolabeled proteins. *J. Colloid Interface Sci.* **143,** 513–526.
3. Piehler, J., Brecht, A., Geckeler, K. E., and Gauglitz, G. (1996) Surface modification for direct immunoprobes. *Biosens. Bioelectron.* **11,** 579–590.
4. Piehler, J., Brecht, A., and Gauglitz, G. (1996) Affinity detection of low molecular weight analytes. *Analytical Chemistry* **68,** 139–143.
5. Piehler, J., Brecht, A., Gauglitz, G., Zerlin, M., Maul, C., Thiericke, R., and Grabley, S. (1997) Label-Free Monitoring of DNA-Ligand Interactions. *Analytical Biochemistry* **249,** 94–102.
6. Birkert, O. and Gauglitz, G. (2002) Development of an assay for label-free high-throughput screening of thrombin inhibitors by use of reflectometric interference spectroscopy. *Anal. Bioanal. Chem.* **372,** 141–147.
7. Birkert, O., Tunnemann, R., Jung, G., and Gauglitz, G. (2002) Label-free parallel screening of combinatorial triazine libraries using reflectometric interference spectroscopy. *Anal. Chem.* **74,** 834–840.
8. Kröger, K., Bauer, J., Fleckenstein, B., Rademann, J., Jung, G., and Gauglitz, G. (2002) Epitope-mapping of transglutaminase with parallel label-free optical detection. *Biosensors and Bioelectronics* **17,** 937 – 944.
9. Hanel, C. and Gauglitz, G. (2002) Comparison of reflectometric interference spectroscopy with other instruments for label-free optical detection. *Anal. Bioanal. Chem.* **372,** 91–100.
10. Wink, T., de Beer, J., Hennink, W. E., Bult, A., and van Bennekom, W. P. (1999) Interaction between plasmid DNA and cationic polymers studied by surface plasmon resonance spectrometry. *Analytical Chemistry* **71,** 801–805.
11. Melendez, J., Carr, R., Bartholomew, D. U., Kukanskis, K., Elkind, J., Woodbury, R., Furlong, C., and Yee, S. (1996) A commercial solution for surface plasmon sensing. *Sensors and Actuators B: Chemical* **35,** 212–216.
12. Cush, R., Cronin, J., Steward, W., Maule, C., Molloy, J., and Goddard, N. (1993) The resonant mirror: a novel optical biosensor for direct sensing of biomolecular interactions, Part I: Principle of operation and associated instrumentation. *Biosens. Bioelectron.* **8,** 347–354.

13. Cunningham, B., Li, P., Lin, B., and Pepper, J. (2002) Colorimetric resonant reflection as a direct biochemical assay technique. *Sensors and Actuators B: Chemical* **81,** 316–328.
14. Baird, C. L. and Myszka, D. G. (2001) Current and emerging commercial optical biosensors. *J. Mol. Recognit.* **14,** 261–268.
15. Rich, R. L. and Myszka, D. G. (2003) A survey of the year 2002 commercial optical biosensor literature. *J. Mol. Recognit.* **16,** 351–382.
16. Zimmermann, B., Hahnefeld, C., and Herberg, F. W. (2002) Applications of biomolecular interaction analysis in drug development. *TARGETS* **1,** 66–73.
17. Engvall, E. (1980) Enzyme immunoassay ELISA and EMIT. *Methods Enzymol.* **70,** 419–439.
18. Karlsson, R. (1994) Real-time competitive kinetic analysis of interactions between low-molecular-weight ligands in solution and surface-immobilized receptors. *Anal. Biochem.* **221,** 142–151.
19. Dillon, P. P., Daly, S. J., Manning, B. M., and O'Kennedy, R. (2003) Immunoassay for the determination of morphine-3-glucuronide using a surface plasmon resonance-based biosensor. *Biosensors and Bioelectronics* **18,** 217 – 227.
20. Tudos, A. J., Lucas-van den Bos, E. R., and Stigter, E. C. (2003) Rapid surface plasmon resonance-based inhibition assay of deoxynivalenol. *J. Agric. Food Chem.* **51,** 5843–5848.
21. Deinum, J., Gustavsson, L., Gyzander, E., Kullman-Magnusson, M., Edström, Å., and Karlsson, R. (2002) A thermodynamic characterization of the binding of thrombin inhibitors to human thrombin, combining biosensor technology, stopped-flow spectrophotometry, and microcalorimetry. *Analytical Biochemistry* **300,** 152–162.
22. Day, Y. S., Baird, C. L., Rich, R. L., and Myszka, D. G. (2002) Direct comparison of binding equilibrium, thermodynamic, and rate constants determined by surface- and solution-based biophysical methods. *Protein Sci.* **11,** 1017–1025.
23. Roos, H., Karlsson, R., Nilshans, H., and Persson, A. (1998) Thermodynamic analysis of protein interactions with biosensor technology. *J. Mol. Recognit.* **11,** 204–210.
24. Markgren, P. O., Schaal, W., Hamalainen, M., Karlen, A., Hallberg, A., Samuelsson, B., and Danielson, U. H. (2002) Relationships between structure and interaction kinetics for HIV-1 protease inhibitors. *J. Med. Chem.* **45,** 5430–5439.
25. Karlsson, R., Kullman-Magnusson, M., Hamalainen, M. D., Remaeus, A., Andersson, K., Borg, P., Gyzander, E., and Deinum, J. (2000) Biosensor analysis of drug-target interactions: direct and competitive binding assays for investigation of interactions between thrombin and thrombin inhibitors. *Anal. Biochem.* **278,** 1–13.
26. Gestwicki, J. E., Hsieh, H. V., and Pitner, J. B. (2001) Using receptor conformational change to detect low molecular weight analytes by surface plasmon resonance. *Anal. Chem.* **73,** 5732–5737.
27. Carrasco, C., Facompre, M., Chisholm, J. D., Van Vranken, D. L., Wilson, W. D., and Bailly, C. (2002) DNA sequence recognition by the indolocarbazole antitu-

mor antibiotic AT2433-B1 and its diastereoisomer. *Nucl. Acids. Res.* **30,** 1774–1781.

28. Hendrix, M., Priestley, S. E., Joyve, G. F., and Wong, C.-H. (1997) Direct observation of aminoglycoside-RNA interactions by surface plasmon resonance. *J. Am. Chem. Soc.* **119,** 3641–3648.
29. Chapman, R. L., Stanley, T. B., Hazen, R., and Garvey, E. P. (2002) Small molecule modulators of HIV Rev/Rev response element interaction identified by random screening. *Antiviral Res.* **54,** 149–162.
30. Li, K., Davis, T. M., Bailly, C., Kumar, A., Boykin, D. W., and Wilson, W. D. (2001) A heterocyclic inhibitor of the REV-RRE complex binds to RRE as a dimer. *Biochemistry* **40,** 1150–1158.
31. Cheng, X., Phelps, C., and Taylor, S. S. (2001) Differential binding of cAMP-dependent protein kinase regulatory subunit isoforms Ialpha and IIbeta to the catalytic subunit. *J. Biol. Chem.* **276,** 4102–4108.
32. Rich, R. L., Day, Y. S., Morton, T. A., and Myszka, D. G. (2001) High-resolution and high-throughput protocols for measuring drug/human serum albumin interactions using BIACORE. *Anal. Biochem.* **296,** 197–207.
33. Yaqub, S., Abrahamsen, H., Zimmerman, B., Kholod, N., Torgersen, K. M., Mustelin, T., Herberg, F. W., Tasken, K., and Vang, T. (2003) Activation of C-terminal Src kinase (Csk) by phosphorylation at serine-364 depends on the Csk-Src homology 3 domain. *Biochem. J.* **372,** 271–278.
34. Myszka, D. G. (2000) Kinetic, equilibrium, and thermodynamic analysis of macromolecular interactions with BIACORE. *Methods Enzymol.* **323**, 325–340.
35. Karlsson, R. and Falt, A. (1997). Experimental design for kinetic analysis of protein-protein interactions with surface plasmon resonance biosensors. *J. Immunol. Methods* **200,** 121 133.
36. Langmuir, I. (1916) The constitution and fundamental properties of solids and liquids. Part I. Solids. *J. Am. Chem. Soc.* **38,** 2221–2295.
37. Langmuir, I. (1918) The adsorption of gases on plane surfaces of glass, mica and platinum. *J. Am. Chem. Soc.* **40,** 1361–1403.
38. Hall, D. R., Gorgani, N. N., Altin, J. G., and Winzor, D. J. (1997) Theoretical and experimental considerations of the pseudo-first-order approximation in conventional kinetic analysis of IAsys biosensor data. *Anal. Biochem.* **253,** 145–155.
39. O'Shannessy, D. J. and Winzor, D. J. (1996) Interpretation of deviations from pseudo-first-order kinetic behavior in the characterization of ligand binding by biosensor technology. *Anal. Biochem.* **236,** 275–283.
40. Hall, D. R., Cann, J. R., and Winzor, D. J. (1996) Demonstration of an upper limit to the range of association rate constants amenable to study by biosensor technology based on surface plasmon resonance. *Anal. Biochem.* **235,** 175–184.
41. Schuck, P. and Minton, A. P. (1996) Analysis of mass transport-limited binding kinetics in evanescent wave biosensors. *Anal. Biochem.* **240,** 262–272.

3

Label-Free Detection of Protein–Ligand Interactions by the Quartz Crystal Microbalance

Andreas Janshoff and Claudia Steinem

Summary

In recent years the quartz crystal microbalance (QCM) has been accepted as a powerful technique to monitor adsorption processes at interfaces in different chemical and biological research areas. In the last decade, the investigation of adsorption of biomolecules on functionalized surfaces turned out to be one of the paramount applications of the QCM comprising the interaction of nucleic acids, specific molecular recognition of protein-receptor couples, and antigen–antibody reactions realized in immunosensors. The advantage of the QCM technique is that it allows for a label free detection of molecules. This is a result of the fact that the frequency response of the quartz resonator is proportional to the increase in thickness of the adsorbed layer. However, in recent years it became more and more evident that quartz resonators used in fluids are more than mere mass or thickness sensors. The sensor response is also influenced by viscoelastic properties of the adhered biomaterial, surface charges of adsorbed molecules and surface roughness. These phenomena have been used to get new insights in the adhesion process of living cells and to understand their response to pharmacological substances by determining morphological changes of the cells. In this chapter we describe a protocol to explore the kinetics and thermodynamics of specific interactions of different proteins such as lectins and annexins with their ligands using receptor bearing solid supported lipid bilayers.

Key Words: Lipid bilayer; gold surface; annexin; lectin; peanut agglutinin; ganglioside; thickness shear mode resonator; bulk acoustic wave sensor.

1. Introduction

Owing to the proportionality between resonance frequency and mass change first analytically described in 1959 by Sauerbrey *(1)*, the quartz crystal micro-

From: *Methods in Molecular Biology, vol. 305: Protein–Ligand Interactions: Methods and Applications*
Edited by: G. U. Nienhaus © Humana Press Inc., Totowa, NJ

balance (QCM) became an invaluable tool as a mass and thickness sensor in gas phase and in vacuum. For the majority of bioanalytical applications it was, however, necessary to follow adsorption processes in an aqueous environment. Only the development of oscillator circuits capable of exciting quartz resonators under liquid load in the 1980s paved the way for the QCM to be used in biological applications *(2)*. Nowadays, quartz resonators are well recognized as biological sensor devices to quantify adsorption processes of biomolecules on functionalized surfaces *(3–5)*. These comprise the interaction of DNA and RNA with complementary strands, specific recognition of protein–ligands by immobilized receptors, the detection of virus capsids, bacteria, mammalian cells *(6)*, and the development of complete immunosensors. The sensitivity of a QCM is mainly determined by the fundamental resonance frequency of the quartz resonator. For the majority of applications, the QCM technique is used to monitor the adsorption of molecules with a molecular mass of more than 10 kDa. However, as is demonstrated in this chapter, in certain cases, the technique may also be applied to detect the interaction of small ligands with proteins by using a competition assay. An indirect monitoring is then used as the mass of the ligands would in principle be too small to be detected with a common QCM.

2. Materials

2.1. Quartz Plates

1. 5 MHz polished AT-cut quartz crystals (plano-plano) (Kristallverarbeitung KVG, Neckarbischofsheim, Germany).
2. Gold (Degussa AG, Hanau, Germany), chromium (BalTec, Balzers, Liechtenstein).
3. Silver conductive paint (RS components, Mörfelden-Walldorf, Germany).

2.2. Lipid Bilayers

1. 1-Palmitoyl-2-oleoyl-*sn*-glycero-3-phosphocholine (POPC), 1-palmitoyl-2-oleoyl-*sn*-glycero-3-phosphoserine (POPS) (Avanti Polar Lipids, Alabaster, AL), G_{M1} (Sigma-Aldrich, Taufkirchen, Germany).
2. Octanethiol (Fluka, Neu-Ulm, Germany).

2.3. PNA-Carbohydrate Interaction

1. Peanut agglutinin (Sigma-Aldrich).
2. β-Gal*p*-(1→3)-GalNAc, β-D-galactose (Sigma-Aldrich).
3. Buffer: 50 m*M* Tris-HCl, 200 m*M* NaCl, pH 7.4.

2.4. Annexin-Ca^{2+} Interaction

1. Porcine heterotetrameric annexin A2 (A2t) is purified from intestine according to a procedure of Gerke and Weber *(7)*. Protein concentration is determined using an extinction coefficient of $\varepsilon_{280} = 0.65\ cm^2\ mg^{-1}$.

2. Calciumchloride dihydrate (Fluka), ethylene glycol-bis(β-aminoethyl) ether-*N,N,N',N'*-tetraacetic acid (EGTA) (Sigma-Aldrich).
3. Buffer: 20 m*M* Tris-HCl, 100 m*M* NaCl, 1 m*M* NaN_3, pH 7.4.

3. Methods

3.1. Active Oscillator Mode—QCM Setup

The experimental setup for the QCM-measurements is rather simple and low-priced **(Fig. 1).** A disk-like plano–plano AT-cut quartz resonator, which is also termed thickness shear mode (TSM) resonator or bulk acoustic wave (BAW) sensor with a diameter of 14 mm and a fundamental resonance frequency of 5 MHz is the core component of the system. The quartz plate is sandwiched between two evaporated disk-like gold electrodes covering an area of 0.3 cm^2 each (*see* **Note 1**). Both gold electrodes are connected to an oscillator circuit from the bottom of the quartz plate via gold spring contacts (*see* **Note 2**). Stable oscillation of a quartz plate only occurs at the resonance frequency of the crystal. If the crystal is incorporated into a feedback loop of an oscillator circuit, it becomes the frequency determining element as its quality factor is very large. Free oscillation of thickness shear mode resonators is restricted to load situations with the phase maximum above zero. At a phase maximum below zero degree phase shift caused by high damping, active oscillation breaks down. A self-made oscillator circuit is used composed of an integrated circuit SN74LS124N from Texas Instruments connected to a frequency counter from Agilent (HP 53181A). Data are collected by a personal computer via RS232 from the frequency counter. The crystal holder is made of teflon and allows the immersion of one of the two quartz surfaces to an aqueous solution. The geometry of the flow cell is designed in a way that a stagnation flow point results, which ensures fast and predictable mass transport to the crystal **(Fig. 1)** *(8)*. An overall volume of 2 mL is pumped through the system by using a peristaltic pump. The crystal and the oscillator circuit are placed in a temperature-controlled chamber, which also serves as a Faraday cage. The temperature is kept constant at 20°C.

3.2. Sensitivity of a QCM

The sensitivity of a quartz plate is defined according to the Sauerbrey equation:

$$\Delta f = -\frac{2Nf_0^2}{A\sqrt{\mu_q \rho_q}}\Delta m = -S_f \Delta m \tag{1}$$

The integral mass sensitivity or Sauerbrey-constant S_f depends on the square of the fundamental frequency f_0. ρ_q is the density and μ_q the piezoelectric stiffened shear modulus of quartz, *A* the electrode area, and *N* the overtone number.

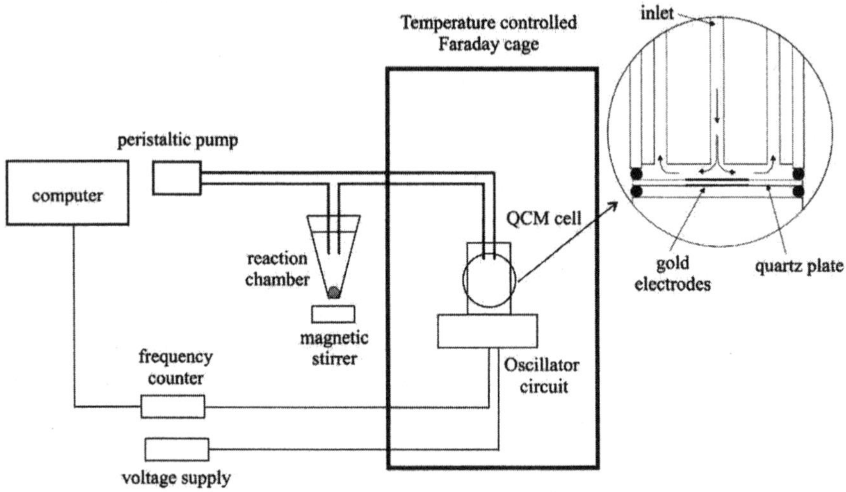

Fig. 1. Experimental setup used for QCM measurements. The magnification depicts the stagnation flow point geometry of the QCM cell.

According to **Eq. 1** a 5 MHz quartz plate exhibits a sensitivity S_f of 0.057 Hz cm^2/ng, whereas practical sensitivities are usually smaller. For plano-plano crystals, Hillier and Ward *(9)* determined a value of 0.036 Hz cm^2/ng for S_f. As a consequence, the mass sensitivities should be determined in a separate experiment prior to the affinity sensing. For instance, electrodeposition of metals like copper gives good results. For the detection of small ligands (ions, small organic molecules) that bind to immobilized proteins the sensitivity is, in most cases, not large enough. It cannot be generally ruled out that small molecule binding cannot be detected in aqueous solution as surface effects such as charge, immobilized water and changes in viscoelasticity might add to the overall observed frequency change and thus might increase the apparent sensitivity (*see* **Note 3**). In most cases, however, for small molecule binding the sensitivity must be increased by increasing the fundamental resonance frequency of the quartz plate or by working with high overtone numbers N. Resonators with a fundamental resonance frequency of up to 27 MHz are frequently used ***(10,11)***. The natural limitation of using quartz plates with even larger frequencies is their thickness, which is inversely proportional to the resonance frequency.

3.3. Kinetic and Thermodynamic Treatment of the Data

The most challenging part of investigating protein–ligand interactions at an interface is data treatment to obtain quantitative parameters. Assuming a reversible binding of the protein to the surface confined receptor, the adsorption kinetics can be described in terms of the following master equation ***(12–14)***:

$$\frac{d\Theta}{dt} = k_{\mathrm{on}}\pi a^2 n_v(\delta)\Phi(\Theta) - k_{\mathrm{off}}\Theta\psi(\Theta), \tag{2}$$

in which Θ denotes the coverage of the surface ranging from 0 to 1, k_{on} and k_{off} are the rate constants for adsorption and desorption of the protein, πa^2 is the area occupied by one protein, $\Phi(\Theta)$ is the surface blocking function, $\psi(\Theta)$ the release function, and $n_v(\delta)$ the protein density (number of proteins per unit volume) at the position of the interface. For the Langmuir case, in which independent adsorption sites are assumed, $\Phi(\Theta)$ is approximated by $1-\Theta$ and $\psi(\Theta) = 1$. This is, however, not realistic for a coverage larger than 0.1.

Because the proteins need to be transported to the surface of the quartz plate in order to overcome the barrier for adsorption we have to consider some aspects of mass transport. The flux towards the surface can be approximated by:

$$j = k_{\mathrm{D}}\left[n_v(\infty) - n_v(\delta)\right] \tag{3}$$

with $n_v(\infty)$, the protein number density in the bulk, $n_v(\delta)$ that at the interface, and k_{D} the transport coefficient that can be approximated by

$$k_{\mathrm{D}} \approx 0.78\left(\left[1.78 + 0.186\,\eta_{\mathrm{k}}^{-1}RV_{\mathrm{m}} + 0.034\left(\eta_{\mathrm{k}}^{-1}RV_{\mathrm{m}}\right)^2\right]D_{\infty}^2V_{\mathrm{m}}R^{-2}\right)^{1/3}$$

for the radial impinging jet assuming small Reynolds numbers (Re < 40) and $k_{\mathrm{D}}=(D_{\infty}/\pi t)^{1/2}$ for diffusion controlled adsorption. D_{∞} is the unrestricted self-diffusion constant of the molecules in solution, η_{k} is the kinematic viscosity of the solution ($\eta_{\mathrm{k}} = 10^{-6}$ m^2/s), V_{m} is the mean linear velocity in the cell, and R the capillary radius. It is not possible to find an analytical solution for the full range of coverage, but it is possible to find approximate solutions for low and high coverage. At very low coverage it is reasonable to use the Langmuir approximation, however, it is necessary to employ a smaller maximum coverage than 1. Following the two-dimensional random sequential adsorption scheme the maximal coverage for spherical particles is 0.547 as determined from Monte Carlo simulations. Because the QCM rarely provides absolute numbers for the coverage it is just good to keep in mind that a real coverage of 1 is physically not possible.

If adsorption is slow as compared to mass transport to the interface and the blocking function can be approximated by $\Phi(\Theta) = 1-\Theta$ and the release function is $\psi(\Theta) = 1$, which is identical with the assumption that binding is noncooperative, it is reasonable to use the Langmuir adsorption isotherm, which provides the easiest approach to obtain numbers for the binding constant. For the adsorption kinetics one then obtains:

$$\Theta = 1 - \exp\left\{-\left(k_{on}c + k_{off}\right)t\right\} \tag{4}$$

and consequently for the isotherm:

$$\Theta = \frac{K_a c}{1 + K_a c} \tag{5}$$

with c, the bulk concentration of the protein in solution and K_a the association constant ($K_a = k_{on}/k_{off}$).

3.4. Immobilization of Lipid Bilayers

Prerequisite for a sensitive detection of the adsorption of a molecule to an interface without interferences from nonspecific adsorption is proper surface functionalization. Here, solid supported membranes are advantageous over a simple derivatization of the surface as lipid bilayers prevent nonspecific adsorption and allow a controlled and highly oriented immobilization of receptor molecules on surface *(15)*. Preparation of solid supported lipid bilayers on gold surfaces of a quartz plate is achieved in two steps. First, a chemisorbed layer of an alkanethiol is deposited on the surface and in the second step a phospholipid monolayer is physisorbed on top of this layer by fusion of large unilamellar vesicles (**Fig. 2**).

3.4.1. Vesicle Preparation

The entire procedure of vesicle preparation is performed above the main phase transition of the lipid mixture. For POPC/ganglioside and POPC/POPS mixtures the procedure can thus be done at room temperature.

1. Pipet chloroform/methanol (1:1, *v/v*) stock solutions of the lipids (typically 10 mg/mL) in a cleaned glass test tube in the desired molar ratio.
2. Slowly evaporate the organic solvent in a stream of nitrogen to obtain lipid films at the bottom of glass test tubes.
3. Remove remaining solvent within the resulting lipid film in vacuum for 2–3 h. These lipid films can be stored at 4°C for at least 1–2 wk.
4. Add buffer to the lipid film leading to a final lipid concentration of 0.5–1 mg/mL and incubate for at least 10 min. After the lipid film has been swollen, vortex the suspension for 30 s, incubate for 5 min in the water bath above the phase transition temperature of the lipid and periodically repeat this procedure four times, which eventually results in the formation of multilamellar vesicles.
5. Large unilamellar vesicles are obtained by the extrusion method using a miniextruder. Press the lipid suspension 31 times through two stacked polycarbonate membranes with a defined pore diameter of 100 nm using a miniextruder

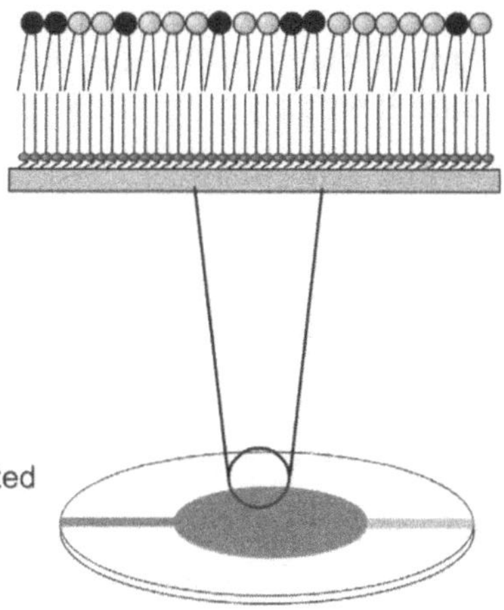

Fig. 2. Schematic representation of a lipid bilayer composed of a first chemisorbed octanethiol and a physisorbed phospholipid monolayer immobilized on a gold electrode of a quartz plate.

(Avestin Europe GmbH, Mannheim, Germany). With this method, unilamellar vesicles with a mean diameter of around 80–120 nm are obtained.

3.4.2. Preparation of a Chemisorbed Octanethiol Monolayer

1. Clean the quartz plate with the evaporated gold electrodes in Piranha solution (70% sulfuric acid, 30% hydrogen peroxide, *v/v*) for 5 min and rinse thoroughly with ultrapure water.
2. Expose the quartz plate for 5 min to high energy argon plasma (plasma cleaner, Harrick Scientific Corporation, New York, NY) to remove organic contaminations.
3. Insert the quartz plate quickly into the quartz holder and add a 1 m*M* ethanolic solution of octanethiol. Incubate the surface for at least 30–60 min to allow for chemisorption of the thiol component on the gold electrode and to ensure complete surface coverage.
4. Rinse with first ethanol and then with the appropriate buffer that is also used for physisorption of the second monolayer obtained by vesicle spreading and fusion (*see* **Note 4**).

3.4.3. Formation of the Second Phospholipid Monolayer

1. Add the unilamellar vesicles immediately after rinsing the octanethiol monolayer with buffer. A final lipid concentration of 0.2 mg/mL is appropriate.
2. Incubate the surface for at least 1 h above the main phase transition of the lipid mixture.

3. After incubation, rinse with buffer to remove non-bound vesicles from solution and from the surface. Make sure that the surface never gets dry to prevent disruption of the formed lipid bilayer.

3.5. Interaction of Peanut Agglutinin With Gangliosides

As a first example, the interaction of the lectin peanut agglutinin (PNA) with ganglioside containing lipid bilayers is described. By means of QCM experiments, the molecular requirements of the carbohydrate ligand for high affinity binding of the lectin can be elucidated ***(16,17)***.

3.5.1. Molecular Structure of the PNA Carbohydrate Ligand

Solid supported membranes allow embedding of different carbohydrate ligands for the lectin PNA, whereas the lipid matrix minimizes nonspecific protein binding. The lectin PNA from *Arachis hypogaea* is composed of four identical subunits each harboring a specific binding site for the carbohydrate structure β-Gal*p*-(1→3)-GalNAc. The specificity of PNA binding to different carbohydrate structures immobilized on the solid support can be quantified by monitoring adsorption isotherms.

1. Prepare lipid bilayers composed of octanethiol and POPC/G_{M1} (4.8 mol% G_{M1}) and octanethiol and POPC/asialo-G_{M1} (4.8 mol% asialo–G_{M1}), respectively, on the gold surface of a quartz plate.
2. Add stepwise different PNA concentrations to the external vial and monitor the QCM response.
3. Read out the equilibrium frequency shift after each protein addition and plot Δf vs the protein concentration in solution.

From the adsorption isotherm, the dissociation constants can be extracted by fitting a Langmuir-adsorption isotherm (**Eq. 5** to the data [**Fig. 3**] [*see* **Note 5**]).

For G_{M1} as membrane confined receptor $\Delta f_{max} = (61 \pm 1)$ Hz and $K_a = (0.83 \pm 0.04) \cdot 10^6 M^{-1}$ are obtained, whereas for *asialo*-G_{M1} K_a reads $(6.5 \pm 0.3)\ 10^6 M^{-1}$ and $\Delta f_{max} = (28 \pm 1)$ Hz. The significant higher binding constant of PNA to asialo-G_{M1} demonstrates that the disaccharide β-Galp-(1→3)-GalNAc is a high affinity receptor for PNA, while the additional negatively charged sialic acid in G_{M1} diminishes the binding affinity of PNA to the saccharide.

3.5.2. Determination of Binding Constants of Different Carbohydrate Structures

A lot of different carbohydrates are available as soluble components but only a few are linked to a lipid moiety so that they can be inserted into a lipid bilayer. Thus, it is advantageous to use a competition assay based on soluble carbohydrates to determine their binding constants to PNA.

The assay is based on the response of a quartz plate upon addition of a defined concentration of PNA to a G_{M1}-doped POPC membrane in the absence

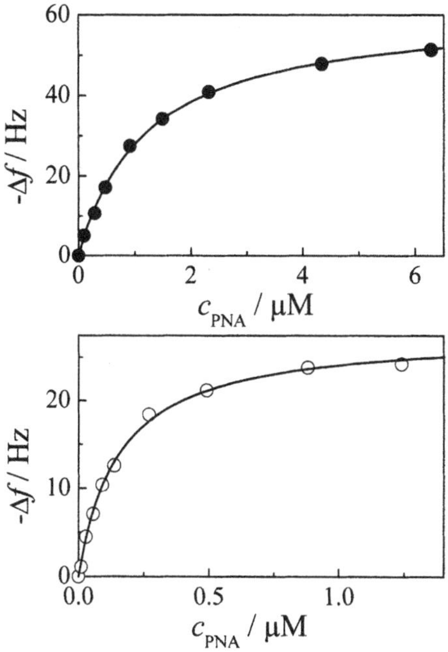

Fig. 3. Adsorption isotherms of PNA on (•) G_{M1} (4.8 mol%) containing OT/POPC-bilayers and on (○) *asialo*–G_{M1} (4.8 mol%) containing OT/POPC layers. The solid lines represent the results of the fitting procedure according to **Eq. 5**. Fitting parameters: (•) $K_a = (0.83 \pm 0.04) \cdot 10^6 M^{-1}$, $\Delta f_{max} = (61 \pm 1)$ Hz; (○) $K_a = (6.5 \pm 0.3) \cdot 10^6 M^{-1}$, $\Delta f_{max} = (28 \pm 1)$ Hz.

and presence of a carbohydrate that is competing with the binding site on the surface.

1. Prepare a lipid bilayer composed of octanethiol and POPC/G_{M1} (4.8 mol% G_{M1}) on the gold surface of a quartz plate.
2. Add 2 μ*M* PNA to a G_{M1}-containing POPC-monolayer, first in the absence, second in the presence of 26.5 m*M* β-D-galactose and third in the presence of 0.265 m*M* β-Gal*p*-(1→3)-GalNAc.
3. Read out the corresponding frequency shifts. Binding of PNA in the presence of the corresponding ligand causes a lower frequency decrease than in the absence of ligands caused by a reduced concentration of accessible PNA binding sites in the bulk solution.

3.5.3. Theory and Data Workup

In order to obtain the intrinsic binding constant $K_{S,i}$ of one binding site of PNA to the corresponding carbohydrate S in solution the following approach can be pursued: PNA harbors four identical subunits each capable of binding a carbohydrate ligand with the same intrinsic binding constant $K_{S,i}$. Five different PNA-carbohydrate complexes have thus to be distinguished: PNA, PNA-S, PNA-S_2, PNA-S_3, and PNA-S_4. Except for PNA-S_4, all species still have free binding sites allowing them to bind to the immobilized ganglioside G_{M1}. The corresponding equilibria at the membrane surface according to the simple Langmuir model can be written as follows:

$$\begin{aligned} 4k_{a,i}c_{PNA}\left(1-\Theta_t\right) &= k_{d,i}\Theta_{PNA} \\ 3k_{a,i}c_{PNA\text{-}S}\left(1-\Theta_t\right) &= k_{d,i}\Theta_{PNA\text{-}S} \\ 2k_{a,i}c_{PNA\text{-}S_2}\left(1-\Theta_t\right) &= k_{d,i}\Theta_{PNA\text{-}S_2} \\ k_{a,i}c_{PNA\text{-}S_3}\left(1-\Theta_t\right) &= k_{d,i}\Theta_{PNA\text{-}S_3} \end{aligned} \tag{6}$$

Θ_t is the total surface coverage and can be expressed as the sum of the coverage of each PNA species:

$$\Theta_t = \Theta_{PNA} + \Theta_{PNA\text{-}S} + \Theta_{PNA\text{-}S_2} + \Theta_{PNA\text{-}S_3} \tag{7}$$

$k_{a,i}$ and $k_{d,i}$ are the intrinsic rate constants of adsorption and desorption of one binding site, respectively. After some algebra the Langmuir equation reads:

$$\frac{\Delta f}{\Delta f_{max}} = \frac{K_{a,i}\left(4c_{PNA} + 3c_{PNA\text{-}S} + 2c_{PNA\text{-}S_2} + c_{PNA\text{-}S_3}\right)}{1 + K_{a,i}\left(4c_{PNA} + 3c_{PNA\text{-}S} + 2c_{PNA\text{-}S_2} + c_{PNA\text{-}S_3}\right)} \tag{8}$$

The intrinsic Langmuir adsorption constant $K_{a,i} = k_{a,i}/k_{d,i}$ can be derived from the Langmuir adsorption constant $K_a = 4K_{a,i}$ assuming that all binding sites of the PNA-carbohydrate complexes exhibit the same binding constant to G_{M1}. The sum in brackets of **Eq. 8** represents the total concentration of binding sites c_{pro} (pro = protomer) capable of interacting with the surface confined receptor:

$$c_{pro} = 4c_{PNA} + 3c_{PNA\text{-}S} + 2c_{PNA\text{-}S_2} + c_{PNA\text{-}S_3} \tag{9}$$

Taking the protomer concentration in solution, equilibrium between the carbohydrate and PNA-binding sites reads:

$$\text{protomer} + \text{S} \rightleftharpoons \text{promotor} - \text{S}$$

The intrinsic binding constant $K_{s,i}$ can be derived from equilibrium:

$$K_{s,i} = \frac{c_{pro\text{-}S}}{c_{pro}c_S} = \frac{c_{pro\text{-}S}}{c_{pro}\left(c_{S,0} - c_{pro\text{-}S}\right)}, \tag{10}$$

where $c_{S,0}$ is the initial concentration of the corresponding water soluble carbohydrate. $c_{pro\text{-}S}$, c_S and c_{pro} are the equilibrium concentrations of the corresponding species. Assuming that the amount of protomers bound to the surface is negligible compared to the total number of protomers ($4c_{PNA,0}$) the intrinsic binding constant $K_{S,i}$ can be calculated from the equilibrium concentration of the unbound protomers c_{pro}:

$$K_{s,i} = \frac{4c_{PNA,0} - c_{pro}}{c_{pro}\left[c_{S,0} - \left(4c_{PNA,0} - c_{pro}\right)\right]} \quad (11)$$

c_{pro} can be directly calculated from the frequency shift Δf.

For PNA binding ($c_{PNA,0}$ = 2 μ*M*) in the presence of $c_{S,0}$ = 0.265 m*M* β-Gal*p*-(1→3)-GalNAc a frequency shift of (13 ± 2) Hz translates into an inhibition constant of $K_{S,i}$ = (20 ± 5)·10^3 M^{-1}. A frequency shift of (11 ± 2) Hz obtained for the adsorption of 2 μ*M* PNA in the presence of 26.5 m*M* β-D-galactose leads to $K_{S,i}$ = (250 ± 50) M^{-1} (*see* **Note 6**).

3.6. Interaction of Annexin A2t With Lipid Bilayers

As a second example, it is described, how the Ca^{2+}-requirement of the protein annexin A2t for binding to acidic lipids can be quantified by the QCM technique. Annexin A2t is a heterotetrameric protein composed of two annexin A2 molecules and one S100 A10 homodimer that binds in a Ca^{2+}-dependent manner to negatively charged phospholipids ***(18)***.

3.6.1. Affinity of Annexin A2t to Lipid Bilayers

As annexin binding to a lipid membrane in the presence of calcium ions is a three-component system, the amount of membrane bound protein depends, besides the membrane composition, on two parameters, the protein and the free Ca^{2+}-concentration. For a simpler data treatment, it is wise to work under conditions where one component is present in excess while the other ones' concentration is varied. To obtain the protein concentration required for maximum surface coverage, first an adsorption isotherm is monitored at a high calcium ion concentration of 1 m*M*, while the annexin A2t concentration is varied. All measurements are done in a buffer composed of 20 m*M* Tris-HCl, 0.1 *M* NaCl, 1 m*M* NaN_3, pH 7.4 with either $CaCl_2$ or EGTA as indicated.

1. Prepare a lipid bilayer composed of octanethiol and POPC/POPS (4:1) on the gold surface of a quartz plate.
2. Add different annexin A2t concentrations to the external vial and monitor the QCM response.
3. Read out the equilibrium frequency shift after each protein addition and plot Δf vs the protein concentration in solution.

From the adsorption isotherm, the protein concentration necessary for maximum surface coverage can be read out, which amounts to at least 0.35 m*M* in this case (**Fig. 4**). This is the concentration that will be used for the experi-

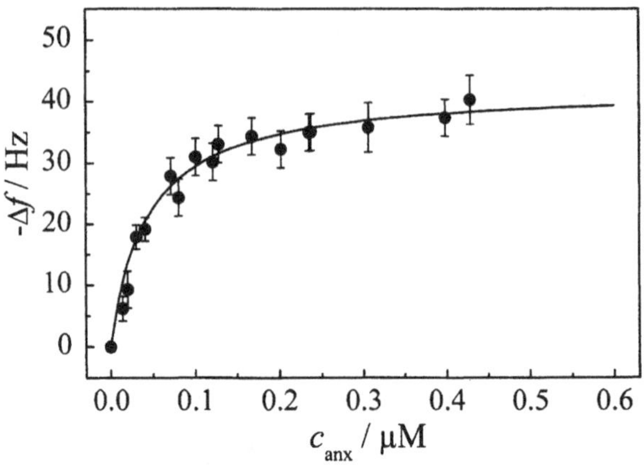

Fig. 4. Adsorption isotherm of annexin A2t to a solid supported membrane composed of octanethiol and a POPC/POPS (4:1) monolayer in the presence of 1 m*M* $CaCl_2$. The solid line is the result of fitting **Eq. 5** to the data with the following result: $K_a = (2.3 \pm 0.3) \cdot 10^7 M^{-1}$, $\Delta f_{max} = (42 \pm 2)$ Hz.

ments in which the Ca^{2+}-concentration is varied and the protein concentration is kept constant.

3.6.2. Determination of Ca^{2+}-Dissociation Constants

Two different procedures are conceivable to quantify the required Ca^{2+}-concentration of annexin A2t binding to solid supported membranes ***(19)***:

1. For this procedure, the protein is first bound to the membrane in the presence of Ca^{2+} followed by a stepwise decrease in Ca^{2+}-concentration by adding EGTA.
   a. Add annexin A2t with a concentration of 0.35 μ*M* to a solid supported membrane in the presence of 1 m*M* $CaCl_2$ and monitor the frequency shift ($\Delta f_{\text{Anx A2t,Ca}^{2+}}$) to control the protein adsorption process.
   b. Rinse with buffer containing 1 m*M* $CaCl_2$.
   c. Add increasing amounts of an EGTA-containing buffer (5 or 50 m*M* EGTA) and monitor the frequency shift time resolved. The stepwise increase in resonance frequency is termed Δf_{EGTA}.
   d. Calculate the free Ca^{2+}-concentration taking the pH and the complex binding constants of Ca^{2+} and EGTA into account ***(20)***.
   e. Plot the apparent coverage

$$\Theta = \frac{\Delta f_{\text{Anx A2t,Ca}^{2+}} - \Delta f_{\text{EGTA}}}{\Delta f_{\text{Anx A2t,Ca}^{2+}}}$$

   as a function of the free Ca^{2+}-concentration.

2. For the second procedure, annexin A2t is added to an EGTA-containing solution, and membrane binding is triggered by stepwise increasing the Ca^{2+}-concentration.
   a. Add annexin A2t to solid supported membranes in the presence of 0.1 m*M* EGTA buffer, which typically leads to a faint frequency shift termed $\Delta f_{\text{Anx A2t, EGTA}}$.
   b. Add increasing amounts of a Ca^{2+}-containing buffer (1 or 10 m*M* $CaCl_2$) and monitor the decrease in resonance frequency indicative of a stepwise protein adsorption up to a final value of $\Delta f_{Ca^{2+},\text{end}}$.
   c. Calculate the free Ca^{2+}-concentration taking the pH and the complex binding constants of Ca^{2+} and EGTA into account ***(20)***.
   d. Plot the apparent surface coverage

$$\Theta = \frac{\Delta f_{\text{Anx A2t,EGTA}} + \Delta f_{Ca^{2+}}}{\Delta f_{\text{Anx A2t,EGTA}} + \Delta f_{Ca^{2+},\text{end}}}$$

vs the free Ca^{2+}-concentration.

3.6.3. Theory and Data Workup

The adsorption of annexin (Anx) to lipid membrane binding sites (B) in the presence of Ca^{2+} can be described by the following equilibrium:

$$\text{Anx} + n\text{Ca}^{2+} + \text{B} \rightleftharpoons \text{AnxCa}_n\text{B}$$

For the dissociation constant the following expression can be formulated:

$$K_d = \frac{c_{\text{anx}} c_{\text{Ca}}^{n} \Gamma_B}{\Gamma_{\text{anxCa}_n\text{B}}} = \frac{c_{\text{anx}} c_{\text{Ca}}^{n} \left(\Gamma_{B,0} - \Gamma_{\text{anxCa}_n\text{B}}\right)}{\Gamma_{\text{anxCa}_n\text{B}}} \tag{12}$$

c_{anx} is the annexin, and c_{Ca} the free Ca^{2+}-concentration. Γ_B denotes the surface concentration of free membrane binding sites, $\Gamma_{\text{anxCa}_n\text{B}}$ the occupied, and $\Gamma_{B,0}$ the total binding sites. With the definition of the apparent surface coverage one can derive the following equation:

$$\Theta = \frac{c_{\text{anx}} c_{\text{Ca}}^{n}}{K_d + c_{\text{anx}} c_{\text{Ca}}^{n}} \tag{13}$$

If the Ca^{2+}-concentration is kept constant while the annexin A2t concentration is varied, **Eq. 13** becomes a function of the annexin concentration:

$$\Theta = \frac{c_{\text{anx}}}{K_d^{*} + c_{\text{anx}}} \tag{14}$$

with $K_d^{*} = K_d / c_{\text{Ca}}^{n}$.

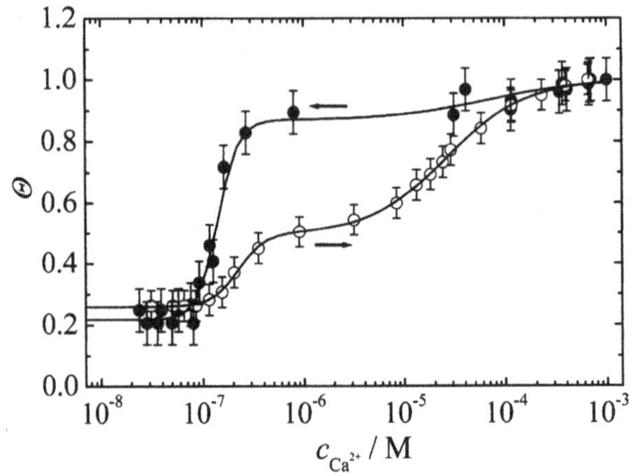

Fig. 5. Ca^{2+}-dependent adsorption and desorption of annexin A2t to a solid supported membrane composed of octanethiol and a POPC/POPS (4:1) monolayer. The values for the surface coverage are obtained from the desorption of annexin A2t as a result of the stepwise addition of EGTA (arrow pointing to the left) to solutions containing 1 m*M* $CaCl_2$ or stepwise addition of $CaCl_2$ (arrow pointing to the right) to an annexin A2t and EGTA containing solution. The total annexin A2t concentration was > 0.35 μ*M* in all experiments. It should be noted that the given surface coverage is a relative value.

If, however, the annexin concentration is kept constant **Eq. 13** is just a function of the Ca^{2+}-concentration and equation can be simplified to:

$$\Theta = \frac{c_{Ca}^{n}}{K_{d}^{\#} + c_{Ca}^{n}} = \frac{c_{Ca}^{n}}{K_{0.5}^{n} + c_{Ca}^{n}} \tag{15}$$

with $K_d^{\#} = K_d / c_{anx}$.

The surface coverage Θ is defined as the relative amount of binding sites occupied by annexin A2t and the total amount of binding sites present on the membrane. $K_{0.5}$ is the Ca^{2+}-concentration, at which half of the proteins are bound to the membrane, and is equivalent to the nth root of the dissociation constant. Using only one value is certainly a simplification, since the result of the Ca^{2+}-titration arises from a complex set of equilibria so that the obtained midpoints ($K_{0.5}$-values) do not represent single equilibrium constants.

Because there are two distinct desorption processes discernable in the curve of **Fig. 5**, a linear combination is used for data evaluation:

$$\Theta = \Theta_1 + \Theta_2 = \frac{c_{Ca}^{n_1}}{K_{d1}^{\#} + c_{Ca}^{n_1}} + \frac{c_{Ca}^{n_2}}{K_{d2}^{\#} + c_{Ca}^{n_2}} \tag{16}$$

Fitting the parameters of **Eq. 16** to the desorption data leads to a value for $K^{\#}_{d1}$ = (140 ± 20) n*M*, n_1 = 4.2 ± 1.0 and $K^{\#}_{d2}$ = (70 ± 20) μ*M*, and n_2 = 0.9 ± 0.5. The same procedure is applied to the adsorption curve. Here, $K^{\#}_{d1}$ reads (230 ± 50) n*M*, n_1 = 3.3 ± 0.9, $K^{\#}_{d2}$ = (32 ± 6) μ*M*, and n_2 = 1.1 ± 0.3. The intermediate surface coverages are quite different for the adsorption and desorption process resulting in a strong hysteresis. Whereas for the adsorption process an intermediate surface coverage of 0.50 ± 0.07 is observed that of the desorption is 0.87 ± 0.07.

4. Notes

1. Gold electrodes are deposited by thermal evaporation through an appropriate mask design on each side of a quartz plate using an evaporation unit (E 306, Edwards, UK). After applying a layer of chromium (10–20 nm) to improve the adhesion of gold on the quartz, the gold layer is deposited subsequently with a final thickness of about 200 nm. Prior to the incubation of the gold surfaces in the self-assembly solution, they are exposed to an argon plasma with high energy (plasma cleaner, Harrick Scientific Corporation, New York, NY) for 5–10 min.
2. Besides the disc-like electrode on both sides of the quartz plate, a small gold area is evaporated at the edge of the plate to connect it with the electrode on the other side by a silver adhesive to allow contacting both electrodes from one side.
3. When using the quartz resonator in aqueous solutions other external parameters influence the shear oscillation, which should be kept in mind ***(21,22)***. The generation of longitudinal waves, which are reflected at the liquid air interface, alter the resonance frequency of the quartz plate. If working in an open crystal holder, periodical changes in the resonance frequency are recorded due to evaporation of the liquid. Changes in ionic strength, dielectric constant of the electrolyte, and viscosity of the liquid, respectively may change the resonance frequency. If the buffer conditions are changed during the experiment, an undesired parasitic frequency shift may occur, dependent on the geometry of the electrodes and conductance of the solution. Surface roughness of the quartz resonator also influences interpretation of adsorption phenomena. Alteration of hydrophilicity upon adsorption can lead to tremendous changes in resonance frequency as rough and hydrophilic surfaces entrap liquids in small cavities thus contributing to the overall mass detected by the device, while hydrophobic cavities are not wetted by the liquid resulting in the inclusion of air or vacuum. This implies that the resonance frequency jumps suddenly to smaller values when changing from a hydrophobic to a hydrophilic surface.
4. If available, analysis of the membrane preparation by impedance spectroscopy is very helpful to evaluate the quality of the lipid bilayer preparation ***(15)***. For data evaluation, an equivalent circuit composed of a capacitance C_m representing the monolayer or bilayer in series to an Ohmic resistance R_e representing the bulk resistance and the wire connections is typically used. Chemisorption of an octanethiol monolayer can be considered as successful, if the capacitance of the octanethiol monolayer is (2.0 ± 0.2) μF/cm^2. For the formed lipid bilayer a capacitance value of (1.1 ± 0.2) μF/cm^2 indicates a successful preparation.

5. **Eq. 5** can be fit to the data in a modified form:

$$\Delta f = \Delta f_{\max} \frac{K_a c}{1 + K_a c}.$$

The maximum frequency shift and the binding constant can be extracted from the isotherm.

6. In principle, this competition assay works with all kinds of protein ligand pairs, however, sensitivity and the practically accessible concentration-regime is better and larger if the binding constant between the partners is not too high. For instance, streptavidin and biotin show rather an all or nothing reaction upon adding biotin to a surface bound streptavidin monolayer attached to the surface via biotinylated lipids. At a certain threshold concentration all of the streptavidin is suddenly released from the surface. This behavior is of course not suited to determine the concentration of the added ligand safely.

Acknowledgments

The authors are very much indebted to invaluable input of Simon Faiß, Eike Lüthgens, Katja Kastl, Alexander Herrig, and Michaela Ross.

References

1. Sauerbrey, G. (1959) Verwendung von Schwingquarzen zur Wägung dünner Schichten und zur Mikrowägung. *Z. Phys.* **155,** 206–222.
2. Nomura, T. and Okuhara, M. (1982) Frequency shifts of piezoelectric quartz crystals immersed in organic liquids. *Analytica Chimica Acta* **142,** 281–284.
3. Janshoff, A., Galla, H.-J., and Steinem, C. (2000) Piezoelectric mass-sensing devices as biosensors - an alternative to optical biosensors? *Angew. Chem. Int. Ed.* **39,** 4004–4032.
4. Janshoff, A. and Steinem, C. (2001) Quartz crystal microbalance for bioanalytical applications. *Sensors Update* **9,** 313–354.
5. Marx, K. A. (2003) Quartz crystal microbalance: a useful tool for studying thin polymer films and complex biomolecular systems at the solution-surface interface. *Biomacromolecules* **4,** 1099–1120.
6. Wegener, J., Janshoff, A., and Steinem, C. (2001) The quartz crystal microbalance as a novel means to study cell-substrate interactions in situ. *Cell Biochem. Biophys.* **34,** 121–151.
7. Gerke, V. and Weber, K. (1984) Identity of p36K phosphorylated upon Rous sarcoma virus transformation with a protein purified from brush borders; calcium-dependent binding to non-erythroid spectrin and F-actin. *EMBO J.* **3,** 227–233.
8. Adamczyk, Z., Siwek, B., Warszynski, P., and Musial, E. (2001) Kinetics of particle deposition in the radial impinging-jet cell. *J. Colloid Interface Sci.* **242,** 14–24.
9. Hillier, A. C. and Ward, M. D. (1992) Scanning electrochemical mass sensitivity mapping of the quartz crystal microbalance in liquid media. *Anal. Chem.* **64,** 2539–2554.

10. Furusawa, H., Murakawa, A., Fukusho, S., and Okahata, Y. (2003) In vitro selection of N-peptide-binding RNA on a quartz-crystal microbalance to study a sequence-specific interaction between the peptide and loop RNA. *Chem Bio Chem* **4,** 217–220.
11. Okahata, Y., Kawase, M., Niikura, K., Ohtake, F., Furusawa, H., and Ebara, Y. (1998) Kinetic measurements of DNA hybridization on an oligonucleotide-immobilized 27-MHz quartz crystal microbalance. *Anal. Chem.* **70,** 1288–1296.
12. Adamczyk, Z. and Werónski, P. (1999) Application of the DLVO theory for particle deposition problems. *Adv. Colloid Interface Sci.* **83,** 137–226.
13. Adamczyk, Z. (2000) Kinetics of diffusion-controlled adsorption of colloid particles and proteins. *J. Colloid Interface Sci.* **229,** 477–489.
14. Adamczyk, Z. (2003) Particle adsorption and deposition: role of electrostatic interactions. *Adv. Colloid Interface Sci.* **100–102,** 267–347.
15. Steinem, C., Janshoff, A., Ulrich, W.-P., Sieber, M., and Galla, H.-J. (1996) Impedance analysis of supported lipid bilayer membranes: a scrutiny of different preparation techniques. *Biochim. Biophys. Acta* **1279,** 169–180.
16. Janshoff, A., Steinem, C., Sieber, M., and Galla, H.-J. (1996) Specific binding of peanut agglutinin to GM1-doped solid supported lipid bilayers investigated by shear wave resonator measurements. *Eur. Biophys. J.* **25,** 105–113.
17. Steinem, C., Janshoff, A., Wegener, J., Ulrich, W.-P., Willenbrink, W., Sieber, M., and Galla, H.-J. (1997) Impedance and shear wave resonance analysis of ligand-receptor interaction at functionalized surfaces and of cell monolayers. *Biosens. Bioelectronics* **43,** 339–348.
18. Gerke, V. and Moss, S. E. (2002) Annexins: from structure to function. *Physiol. Rev.* **82,** 331–371.
19. Ross, M., Gerke, V., and Steinem, C. (2003) Membrane composition affects the reversibility of annexin A2t binding to solid supported membranes: a QCM study. *Biochemistry* **42,** 3131–3141.
20. Stockbridge, N. (1987) EGTA. *Comput. Biol. Med.* **17,** 299–304.
21. Bandey, H. L., Martin, S. J., Cernosek, R. W., and Hillmann, A. R. (1999) Modeling the responses of thickness-shear mode resonators under various loading conditions. *Anal. Chem.* **71.**
22. Lucklum, R. and Hauptmann, P. (2003) Transduction mechanism of acoustic-wave based chemical and biochemical sensors. *Meas. Sci. Technol.* **14,** 1854–1864.

4

Measurement of Solvent Accessibility at Protein–Protein Interfaces

Jeffrey G. Mandell, Abel Baerga-Ortiz, Arnold M. Falick, and Elizabeth A. Komives

Summary

Methods are presented for monitoring solvent accessibility of protein–ligand and protein–protein interfaces. The kinetics of solvent accessibility at the protein–protein interface is monitored by amide hydrogen/deuterium ($H/^2H$) exchange detected by matrix assisted laser desorption/ionization time-of-flight (MALDI-TOF) mass spectrometry (MS). A straightforward theoretical analysis is presented for determining the concentration of a weakly binding ligand that is required for achieving a situation in which the receptor is essentially 100% bound, and this is verified by control experiments. We show that when the receptor is essentially 100% bound it is possible to distinguish amide exchange as a result of solvent accessibility at the interface from amide exchange caused by complex dissociation. Methods are also presented for the measurement of tightly bound complexes of large interactions such as antibody-antigen complexes. Quantitation of the number of amides sequestered at the interface can be related to the number of H_2O molecules excluded from the interface.

Key Words: Amide $H/^2H$ exchange; MALDI-TOF mass spectrometry; hydration.

1. Introduction

We have recently reported amide exchange experiments that probe for solvent accessibility changes at protein–protein interfaces ***(1–4)***. Although details of individual experimental designs are not presented here, we have taken several different approaches depending on the system. In one case, an antibody epitope was mapped. This study involved two large proteins that could not be analyzed simultaneously and methods were devised for separating and analyz-

From: *Methods in Molecular Biology, vol. 305: Protein–Ligand Interactions: Methods and Applications*
Edited by: G. U. Nienhaus

ing one protein *(3)*. In other cases, the on-exchange or off-exchange experiment may be preferable depending on whether the complex readily forms and dissociates *(4,5)*.

Amide exchange experiments have traditionally been performed to understand protein folding and unfolding and the amide exchange rates related to hydrogen bond formation *(6)*. In the case of protein–protein interfaces, there are not usually hydrogen bonds forming across the interface, it is rather a case of decreased solvent accessibility which can arise because of a number of factors including side chain interactions, decreased loop mobility, etc. Because it is known that the amides in the interior of a protein exchange more slowly than those on the surface, it is nearly always the more rapidly exchanging surface amides that require monitoring for protein–protein interface studies. Typically, amides on the surface of a protein exchange within seconds to minutes. This exchange rate is slower than that for unstructured peptides *(7)* but relatively faster than the core of the protein *(8)*.

It is easy to see that the protein–protein binding equilibrium constant must play a role in whether or not interface protection is observed. This is because if the complex is dissociating and reassociating rapidly, then the amides will have time to exchange when the proteins are unbound, and interface protection will not be observed. To address this issue, some knowledge of the binding equilibrium constant is required in advance. It is also possible that H_2O molecules will be able to access part of the interface albeit to a reduced degree, and it is difficult to understand how and when it is best to observe these partly excluded amides. In our experience, there are amides at the protein–protein interface that are completely solvent excluded as long as the two proteins remain bound *(2,9)*. For these completely solvent excluded amides, the observed hydrogen/deuterium ($H/^2H$) exchange rate is controlled by the protein–protein dissociation and association rates as well as the intrinsic rate of hydrogen exchange:

$$[R_D{\cdot}L] \underset{k_a}{\overset{k_d}{\rightleftarrows}} R_D + L \xrightarrow{k_{ex}} R_H \tag{1}$$

Where R_H is protonated receptor, R_D is deuterated receptor, L is ligand, k_{ex} is the intrinsic amide exchange rate (min^{-1}) for amides in the uncomplexed receptor, k_d is the rate of dissociation of the complex (min^{-1}), and k_a is the rate of association of the proteins undergoing complexation ($M^{-1}\,\text{min}^{-1}$). $H/^2H$ exchange at a protein–protein interface could occur because of solvent accessibility at the interface or because of exchange that occurs when the complex dissociates. A theoretical analysis shows that under the conditions of the $H/^2H$ exchange experiment presented here (protein concentration, 3 μ*M*), if the K_d is less than 2 n*M*, a 1:1 ratio of ligand to receptor is sufficient to keep essentially 100% of the receptor bound throughout the experiment. If the K_d is 10–100 n*M*, then higher ratios of ligand to receptor are required to study $H/^2H$ exchange kinetics at the interface in the bound complex, and the interplay between the binding kinetics and exchange kinetics becomes important *(2)*. In these cases, a full knowledge of the binding kinetics is required prior to quantitative interpretation of the data. For protein–protein interactions that are in rapid equilibrium, the amides excluded at the interface should not show pH-dependent exchange rates *(2)*.

2. Materials

1. **Reagents.** Deuterium oxide (99.996%) was purchased from Cambridge Isotope Laboratories (Andover, MA). All other reagents were of the highest purity possible. Polypropylene 0.5 mL microcentrifuge tubes and thin-walled PCR tubes (0.5 mL) were purchased from USA/Scientific (Ocala, FL).
2. **Proteins.** Purified proteins should be exchanged into a low-salt buffer in which they are stable. Portions of 400–700 pmol should be prepared by centrifugal concentration using Centricon (Millipore Corp., Bedford, MA) and stored either lyophilized or concentrated depending on stability. Immobilized pepsin (on crosslinked 6% agarose, 2–3 mg of pepsin/mL of gel, obtained from Pierce Chemicals, Rockford, IL) was used to cleave the proteins into peptides of 8–20 amino acids *(10)*.
3. **MALDI-TOF Matrix.** The matrix used was 5 mg/mL α-cyano-4-hydroxycinnamic acid (Sigma Chemicals, St. Louis, MO) which was recrystallized once from ethanol. Matrix solution contained 1:1:1 acetonitrile, ethanol, and 0.1% trifluoroacetic acid and was adjusted to pH 2.2 with 2% trifluoroacetic acid using an Inlab 423 pH electrode (Mettler Toledo Inc., Wilmington, MA). The matrix solution was chilled on ice for at least 2 h prior to use in the experiments and was freshly prepared for each experiment.
4. **Solutions and Buffers.** 2% (w/v) TFA mix 0.75 mL of sequencing grade TFA with 49.25 mL MilliQ H_2O. This solution can be kept for 1 mo in a sealed glass bottle stored at 4°C.
   0.1% (w/v) TFA mix 2.5 mL of 2% TFA solution with 47.5 mL MilliQ H_2O. This solution should be made fresh daily.
   Deuterated buffers are made by preparing 10 mL of the buffer in H_2O and adjusting the pH as usual. This solution is then freeze-dried (lyophilized) and resuspended in the same volume (10 mL) of D_2O. The lyophilization process should be repeated a second time to prepare a 99% deuterated buffer solution.
   Phosphate buffered saline for storage of the antibody beads is 50 m*M* NaH_2PO_4 pH 6.6, 50 m*M* NaCl containing 0.2% NaN_3 (w/v).

3. Methods

3.1. Overall Scheme

Figure 1 shows the overall scheme of the on-exchange and off-exchange experiments. In order to quantitatively relate the amount of amide exchange at a protein–protein interface to the number of excluded H_2O molecules, it is nec-

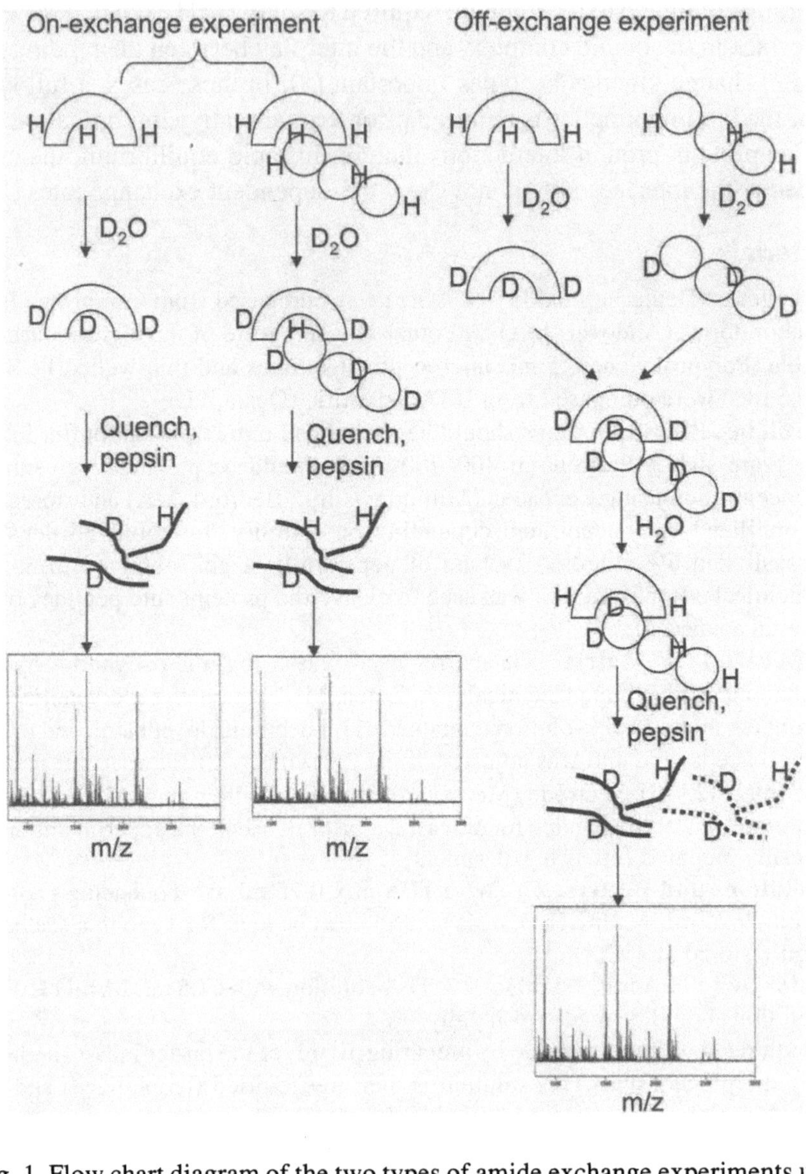

Fig. 1. Flow chart diagram of the two types of amide exchange experiments used to study protein–protein interfaces. In the on-exchange experiment, the protein–protein complex shows a region in which less deuterium is incorporated when compared to control experiments using each protein alone. In the off-exchange experiment, each protein is first allowed to incorporate deuterium, and then the complex is formed and deuteriums are off-exchanged by dilution back into H_2O. In the off-exchange experiment on the protein–protein complex, the presence of remaining deuterons after off-exchange compared with control experiments using each protein alone indicates the interface.

essary to determine that the amides that show differences between the bound and free states are on the surface of the protein. This requires comparison of the numbers of amides in each peptide that rapidly exchange in on-exchange experiments to those in off-exchange experiments. For interface mapping, either experiment suffices, however the additional information provided by the off-exchange experiments helps distinguish between interface protection and conformational changes.

3.2. Experimental Setup

Prior to carrying out an amide exchange experiment, there are several preparations that need to be undertaken. First, it is necessary to determine the pepsin digestion conditions. Then it is necessary to be able to reproducibly prepare protein samples of known pH and these need to be concentrated enough that the pepsin digest products can be reliably observed in the mass spectrometer. For matrix assisted laser desorption/ionization time-of-flight (MALDI-TOF), the final concentration of peptides needs to be 1 pmol/µL. Finally, the identity of the peptides produced in the pepsin digest need to be determined. The methods for these preparation steps will be described first and then the considerations of the on- and off-exchange experiments will be addressed. Finally, the data analysis procedures will be described.

3.2.1. Pepsin Digestion and Identification of Peptic Peptides

Pepsin digestion was carried out with immobilized pepsin (25 µL pepsin beads, 50 µL pepsin slurry. The beads were washed twice in a 1.5 mL eppendorf tube with 1 mL of chilled 0.1% TFA by vortexing and subsequent centrifugation for 2 min at 7000*g* immediately before use. The pepsin beads can be washed and re-used approximately five times, but begin to lose activity 4 h after acidification. Quenched protein samples (120–130 µL) were added to the pepsin and digested for 10 min on ice with mixing every minute. The immobilized pepsin was removed by centrifugation for 1 min at 14,000*g* in a chilled eppendorf centrifuge, and the supernatant was transferred to a 0.5 mL thin-walled microcentrifuge tube, aliquotted and immediately frozen in liquid N_2. Samples were quenched, digested, and frozen in under 15 min. Digested samples can be stored at –80°C for up to 2 wk before analysis with no loss of deuteration. Others have shown that artifacts can be introduced if deuterated proteins are quenched and frozen prior to digestion.

It is important to obtain the identity of the pepsin digest products in a reliable manner. Several bioinformatics tools and mass spectrometry data analysis software packages provide the option of entering peptide masses along with a protein sequence and these purport to identify pepsin digest peptides. These algorithms are only correct some 85% of the time as they are based on the most common but not the only pepsin cleavage sites. Therefore, in order to obtain reliable peptide identities, we use the following approach. First, the pepsin digest is performed *exactly* as it will be in the experiment (i.e., cold quench solution, pepsin, matrix solution and MALDI target plate). This mixture is then combined with mass standards to varying ratios and analyzed. The spectrum showing the best 1:1 ratio of standards to pepsin digest peptides is then internally calibrated based on the added standards. The masses of the peptides in this spectrum are accurate to within 20 ppm on a typical MALDI-TOF instrument that is well-tuned. These masses can then be used to search *all* of the protein sequence to find those peptides for which there is only one possible match within the 20 ppm cut-off. All other peptides in the mixture must be identified by MS/MS sequencing. This is best done by separating the pepsin digest mixture on a C18 analytical HPLC column and spotting each of the peaks. Typically only about half of the total peptides in the mixture are actually observed in the mass spectrum of the mixture, and there is no point in identifying those that are not observed. Each of the peptides in the mass spectrum that require identification is found in one of the HPLC separated fractions and then can be readily sequenced from this purified fraction. An HPLC separation of approximately 1 nmol of protein will yield enough peptide to sequence by post-source decay sequencing, C-terminal carboxypeptidease Y sequencing (these are MALDI-based sequencing methods) or by ESI MS/MS on an ion trap or tandem Q-TOF mass spectrometer. It is important to note that while pepsin cleavage is not predictable, it is completely reproducible so that sequencing need only be carried out once prior to all the amide exchange experiments.

3.2.2. Mass Spectrometry

In the experiments presented here, MALDI-TOF mass spectrometry (MS) was used. We developed the methodology for measurement of amide exchange by MALDI-TOF because it is easier to do, and no modification of commercially available instruments is required ***(11)***. MS to measure amide exchange was pioneered by David Smith in the early 1990s using electrospray ionization ***(10)***. The same results can be obtained by measurements on an electrospray mass spectrometer and similarly high resolution data can now be obtained using a tandem quadrupole time-of-flight mass spectrometer. MALDI-TOF spectra were acquired on a Voyager DE STR (Applied Biosystems, Framingham, MA).

1. Chill the MALDI targets overnight at 4°C in a plastic case to prevent condensation of atmospheric H_2O prior to use.
2. Quickly (< 30 s) defrost the frozen samples (one at a time) to 0°C by warming in the hand and *flicking* the tube just until a small fragment of ice remains.
3. Place the sample in a pre-chilled eppendorf tube rack at 4°C.

4. Immediately remove a small portion, typically 5 µL, mix with an equal volume of pre-chilled matrix solution, and spot 1 µL onto the chilled MALDI target.
5. Immediately place the target in a dessicator under a moderate vacuum such that the spot dries in 1 min. Slow drying in moderate vacuum was found to improve sample analysis, presumably because of improved crystal growth.
6. Transfer the chilled, dried plate as quickly as possible (< 10 s) to the mass spectrometer.
7. Analyze one sample at a time on a single target plate so that each sample is treated identically and experiences the same amount of artifactual back exchange, thereby avoiding the necessity of correcting for back exchange occurring during analysis. No more than 3 min should elapse from the time of defrosting to completion of the MALDI-TOF analysis with the bulk of the time being the 1 min of slow drying and the 1.5 min that elapses between the time the target is loaded into the spectrometer and when the first laser shot occurs.
8. Collect 256 scans and average. A complete analysis of the behavior of deuterated control peptides under the conditions of this MALDI-TOF experiment has been presented ***(11)***.

3.2.3. pH of $H/^2H$ Exchange Mixtures

Amide exchange rates are strongly dependent on pH and therefore control of the pH throughout the amide exchange experiments is critical to obtaining high quality data. Measurements of pH were made on nondeuterated mock solutions to avoid electrode isotope effects. We tried several electrodes to measure volumes of less than 50 µL, but these were not very reliable. Therefore, even when deuteration was performed on 10 µL samples, prior to the first experiment, the pH of five such samples combined together was measured in order to know exactly the pH of the deuteration solution. Measurements of solutions of >50 µL was accurately performed using an Inlab 423 pH electrode (Mettler Toledo, Inc.). Deuterated buffer solutions were prepared from 1 *M* stock buffers (in H_2O) in exactly the same manner as the mock solutions.

3.3. Measurement of the Rate of Incorporation of Deuterons Into Surface Amides of Proteins (On-Exchange)

The kinetics of incorporation of solvent deuterons are a good measure of the solvent accessibility of each region of a protein, although care must be taken in interpreting comparisons between proteins and in using such information to make conclusions about the *folded state* of a protein. Depending on the stability of the protein at relatively high concentrations, the starting protein is either prepared lyophilized, or as a solution at a concentration of 100 µ*M*. The protein can be prepared in any buffer, but it is important to remember that if pure D_2O is added, the starting protein must contain 10X buffer. Otherwise, deuterated buffer should be prepared (*see* **Subheading 3.4.**). The concentration of the protein in the on-exchange step needs to be at least 10 μ*M* so that after a 1:10 quench the final concentration is 1 pmol/μL for spotting on the MALDI target.

1. Prepare samples after deuteration for varying lengths of time (typically 0–30 min). Samples should be prepared in triplicate for high statistical significance.
2. Quench each sample by addition of 120 μL H_2O (0°C) containing approx 5 μL 2% TFA to give a final pH of 2.2 (note: the first time the experiment is performed, a trial sample must be prepared so that the exact amount of 2% TFA required to bring the final pH to 2.2 can be determined exactly).
3. Digest each sample at 0°C for 1–10 min with immobilized pepsin, aliquot, and freeze in liquid N_2 as described previously.
4. Analyze the data after collecting all the mass spectra for several different times of deuteration. The protein–protein interface is identified as the region of the protein that shows decreased solvent accessibility in the complex as compared to solvent accessibility of these same regions in the separated proteins.

3.4. Measurement of the Rates of Off-Exchange of Deuterons

The rates of off-exchange of deuterons from protein–protein interfaces can provide additional information about the solvent accessibility of the interface *(2)*. The experiment includes measurement of the on-exchange of deuterium into each of the proteins involved in the interaction. After on-exchange is measured, a suitable set of times of on-exchange are chosen and these samples are then allowed to complex, and are diluted into H_2O for the off-exchange part of the experiment. With both on-exchange and off-exchange properties of the interface available, interpretations of conformational changes vs interface protection are more easily made.

3.4.1. Off-Exchange From the Protein–Protein Interface

1. Perform the on-exchange experiment as described above on the separate proteins. The on-exchange data gives information about how solvent accessible the various surface regions of the protein are.
2. The off-exchange experiment is initiated in the same way as the on-exchange. Incubate each protein separately in D_2O (typically in 6 μL) for a defined time (we typically use 8 min).
3. Combine the two protein solutions (12 μL total) in a thin-walled eppendorf tube (these are usually sold for use with the polymerase chain reaction) and allow them to complex for 2 min.
4. Initiate off-exchange by diluting 1:10 dilution with 120 μL H_2O at 25°C (note it is important to check that the pH of each starting protein solution is the same so that when they are mixed and diluted the pH does not change). The off-exchange times should extend at least as long as the on-exchange (we typically do varying times (1–30 min.).
5. Quench the off-exchange reaction by plunging the eppendorf tube into ice water and by rapid addition of a pre-determined amount of 2% trifluoroacetic acid

(appprox 5 μL) to bring the pH down to 2.2. The precise amount of TFA required for quenching should be determined by titration prior to each experiment.
6. Digest the proteins with pepsin and freeze as already described for the on-exchange experiment.
7. Thaw and analyze each sample by MALDI-TOF MS as already described. In our experience, if the protein complex has a total molecular weight greater than 80 kDa, it is likely that some of the peptides will overlap and some quantitative data will not be lost as a result. It is still possible to follow the off-exchange of each peptide qualitatively, and if differences are seen in a peptide that is overlapping, it is worthwhile to attempt some separation scheme. One method we have used is to quickly adsorb 10 μL of the sample to a zip tip and then elute with 0.1% TFA containing 20% followed by 50% and then 100% acetonitrile. This does not result in significant time lost in analysis and often the peptides separate and the deuteration can then be quantitated.
8. Perform control experiments in which each protein is analyzed separately in the off-exchange. These experiments should reproduce the same pH and volume of dilution conditions as for the experiments on the complex.

3.4.2 Off-Exchange With Removal of One of the Proteins

In one case, we mapped an antibody epitope. This required removal of the antibody prior to analysis of the protein in which the epitope was found because the antibody contributed a large number of peptide fragments to the mass spectrum ***(3)***. For this approach, the following protocol can be used.

1. Covalently link the antibody to protein G agarose beads (Sigma Chemicals) by cross-linking with 20 m*M* dimethylpimelimidate (DMP, Pierce, Rockford IL) according to the manufacturers instructions.
2. Wash the beads extensively with borate buffer and then with a solution of 10 m*M* glycine pH 1.7 to eliminate the mAb that did not get covalently bound to the beads according to the manufacturers instructions.
3. Exchange the buffer for the one you want to use in your experiment, we used phosphate-buffered saline, pH 6.6
4. Store the antibody beads in 30 μL aliquots.
5. Perform control experiments to ascertain the nonspecific binding of the protein–protein G beads, and to ascertain the binding capacity of the mAb beads.
6. The epitope mapping experiment is a variation of the off-exchange experiment. The first step is to resuspend the binding protein in 3 μL deuterated buffer for 10 min to allow deuteration of surface amides.
7. At the same time, resuspend the mAb beads (30 μL) in 270 μL of deuterated buffer for 10 min, centrifuge and decanted to 30 μL.
8. Mix the MAb beads and binding protein together for 10 min (the volume should be 33 μL).
9. Perform the off-exchange by diluting the mixture into 270 μL of H_2O.
10. Centrifuge the beads, discard the supernatant, and resuspend the beads (approx 30 μL) in 270 μL H_2O.

11. Centrifuge again and decante to 30 µL. The time from the first dilution into H_2O until completion of the final centrifugation should be 2 min. During the 2 min off-exchange is occurring in the complex.
12. Perform a series of experiments with longer off-exchange times by allowing additional time after the second resuspension.
13. Quench each reaction by mixing with 30 µL of a chilled (4° C) solution of equal parts of 0.1% TFA pH 2.2 and 1-propanol. This solution not only quenches the exchange but also elutes the binding protein from the mAb, which is covalently attached to the beads and is removed by centrifugation and discarded.
14. Mix the supernatant (30 µL) with 100 µL of a slurry of immobilized pepsin and process as already described.
15. Perform a control experiment (no-mAb) by resuspending the binding protein in 3 µL of D_2O and subjecting it to back exchange with 27 µL of H_2O. The rest of the experiment is performed exactly the same way as already described for the MAb experiment. This control experiences a 90% dilution of the D_2O with H_2O while the complex sample on the beads experiences a 99% dilution. The difference is dilution is corrected for during data analysis (*see* below).

3.5. Data Analysis

The mass spectra are displayed and analyzed using the GRAMS-MS software because this software allows for stacking the plots and facilitates accurate assignment of the monoisotopic peak of the mass envelopes (**Fig. 2**). If the GRAMS-MS software is not available, it is essential to devise a way in the alternative software for stacking the plots. This is because it is otherwise difficult to accurately calibrate the spectra in which a substantial amount of deuterium incorporation into the peptides has occurred. For some deuterated samples the monoisotopic peak was not present so higher mass peaks of the same envelope were used; these peaks were identified in the uncalibrated (other than instrument calibration) stacked plots. Then these peaks were used for internal calibration after appropriate unitary additions. The average mass of a peptide was calculated by determining the centroid of its isotopic envelope using the CAPP software written by Mandell *(11)*. The difference between the average masses of the deuterated and nondeuterated peptide gave the raw amount of deuterium incorporated.

For MALDI-TOF $H/^2H$ exchange data, two corrections are required to convert the raw data into the number of amide deuterons incorporated. First, a residual amount of deuterium remains in the dried spot on the MALDI-TOF target plate, and this residual amount labels all of the rapidly exchanging sites in each peptide. To correct for this, it is first necessary to compute the amount of residual deuterium from all of the dilutions in the experiment. For example, if the experiment involves mixing 100% D_2O, 95:5 with concentrated protein, then a 1:10 dilution into pepsin and a 1:1 dilution into matrix, then the residual amount of deuterium is 2.5%. The sum of all rapidly exchanging positions

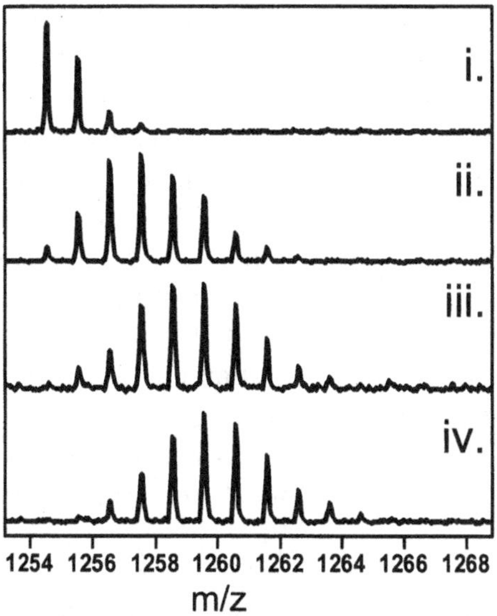

Fig. 2. For the on-exchange experiment, the proteins separately or in complex are allowed to incubate in deuterated buffer for varying lengths of time prior to quenching, digesting, and mass spectrometric analysis. The MALDI-TOF mass spectrum contains many peptides that cover much of the sequence of the entire protein. Here is shown the mass envelope of one peptide from MALDI-TOF mass spectra taken after increasing times of deuteration (i–iv are 0, 1, 2, 5 min, respectively). One can see that the mass envelope broadens somewhat, and increasing deuterium incorporation is observed over the first few minutes. Partial incorporation over the first several minutes and then a slowing of incorporation is typical for a region of the protein that is partly surface-exposed.

(N-term-3; C-term-1; Asp, Glu, Cis, His, Lys, Ser, Thr, Tyr, Trp-1; Asn, Gln-2; Arg-4) in each peptide is multiplied by the residual deuterium, and this amount is subtracted from the raw amount of deuterium incorporated to give the amount of deuterium present at amide positions.

The raw amount of deuterium present at amide positions must then be corrected for back exchange, in order to determine the real amount of incorporation into amide positions during the labeling period. In the MALDI-TOF experiment, all of the peptides are analyzed simultaneously, so a single back-exchange correction term is used for all the peptides. This is typically obtained from a sample that is deuterated for 24 h and subsequently treated in the exact same manner as the other experimental samples. For globular proteins, not all regions will be completely deuterated after 24 h, but some will. The amount of deuterium is determined for each peptide as already described, and those peptides that have incorporated the most deuterium per amide are used in the analysis. For example, if the most deuterated peptides had 71% of the amides deuterated, then the back exchange that occurred was 29%. The amount of deuterium incorporated into amide positions is corrected for back exchange by dividing by the fraction most deuterated so that if a peptide has five deuterons incorporated, and the amount of back exchange that occurred was 29%, the real amount of amide deuteration is 5/0.71 or seven amide deuterons. It is important to remember that if the peptide had 14 amides, 50% of the peptide is deuterated, but it equally possible that half of the peptide is 100% deuterated or that all of the peptide is 50% deuterated.

Kinetic plots of deuterium incorporation into amides during on-exchange fit best to a two-exponential model accounting for deuterons exchanging at a very rapid rate (fully solvent accessible amides) and an intermediate rate (amides with reduced solvent accessibility in the folded protein) using the following equation:

$$D=N_{fast}(1-e^{-k_{fast}t})+N_{slow}(1-e^{-k_{slow}t}) \quad (2)$$

where D is the total number of deuterons at time, t, N_{fast} is the number of deuterons exchanging at the fast rate, k_{fast}, and N_{slow} is the number of deuterons exchanging at the slow rate, k_{slow}. The fit was implemented in KaleidaGraph 3.0 (Synergy Software, Inc.). The rapidly-exchanging protons had all exchanged by the first time point, so k_{fast} was fixed. Other floating parameters (N_{fast}, N_{slow}, and k_{slow}) were completely insensitive to changes in the value of k_{fast} from 10 min^{-1} to 100min^{-1}, so k_{fast} was set to 30 min^{-1}, the median of the amide exchange rates for the rapidly exchanging amides in the mannose permease domain P13 from *E. coli* ***(12)***.

Kinetics of off-exchange typically fit best to either a bi- or tri-exponential model. The different rates reflect those amides that were not protected from solvent in the complex and remained fast exchanging (set to 30 min^{-1}), those partially excluded and exchanging at an intermediate rate (typically 0.1–0.5 min^{-1}), and those completely excluded that exchange at a slow rate (typically 0.04– < 0.01 min^{-1}). The bi-exponential model was used when no improvement in the fit was observed when the bi- and tri-exponential models were compared. In these cases, typically no deuterons were found to exchange at the intermediate rate. The tri-exponential equation is

$$D = N_{fast}e^{-k_{fast}t} + N_{inter}e^{-k_{intermed}t} + N_{slow}e^{-k_{slow}t} \quad (3)$$

where D represents the number of deuterons at time t, N_{fast} represents the number of fast-exchanging deuterons, k_{fast} is the fast-exchange rate, N_{inter} is the number of intermediate-exchanging deuterons, $k_{intermed}$ is the intermediate-exchange rate, N_{slow} is the number of slow-exchanging deuterons, k_{slow} is the slow-exchange rate. The fast rate k_{fast} was again fixed at 30 min^{-1} and the other variables were not sensitive to this fast rate.

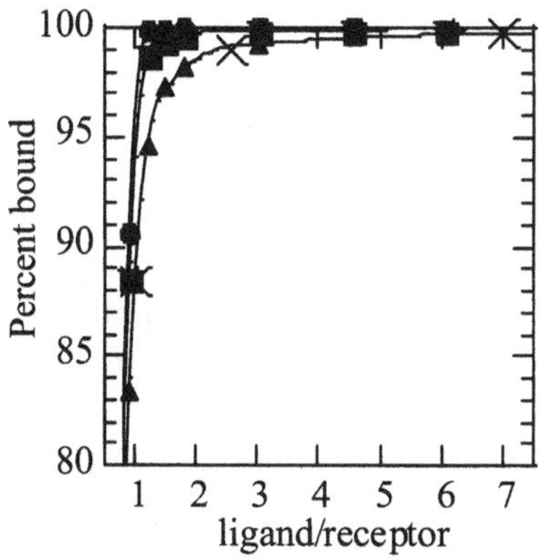

Fig. 3. Plot of the relationship between the concentration of ligand and the percent of protein bound to the ligand in a typical amide exchange experiment. Theoretical curves for K_d = 0.1 n*M* and 1 n*M* were essentially indistinguishable (●). To achieve 100% bound, ligand:receptor ratios greater than 1:1 are required for K_d = 10 n*M* (■) and higher. The curve for K_d = 50 n*M* (▲) shows that for these weaker binding affinities, ratios greater than 5:1 are required to achieve 100% bound. Actual experimental data showing the amount of deuterium retained at the interface for the thrombin-TMEGF45 interaction, which has a K_d close to 100 n*M*, fall closest to the 50 n*M* curve. The experimental data were measured for thrombin:TMEGF45 ratios of 1:1, 2.8:1 and 7:1 (X) and the percent bound was determined from assuming that the amount of deuterium retained after 1 min off-exchange was 100%.

4. Notes

1. It is important to note that the results of these types of experiments are extremely difficult to interpret in the absence of a structure of at least one of the proteins present in the complex.
2. In one case, we studied the binding of TMEGF45 to thrombin which is a weak interaction (the K_d is 120 n*M*). This interaction showed increasing amounts of deuterium retained at the interface as the ratio of TMEGF45 to thrombin was increased from 2.6:1 to 7:1 corresponding to 98.77 and 99.78 % bound respectively (**Fig. 3**). For the 2.6:1 ratio, the complex was apart for 1.5 s during the 2 min off exchange time, and this was enough to decrease the observed deuteration compared with the 0.3 s apart at a 7:1 ratio. To ensure the observed $H/^2H$ exchange was for amides at the interface of the protein complex, and not due to complex dissociation, a ratio of TMEGF45 to thrombin of 7:1 was used for the quantitative measurement of exchange rates. At this ratio, the concentration of TMEGF45 was 24 μ*M* and $k_a \cdot [TM] = 30{,}000\ min^{-1} >> k_{ex}$. The possibility of nonspecific binding arises with high ratios of ligand:receptor, so comparisons were made with lower ratios of TMEGF45:thrombin (data not shown). The same surface regions of thrombin showed slowed exchange in the complex at ratios of 7:1 and 2.6:1, consistent with a lack of nonspecific binding.

3. In order to try to differentiate partially excluded regions from the solvent inaccessible *core* of the interface, we measured the off-exchange at two different pHs (6.6 and 7.9). Those regions of the interface that showed a decrease in solvent accessibility at pH 6.6 (where the amides exchange more slowly) but not at pH 7.9 (where the amides exchange more quickly) were judged to be partially solvent accessible. Amides were observed not to exchange at both pHs for up to 30 min when the protein–protein complex was under 100% bound conditions.
4. It is often difficult to differentiate between solvent accessibility changes that arise from conformational changes and those that arise from interface protection. To do this we required that the amides that were protected be in regions that on-exchanged rapidly (indicative of surface location). For thrombin, we found surface loop regions that incorporated deuterium rapidly at both pH. Other regions spanned surface as well as partly buried sites. These showed rapid deuteration at pH 7.9, but also contained a few amides that became deuterated more slowly at pH 6.6 indicating that the region was not completely solvent accessible. Finally, some regions incorporated deuterium slowly at both pHs. In the structure of thrombin, these regions had the least surface solvent accessibility as determined by calculating the solvent accessible surface area of each individual region using GRASP *(13)*. In the off-exchange experiment, we could observe amides that on-exchanged slowly, and off-exchanged more slowly still. These were not immediately assigned to interface regions unless the region was adjacent to a region that was clearly identified as interface. This assessment requires knowledge of the structure of the protein.

References

1. Mandell, J. G., Falick, A. M., and Komives, E. A. (1998) Identification of protein–protein interfaces by decreased amide proton solvent accessibility. *Proc. Nat. Acad. Sci. USA* **95,** 14,705–14,710
2. Mandell, J. G., Baerga-Ortiz, A., Akashi, S., Takio, K., and Komives, E. A. (2001) Solvent accessibility of the thrombin-thrombomodulin interface. *J. Mol. Biol.* **306,** 575–589.
3. Baerga-Ortiz, A., Hughes, C. A., Mandell, J. G., and Komives, E. A. (2002) Epitope mapping of a monoclonal antibody against human thrombin by H/D-exchange mass spectrometry reveals selection of a diverse sequence in a highly conserved protein. *Protein Sci.* **11,** 1300–1308.
4. Anand, G. S., Law, D., Mandell, J. G., Snead, A. N., Tsigelny, I., Taylor, S. S., Ten Eyck, L., and Komives, E. A. (2003) Identification of the Protein Kinase A Regulatory RI -Catalytic Subunit Interface by Amide H/^{2}H Exchange and Protein Docking. *Proc. Nat. Acad. Sci. USA* **100,** 13,264–13,269.
5. Hughes, C. A., Mandell, J. G., Anand, G. S., Stock, A. M., and Komives, E. A. (2001) Phosphorylation causes subtle changes in solvent accessibility at the interdomain interface of methylesterase CheB. *J. Mol. Biol.* **307,** 967–976.
6. Englander, S., Mayne, L., Bai, Y., and Sosnick, T. (1997) Hydrogen exchange: the modern legacy of Linderstrøm-Lang. *Protein Science* **6,** 1101–1109.

7. Bai, Y., Milne, J. S., Mayne, L., and Englander, S. W. (1993) Primary structure effects on peptide group hydrogen exchange. *Proteins* **17,** 75–86.
8. Dharmasiri, K. and Smith, D. L. (1996) Mass spectrometric determination of isotopic exchange rates of amide hydrogens located on the surfaces of proteins. *Analytical Chemistry* **68,** 2340–2344.
9. Baerga-Ortiz, A., Bergqvist, S. P., Mandell, J. G., and Komives, E. A. (2004) Two different proteins that compete for binding to thrombin have opposite kinetic and thermodynamic profiles. *Protein Sci.* **13,** 166–176.
10. Zhang, Z. and Smith, D. L. (1993) Determination of amide hydrogen exchange by mass spectrometry: a new tool for protein structure elucidation. *Protein Sci.* **2,** 522–531.
11. Mandell, J. G., Falick, A. M., and Komives, E. A. (1998) Measurement of amide hydrogen exchange by MALDI-TOF mass spectrometry. *Analytical Chemistry* **70,** 3987–3995.
12. Gemmecker, G., Jahnke, W., and H., K. (1993) Measurement of fast proton exchange rates in isotopically labeled compounds. *J. Am. Chem. Soc.* **115,** 11,620–11,621.
13. Petrey, D. and Honig, B. (2003) GRAPZ: visualization, surface properties, and electrostatics of macromolecular structures and sequences. *Methods Enzymol.* **374,** 492–509.

5

Hydrophobic Interaction Chromatography

Harnessing Multivalent Protein–Surface Interactions for Purification Procedures

Herbert P. Jennissen

Summary

Hydrophobic interaction chromatography (HIC) is one of the basic purification procedures in the biosciences. However, because of its complexity it has not gained the same foothold in the methodological repertoire of protein chemistry as has affinity chromatography or ion exchange chromatography. This is mainly a result of the lack of a general optimization procedure for the reversible adsorption and elution of a novel protein to be purified. Further problems arise from the fact that most commercial hydrophobic adsorbents are inadequate for an ideal performance in downstream processing procedures, because these media are too hydrophobic and elution of proteins in their native state is often impossible. Therefore, as in the 1970s a bioscientist of today has to be capable of synthesizing a small library of hydrophobic gels from which he or she can then select and optimize the ideal matrix for their special needs. In addition, a general optimization method employing the critical hydrophobicity concept has now been devised that should allow the application of HIC methodology to many hitherto unpurified proteins. In this chapter, the reader is first introduced to the theoretical background (multivalence, negative cooperativity, adsorption hysteresis) of the binding of protein ligands to hydrophobic supports, so that they will be capable of independently adapting HIC to a novel protein. Then a simple nontoxic method is described for the synthesis of HIC-gel libraries consisting of a homolgous series of uncharged alkyl-Sepahroses of three chain lengths (butyl, pentyl, and hexyl Sepahrose) prepared with different degrees of substitution. From this series a critical hydrophobicity gel can then be selected and employed for critical hydrophobicity HIC. A detailed example for the chromatography of human fibrinogen is given that has been employed as a one-step procedure for the purification of fibrinogen from human plasma.

From: *Methods in Molecular Biology, vol. 305: Protein–Ligand Interactions: Methods and Applications*
Edited by: G. U. Nienhaus © Humana Press Inc., Totowa, NJ

Key Words: Adsorption hysteresis; alkyl agarose library; carbonyl diimidazole; critical hydrophobicity HIC; hydrophobic interaction chromatography (HIC); library of HIC gels; ligand definition; multivalence; negative cooperativity; lattice-site binding function; bulk ligand binding function.

1. Introduction

1.1. Ligands in Chromatographic Systems

Chromatographic methods based on a binding reaction between a chromatographic matrix and a protein can be classed into the large group of *protein–ligand interactions*. However, because the usage of the word *ligand* differs significantly in the literature a consensus definition of this word should form the basis of its use in this chapter. According to the Nomenclature Committee of the IUBMB as of 1992 *(1)* a ligand is defined as follows: "If it is possible or convenient to regard part of a polyatomic molecular entity as central, then the atoms, groups or molecules bound to that part are called ligands." In general usage, the polyatomic central molecular entity is large in comparison to the smaller peripheral ligands. In the case of insulin binding to its receptor or a coenzyme to its apoenzyme things appear simple, although in the case of the insulin-receptor interaction the ligand is a protein. In the latter case the *protein–ligand interaction* reduces to a protein–protein interaction. According to the IUBMB definition, systems in which a protein is adsorbed to a solid surface or a chromatographic support, the protein corresponds to the ligand, and the solid surface or chromatographic support to the polyatomic central entity. In anticipation of the IUBMB consensus definition we defined a ligand for adsorption studies in 1979 as "a solute molecule, e.g., protein capable of noncovalent interactions with the surface of the agarose matrix or a biological superstructure" *(2)*. In fact, we have been using the term protein ligand synonymously for adsorbate since 1976 *(3)*. Consequently, it can be calculated that one bead (= central entity) of Sepharose 4B can maximally bind approx 4.8×10^{10} molecules of phosphorylase *b* (= ligand) *(4)*. In accordance with the IUBMB definition protein molecules with the properties of an adsorbate will generally be defined as ligands in this chapter. In distinction, immobilized molecules of small molecular weight (e.g., alkyl moieties) on chromatographic supports will be called alkyl residues *(3)*.

1.2. Protein–Surface Lattice Interactions

Today, it is generally accepted that a protein molecule adsorbing to a solid surface covered with immobilized alkyl residues will bind to several residues at the same time *(3,5)*. This was unclear early in the 1970s. The corresponding adsorption mechanism has been termed multivalent binding *(3,5)* on a two-dimensional lattice of binding sites (**Fig. 1**, *see* **ref. 6**). Valence was defined

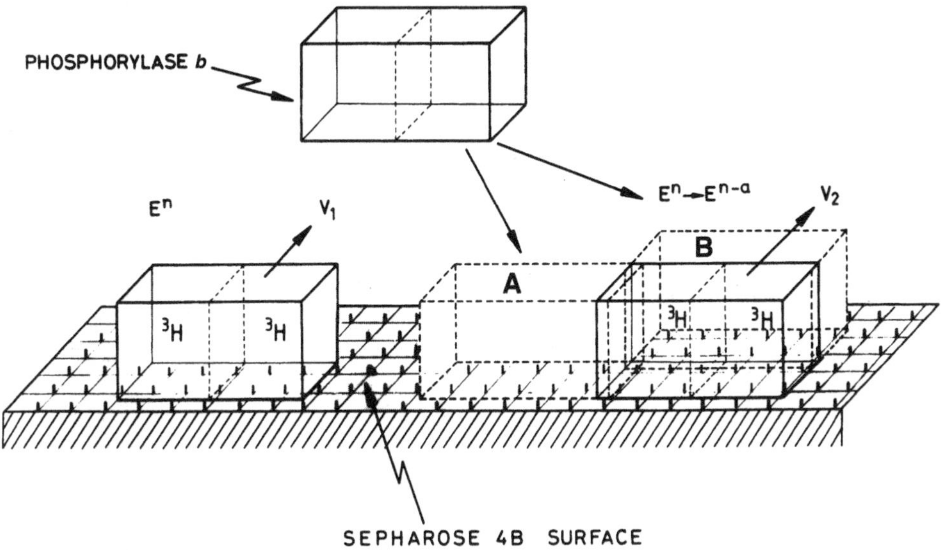

Fig. 1. Schematic model of the multivalent binding of a protein ligand (phosphorylase *b*) on a planar lattice of alkyl binding sites. The hatched area corresponds to the agarose surface showing the two-dimensional lattice of butyl residues as biding sites (vertical dashes). The box is a schematic representation of the dimer phosphorylase *b* (dimensions 6.3 × 6.3 × 11.6 nm). The scheme also displays lateral interactions between a ^{3}H-labelled enzyme and an unlabelled species where the labelled molecule is displaced from the surface. For further details *see* **ref. *19*** and the text.

(2) as "the no covalent force (better: energy of binding) resulting from the interaction of one lattice binding site (e.g., alkyl residue) on the matrix with a corresponding finite area (e.g., hydrophobic binding site) of the protein." Multivalent binding occurs above a critical surface concentration of alkyl residues, when the protein covers a surface area on the adsorbent containing more than one immobilized alkyl residue. The surface area covered by the protein, i.e., the footprint of the protein is called a binding unit *(7)* and contains several lattice binding sites, i.e., immobilized alkyl residues. From **Table 1** it is clear that all protein chromatography methods based on noncovalent interactions in the range of 4–10 kJ/mol will be multivalent requiring a critical surface concentration as a threshold for binding *(7)*.

On the basis of multivalence, which implies a specific geometry of sites, protein adsorption on 2D-lattices can compared to a molecular recognition process. Molecular recognition goes back to Fischer who devised the lock-and-key model *(8)*. The lock-and-key model has lost little of its heuristic potential, but has been developed further via the complementarity concept of Pauling *(9)* and

Table 1
Classification of Chromatography Methods According to Chemical Bonds and Interactions

Bond type/ interaction	Length nm	Bond strength (kJ/Mol)		Force Law	Binding force N	Chromatography (thermodynamics)	Reference
		In vacuum	In water				
Covalent b.	0.15	377	377		10^{-8}	Covalent chromatography (enthalpic)	*(27)*
Coordinative b. (monocoordinated)	0.19	20–30	20–30			Metall-chelate chromatograpy (enthalpic)	*(28)*
Ionic b.	0.25	335	4	$-\frac{q_1 - q_2}{r^2}$		Ion exchange chromatography (enthalpic)	*(29,30)*
Hydrogen b.	0.3	16	4			Hydroxy apatite chromatography (enthalpic)	*(30–32)*
Van der Waals interaction.	0.2	4	4	$-\frac{\alpha^2}{r^6}$	10^{-11}	Enthalpic	
Hydrophobic interaction.	0.2–0.3	—	8 –11	$\exp^{-D/Do}$		Hydrophobic interaction chromatography (entropic)	*(33,34)*
Charge transfer inter.	~ 0.2	4	4			Charge transfer chromatography (enthalpic)	*(35)*

[a] The different chromatographies depicted are covalent chromatography ***(27)***, metal chelate chromatography ***(28)***, ion exchange chromatography ***(29,30)***, hydroxyapatite chromatography ***(30–32)***, hydrophobic interaction chromatography ***(33,34)*** and charge transfer chromatography ***(35)***. Some of the data was taken from **ref. *36***.

the induced fit or hand-in-glove model of Koshland *(10)*. On the basis of this work, Katchalski-Katzir *(11)* has suggested three main parameters in protein recognition: 1) complementarity (lock-and-key), 2) specificity, and 3) dynamics (induced fit). Specificity was wonderfully characterized by Katchalski-Katzir in connection with a lecture held in Gwatt, Switzerland in 1993 (*see* **ref. *12***): "If there is only one protein and one ligand there is no specificity. Specificity is the selection of one ligand among 1000 ligands." In essence, this means that specificity corresponds to a good signal-to-noise ratio.

How can a recognition reaction occur on a two-dimensional lattice of binding sites? As was pointed out previously *(3,5)*, the classical lock-and-key model cannot be directly applied to protein adsorption on alkyl agaroses. This is precluded by multivalence. However, if a modern two-dimensional key is taken into account, a similarity of such keys to a protein–lattice interaction cannot be denied. Two-dimensional plastic keys in credit card format in use for public garages are shown in **Fig. 2**. They differ in the number and geometric distribution of buttons on the card surface leading to a specific recognition in the lock opening. Similarly, one can imagine the differential binding of proteins to a two-dimensional lattice of binding sites capable of discriminating the binding of two proteins. Thus, a specific recognition of proteins in the sense of Katchalski-Katzir *(11)* on two dimensional lattices of binding sites is indeed possible and forms the basis of the chromatographic purification of proteins *(7)*. As has been recently pointed out *(6)*, three parameters govern the adsorption of a protein to a homologous series of hydrophobic alkyl Sepharoses: 1) the chain-length, 2) the surface concentration of immobilized alkyl residues, and 3) the salt concentration. On the basis of the first two fundamental parameters determining protein adsorption on natural and artificial surfaces one may distinguish two categories of protein adsorption isotherms: (a) a lattice-site binding function, and (b) a bulk ligand binding function *(4,13–15)*. The equations describing these binding functions have been derived previously and are similar power functions as the so-called Hill equation *(3,4,14)*.

The lattice site binding function governs the binding of an immobilized residue, constituting a surface lattice-site, to a complementary site (patch, pocket) on the protein adsorbed from the bulk solution. Protein adsorption is multivalent (*see* **Fig. 1**). The more lattice sites that simultaneously interact with a protein molecule, the higher the affinity of binding will be *(16)*. Protein adsorption, therefore, increases as a function of the surface concentration of lattice sites (at a constant equilibrium concentration of free bulk protein) according to the following equation *(3,4,17)*:

$$\frac{\theta_S}{1-\theta_S} = K_S\left(\Gamma_r^S\right)^{n_s} \qquad (1)$$

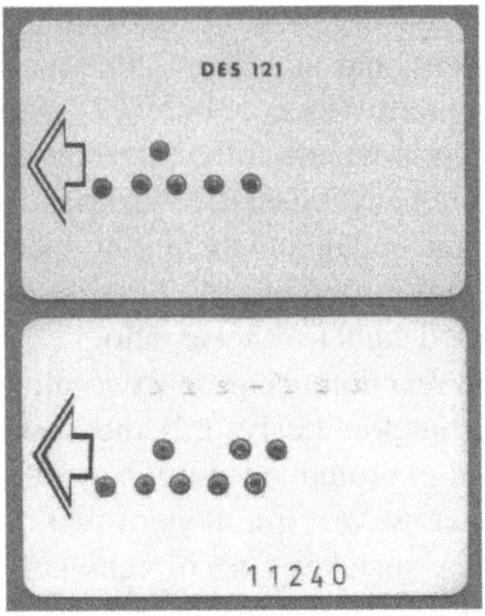

Fig. 2. Modern two-dimensional keys in credit-card format. The small black buttons on the surface allow a specific recognition in the lock. In a similar manner a molecular recognition of a protein ligand can occur on a two-dimensional lattice of binding sites of a HIC support. The size of the keys is: 84 × 54 × 1 mm. For further details *see* **Fig. 1** and the text.

where θ_S is the fractional saturation of binding units with protein on the adsorbent as a function of the surface concentration of lattice sites in a binding unit. In essence, this means that θ_S corresponds to the fractional saturation of the involved protein with immobilized alkyl residues ***(4)***. As the protein is saturated with immobilized residues the affinity to the surface increases and the binding units fill up. Γ_r^S is the surface concentration of lattice sites (i.e., immobilized residues), K_S is the lattice site adsorption constant, and n_S is the lattice site adsorption coefficient. For the case $n_S = 1$, **Eq. 1** reduces to a rectangular hyperbola. Saturation of binding occurs when the complementary binding sites of the protein in a binding unit on the adsorbent surface cannot accommodate further lattice sites ***(17)***. The corresponding half-saturation constant $K_{S,0.5}$ relates to K_S according to: $K_{S,0.5} = (K_S)^{1/n_S}$.

The bulk ligand (i.e., protein) binding function, which corresponds to a classical protein adsorption isotherm, can be described by an analogous equation ***(3,4,17)***:

$$\frac{\theta_B}{1-\theta_B} = K_B\,[P]^{n_B} \tag{2}$$

where θ_B is the fractional saturation of binding units on the surface at a constant lattice site concentration with the bulk protein as independent variable. In this case θ_B corresponds to the fractional saturation of gel surface with protein. $[P]^{n_B}$ is the bulk protein concentration at equilibrium, K_B is the bulk-ligand adsorption constant and n_B the bulk-ligand adsorption coefficient. For the case $n_B = 1$, **Eq. 2** also reduces to a rectangular hyperbola (Langmuir isotherm). Saturation occurs, when the solid phase surface cannot accommodate additional protein molecules. The half-saturation constant $K_{B,0.5}$ in this case relates to K_B according to: $K_{B,0.5} = (K_B)^{1/n_B}$ ***(3,4,13)***.

Considering chromatographic systems based on alkyl modified surfaces the lattice site binding function adequately describes the above surface concentration parameter, which in the range of C1 to C6 corresponds to a sigmoidal binding curve displaying positive cooperativity ***(3,5,18)***. On the other hand, a classical binding isotherm monitored to determine the binding capacity of a defined gel corresponds to the bulk ligand binding function displaying negative cooperativity ***(3,19)*** and adsorption hysteresis ***(2,20,21)***. Negative cooperativity implies that the affinity of the adsorbent for the protein decreases as the adsorbent is saturated ***(3)***. Thus, HIC gels will always contain a mixture of high and low affinity sites by nature of multivalence and overlapping binding units ***(3)***. Protein molecules binding to high-affinity binding units will often be forced into conformational changes resulting in a further increase in affinity and adsorption hysteresis ***(20)***. This may lead to a surface denaturation of the protein and so-called *irreversible adsorption* (*see* **ref. *4***). In order to avoid the drawbacks of multivalence, negative cooperativity and adsorption hysteresis on HIC-gels the critical hydrophobicity method of hydrophobic interaction chromatography (i.e., critical hydrophobicity HIC) was developed.

2. Materials

2.1. Chromatography

1. Sepharose 4B (Amersham Bioscience).
2. Carbonyl diimidazole (Acros, Fisher Scientific GmbH, Schwerte Germany, or Sigma, Munich, Germany).
3. Dried acetone (Merck, Darmstadt, = 0.01% H_2O).
4. Pure acetone (Baker, Deventer, NL, = 0.5% H_2O).
5. n-Alkyl amines (butyl, pentyl and hexyl amine) Acros (Fisher Scientific GmbH).
6. $[1-^{14}C]$ n-Butylamine hydrochloride (Biotrend, Köln, Germany).
7. $[1-^{14}C]$ n-Hexylamine hydrochloride (Biotrend).
8. Scintillation cocktail Ecolume (ICN Biochemicals Inc., Irvine CA).
9. Low volume Plexiglas columns (1 cm i.d. x 12 cm height).
10. Larger volume Plexiglas column (2 cm i.d. x 15 cm height).
11. Peristaltic pump.

2.2. Proteins

1. Crude extract.
2. Purified proteins (phosphorylase *b*, calmodulin, Fibrinogen).

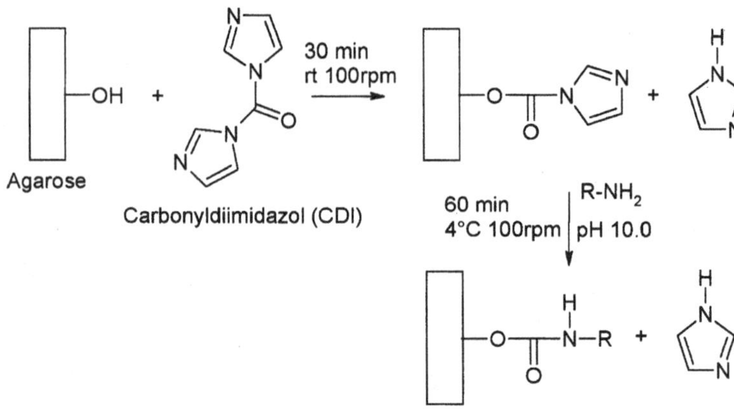

Fig. 3. Activation of Sepharose 4B with carbonyl diimidazole for the coupling of primary amines. For further details *see* **refs. *6,37,38***.

2.3. Buffers and Solutions

1. Buffer A: containing 10 m*M* *tris*-(hydroxymethyl)-aminomethane/maleate, 5 m*M* dithioerythritol, 1.1 *M* ammonium sulfate, 20% sucrose, pH 7.0.
2. Buffer B: 20 m*M* Tris-HCl, 1 m*M* $CaCl_2$, pH 7.0.
3. Alkaline SDS solution: 0.1 *M* NaOH, 1% SDS.
4. Buffer C: buffer B + 0.3 *M* NaCl.
5. Buffer D: 20 m*M* Tris-HCl, 0.3 *M* NaCl, 10 m*M* ethylene glycol-*bis*-(β-amino-ethylether)*N,N,N',N'*-tetraacetic acid, i.e., EGTA, pH 7.0.
6. Buffer E: 50 m*M* Tris-HCl, 150 m*M* NaCl, 1 m*M* EGTA, pH 7.4.
7. Buffer F: 1 m*M* sodium β-glycerophosphate pH 7.0.
8. Buffer G: 50 m*M* Tris-HCl, 1500 m*M* NaCl, 1 m*M* EGTA, pH 7.4.

3. Methods

3.1. Synthesis of Homologous Hydrophobic Gel Libraries

The stability of Sepharose 4B in acetone solution was shown in **ref. *22***. The employment of dry acetone solutions for the activation of functional surface groups with carbonyl diimidazole (*see* **Fig. 3** for reaction scheme) was shown in **ref. *23***.

3.1.1. Activation of Sepharose 4B With 20 mg/mL Incubation Mixture of CDI

Twenty-five grams wet weight Sepharose 4B (= unit of volume) ***(24)***. Wash procedure by suction on Büchner funnel:

1. 20 vol H_2O.
2. 20 vol H_2O/acetone 3:1.

3. 20 vol H_2O/acetone 1:3.
4. 20 vol pure acetone (Baker, Deventer, NL, = 0.5% H_2O).
5. The acetone is removed by suction leading to a white dry powder of Sepharose 4B.
6. To approx 0.8 g dry Sepharose powder (~ 25 mL packed gel) is added 1.5 g carbonyl diimidazole in 2 vol (50 mL) dried acetone (Merck, = 0.01% H_2O) under gentle stirring.

After 30 min of activation at room temperature (rt) the gel suspension is again washed on Büchner funnel.

7. 20 vol H_2O/acetone 1:3.
8. 20 vol H_2O/acetone 3:1.
9. 20 vol H_2O.
10. The resulting 25 g wet weight Sepharose is added to 2 vol 2 *M* alkyl amine pH 10.0 (which may contain 0.25 µCi ^{14}C-alkyl amine) and incubated for 60 min at 4°C.

The resulting alkyl Sepharose 4 B is then washed on Büchner funnel:

11. 20 vol 0.01 *M* NaOH.
12. 20 vol 0.02 *M* HCl.
13. 20 vol H_2O.
14. The resulting 25 g wet weight alkyl Sepharose is taken up in about 2 vol of H_2O to which a small amount of NaN_3 is added and stored at 4°C.

3.1.2. Hydrolysis and Analysis of ^{14}C-Alkyl-Sepharose (**24**)

1. One milliliter packed gel of ^{14}C-alkyl Sepharose is added to 1 vol 32% HCl and heated to 50°C for 45 min.
2. The reaction is stopped and the solution neutralized by addition of 0.20 mL 10 *N* NaOH.
3. The neutralized hydrolysate solution (approx 2 mL) is added to 15 mL Scintillation cocktail Ecolume for scintillation counting (β-counter) and quantification (quench correction should not be forgotten, *see* **Notes 1–6**).

3.2. Measurement of Protein Adsorption on Columns

The following four protocols show the use of high surface area adsorbents. In each protocol the protein concentration in solution needs to be determined after adsorption took place. Protein concentration can be measured directly, according to the method of Lowry et al. *(25)*, by measuring protein's UV absorbance or fluorescence, using ^{125}I- or FITC-labels, or, in the case of adsorption of enzymes, by measuring a protein's enzymatic activity.

3.2.1. Saturating Sample-Load Method

This method can be employed for all sorts of gel particles (low density) and also for powders of glass, inorganic material or metal. The saturation sample-load method should be employed if one wants to compare the *saturation capacity* (Qs = Ce/Co = 1.0, where Ce and Co are the protein concentrations in effluent and sample respectively, *see* below) of different gels or powders at a defined bulk protein concentration. The methods employed for the measurement of adsorption of the enzyme phosphorylase *b* to butyl-Sepharose 4B particles at 5°C, described briefly below, can be found in more detail in **refs.** ***2,3,19***.

Preparation of protein solution. Before use, the protein, e.g., phosphorylase *b*, is extensively dialyzed against the adsorption buffer A *(2)*. Sucrose is included in buffer A to minimize nonspecific adsorption to the agarose gel backbone. Before adsorption the substituted beaded or particulate adsorbent is equilibrated with adsorption buffer A. Low volume Plexiglas columns (1 cm i.d. × 12 cm height) containing 1–3 mL packed gel or larger volume columns (with a large cross section for fast flow) of the dimensions 2 cm i.d. × 15 cm height filled with 10–20 mL packed gel can be employed at flow rates of 5–10 cm/h. A purified protein solution (phosphorylase *b*) or a crude extract is applied by pump or gravity to the columns until no more enzyme is adsorbed, i.e., until the protein concentration in the run-through is identical to that in the applied sample. The columns are then washed with a 10- to 50-fold buffer volume until significant protein can no longer be detected in the run-through. Elution of protein is either accomplished by specific affinity agents, salts ***(24)***, or, for protein balance studies, by denaturants, e.g., urea, SDS applied to the column in buffer. The adsorbed amount of protein is calculated from difference measurements of the amount applied and the amount in the run-through, or by elution of the adsorbed protein by a strong denaturing detergent mixture such as 0.1 *M* NaOH, 1% SDS and subsequent determination of the protein amount in this eluent.

3.2.2. Limited Sample-Load Method

This method can also be employed for all sorts of gel particles and for powders of glass, inorganic material or metal. The limited sample-load method is a good screening procedure for a series of adsorbents or protein samples. In this method a defined amount of protein (e.g., 1 mg) in a defined sample volume is applied to each column under identical conditions. The applied nonsaturating amount is dimensioned so as to be 100% adsorbed (i.e., no protein in run-through) on an adsorbent displaying the expected maximal affinity and capacity. The method allows a comparison of *relative adsorption capacities and affinities* of adsorbents on a quantitative chromatographic basis. This method, employed for quantitative evaluation of the adsorption of calmodulin, fibrinogen and peptides to beaded agarose adsorbents of varying hydrophobicity, is exemplified below.

Example fibrinogen (*see* **ref.** ***26***). Quantitative hydrophobic adsorption chromatography of fibrinogen is performed on the same column type containing 2 mL of packed gel. The gel is washed and equilibrated with 20 vol buffer E

(50 m*M* Tris-HCl, 150 m*M* NaCl, 1 m*M* EGTA, pH 7.4) at 5–10 cm/h. As sample 1 mg purified fibrinogen is applied in a volume of 1 mL (in buffer E) and fractions of 1.5 mL are collected. The column is then washed with 15 mL buffer E and eluted with 7.5 *M* urea and at high hydrophobicity of the gel with 1% SDS ***(26)***.

Example calmodulin (*see* **ref. *26***). In the case of calmodulin quantitative adsorption, chromatography is performed on a column (0.9 × 12 cm) containing 2 mL of packed gel of various alkyl agaroses. The gel was washed and equilibrated with 20 vol buffer B. 1 mg of purified calmodulin is applied in a sample volume of 1 mL (in buffer B) and fractions of 1 mL are collected. The column is then washed with 9 mL buffer B and then with 9 mL buffer C. The adsorbed calmodulin is eluted with buffer D. The EGTA chelates the calmodulin-bound Ca2+, thereby changing the conformation of calmodulin to a more hydrophilic species leading to elution. For quantifying tightly bound, i.e., *irreversibly bound*, calmodulin on highly hydrophobic columns elution is continued with 7.5 *M* urea and finally with 1% SDS added to the adsorption buffer ***(26)***.

Example tryptophan tripeptide (*see* **ref. *26***). Quantitative hydrophobic adsorption chromatography of a tripeptide, Trp-Trp-Trp (Paesel and Lorei, Frankfurt, Germany, 98.5% purity) is performed on a column (0.9 × 12 cm) containing 2 mL of packed gel. The gel is washed and equilibrated with 20 vol. buffer F. As sample 1 mg Trp-tripeptide is applied in a volume of 1 mL (in buffer F) and fractions of 1.5 mL are collected. The column is then washed with 15 mL buffer F and eluted with 1% SDS ***(26)***. Calmodulin, fibrinogen and the tripeptide in the obtained fractions are measured directly according to the method of Lowry et al. ***(25)*** employing BSA (all other proteins) or the Trp-tripeptide (for the peptide only) as standard.

In the examples of quantitative analytical chromatography as above the experiments can be performed at room temperature, good flow is usually achieved by gravity and only fresh, nonregenerated gel should be employed for each experiment.

3.3. Critical Hydrophobicity Hydrophobic Interaction Chromatography

High yields in hydrophobic interaction chromatography can only be obtained if the protein to be purified is fully excluded from the gel under elution conditions that are as physiological as possible, i.e., at low ionic strength. This means that the gel should be fully non-adsorbing under these conditions. The strategy for optimizing a hydrophobic alkyl Sepharose for a chromatographic purification step is outlined in **Table 2**. The method involves four basic steps of which steps one and two are performed at low or physiological ionic strength (i.e., I = 0.15): 1) the selection of an appropriate chain length, 2) the determination of the critical surface concentration of alkyl residues (critical hydrophobicity), 3)

Table 2
Strategy for Application of Critical Hydrophobicity Procedure

1. Application of protein sample (0.5–1 mg/mL packed gel) at a low salt concentration. Selection of appropriate alkyl chain length from homologous chain length library (~ 20 mmol/mL packed gel) i.e., that gel which adsorbs approx 50% (chain-length parameter).
2. Variation of the surface concentration (gel library) for determination of the critical hydrophobicity of above selected gel (surface concentration parameter).
3. Determination of the optimum high salt concentration (e.g., 1–5 *M* NaCl) for binding the protein sample on the critical hydrophobicity support (salt parameter).
4. Application of protein solution to critical hydrophobicity support at specified high salt concentration and elution by inverse salt gradient.

Table 3
Synthesis of a Surface Concentration Series of ^{14}C-Butyl Sepharoses by the Carbonyl Diimadazole (CDI) Method [a]

Carbonyl diimazole (mg/mL)	Degree of substitution (mmol/mL) packed gel
0.39	1.0
0.84	2.0
2.2	5.0
4.4	10.0
6.5	15.9
9.0	20.0
9.5	25.6
14.4	30.0
18.0	41.2

[a] For further details *see* **ref. 6** and the text.

the determination of the minimal salt concentration (e.g., NaCl) necessary for a complete adsorption of the protein, and 4) chromatography: adsorption of the protein on critical hydrophobicity gel at the determined high salt concentration and elution by a negative salt gradient. Before the optimization can begin, a library of hydrophobic gels must be synthesized. As a rule gels with immobilized butyl, pentyl, and hexyl residues (Seph-C4, Seph-C5, Seph-C6) suffice.

3.3.1. Gel Libraries of Seph-Cn Containing Different Immobilized Residue Concentrations

A typical surface concentration series of Seph-C4 synthesized by the CDI method is shown in **Table 3**. If potted graphically it can easily be seen that the degree of substitution (i.e., surface concentration, *see* **ref. 4**) increases linearly with the concentration of CDI in the incubation mixture. For synthesizing three series (Seph-C4 to Seph-C6) the activated CDI-Sepharose (acetone dried powder) can be divided into there equal portions which are rehydrated separately and incubated with the corresponding butyl-, pentyl-, or hexyl-amine solution, respectively. To control the degree of substitution at least one series should be isotopically labeled with a ^{14}C-alkyl amine.

3.3.2. Determination of Critical Hydrophobicity

In the next step the three series are tested for protein binding by the limited sample load method (see above). As a rule of the thumb, an intermediate degree of substitution (e.g., approx 20 µmol/mL packed gel of each series is tested for binding. The alkyl Sepharose binding ca. 50% of the protein load is then selected for determination of the critical hydrophobicity. A typical example for such an experiment is shown in **Table 4** for the adsorption of fibrinogen to library of homologous Seph-Cn gels. On Seph-C4 it can be seen that no fibrinogen is bound to the 21.7 µmol gel. In the Seph-C5 series the 22 µmol gel binds 68% of the applied fibrinogen and in the Seph-C6 series the 17.2 µmol gel already binds 96% of the fibrinogen. Thus, the Seph-C5 series is selected in separate experiments (not shown) and the critical hydrophobicity is determined to about 13.0 µmol/mL packed gel. A second example is shown in **Table 5** for the protein calmodulin which develops a hydrophobic surface in the presence of Ca^{2+}. As shown in **Table 5,** calmodulin did not bind at all to Seph-C4. Therefore, binding was tested on a series of Seph-C6 with the 7.7 µmol gel already binding 83% of the applied amount. Calmodulin is a special case since the hydrophobicity of the protein can be changed at will by adding or deleting Ca^{2+} ions. Even on high hydrophobicity gels there is little danger of *irreversible* adsorption because elution can be facilitated by EGTA which chelates Ca^{2+}. Thus, in this special case calmodulin can be nicely purified (not shown) on a gel above its critical hydrophobicity (i.e., 4.3 µmol/mL packed gel in the presence of Ca^{2+}) on the support with 7.7 µmol/mL packed gel. Finally, the tripeptide of tryptophan can be shown to bind weakly to a Seph-C6 of 41 µmol/mL packed gel. Under isocratic elution conditions *(26)* the tripeptide is significantly retarded by the column. Thus, below the critical hydrophobicity as defined here the alkyl agaroses can also be employed under isocratic conditions for the purification of proteins. Because of negative cooperativity and adsorption hysteresis an ideal linear chromatography will generally not be possible *(4)*.

3.3.3. Critical Hydrophobicity Chromatography

Figure 4 shows a chromatographic run of purified human fibrinogen on a critical hydrophobicity pentyl Sepharose gel. Fibrinogen in an amount of 55 mg

Table 4
Quantitative Hydrophobic Interaction Chromatography of Fibrinogen on Alkyl-Agaroses [a]

Type of Gel	Degree of substitution (mmol/mL) packed gel	Fibrinogen: Applied (mg)	Fibrinogen: Excluded (unbound) (mg)	Fibrinogen: Bound (mg)	Fibrinogen: Eluted (bound) (mg)	Total yield (%)
Seph-C4	4.4	1.0	0.891	0.109	0	89
	15.3	1.0	0.886	0.114	0	89
	21.7	1.0	0.920	0.080	0	92
	36.8	1.0	0.878	0.122	0.030	91
	46.8	1.0	0.493	0.506	0.366	89
Seph-C5	3.0	1.0	0.887	0.113	0	89
	8.4	1.0	0.890	0.110	0.030	92
	12.1	1.0	0.899	0.101	0	90
	22.0	1.0	0.318	0.682	0.446	77
	41.6	1.0	0	1.00	0.722	73
Seph-C6	1.9	1.0	0.808	0.192	0	81
	4.3	1.0	0.828	0.172	0	83
	7.7	1.0	0.296	0.704	0.545	85
	11.0	1.0	0.049	0.951	0.712	77
	17.2	1.0	0.036	0.964	0.795	84
	42.0	1.0	0.014	0.986	0.769	79

[a] The data were derived from column experiments as described in methods and the text. Fibrinogen (1 mg in 1 mL sample volume) was applied to 2 mL packed gel in buffer E. The column was then washed with the same buffer and eluted with 7.5 *M* urea. The exclude amount of fibrinogen corresponds to the amount washed from the column in buffer E. The total yield corresponds to the amount recovered in the buffer wash and after elution with urea as percent of amount applied . The difference between amount applied and amount recoverd is the protein fraction which cannot be eluted ("irreversibly" bound protein). For further details and the definition of gel types *see* **ref. *26***, Methods and the text.

is applied to a 20 mL packed gel column (Seph-C5, 13.6 μmol/mL packed gel) in the presence of 1.5 *M* NaCl which was determined as the ideal salt concentration in separate experiments (salt parameter). It can be seen, that no fibrinogen is in the run-through. The adsorbed fibrinogen can then be eluted by a negative salt gradient with a recovery of 86% (48 mg). Similarly, fibrinogen can be purified in a single step from human plasma ***(6)***.

4. Notes

1. It is of prime importance that the restrictions in connection with the water content of the acetone (pure and dried acetone) are adhered to. If the water content of the

Table 5
Quantitative Hydrophobic Interaction Chromatography of Calmodulin on Alkyl-Agaroses [a]

		Calmodulin				
Type of Gel	Degree of substitution (μmol/mL) packed gel	Applied (mg)	Excluded (mg)	Bound (mg)	Eluted (mg)	Total yield (%)
Seph-C6	1.9	1.0	0.96	0.04	0	96
	4.3	1.0	0.90	0.10	0.02	92
	7.7	1.0	0.17	0.83	0.60	87
	17.2	1.0	0.19	0.81	0.68	87

[a] The data (taken from **ref. *2***) were derived from column experiments as described in methods and the text. Calmodulin was applied in buffer B (+ Ca++), the column was then washed with buffer C (+ 300 m*M* NaCl) and finally eluted with buffer D (+ EGTA). The excluded amount of calmodulin corresponds to the amount washed from the column in buffers B and C. The eluted amount corresponds to the amount eluted with buffer D. The total yield corresponds to the amount recovered in buffers B–D as percent of amount applied. The difference between amount applied and amount recoverd is the non-EGTA elutable fraction ("irreversibly" bound protein) which can however be eluted by detergents. For further details and the definition of gel types, *see* Methods, the text and **ref. *26***.

acetone is too high (> 0.5%) the CDI method will not work reproducibly. If costs are no problem then only dried acetone (= 0.01% H_2O) should be used for all steps involving CDI and acetone dried gels and refreshed regularly.

2. In the case of ^{14}C-scintillation counting the quench corrections can be omitted if all solutions give approximately the same quenching. This can be achieved by hydrolyzing unsubstituted Sepharose 4 B for use in the controls and the standard curves for determining the specific radioactivity of the alkyl amine. If the quenching in the samples varies widely, then a quench correction curve has to be prepared.
3. The scintillation counting (3–5 counting rounds) should be done rapidly after the hydrolysis of the gel since the gel/scintillation mixture is unstable and will deteriorate (milky appearance) after about 1 d.
4. It should be noted that hydrophobic interactions have positive temperature coefficients. Thus there will generally be a higher affinity of the protein for the gel at room temperature compared to the 5°C in a cold room, where a protein may not bind under the same conditions.
5. NaCl has been suggested as the salt of choice in critical hydrophobicity HIC because it has low salting-out properties. A strongly salting-out neutral salt like ammonium sulphate may lead to the formation of heterologous protein aggregates on a column run with a crude extract which will lower the achievable purification factor. If adsorption of the selected protein cannot be achieved by 3–4 *M* NaCl then ammonium sulphate should be tried which otherwise is a very gentle salt for proteins and inhibits proteolysis.

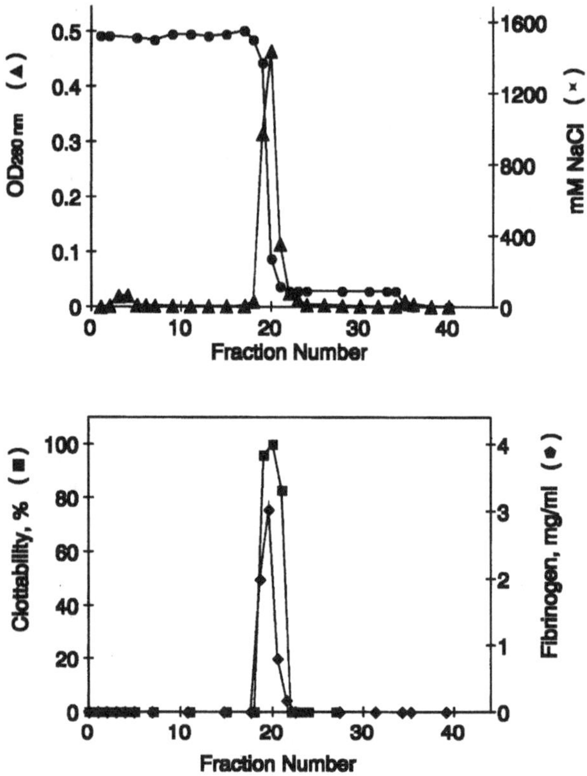

Fig. 4. Hydrophobic interaction chromatography of purified human fibrinogen on critical hydrophobicity Seph-C5 (13.6 mmol/mL packed gel). As sample 2.1 mL (26.4 mg/mL) purified human fibrinogen (Behring) was applied to a column (1.6 cm i.d. × 10 cm) filled with 20 mL packed pentyl Sepharose (13.6 mmol/mL packed gel) in buffer G. The flow rate was approx 70 mL/h and fractions of 8 mL were collected. The adsorbed fibrinogen was eluted in a single step with buffer E which leads to a steep negative gradient. The peak fractions of the eluted fibrinogen had a clottablility more than 96%.

6. Storage: Sepharose 4B and CDI-Sepharose can be stored for several days in the acetone dried white powder form in a desiccator without loss of structure or activity at room temperature. The alkyl agarose libraries are stable without leaching for years in aqueous bacteriostatic/bactericidal solutions at 5°C.

References

1. IUPAC-IUBMB Joint Commission on Biochemical Nomenclature (JCBN) & Nomenclature Commission of IUBMB (NC-IUBMB) (1992) Ligands in biochemistry and inorganic chemistry. *Eur. J. Biochem.* **204,** 1–3.

2. Jennissen, H. P. and Botzet, G. (1979) Protein binding to two-dimensional hydrophobic binding-site lattices:adsorption hysteresis on immobilized butyl-residues. *Int. J. Biol. Macromol.* **1,** 171–179.
3. Jennissen, H. P. (1976) Evidence for negative cooperativity in the adsorption of phosphorylase b on hydrophobic agaroses. *Biochemistry* **15,** 5683–5692.
4. Jennissen, H. P. (1981) Immobilization of residues on agarose gels: Effects on protein adsorption isotherms and chromatographic parameters. *J. Chromatogr.* **215,** 73–85.
5. Jennissen, H. P. (1976) Multivalent adsorption of proteins on hydrophobic agaroses. *Hoppe-Seyler's Z Physiol. Chem.* **357,** 1201–1203.
6. Jennissen, H. P. (2000) Hydrophobic interaction chromatography: the critical hydrophobicity approach. *Int. J. Bio-Chromatography* **5,** 131–163.
7. Jennissen, H. P. (1978) Multivalent interaction chromatography as exemplified by the adsorption and the desorption of skeletal muscle enzymes on hydrophobic alkyl-agaroses. *J. Chromatogr.* **159,** 71–83.
8. Fischer, E. (1894) Einfluß der configuration auf die wirkung der enzyme. *Ber. Dt. Chem. Ges.* **27,** 2985–2993.
9. Pauling, L. (1946) Molecular architecture and biological reactions. *Chem. Eng. News* **24,** 1375–1377.
10. Koshland, D. E. (1994) The key-lock theory and the induced fit theory. *Angew Chem. Int. Ed. Engl.* **33,** 2375–2378.
11. Katchalski-Katzir, E. (1983) Some general considerations on the recognition by and of proteins. In *Affinity Chromatography and Biological Recognition* (Chaiken, I.M., Wilchek, M., and Parikh, I., eds.). Academic Press, Inc., London, pp. 7–26.
12. Eberle, A. N., Fischer, E. A., and Parikh, I. (1995) Preface to proceedings of 1th symposium on biorecognition and affinity technology in Gwatt Switzerland 1993. *J. Mol. Recognit.* **8,** 1–2.
13. Jennissen, H. P. (1988) General aspects of protein adsorption. *Makromol. Chem. Macromol. Symp.* **17,** 111–134.
14. Jennissen, H. P. and Botzet, G. (1993) The binding of phosphorylase kinase to immobilized calmodulin. *J. Mol. Recognit.* **6,** 117–130.
15. Jennissen, H. P. (1989) Proteinadsorption auf alkylsubstituierten Oberflaechen. *Berichte der Bunsen-Gesellschaft fuer Physik. Chemie* 949–956.
16. Jennissen, H. P. (1981) The binding and regulation of biological active proteins on cellular interfaces: model studies of enzyme adsorption on hydrophobic binding site lattices and biomembranes. *Adv. Enzyme Regul.* **19,** 377–406.
17. Jennissen, H. P. (1977) Multivalent interactions on hydrophobic agaroses. *Hoppe-Seyler's Z Physiol. Chem.* **358,** 255.
18. Jennissen, H. P. (2002) Hydrophobic interaction chromatography. In *Nature Encyclopedia of Life Sciences*. Nature Publishing Group, London, pp. 353–361.
19. Jennissen, H. P. (1986) Protein binding to two-dimensional hydrophobic binding-site lattices: Sorption kinetics of phosphorylase *b* on immobilized butyl residues. *J. Colloid. Interface Sci.* **111,** 570–586.

20. Jennissen, H. P. (1985) Protein adsorption hysteresis. In *Surface and Interfacial Aspects of Biomedical Ploymers, Vol. 2, Protein Adsorption* (Andrade, J. D., ed.). Plenum Press, New York, pp. 295–320.
21. Jennissen, H. P. (1987) General aspects of protein adsorption. *Biol. Chem. Hoppe-Seyler* **368,** 751–752.
22. Demiroglou, A. and Jennissen, H. P. (1990) Synthesis and protein-binding properties of spacer-free thioalkyl agaroses. *J. Chromatogr.* **521,** 1–17.
23. Chatzinikolaidou, M., Laub, M., Rumpf, H. M., and Jennissen, H. P. (2002) Biocoating of electropolished and ultra-hydrophilic titanium and cobalt chromium molybdenium alloy surfaces with proteins. *Materialwiss. Werkstofftech.* **33,** 720–727.
24. Jennissen, H. P. and Heilmeyer-Jr, L. M. G. (1975) General aspects of hydrophobic chromatography. Adsorption and elution characteristics of some skeletal muscle enzymes. *Biochemistry* **14,** 754–760.
25. Lowry, O. H., Rosebrough, N. J., Farr, A. L., and Randall, R. J. (1951) Protein measurement with the folin phenol reagent. *J. Biol. Chem.* **193,** 265–275.
26. Jennissen, H.P. and Demirolgou, A. (1992) Base-atom recognition in protein adsorption to alkyl agaroses. *J. Chromatogr.* **597,** 93–100.
27. Rydén, L. and Carlsson, J. (1989) Covalent chromatography. In *Protein Purification* (Janson,J.-C. and Rydén, eds.). VCH Publishers, Inc., New York, NY, pp. 252–274.
28. Mrabet, N. T. and Vijayalakshmi, M. A. (2002) Immobilized metal-ion affinity chromatography. In *Biochromatography* (Vijayalakshmi,M.A., ed). Taylor and Francis, London, pp. 272–294.
29. Boschetti, E. and Girot, P. (2002) Ion exchange interaction chromatography. In *Biochromatography* (Vijayalakshmi, M .A., ed.). Taylor and Francis, London, pp. 24–45.
30. Karlsson, E., Rydén, L., and Brewer, J. (1989) Ion exchange chromatography. In *Protein Purification* (Janson,J.-C. and Rydén, eds), VCH Publishers, Inc., New York, pp. 107–148.
31. Misra, D.N. (1999) Adsorption from solutions on synthetic hydroxyapatite: non-aqueous vs. aqueous solvents. *J. Biomed. Mater. Res.* **48,** 848–855.
32. Posner, A. S. (1985) The structure of bone apatite surfaces. *J. Biomed. Mater. Res.* **19,** 241–250.
33. Israelachvili, J. and Pashley, R. (1982) The hydrophobic interaction is long range, decaying exponentially with distance. *Nature* **300,** 341–342.
34. Jennissen, H. P. (2002) Hydrophobic (interaction) chromatography of proteins. In *Biochromatography* (Vijayalakshmi, M. A., ed.). Taylor and Francis, London, pp. 47–71.
35. Porath, J. (1989) Electron-donor-acceptor chromatography (EDAC) for biomolecules in aqueous solutions. In *Protein Recognition of Immobilized Ligands* (Hutchens,T.W., ed.). Alan R. Liss, Inc., New York, pp. 101–122.
36. Alberts, B., Bray, D., Lewis, J., Raff, M., Roberts, K., and Watson, J. D. (1989) *Molecular Biology of the Cell.* Garland Publications, Inc., New York, pp. 88–92.

37. Bethell, G. S., Ayers, S., Hancock,W. S., and Hearn, M. T. W. (1979) A novel method of activation of cross-linked agaroses with 1,1'-carbonyldiimidazole which gives a matrix for affinity chromatography devoid of additional charged groups. *J. Biol. Chem.* **254,** 2572–2574.
38. Zumbrink, T., Demiroglou, A., and Jennissen, H. P. (1995) Analysis of affinity supports by 13C-CP/MAS-NMR spectroscopy: application to carbonyldi-imidazole- and novel tresyl chloride-synthesized agarose and silica gels. *J. Mol. Recognit.* **8,** 363–373.

6

Sedimentation Velocity Method in the Analytical Ultracentrifuge for the Study of Protein–Protein Interactions

Claus Urbanke, Gregor Witte, and Ute Curth

Summary

Sedimentation analysis in the analytical ultracentrifuge can be employed to detect macromolecular interactions. Whenever two molecules interact the mass of the resulting complex is increased and this is reflected in the sedimentation behavior. In this chapter we discuss how this phenomenon can be utilized to determine quantitative parameters of an interaction. An example, interaction of single-stranded DNA binding protein with a subunit of DNA polymerase III holoenzyme is given together with a thorough treatment of the relating theory and a description of evaluation algorithms.

Key Words: Analytical ultracentrifugation; binding constants; binding stoichiometry; protein–protein interaction; sedimentation rate.

1. Introduction

1.1. Analytical Centrifugation

Whenever a gravitational field is exerted upon a particle (e.g., a protein in solution) a force results that is proportional to the mass of the particle and the strength of the gravitational field. This force will provoke a movement of the particle. Such a movement in principle can be induced by the earth's gravitational field, but for all practical purposes on a molecular scale this field is far too weak to cause any measurable effect. Thus, strong gravitational fields are created by spinning the sample in a centrifuge.

In the 1920s, a big problem of the then uprising biochemistry was the determination of the molecular mass of a protein. The proteins that had been isolated by then withstood all attempts to elucidate their molecular nature.

From: *Methods in Molecular Biology, vol. 305: Protein–Ligand Interactions: Methods and Applications*
Edited by: G. U. Nienhaus © Humana Press Inc., Totowa, NJ

Svedberg *(1)* at Upsala tried to measure the mass of myoglobin and hemoglobin by spinning a solution at high speeds and observing the concentration changes while the centrifuge was running. The name for this method was termed *analytical ultracentrifugation* by virtue of the fact that the contents of the centrifuge tube could be analyzed *in situ* while running the centrifuge, and because the centrifuges of Svedberg ran at that time at unbelievably high speeds of up to 20,000 revolutions per minute (rpm). In 1926, he succeeded in determining the molar mass of hemoglobin with this method *(1)*

The basic principle and the underlying theory of analytical ultracentrifugation have not changed considerably since then. A recent review of the use of analytical ultracentrifugation in protein science has been given by Lebowitz et al. ***(2)***. There are two basically different kinds of experimental techniques available with the analytical ultracentrifuge, 1) sedimentation velocity and 2) sedimentation-diffusion equilibrium. Both techniques can be used to measure interactions between molecules. Because sedimentation-diffusion equilibrium methods are mainly used to study self assembly of proteins, this chapter will focus on the use of sedimentation velocity experiments in analyzing heterologous associations.

If in a sedimentation experiment two macromolecules form a complex, the molecular mass of such a complex is larger than that of its constituents. If no dramatic increase in the frictional coefficient occurs, such complex will sediment faster than any of its constituents. Thus, an increase in sedimentation speed in a mixture of two proteins is proof for their interaction. An example for this simple approach to detect protein–protein interactions is the binding of Internalin A of *Listeria monocytogenes* to the epithelial plasma membrane protein E-cadherin, one of the main steps in *Listeria* infection ***(3)***. Because in an ultracentrifugation experiment the different constituents of such a solution of interacting macromolecules can be separated, a more quantitative analysis is possible leading to binding isotherms, binding stoichiometries and binding constants. The theoretical background for such a procedure is given at the end of this chapter. Before that the authors will give a description of the instrument and will describe an example of such an interaction analysis.

1.2. The Analytical Ultracentrifuge

An analytical ultracentrifuge is basically a combination of a centrifuge with a photometric detection unit allowing the observation of the contents of the centrifuge tube (here called cell) during the run. At present, there seems to be only a single brand of analytical ultracentrifuge on the market, the Beckman/Coulter XL-A/XL-I. This centrifuge can be supplied with photometric (absorption, XL-A) as well as with interometric (Rayleigh interferometer, XL-I) detection. **Figure 1** schematically depicts the main components of the photometric

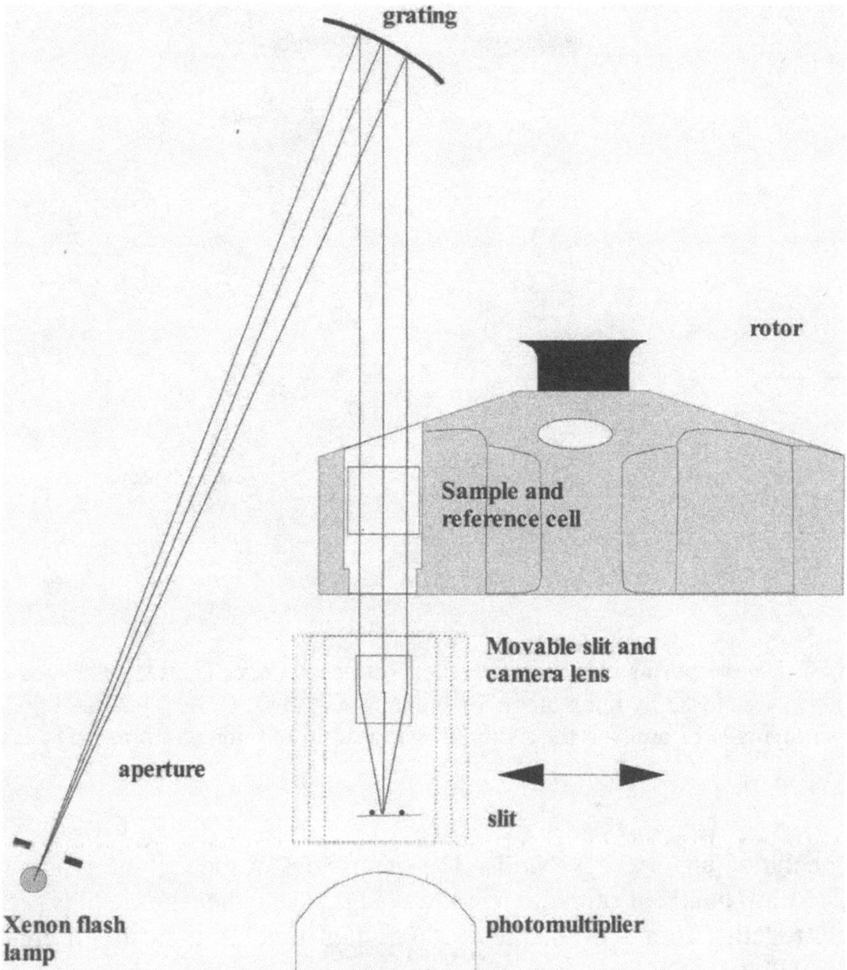

Fig. 1. Schematic drawing of the absorbance detector in an analytical ultracentrifuge. Light from the Xenon flashlamp is projected parallel by a toroidal grating onto the spinning rotor. Below the rotor there is a camera lens and a narrow slit which move above a photomultiplier to scan an image of the cell. Whenever a sector of the measuring cell (cf. **Fig. 2**) is exactly aligned with the beam of light the lamp is ignited. During a scan of the cell the transmission of reference and sample sector are measured alternately and converted to absorption readings.

detection in this analytical ultracentrifuge. The measuring cell of the centrifuge contains two compartments, one for the sample and the other for the reference buffer. The compartments are sector shaped so that the side walls coincide with straight lines coming from the center of revolution to avoid

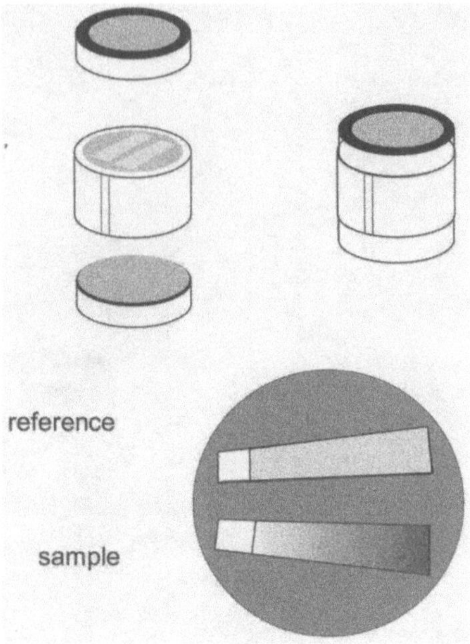

Fig 2. The measuring cell of an analytical ultracentrifuge. The sector-shaped compartments are closed by thick quartz or sapphire windows. Optical path length of the compartments is 12 mm and the sectors are approx 15 mm long holding up to 450 μL sample.

sedimentation against these walls. The other sides of the cell are made up of thick (5 mm) quartz or sapphire windows (**Fig. 2**), and the whole cell is pressed tightly together in a 1 in. diameter aluminium tube. A typical cell contains approx 400 μL of sample and reference each, and should show an absorption between 0.1 and 1 at the desired wavelength. For protein work at 280 nm this corresponds to a concentration of 0.1 to 1 mg/mL. Working at lower concentrations is possible if one uses the peptide absorption around 220 nm, although noise and interference from buffer UV absorption increase considerably. At present, two types of rotor are available, a 4 place rotor that can run up to 60,000 rpm and an 8 place rotor restricted to 50,000 rpm. Because one of the places in the rotor is occupied by a counterbalance, three or seven samples can be run simultaneously.

At the beginning of a sedimentation experiment the sample sector is filled homogeneously with the sample. Only special cell types allow overlaying or underlaying of different solutions in the sector. When sedimentation proceeds a boundary is formed between pure solvent and the solution containing the macromolecule, and the movement of this boundary can be observed by

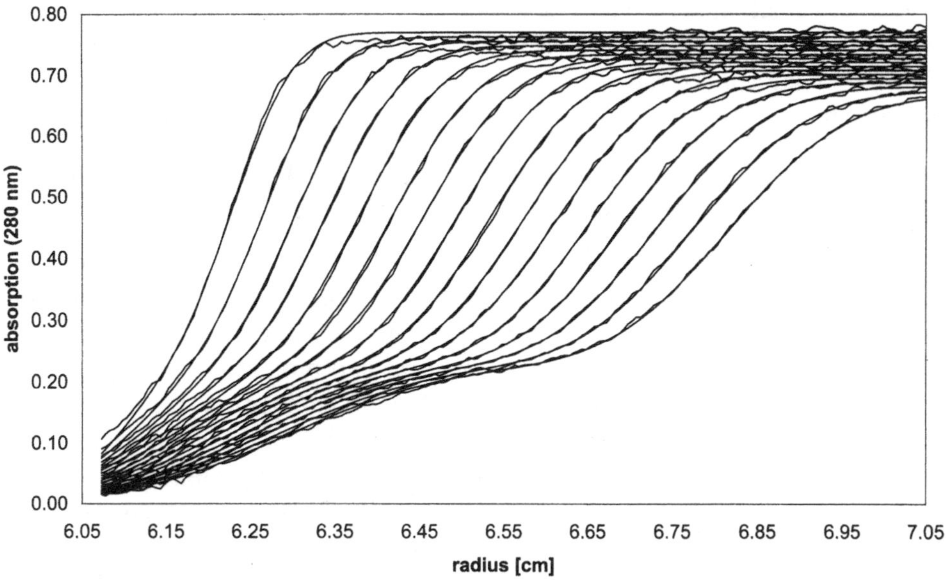

Fig. 3. Cosedimentation of 15 μ*M* χ protein and 2,5 μ*M* *Eco*SSB (tetramer). Smooth lines are theoretical sedimentation profiles calculated for two sedimenting boundaries with Gaussian shape (cf. **Eq. 14**) with parameters $A^P_{slow} = 0.25$, $A^P_{fast} = 0.57$ and $s_{slow} = 1.78$, $s_{fast} = 5.08$.

repeatedly scanning the cell absorption. **Figure 3** shows an example of such a sedimentation experiment with interacting proteins forming two sedimenting boundaries (details described later).

2. Materials and Methods

2.1. Setting Up an Experiment

If setting up an experiment for protein–ligand or protein–protein interaction several prerequesites have to be fulfilled. First, the sedimentation coefficient of the complex boundary must be larger than the sedimentation coefficient of the slower moving species. This would allow the separation of complex and at least one of the interacting species. For protein–protein interaction the changes in sedimentation coefficient caused by complex formation will be large whenever the two proteins are of similar size. For a standard approach to such a protein–protein interaction study, one should take the larger protein and titrate it with the smaller one. Quantitative evaluation of the experiment requires concentration of the smaller protein to be large enough that unbound protein can be observed (*v.i.*). For a reliable binding isotherm it may be necessary to do two runs in an 8 place rotor yielding up to 14 data points.

If a very small ligand binds to a protein the changes in sedimentation rate will be too small to be detected. In such a case, the evaluation procedure does not change but the concentration of the small ligand must be determined after the complex has sedimented requiring enough absorption signal from the free ligand.

2.2. Example: Interaction of Single-Stranded DNA Binding Protein With the χ (v.s.) Subunit of DNA Polymerase III Holoenzyme

Single-stranded DNA binding proteins (SSB) are defined as proteins that bind specifically to DNA in its single-stranded conformation (ssDNA) and do not recognize double-stranded DNA. They show no direct catalytic function. The major role of these proteins is to stabilize and protect DNA in its single-stranded conformation and to prepare ssDNA for replication, recombination, or repair. Shortly after the discovery of such SSB in *E. coli* it was recognized that besides binding ssDNA these proteins recruit other proteins, e.g., to the replication machinery. One of the proteins that *Eco*SSB recognizes is χ protein from the *E. coli* DNA polymerase III replisom. Analytical ultracentrifugation has been used to quantify this protein–protein interaction and a detailed experimental procedure for obtaining the published result ***(4)*** is described later.

2.2.1. Procedure

*Eco*SSB and χ protein were prepared from overproducing strains as described previously ***(4)***. The proteins were dialysed extensively against a buffer containing 0.3 *M* NaCl, 0.02 *M* KPi pH 7.4. Protein concentration was determined spectrophotometrically at 280 nm using a molar extinction of 113,000 M^{-1}cm^{-1} for *Eco*SSB and 29,280 M^{-1}cm^{-1} for χ protein. AUC samples were prepared by mixing 2.5 μ*M* *Eco*SSB (tetramer) with different amounts of χ protein in 400 μL of the buffer (*v.s.*). Fourteen different samples (including free *Eco*SSB) were spun at 45,000 rpm in the analytical ultracentrifuge in two runs with seven samples each. Absorbance traces were recorded for each sample at time intervals of approx 570 s. A typical result is shown in **Fig. 3**.

For evaluation the sedimentation traces of each sample were fitted using **Eq. 14. Figure 4** shows that the sedimentation coefficient of the fast moving boundary clearly increases with increasing amounts of χ protein. The plateau absorbances of the fast and slow moving boundaries are corrected with the absorbancies for the samples measured spectrophoto metrically before the run. From these values the concentration of free and bound χ protein are calculated.

2.2.2. Results

Figure 5 shows the resulting binding isotherm. Nonlinear least square fitting of a model where *n* molecules χ bind to an *Eco*SSB tetramer yields an affinity constant of $K = 2.5 \cdot 10^5\ M^{-1}$ and a binding stoichiometry $n = 4.1$. The

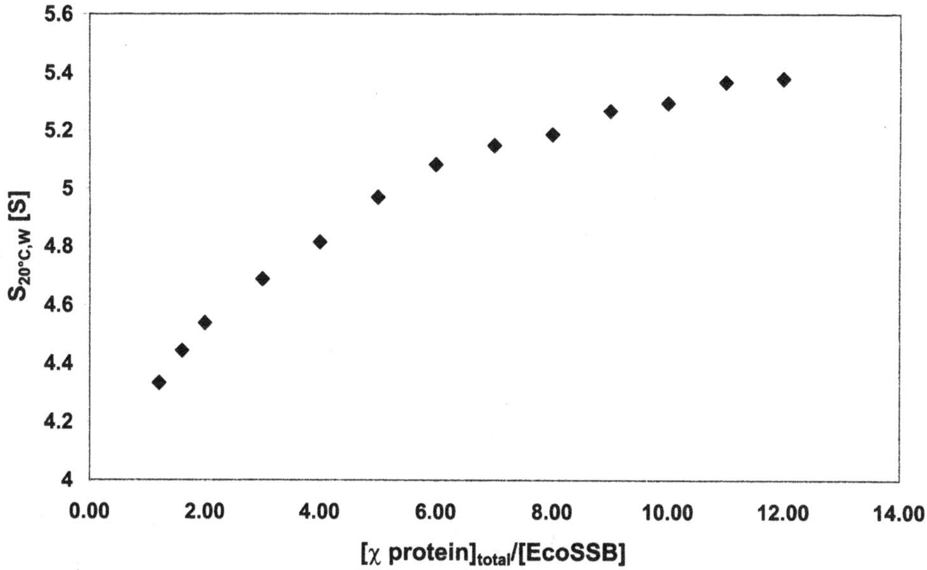

Fig. 4. Cosedimentation of 2.5 μ*M* *Eco*SSB (tetramer) and different concentrations of χ (*v.s.*) protein. The sedimentation coefficient of the faster moving boundary increases with increasing amounts of ligand.

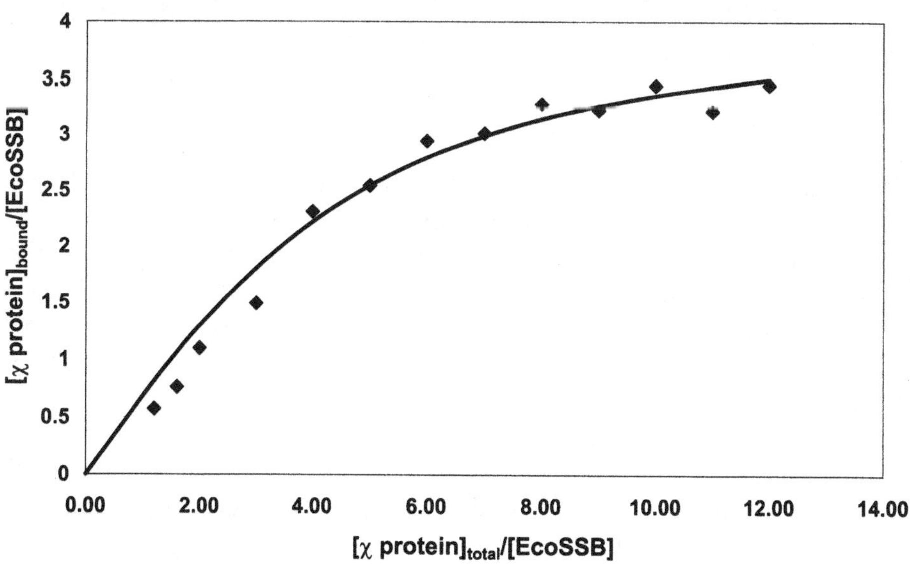

Fig. 5. Cosedimentation of 2.5 μ*M* *Eco*SSB (tetramer) and different concentrations of χ (*v.s.*) protein. Binding isotherm calculated from the amplitudes of the slow and fast moving boundaries. The smooth line represents a binding isotherm calculated with $n = 4.1$ and $K = 2.5 \cdot 10^5\ M^{-1}$.

stoichiometry is in good accordance with the fact that *Eco*SSB is made up from four identical subunits each carrying a binding site for χ protein. The binding constant is moderate probably indicating the dynamic nature of the interaction.

3. Theory and Evaluation Algorithms

3.1. Sedimentation Velocity of a Single Species

In a centrifuge spinning at a speed of U rpm the gravitational field Z_{rel} at a distance r (m) from the center of rotation in units of the earth's gravitational field (1g = 9.8 m/s^2) is given by

$$Z_{rel} = 1.119 \cdot 10^{-3} \cdot r \cdot U^2 \, [g] \tag{1}$$

A particle of mass m (kg) suspended in a solvent of density ρ (kg/m^3) and having a partial specific volume $\bar{v}$ (m^3/kg) in such a field experiences a force given by:

$$F_{gravitation} = m \cdot (1 - \bar{v} \cdot \rho) \cdot \left(\frac{2 \cdot \pi \cdot U}{60} \right)^2 \cdot r = m \cdot (1 - \bar{v} \cdot \rho) \cdot \omega^2 \cdot r \tag{2}$$

with ω (s^{-1}) being the angular velocity of the rotor. The term $m \cdot (1 - \bar{v} \cdot \rho)$ is called the reduced mass of the particle and is the actual mass corrected for buoyancy. This force will accelerate the particle but as the speed v (m/s) of the particle increases, a frictional force proportional to the speed, $F_{friction} = f \cdot v$ and opposing the gravitation will increase until both gravitational acceleration and friction cancel and the particle will travel at constant speed:

$$v = \frac{m \cdot (1 - \bar{v} \cdot \rho)}{f} \cdot \omega^2 \cdot r = s \cdot \omega^2 \cdot r \tag{3}$$

This canceling of forces will be reached after a very short time, much less time than it takes the particle to travel over its own diameter. The sedimentation constant s (s) is a property of the particle and the viscosity η (Pa · s) and density ρ of the solvent. It is usually given in units of *Svedberg* (S) with 1S = 10^{-13} s. To make sedimentation constants comparable the properties of the solvent (indicated by T,S) are corrected to pure water at 20° C (indicated by 20°C,W)

$$s_{20^\circ C,W} = s_{T,S} \cdot \frac{\eta_{T,S}}{\eta_{20^\circ C,W}} \cdot \frac{1 - \bar{v} \cdot \rho_{20^\circ C,W}}{1 - \bar{v} \cdot \rho_{T,S}} \tag{4}$$

From the movement of the sedimenting boundary (cf. **Fig. 3**) the sedimentation coefficient can be evaluated by

$$\frac{\partial \ln \bar{x}}{\partial t} = s \cdot \omega^2 \tag{5}$$

with $\bar{x}$ being the position of the boundary, usually the point of inflection or the position at half height.

3.2. Sedimentation of Interacting Molecules

However, for a more quantitative interpretation the simple approach of observing the sedimentation coefficient alone is insufficient and a thorough treatment of the transport processes in the analytical ultracentrifuge is necessary. Lamm *(5)* has derived a differential equation that completely describes the behavior of a single species k in a sector-shaped centrifuge cell with operators

$$\frac{\partial c_k}{\partial t} = D_k \Omega_k - S_k \Psi_k \text{ with operators } \Omega = \frac{\partial^2 c}{\partial x^2} - \frac{1}{x}\frac{\partial c}{\partial x} \text{ and } \Psi = x \cdot \frac{\partial c}{\partial x} + 2 \cdot c \qquad (6)$$

where D_k is the diffusion coefficient and c_k the concentration of species k. This equation can be extended to the treatment of interacting species. The following theory follows essentials of a similar derivation given earlier *(6)*. If a ligand, A, interacts with a macromolecule, B, containing n equivalent binding sites, the law of mass action together with the appropriate statistical factors gives for the concentration of each complex where i molecules of A are bound to B:

$$nA + B \rightleftharpoons + AB + (n-1)A \rightleftharpoons \cdots A_iB + (n-1)A \cdots \rightleftharpoons A_nB$$

$$C_{A_ib} = (K \cdot C_A)^i \cdot C_B \cdot \prod_{j=1}^{i}\left(\frac{n+1}{j} - 1\right) = \Lambda_i \cdot C_B \qquad (7)$$

Note that Λ_i a pure number that only depends on the (local) concentration of A. This law of mass action can be combined with Lamm's differential equation *(6)* if the equilibration is fast enough so that the law of mass action holds during the whole sedimentation experiment. In numerical simulations *(7)* it could be shown that this condition is well fulfilled whenever the relaxation time of the equilibration is shorter than 1/10 of the time it takes the fastest molecule to sediment from the meniscus to the bottom:

$$\begin{aligned} \frac{\partial c_B}{\partial t} + \sum_{i=1}^{N} \frac{\partial c_{A_iB}}{\partial t} &= \left(1 + \sum_{i=1}^{N} \Lambda_i\right) \cdot \frac{\partial c_B}{\partial t} \\ &= D_B \cdot \Omega_B + \sum_{i=1}^{N} D_{A_iB} \cdot \Omega_{A_iB} - s_B \cdot \omega^2 \cdot \psi_B - \sum_{i=1}^{N} s_{A_iB} \cdot \omega^2 \cdot \psi_{A_iB} \\ &= \Omega_B \cdot \left(D_B + \sum_{i=1}^{N} D_{A_iB} \cdot \Lambda_i\right) - \psi_B \cdot \omega^2 \left(s_B + \sum_{i=1}^{N} s_{A_iB} \cdot \Lambda_i\right) \end{aligned} \qquad (8)$$

In case the free ligand A is sedimenting slower than the macromolecule B and its concentration is in excess over the concentration of B, all Λ_i of the above equation will be constant in the region where B is present and only two sedimenting boundaries will move through the cell. The slower moving bound-

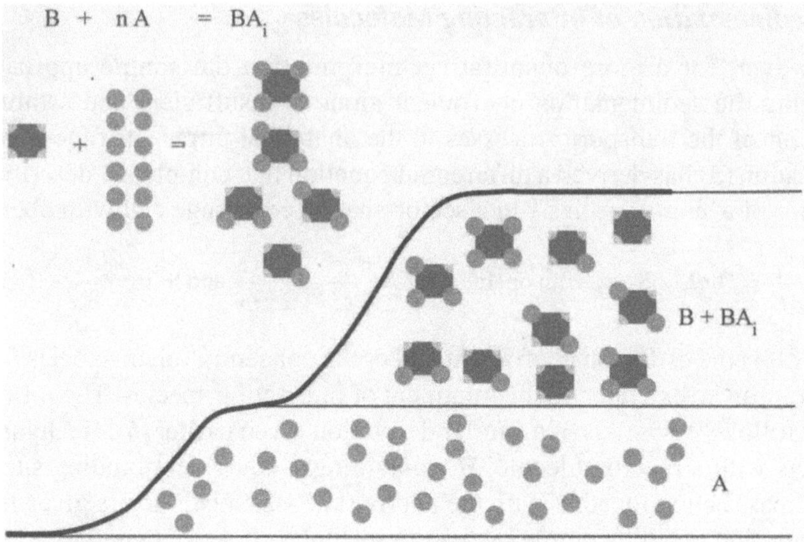

Fig. 6. Cosedimenting interacting molecules. The binding equilibrium between a protein having four identical binding sites for a ligand is depicted in the upper left corner. The sedimentation trace shows two sedimenting boundaries, with the slower one made up by free ligand and the faster one by free protein and all kinds of complexes. Note, that the faster sedimenting species all sediment in presence of an excess of free ligand. This ensures complete equilibration in the region of the faster sedimenting boundary.

ary will depict the sedimentation of free ligand A, whereas the faster moving boundary contains all complexes and free macromolecule B. **Figure 6** depicts such a situation. While the movement of the slower sedimenting boundary is described by the sedimentation (s_{slow}) and diffusion (D_{slow}) coefficients of the free ligand A the faster sedimenting boundary moves with Λ_i weighted average transport coefficients:

$$D_{fast} = \frac{D_B + \sum_{i=1}^{n} D_{A_i B}\Lambda_i}{1 + \sum_{i=1}^{n} \Lambda_i} \text{ and } s_{fast} = \frac{s_B + \sum_{i=1}^{n} s_{A_i B}\Lambda_i}{1 + \sum_{i=1}^{n} \Lambda_i} \quad (9)$$

For an analysis of binding constants and stoichiometries, however, it is necessary to determine the concentrations of bound and nonbound ligands while the actual value of sedimentation and diffusion coefficients can be ignored.

The determination of these concentrations would be fairly straightforward if one would determine the plateau concentrations of the both sedimenting boundaries. Unfortunately, these plateau concentrations decrease when the boundary progresses through the cell. For a single sedimenting boundary this dilution can be described by:

$$\frac{A_p(t)}{A_p(0)} = \left(\frac{x_m}{x}\right)^2 \tag{10}$$

where $A_P(0)$ and $A_P(t)$ are the absorbances of the plateau and x_m and $\bar{x}$ the positions of the boundary at the beginning of the experiment (0) and at time t, respectively. The velocity of the sedimenting boundary as described earlier then reads:

$$\frac{\partial \bar{x}}{\partial t} = s \cdot \bar{x} \cdot \omega^2 \cdot t \text{ or } \frac{\partial \ln \bar{x}}{\partial t} = s \cdot \omega^2 \text{ or } \bar{x} = x_m \cdot e^{s \cdot \omega^2 \cdot t} \tag{11}$$

and the plateau absorbance of a single band becomes:

$$A_P(t) = A_P(0) \cdot e^{-2 \cdot s \cdot \omega^2 \cdot t} \tag{12}$$

For all practical purposes, the shape of a sedimenting boundary can well be described by an integrated Gaussian distribution:

$$G(x) = \frac{G(0)}{\sigma\sqrt{2 \cdot \pi}} \int_{-\infty}^{x} e^{-\frac{1}{2}\left(\frac{y - x_0}{\sigma}\right)^2} dy \tag{13}$$

and substituting the height G_0 and the position x_0 of the boundary by the above definitions the sum of two sedimenting boundaries, e.g., measured in an interaction experiment is given by

$$\begin{aligned} A(x,t) = {} & \frac{A^P_{slow}(0) \cdot \mathrm{e}^{-2 \cdot s_{slow} \cdot \omega^2 \cdot t}}{\sigma_{slow}(t)\sqrt{2 \cdot \pi}} \int_{-\infty}^{x} e^{-\frac{1}{2}\left(\frac{y - x_m \cdot \mathrm{e}^{-2 \cdot s_{slow} \omega^2 \cdot t}}{\sigma_{slow}(t)}\right)^2} dy \\ & + \frac{A^P_{fast}(0) \cdot \mathrm{e}^{-2 \cdot s_{fast} \cdot \omega^2 \cdot t}}{\sigma_{fast}(t)\sqrt{2 \cdot \pi}} \int_{-\infty}^{x} e^{-\frac{1}{2}\left(\frac{y - x_m \cdot \mathrm{e}^{-2 \cdot s_{fast} \omega^2 \cdot t}}{\sigma_{fast}(t)}\right)^2} dy \end{aligned} \tag{14}$$

where $A(x,t)$ is the measured absorbance, A^P_{slow} (0) and A^P_{fast} (0)are the absorbances at the beginning of the experiment and s_{slow} and s_{fast} are the sedimentation coefficients of free ligand A and the complex containing boundary, respectively. In a global analysis these four parameters are extracted from a series of boundaries where $\sigma_{slow}(t)$ and $\sigma_{fast}(t)$ are the widths of these boundaries at the different times t and must be fitted locally.

Because the photometer of an analytical ultracentrifuge only yields relative absorbances *(8)*, these values have to be corrected by the known absorbance of the sample measured in a standard spectrophotometer or calculated from the composition of the sample before the run. If the absorbances measured for both boundaries in the analytical ultracentrifuge are termed $A_{AUC}^{slow} = A_{slow}^{P}(0)$ and $A_{AUC}^{fast} = A_{fast}^{P}(0)$ and the initial absorbance of component A and B are A_A and A_B these corrected absorbances with $A_{total} = A_A + A_B$ are given by:

$$A_{corr}^{slow} = \frac{A_{AUC}^{slow}}{A_{AUC}^{slow} + A_{AUC}^{fast}} \cdot A_{total} \text{ and } A_{corr}^{fast} = \frac{A_{AUC}^{fast}}{A_{AUC}^{slow} + A_{AUC}^{fast}} \cdot A_{total} \tag{15}$$

Because free ligand A is represented by the slow moving boundary and the fast moving boundary represents the sum of all of B and bound ligand A the absorbances of free and bound ligand A can be calculated as:

$$A_A^{free} = A_{corr}^{slow} \text{ and } A_A^{bound} = A_{corr}^{fast} - A_B \tag{16}$$

using the known molar absorbances of A and B, ε_A and e_B, the concentrations of free and bound A,

$$c_A^{bound} = \frac{A_A^{bound}}{\varepsilon_A}$$

as well as the concentration of remaining binding sites on B, $n \cdot c_B^{total} - c_A^{bound}$, can be put into the law of mass action:

$$K = \frac{c_A^{bound}}{c_A^{free} \cdot \left(n \cdot c_B^{total} - c_A^{bound}\right)} \text{ or } \frac{c_A^{bound}}{c_B^{total}} = \frac{n \cdot K \cdot c_A^{free}}{1 + K \cdot c_A^{free}} \tag{17}$$

If such measurements are made at different concentrations of A and B a binding isotherm can be constructed by, e.g., plotting

$$c_A^{bound} / c_B^{total} \text{ vs } \left(c_A^{bound} + c_A^{free}\right) / c_B^{total} = c_A^{total} / c_B^{total} \tag{18}$$

and values for K and n can be fitted to this binding isotherm.

References

1. Svedberg, T. and Fåhraeus, R. (1926) A new direct method for the determination of the molecular weight of the proteins. *J. Am. Chem. Soc.* **48,** 430–438.
2. Lebowitz, J., Lewis, M. S., and Schuck, P. (2002) Modern analytical ultracentrifugation in protein science: a tutorial review. *Protein Sci.* **11,** 2067–2079.
3. Schubert, W. D., Urbanke, C., Ziehm, T., Beier, V., Machner, M. P., Domann, E., Wehland, J., Chakraborty, T., and Heinz, D. W. (2002) Structure of internalin, a major invasion protein of Listeria monocytogenes, in complex with its human receptor E-cadherin. *Cell* **111,** 825–836.

4. Witte, G., Urbanke, C., and Curth, U. (2003) DNA polymerase III chi subunit ties single-stranded DNA binding protein to the bacterial replication machinery. *Nucleic Acids Res.* **31,** 4434–4440.
5. Lamm, O. (1929) Die differentialgleichung der ultrazentrifugierung. *Arkiv för matematik, astronomi och fysik* **21B No. 2,** 1–4.
6. Krauss, G., Pingoud, A., Boehme, D., Riesner, D., Peters, F., and Maass, G. (1975) Equivalent and non-equivalent binding sites for tRNA on aminoacyl-tRNA synthetases. *Eur. J. Biochem.* **55,** 517–529.
7. Urbanke, C., Ziegler, B., and Stieglitz, K. (1980). Complete evaluation of sedimentation velocity experiments in the analytical ultracentrifuge. *Fres. Z. Anal. Chem.* **301,** 139–140.
8. Beckman Model XL-A Analytical Ultracentrifuge (1995) Instruction Manual. Beckman Instruments, Inc., Palo Alto, CA.

7

Protein–Ligand Interaction Probed by Time-Resolved Crystallography

Marius Schmidt, Hyotcherl Ihee, Reinhard Pahl, and Vukica Šrajer

Summary

Time-resolved (TR) crystallography is a unique method for determining the structures of intermediates in biomolecular reactions. The technique reached its mature stage with the development of the powerful third-generation synchrotron X-ray sources, and the advances in data processing and analysis of time-resolved Laue crystallographic data. A time resolution of 100 ps has been achieved and relatively small structural changes can be detected even from only partial reaction initiation. The remaining challenge facing the application of this technique to a broad range of biological systems is to find an efficient and rapid, system-specific method for the reaction initiation in the crystal. Other frontiers for the technique involve the continued improvement in time resolution and further advances in methods for determining intermediate structures and reaction mechanisms. The time-resolved technique, combined with trapping methods and computational approaches, holds the promise for a complete structure-based description of biomolecular reactions.

Key Words: Time-resolved macromolecular crystallography; Laue diffraction; intermediate states; reaction mechanism; SVD.

1. Introduction

Studies of macromolecules by X-ray diffraction technique over the last few decades provided an enormous wealth of information on three-dimensional structures of protein, DNA and RNA molecules. These average, static three-dimensional pictures of macromolecules offer a significant insight into their function. However, to fully understand how these molecules perform their function one has to watch them in action, along a reaction path that often involves short-lived intermediates. Many such reactions are possible in crys-

From: *Methods in Molecular Biology, vol. 305: Protein–Ligand Interactions: Methods and Applications*
Edited by: G. U. Nienhaus © Humana Press Inc., Totowa, NJ

tals as they typically contain a large percentage of solvent, about 40 to 60%. The solvent forms channels and hydration shells around protein molecules and facilitates dynamic processes such as diffusion and binding of substrates and other ligands, turn-over in enzyme crystals, and conformational change in response to absorption of light in photoreceptors. Crystallography can, therefore, also play an important role in visualizing structures of reaction intermediates and elucidating reaction mechanism. The term time-resolved X-ray crystallography (TRX) is sometimes used in a broad sense to encompass a variety of methods that investigate structural intermediates. Trapping methods are often used to extend the lifetime of intermediate species and increase their peak concentration. Physical trapping can be accomplished by conducting a reaction at low temperature (freeze-trapping). Or alternatively, a reaction can be initiated at room temperature and a particular intermediate trapped at a certain time delay after the reaction initiation by rapid cooling (trap-freezing). Chemical trapping can be achieved by a pH change, by site-directed mutagenesis, or by chemical modification of the substrate or cofactor. Although the trapping strategy alters to some degree the reaction and structures of intermediates that are investigated, such methods are nevertheless attractive. These experiments can be conducted using the standard monochromatic oscillation technique and important information about reaction pathways and intermediates can be obtained ***(1–4)***. A more direct, but also technically more challenging strategy to study intermediates, is to utilize the TRX in a strict sense of the term and to follow the unperturbed reaction as it evolves at room temperature. We will focus here on the principles, challenges, and applications of such, more narrowly defined TRX.

TR crystallography and TR spectroscopy share the same goal: revealing the reaction mechanism and structural characteristics of intermediate states. The main advantage of TRX is that detailed structural information is obtained directly and globally, for the entire molecule. Spectroscopic techniques, such as visible or UV absorption, resonance Raman, and IR spectroscopy can be very sensitive to small structural changes but typically provide local structural information indirectly. In some cases, like with the IR amide I band, overall global information is obtained. Although the time course of spectroscopic changes reflects the time course of tertiary structural changes ***(5–12)***, it is often difficult to unambiguously link spectroscopic changes to particular and specific underlying structural changes, and to relate the amplitude of spectroscopic change to the extent of structural change. However, the two techniques are complementary and taken together provide a more complete insight into structural changes and reaction mechanism.

In TRX experiments, one triggers a reaction in molecules in the crystal and uses X-ray pulses to probe structural changes at various time delays following the start of the reaction. The time resolution of the experiment is determined by either the duration of the triggering process or the duration of the X-ray probe pulse, whichever is longer. Experiments require synchrotron X-ray radiation. Synchrotrons are pulsed X-ray sources with a typical X-ray pulse duration of about 100 ps. When longer X-ray exposures are sufficient, a train of X-ray pulses of appropriate duration can be used (*see* **Subheading 3.3.**). The best time resolution achieved to date matches the duration of a single X-ray pulse ***(13)***. Very high X-ray flux is needed to obtain data of necessary quality with ns and sub-ns time resolution. Third-generation synchrotron sources ***(14)*** provide the most intense X-ray pulses to date. Such sources are the Advanced Photon Source (APS) (USA), European Synchrotron Radiation Facility (ESRF) (France), and SPring-8 (Japan).

Experiments that require subsecond time resolution have to be conducted at synchrotron facilities with polychromatic beam capability. It is impossible to record the integral intensity of a reflection in such a short exposure time using the conventional monochromatic method where the crystal has to be rotated/oscillated. The polychromatic, Laue X-ray diffraction technique is used instead, where the crystal is stationary ***(15)***. Bending magnet beamlines are sufficient for experiments that require ms time resolution, whereas insertion device beamlines ***(14)*** are necessary for sub-ms resolution.

A comprehensive review of the present state of the Laue technique as well as examples of its application to static and TR studies can be found in **ref. *15***. The technique has reached a mature stage. Most of the problems in Laue data processing, that limited the use of this technique in the past, have been solved. Structure factor amplitudes from static Laue experiments equal in quality those from standard monochromatic oscillation measurements.

Reaction triggering is a crucial and critical part of TRX experiments and will be discussed in more detail in **Subheadings 3.2.** and **3.4.** Ideally, the triggering is accomplished rapidly in all molecules in the crystal, in a time interval much shorter than the lifetime of the intermediate that is investigated and the duration of the X-ray pulse used to probe the intermediate. In reality, triggering occurs only in a fraction of molecules. One of the experimental goals is to maximize this fraction. The fastest method for triggering a reaction in the crystal is to use ultrashort, fs to ns, laser pulses. The method can be applied to photosensitive molecules that undergo structural changes upon the absorption of light by an embedded chromophore. Examples, extensively studied by TR spectroscopy, include heme proteins ***(13,16–18)***, bacteriorhodopsin ***(19)***, photoactive yellow protein ***(5,20–22)***, and other photoreceptors ***(23)***. We will focus here mainly on TRX studies of such inherently photosensitive proteins.

The readout speed of present large area X-ray detectors used for macromolecular crystallography does not permit to follow the subsecond reactions in real time after a single reaction initiation. Collecting a diffraction image at each time delay, therefore, requires a separate reaction initiation event. In addition, multiple diffraction images at different crystal orientations are needed for a complete data set at each time delay (*see* **Subheading 3.4.**). It is therefore greatly advantageous if the reaction is reversible and the system restores itself to the initial state after a relatively short period of time. For such systems, a single crystal can be used for collecting complete or even multiple data sets. Processes where the triggering step or the reaction itself is irreversible can also be studied in favorable cases, but a new crystal is required for each X-ray exposure.

The result of a pump-probe TRX measurement is a four-dimensional data set, consisting of a time series of structure factor (SF) amplitudes, $|\mathbf{F}(hkl,t)|$. Because the goal is to determine how the known initial structure is changing in time following the reaction initiation, we are actually interested in the difference between the time-dependent SF amplitudes $|\mathbf{F}(hkl,t)|$ and the SF amplitudes corresponding to the initial state, $|\mathbf{F}(hkl)|$. Results are typically presented in real space (x,y,z) rather than reciprocal (h,k,l) space. Time-dependent difference electron density (DED) maps, $\Delta\rho(t)$, are calculated on a three-dimensional grid across the crystal unit cell, with coefficients $\Delta F(h,k,l,t) = |\mathbf{F}(hkl,t)|-|\mathbf{F}(hkl)|$ and phases obtained from the structure of the initial state (*see* **Subheading 3.5.2.**). DED maps are analogous to difference spectra in TR spectroscopy. Negative densities in such maps represent the loss of electrons and positive densities the gain of electrons. A DED that results from a structural change varies relatively smoothly with time when determined with a sufficiently high signal-to-noise ratio (SNR).

Many chemical reactions are considered *simple* ***(24,25)*** and can be represented by a chemical kinetic model where the reaction proceeds along a reaction coordinate and involves a set of discrete intermediate states I_j. These states correspond to the minima on the potential energy surface of the system. They are separated by well-defined energy barriers and the interconversion between them follows exponential behavior. To fully describe the kinetic mechanism one then needs to determine the number of intermediate states, the pathways by which they interconvert, and the rate coefficients of their interconversion (**Fig. 1**). When these are known, the kinetics of the ensemble of molecules in the crystal can be described. Following the reaction initiation, molecules populate intermediate states according to fractional (normalized) concentrations, $I_j(t)$, which are governed by a system of coupled differential equations that describes a given mechanism (**Fig. 2**). Each molecule crosses the energy barrier between the states I_j at random and independently from other molecules, and spends a short time in the transition state (at the top of the barrier) as compared to the residence time in I_j states. An intermediate state can in prin-

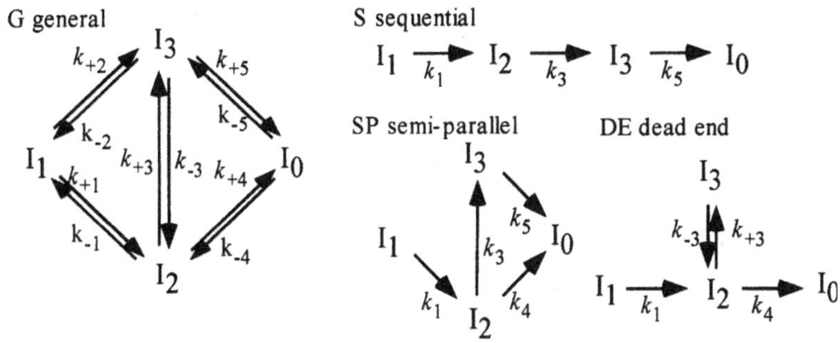

Fig. 1. General mechanism (G) with three intermediate states: I_1, I_2, and I_3. The final state is I_0. In a reversible system this is also the initial state. Three simple mechanisms (S, SP, DE) based on three intermediates are also shown.

A

$$\frac{dI_1}{dt} = -k_1 I_1$$

$$\frac{dI_2}{dt} = k_1 I_1 - k_{+4} I_2 - k_{+3} I_2 + k_{-3} I_3$$

$$\frac{dI_3}{dt} = k_{+3} I_2 - k_{-3} I_3$$

$$I_1 + I_2 + I_3 + I_0 = 1$$

B

$$I_1 = P_{11} e^{\lambda_1 t} + P_{12} e^{\lambda_2 t} + P_{13} e^{\lambda_3 t}$$

$$I_2 = P_{21} e^{\lambda_1 t} + P_{22} e^{\lambda_2 t} + P_{23} e^{\lambda_3 t}$$

$$I_3 = P_{31} e^{\lambda_1 t} + P_{32} e^{\lambda_2 t} + P_{33} e^{\lambda_3 t}$$

Fig. 2. Chemical kinetic model: fractional concentrations of intermediates for mechanism DE in **Fig. 1**. **(A)** Three differential equations describe the mechanism. I_1, I_2, I_3: time-dependent fractional concentrations of molecules populating the intermediate states I_1, I_2, and I_3. The forth equation (sum of all fractional concentration is 1) determines the concentration of the I_0 state. **(B)** General solution of differential equations in (A). Relaxation rate coefficients λ_i are the same for all intermediates and exclusively depend on the rate coefficients k (inverse of $|\lambda_i|$ are relaxation times, $\tau_i = 1/|\lambda_i|$). The λ_i are directly observable, the k are not. The pre-exponentials P_{ji} depend on initial conditions and λ_i, and therefore also depend on the rate coefficients k (*see* ***25a*** for illustrative examples). The three exponential terms in the expression for each intermediate are called transients.

ciple be significantly populated at some time delay, depending on the values of the interconversion rate coefficients (**Fig. 3**). Structural changes can be visualized as hopping of molecules between structurally distinct intermediate states I_j. In other words, what changes in time are the fractional concentrations of molecules in various intermediate states. As a result, TRX experiments do not produce a movie of continuously changing structure, but a set of discrete structures corresponding to the intermediate states.

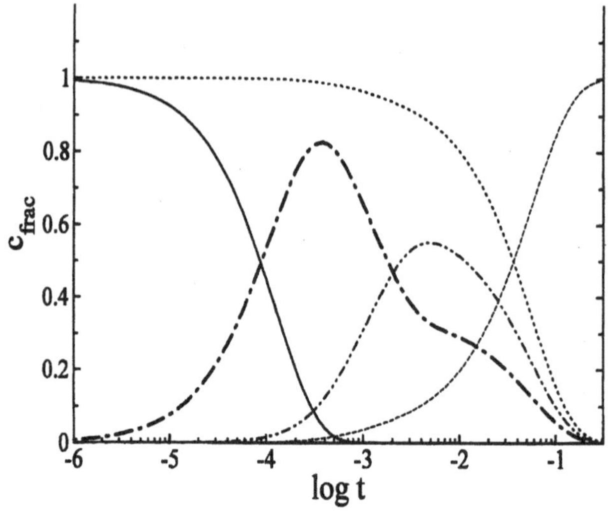

Fig. 3. Calculated fractional concentrations of molecules populating the three intermediate states: I_1 (solid line), I_2 (dot-dashed line), I_3 (double dot-dashed line), and the final state I_0 (dashed line) for the mechanism DE shown in **Figs. 1** and **2**. Reaction triggering is assumed instantaneous and all molecules are in state I_1 at $t = 0$. The sum of I_1, I_2, and I_3 is shown as dotted line. Rate coefficients: $k_1 = 8000\ s^{-1}$, $k_{+3} = 500\ s^{-1}$, $k_{-3} = 300\ s^{-1}$, $k_{+4} = 50\ s^{-1}$. Note a logarithmic time scale.

How does the chemical kinetic model apply to the results of TRX measurements? The DED at each grid point m in a DED map at a given time delay t after the reaction initiation, $\Delta\rho(t)_m$, can be represented by **Eq. 1**:

$$\Delta\rho(t)_m = \sum_{j=1}^{N} I_j(t) \cdot \Delta\rho_{j,m} \tag{1}$$

where $I_j(t)$ is the time-dependent fractional concentration of molecules in the intermediate state I_j, and $\Delta\rho_{j,m}$ is the DED corresponding to a pure intermediate state I_j. Note that the electron densities, but not the measured SF amplitudes $|\mathbf{F}(hkl,t)|$, are proportional to the fractional concentrations ***(26,27)***. Therefore, the analysis should be performed in real space (electron density), rather than reciprocal space (SF amplitudes). Also, note that in general at most delay times several intermediates may be present **(Fig. 3)** and measured DEDs represent a mixture of these intermediate states. Methods for separating the measured mixtures of states $\Delta\rho(t)$ into DEDs $\Delta\rho_j$ of the component intermediate states, and ultimately determining the structures of these intermediate states, have been developed recently ***(27–29)*** and are described in detail in **Subheading 3.5.**

2. Materials

2.1. Time-Resolved X-Ray Diffraction Experiments

1. Crystals.
2. Thin-walled glass or quartz capillaries for crystal mounting and other crystal mounting tools *(30,31)*.
3. A crystal-cooling device (such as the FTS Air-Jet crystal cooler) to maintain constant crystal temperature.
4. Suitable caged compounds for triggering the reaction in crystals that are not inherently photosensitive.
5. A laser for reaction initiation. The following has to be considered when deciding what type of laser to use: wavelength, pulse energy, repetition rate, and pulse duration. The choice will depend on the properties of the sample and the desired time resolution.
6. Optical fibers and/or other optics necessary for delivering the laser light to the sample.
7. A microspectrophotometer for measuring crystal absorption spectra and preliminary TR spectroscopy on crystals. Appropriate light sources for monitoring absorption and reaction initiation have to be considered depending on the absorption properties of the sample.
8. Polychromatic synchrotron X-ray source.
9. Fast and slow X-ray shutters for selecting and isolating individual X-ray pulses or pulse trains.
10. Timing electronics for synchronization of laser and X-ray pulses. The synchronization jitter has to be smaller than the laser or X-ray pulse duration, whichever is longer. Delay generators are needed for adjusting the time delay between laser and X-ray pulses. PIN diodes are generally used for detecting laser and X-ray pulses at the sample location and a fast oscilloscope for measuring the time delay between these pulses.
11. A large area detector for recording diffraction patterns. Image plate or CCD detectors are required to provide sufficient sensitivity.

2.2. Processing and Analysis of Time-Resolved Data

1. Laue data processing. Several software packages are available: LaueView *(32,33)*, Daresbury Laboratory Laue Software Suite (Lauegen, Lscale) *(34,35)*, Leap *(36)*, PrOW *(37)*, Precognition *(38)*.
2. Analysis of DED maps. Integration of DED features: Probe *(16)*; SVD analysis: SVD4TX, GetMech *(27,28)*.

3. Methods

3.1. Sample Preparation

Crystal requirements for TR Laue experiments are more stringent than for the standard, monochromatic oscillation data collection. The Laue technique is more sensitive to crystal mosaicity. Laue diffraction spots become elongated

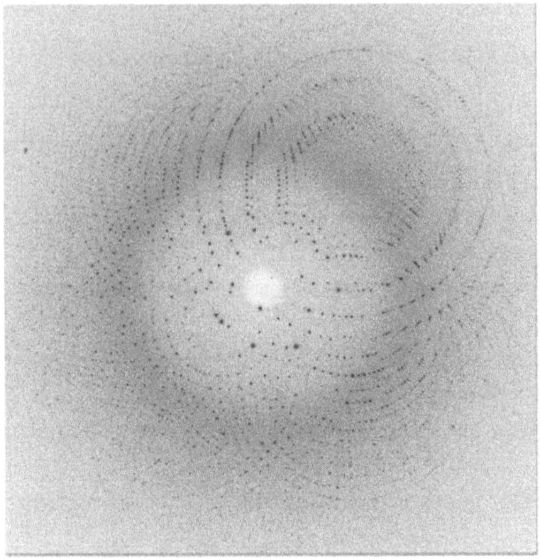

Fig. 4. A Laue diffraction image from a protein crystal collected on the MAR345 image plate detector at the APS 14-ID beamline.

when the mosaicity is increased. In most cases diffraction spots also become more elongated as a result of the photo-activation. In addition, the crystal will degrade as a result of radiation damage. The diffraction spots become more streaky and weaker, and the resolution to which crystal diffracts degrades during the experiment. The elongation of diffraction spots causes additional spatial overlap in the already crowded Laue diffraction images (**Fig. 4**) and spot integration becomes less accurate. Therefore, crystals need to be screened based on mosaicity and on the additional elongation of the diffraction spots when crystals are photo-activated. A compromise between the extent of photo-activation and the spot elongation often becomes necessary. The Laue processing software can handle successfully some degree of spot elongation (approx 4:1 ratio of spot length to spot width). However, one should keep in mind that the SNR is generally inversely proportional to the spot elongation for a given extent of photo-activation.

Crystal size consideration is very important. On the one hand, one needs large crystal volume to measure diffraction intensities with a sufficient SNR when using very short X-ray exposures. On the other hand, crystals cannot be too large as they will also become too optically thick at the wavelength used for reaction triggering (*see* **Subheadings 3.2.** and **3.4.**). In practice, this limits the thickness of the crystals in most cases to about 200 μm, unless special consideration is given to the geometry (*see* **Subheading 3.4.**).

Unlike standard diffraction experiments where crystals are routinely investigated at cryogenic temperatures (around 100 K), TR experiments probe structural changes at room temperature (around 293 K). Crystals are therefore more susceptible to radiation damage by X-rays. Crystals have to be mounted in thin-walled glass or quartz capillaries. Details on practical aspects of mounting crystals in capillaries can be found in the literature ***(30,31)***. Crystals should not be mounted too wet or too dry. A crystal embedded in too much liquid will move more easily when exposed to laser pulses, whereas a dry crystal will degrade faster because it will dry out further when repeatedly exposed to laser pulses. Crystals are typically maintained at a constant temperature by a dry gas stream. It is important to avoid temperature gradients across the capillary by using short capillaries and embedding the entire capillary into the gas stream. Otherwise, condensation or evaporation of the liquid at the crystal location may occur, which will degrade the crystal diffraction quality.

3.2. Reaction Triggering

A rapid and uniform reaction initiation, without a perturbing effect on the crystal or the reaction and with the highest possible fraction of activated molecules, is critical to the success of a TR experiment. Careful assessment and evaluation of the reaction initiation options is therefore essential. Depending on the nature of the reaction and reaction rates, triggering can be done by a rapid change in the concentration of substrates, cofactors, protons or electrons, by photo-activation of a native chromophore or a stable precursor of a substrate or cofactor, or by a rapid change in temperature or pressure.

The simplest method of concentration change is diffusion of substrates, cofactors or protons into a crystal in a flow cell. The method can be used to trigger a single reaction event ***(39–41)*** or a multiple turn-over, steady-state accumulation of an intermediate ***(42–44)***. However, the diffusion process is very slow ranging from seconds to many minutes ***(3,45)***, and only the slowest reactions can be triggered synchronously. These reactions can then be investigated either by Laue or monochromatic X-ray diffraction technique. Flash cooling following diffusion can also be instrumental in capturing the various stages of the reaction ***(39)***.

The fastest method for reaction triggering is photo-initiation using ultrashort laser pulses (fs to ns). Naturally suitable samples for this method are photosensitive proteins ***(13,16–23)***. Short laser pulses can also be used to induce a temperature jump in the crystal for studies of the initial steps of thermal unfolding in proteins ***(46,47)***.

For proteins that do not contain a chromophore and are not inherently photosensitive, phototriggering can be used in combination with caged or photolabile compounds. Caged compounds are inert precursors of substrate or cofactor molecules. Illumination with UV light can liberate the active molecule and trigger the reaction in the crystal ***(45,48,49)***. A number of reactions have been triggered successfully on the μs to ms time scale using caged compounds ***(50–53)***. Very efficient and fast (sub-μs) caging groups are available now that can be incorporated into the hydroxyl, carboxyl, phosphoryl, and amide groups of a wide range of compounds. The compounds exist for photorelease of caged nucleotides, divalent cations (like magnesium and calcium), protons, neurotransmitters, and several amino acids.

Several concerns need to be addressed regarding the photo-initiation. The optical densities of photo-active crystals are often very high. The front layer of the crystal can therefore be easily over-saturated by light whereas the inner part of the crystal remains underexposed. For this reason, the wavelength of the laser light has to be selected such that the optical density is low (<0.5). This typically means that the wavelength is not at the absorption maximum, but on rather weak tails of the absorption bands. Another method to trigger the reaction more uniformly throughout the crystal is to illuminate the crystal from multiple sides.

It is beneficial if the optical density at the excitation wavelength is not only low for the initial state, but also for the intermediate state that forms during the laser pulse. Similarly, the use of shorter laser pulses will prevent the absorption of light by intermediates that form past the laser pulse duration. In both cases, the heating of the crystal will be reduced and a possible undesired photoactivation of the intermediate states is minimized.

The laser pulse energy delivered to the crystal has to be considered carefully. Too low energy will prevent sufficient photo-initiation, whereas too high energy may be damaging. Given the large number of molecules in the crystal (10^{12}–10^{14}), 10–100 μJ/pulse is needed just to match the number of absorbed photons to the number of molecules, assuming all delivered photons are absorbed. Because a relatively low optical density of the crystal is needed for a uniform photo-initiation, a significant fraction of photons will not be absorbed. Also, the laser beam size is typically larger than the crystal for easier alignment and overlap of the X-ray and laser beams (**Subheading 3.4.**), and to ensure a uniform reaction initiation across the crystal. This again reduces the number of photons that are actually absorbed and utilized for reaction initiation and requires higher pulse energy to be delivered. For example, for a crystal of $0.3 \times 0.3 \times 0.2$ mm^3 size containing 10^{14} molecules, with 0.2 OD at the excitation wavelength and the laser beam size of 0.6 mm diameter, only about 10% of the laser energy will be absorbed. As a result, about 1 mJ/pulse, rather than 100 μJ /pulse, has to be delivered to match the number of absorbed photons and number of molecules in the crystal. The temperature increase in the crystal due to the absorbed light is estimated to 1–2 K in this case. The quantum yield (the number of reaction events per photon absorbed) also needs to be taken into account when determining the appropriate laser pulse energy. Other aspects of maximizing the reaction initiation in practice are discussed in **Subheading 3.4.**

When using caged compounds, similar considerations to those mentioned previously regarding the laser wavelength, pulse energy, and illumination apply. Caged compounds typically require intense UV light for activation ***(49)***, and crystal damage is a concern. For more efficient activation of caged compounds with low quantum efficiency, cryophotolysis has been proposed ***(4)*** in combination with temperature-controlled crystallography (rapid warm up/cool down cycle) to trap intermediates.

Preliminary spectroscopic measurements in crystals proved to be a very useful tool to characterize the purity of the sample, to monitor the extent of reaction initiation while optimizing the laser pulse energy, and to monitor the reaction progress in the chromophore region as a guide to TR crystallographic measurements. Several compact microspectrophotometers for on-line or off-line use have been described ***(54–57)***. The Hadfield-Hajdu design ***(55)*** is commercially available (www.4Dx.se).

3.3. Experimental Setup

3.3.1. X-Ray Source

Successful ns TR experiments have been conducted at two beamlines: ID09, ESRF and 14-ID, APS ***(16–18,20,22,28,58)***. The first sub-ns experiment (150 ps time resolution), conducted at the ID09 beamline, has been reported recently ***(13)***. Slower, μs to ms, experiments have also been conducted at the beamlines X26C (National Synchrotron Light Source, Brookhaven National Laboratory) and BL44B2 and BL40XU (SPring8).

At a synchrotron X-ray source, electrons emitted by an electron gun are first accelerated in a linear accelerator (linac), and then in a cycling booster synchrotron to their final energy of several GeV (7GeV at the APS). These high-energy electrons are injected into a large storage ring (1104 m circumference at the APS) where they orbit at a constant energy inside a vacuum chamber for many hours. Their revolution time is 3.683 μs at the APS storage ring. The electrons are bunched in the storage ring and produce intense X-ray pulses of approx 100 ps duration when passing through the bending magnets or insertion devices located around the storage ring ***(14)***. Insertion devices are arrays of magnets that force the electrons to follow a wavy trajectory, which results in more intense X-ray radiation than that produced by bending magnets.

Storage rings are typically operated with a uniform distribution of the electron bunches and thus produce a train of uniformly and quite closely spaced X-ray pulses. Standard operating modes at the APS are 324 or 24 uniformly

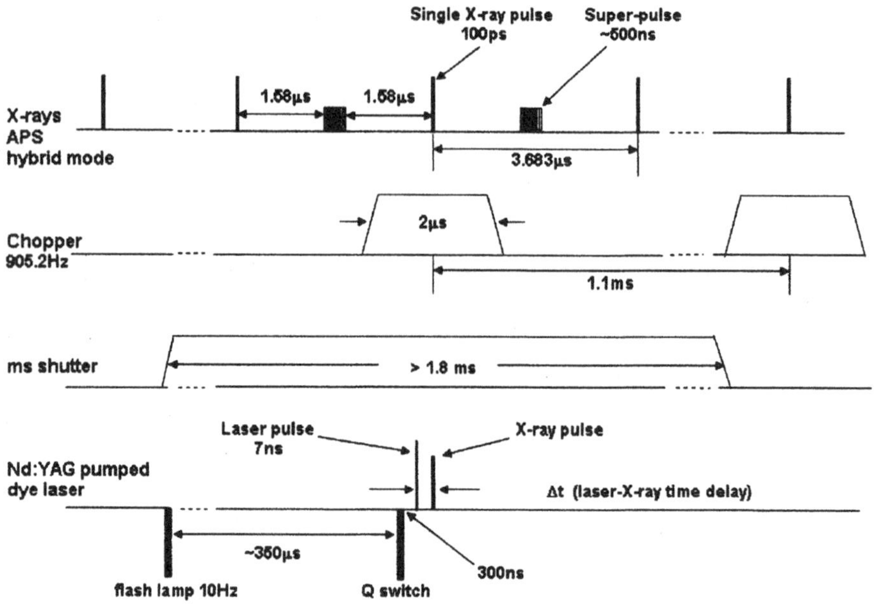

Fig. 5. Timing schematics for TR experiments at the 14-ID beamline, APS. Two X-ray shutters operate in series, synchronized to the X-ray pulse train, to select an X-ray pulse or a short pulse train. Laser firing is also synchronized with the X-ray pulses.

spaced single bunches (11.4 ns and 153 ns bunch spacing, respectively). X-ray shutters are used to isolate single X-ray pulses or super-pulses (trains of single 100 ps X-ray pulses with µs-ms duration). The fastest X-ray shutters ***(59,60)*** have open-time of 1–2 µs and can isolate a single X-ray pulse if the adjacent pulses are separated by more than 0.5–1 µs. This is the case in special operating modes of storage rings, such as the single bunch mode (ESRF) or the hybrid mode (ESRF and APS). In the single bunch mode only one electron bunch is circulating in the storage ring (2.816 µs spacing between consecutive X-ray pulses at the ESRF). In the APS hybrid mode, a single bunch is on one side of the ring while a train of closely spaced bunches (super-bunch) is located on the opposite side. The spacing between the single X-ray pulse and the adjacent super-pulse is 1.58 µs at the APS (**Fig. 5**).

At the polychromatic undulator beamlines of the third generation synchrotron sources, 10^9–10^{10} photons are delivered to a 200 × 200 µm^2 sample in a single X-ray pulse. This is, however, not sufficient to record a diffraction image with a SNR that is required to accurately determine small difference SF amplitudes. Signal averaging is required and 10–100 single X-ray pulse exposures are typically accumulated prior to the detector readout for each diffraction image (*see* **Subheading 3.4.**).

3.3.2. Synchronization and Timing of Laser and X-Ray Pulses

We will describe here the current experimental set-up at the APS beamline 14-ID, built and operated by the Consortium for Advanced Radiation Sources, The University of Chicago, Chicago, IL. For a detailed description of the ESRF beamline ID09 *see* Schotte et al. ***(60)***.

A series of shutters is used to select a single X-ray pulse from the continuous stream of pulses (**Fig. 5**). The fast rotating μs shutter (chopper) has a maximum rotation frequency of 905.2 Hz and a minimum open-time of 2 μs (FWHM). The phase of the chopper relative to the X-ray pulse train is adjustable so that the chopper can transmit either a single, 100 ps X-ray pulse or a 500 ns super-pulse in the APS hybrid mode. The slower ms shutter, with a minimum open-time of 1.8 ms (FWHM), isolates a single opening of the chopper. Two additional, even slower shutters (so-called *heat-load* shutters) are used to prevent excessive heating of the chopper by polychromatic radiation.

Two tunable ns lasers are available for reaction photo-initiation: a Nd:YAG pumped dye laser (Continuum Powerlite 8010/ND6000) and a more easily tunable Nd:YAG pumped OPO laser (Opotek Vibrant). The pulse duration for the two lasers is 7 ns and 4 ns, respectively. The wavelength range accessible with these lasers is 420–850 nm and 240–620 nm, respectively. A timing module (in-house design) provides synchronization of the laser and X-ray pulses. The actual synchrotron RF signal is used as a synchronizing *clock* for both the timing module and the chopper controller. The timing module generates a 10 Hz signal for necessary continuous operation of the Nd:YAG laser flash lamps and also provides the trigger for laser firing. Both signals are sent to the laser via a very accurate time-delay generator (Stanford Research DG535) that facilitates adjustment of the time delay between the laser and X-ray pulses in the range from 1 ns to many seconds.

3.3.3. Sample Environment

The ns lasers are located in the remote laser laboratory. Laser light is passed to the X-ray station and delivered to the sample via an optical fiber. Fibers with a relatively large diameter are used (0.6–0.9 mm) because the laser pulse energy of a few mJ (*see* **Subheading 3.2.**) has to be coupled into the fiber. The coupling is accomplished by focusing the laser light at the tip of the fiber, while exposing only the fiberglass core to the light to prevent damage of the fiber.

The standard arrangement of components in the sample area is shown in **Fig. 6**. The laser beam is focused onto the sample by a small collimating/focusing assembly (L) that provides approx 1:1 focusing. Another optical fiber (not shown) is mounted opposite to the shown fiber for a more efficient reaction photo-initiation (*see* **Subheadings 3.2.** and **3.4.**). The optical paths are adjusted so that the light from both fibers reaches the sample simultaneously (± 0.5 ns).

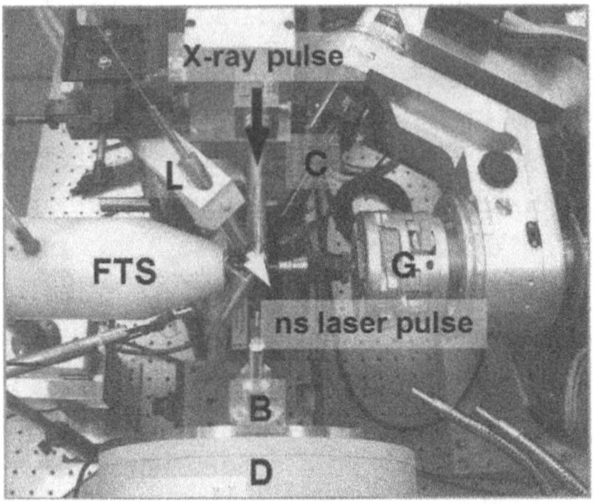

Fig. 6. Sample environment at the 14-ID beamline, APS. The sample capillary is mounted on the goniometer head (G) of a standard kappa diffractometer. The camera for viewing the magnified image of the crystal (C) is also mounted on the diffractometer. The sample is maintained at a constant temperature by a gas stream from the cooling system (FTS). The X-ray beam stop (B) and the MARCCD area detector (D) are also shown.

Two avalanche photodiodes are used to detect the laser and X-ray pulses separately as signal strengths vary significantly. Both are located very close to the sample (not indicated in **Fig. 6**). A diode with a Be window is used to detect the X-ray pulses only. The second diode detects predominantly laser pulses (much stronger signal). The time-offset between the two diodes is fixed and known. The signal from these diodes is sent to a fast oscilloscope (500 MHz Tektronix TDS744A) for measurement of the time delay between the pulses.

3.4. X-Ray Data Collection

3.4.1. Reaction Triggering Considerations

A general scheme of TR X-ray data collection can be described as follows. A single laser pulse is directed to the desired volume of the protein crystal and after a well defined time delay a single X-ray pulse illuminates the laser-irradiated volume. The diffracted X-rays are recorded on a large area detector. Because the signal of interest is the difference between the diffraction before and after the photo-activation, the alignment of the crystal, the X-ray beam, and the laser beam is critical. First, the center of rotation of the diffractometer is brought into the X-ray beam path. The laser beam is then focused at the center of rotation. Finally, the crystal is placed at the center of rotation. The crystal therefore stays at the same position during the data collection, at the intersection of the laser and X-ray beams.

As mentioned in **Subheading 3.3.**, even the most powerful, third-generation synchrotron sources require multiple pump-probe cycles for ns or sub-ns time resolution. Between each pump-probe cycle there must be a wait-time for the original state to be restored in the crystal. The wait-time depends on the cycling time of the sample and the time needed to dissipate the heat deposited by the laser pulse. If the wait-time is too short, the heat may not dissipate before the next pump-probe cycle and the temperature of the sample might creep up with repeated pump-probe cycles. For this reason, the repetition rate is usually only a few Hz even when the cycling time of the protein is faster.

When a crystal is exposed to laser pulses, the increase in crystal mosaicity and crystal motion are generally observed and manifested as elongation (streakiness) of the diffraction spots. The increase in mosaicity can be observed even with a single X-ray pulse exposure whereas the crystal motion is most clearly evident with multiple exposures. The main cause for the crystal motion is thought to be heating related. To reduce the motion, a sufficient wait-time between cycles, as well as minimizing the number of pump-probe cycles should be considered. Also, immobilizing the crystal inside the capillary is highly beneficial. This has recently been successfully accomplished for a variety of crystals by using either glue (standard 5-min epoxy, for example) or a thin coat of polyvinyl to attach the crystal to the capillary ***(61)***.

The X-ray exposure should be maximized without compromising the time resolution for a given time delay. Whenever possible, a train of X-ray pulses should be used for the pump-probe cycle rather than a single X-ray pulse, because more X-ray photons are used per cycle and an image can be obtained with a smaller number of cycles. This has two advantages. First, the total time for the experiment is reduced. Second, the number of laser shots and therefore the crystal damage by the laser pulses is reduced.

The spatial extent of structural changes as well as the overall fraction of photo-activated molecules is generally small. It is therefore imperative to maximize the signal and minimize the background. The SNR of the difference signal is roughly proportional to the square root of the total number of X-ray photons and to the overall fraction of photo-activated molecules, which in turn is proportional to the absorbed energy of the laser pulse (other conditions being equal). Therefore, laser beam size and X-ray beam size should be adjusted to maximize the fraction of excited molecules probed by the X-ray beam. The size of the X-ray beam is the same or slightly smaller than the size of the crystal in order to optimize the SNR. If the X-ray beam is larger than the crystal, the background from the capillary and the liquid around the crystal is increased. On the other hand, if the X-ray beam is too small, a part of the crystal will not contribute to the diffraction. The X-ray beam is typically smaller than the laser beam as the laser beam is generally larger than the crystal (*see* **Subheading 3.2.**).

Larger crystals are preferred because they result in better diffraction, and therefore better difference signal. Efficient photo-activation, however, becomes a problem if the wavelength cannot be tuned to a region where the optical density is sufficiently low (*see* **Subheading 3.2.**). In cases where optically thick crystals have to be used, a small X-ray beam can be used to probe only the surface layer of the crystal that is illuminated by laser light. It is important that the surface layer stays at the intersection of the laser and X-ray beams for all crystal orientations during the data collection.

To minimize systematic errors, ideally all time delays should be collected using a single crystal. In practice, however, this is often not possible because of crystal damage by the X-ray and laser pulses. This limits the maximum number of images that can be collected from one crystal (*see* **Subheading 3.4.3.**). When crystal morphology allows, using long but relatively thin crystals with smaller laser and X-ray beams can be beneficial. The crystal can be translated and the previously unexposed part of the crystal can be used for subsequent data collection.

The laser pulse energy has to be optimized to maximize the photo-activation while preventing the crystal damage and excessive spot elongation (*see* **Subheading 3.1.**). Typical pulse energies are of the order of several mJ for the laser beam size of 0.6–0.9 mm (diameter). When a time series of data sets is collected, the laser pulse energy should be held constant for all time delays. A failure in this regard will introduce systematic errors as the extent of photo-activation will differ for different time delays.

3.4.2. X-Ray Beam Considerations

Undulators, that have an energy bandpass significantly narrower than wigglers, proved to be better X-ray sources for Laue and TR crystallography ***(15,62,63)***. In addition to the reduced polychromatic background and spatial overlap in the Laue images, the higher peak power of undulators also contributes to the improved data quality. The APS Undulator A, used on the 14-ID beamline, has an energy bandpass of approx 1 keV ($\Delta E/E \sim 10\%$, FWHM), compared to the bandpass of approx 10 keV of the APS Wiggler A.

X-ray wavelength and flux are determined by the undulator gap. The Undulator A at 14-ID beamline is typically operated with the energy of the first harmonic in the range of 11.3 to 13.8 keV (0.9 to 1.1Å). Because it is critical that the X-ray flux is maximized for TR experiments, the choice of the wavelength is only the secondary factor. It should be noted, however, that shorter wavelengths cause less radiation damage ***(64)*** and this can therefore be important for room temperature studies.

3.4.3. Data Collection Strategies

In TR Laue crystallography, a complete data set is a multidimensional data matrix: three traditional reciprocal space dimensions and the additional dimension of time. At a particular crystal orientation and time delay, the Laue diffraction pattern contains diffraction spots only for a part of the reciprocal space. Laue images at many different crystal orientations have to be collected to completely sample the reciprocal space at each time delay. The angular step through the reciprocal space depends on the bandpass of the X-ray source. With the APS Undulator A, 2–3° is sufficient. The number of time-points needed and their distribution in time ultimately depends on the reaction that is investigated but a good starting point is to collect three to five points per time decade in the region of interest, equally spaced in logarithmic time.

The beamtime is quite limited at synchrotrons in general. In addition, the special operating modes required for TR experiments are available for only a very limited amount of time. Therefore, one should make a considerable effort to use beamtime efficiently. The mode of experiments and data collection strategy depends on the goal of the experiment. If crystals are being checked for suitability for TR Laue experiments, snapshot images are collected without the laser excitation to evaluate the crystal mosaicity. If the suitability has already been established and real TR data is to be obtained, one often faces two possibilities depending on the choice of the fast variable: crystal orientation or time delay. The most straightforward data collection scheme is to scan the entire angular range at a fixed time delay and subsequently move to another time delay. Because only a limited number of images can be obtained from one crystal as a result of radiation damage, only single time-point or at most a few time-points can be obtained from one crystal. Therefore, in this experimental scheme, laser intensity fluctuation and crystal-to-crystal variation between time-points introduce a systematic error that has a detrimental effect on the accurate determination of the time constants when the entire series of time delays is examined. Nevertheless, this collection strategy is often chosen if the goal of the experiment is to collect preliminary data at a few chosen time delays in order to establish the presence of the signal and determine the photo-activated fraction of molecules. In addition, a highly redundant data at a desired time delay can be obtained most effectively this way by merging multiple data sets collected at the same time delay ***(58)***.

The problem of systematic errors described previously can be circumvented by an alternative data collection scheme where the time delay, rather than the angular setting, is the fast variable. The time range is scanned at a fixed angular setting, which is then advanced for a repeated time scan at the new crystal orientation. A shortcoming of this method is that one crystal may not be sufficient to obtain data for all crystal orientations and all time delays as a result of radiation damage issues. Fortunately, our experience suggests that data from multiple crystals can be merged successfully. Therefore, data sets that are partial in reciprocal space but which cover all time delays can be collected from several crystals and merged to obtain a time series of complete data sets. This constitutes a better method for accurate determination of time constants associated with the formation and decay of intermediates.

Another shortcoming of this approach of data collection is the potential all or nothing aspect. If the entire reciprocal space for the series of time delays is not completed, the data may turn out to be useless. One may run out of beamtime or crystals. This becomes a more serious problem with lower X-ray flux and space groups with low symmetry like monoclinic, where an angular range of 180° has to be collected. To minimize the risks but complete the entire time series that is needed, one can collect several shorter time series rather than one long series. In this case, one time delay common to all time series is desirable as it will greatly help to account for any variation in the extent of photoactivation among different time series.

Because the actual signal in the TR data is the change caused by the reaction initiation, a reference data should be collected and subtracted from the data at positive time delays (so-called light data). It may be possible to use a model from a static measurement as a reference without collecting experimental reference data. However, our experience shows that an experimental reference data collected without light activation (so-called dark data) is superior in reducing systematic error and yield a better SNR. The best way to minimize the systematic errors is to interleave dark and light images for single-time delay if crystal orientation is the fast variable, and to collect dark plus light time series if time delay is the fast variable. This way, any progressive degradation of the crystal will have a minimal effect in the final difference signal. In addition, a negative time delay instead of dark data can be used as a reference. This has an advantage of reducing any systematic error due to potential accumulation of photoproducts during the data acquisition.

3.5. Data Processing and Analysis

Processing of TR data proceeds in three steps: 1) a time series of structure factor amplitudes is derived from diffraction images; 2) a time series of difference electron density maps is derived; and 3) structures of intermediate states are determined. The three steps are explained in **Subheadings 3.5.1.** to **3.5.9.**

3.5.1. Reducing Laue Data

The goal of data reduction is the extraction of accurate diffraction spot intensities or SF amplitudes from the raw images (**Fig. 4**). The major steps involved are indexing, integration, scaling, and deconvolution of harmonic overlaps *(15)*. The first three steps are challenging for the polychromatic, Laue method, whereas the fourth step is actually unique for this method. Specialized software is used for processing of Laue data (*see* **Subheading 2.**). The software must cope with the following problems specific for the Laue method:

1. In crowded Laue diffraction patterns probability is high that the reflections spatially overlap. Spot profiles are used to separate these overlaps ***(32,37)***. The profiles are derived from well separated, nonoverlapping reflections. By using the narrow bandpass undulators as X-ray sources, the spatial overlap problem is significantly reduced ***(62,63)***.
2. The intensity of the polychromatic incident radiation varies as a function of wavelength. In addition, the scattering power of the crystal as well as other parameters such as the detector sensitivity are wavelength dependent ***(26)***. When the orientation of the crystal is changed, a given reflection may be stimulated at a different wavelength. Consequently, intensity collected from this reflection or its symmetry mates may vary strongly with the orientation of the crystal. Therefore, the intensities must be brought to a common scale before they can be merged. This procedure is known as wavelength normalization and results in a so-called λ-curve, which represents all wavelength dependent effects as seen by the detector ***(32,35)***.
3. The polychromatic X-ray beam may excite at the same time reflections whose indices are multiples of a common basic hkl triplet. For example, the reflections with indices 2 4 0 and 3 6 0 both contain a multiple of the basic 1 2 0 triplet and are, therefore, harmonic to each other. Harmonic reflections lie on a radial line in the reciprocal space (starting at the origin), scatter in exactly the same direction and overlap precisely at the detector ***(33)***. As these reflections correspond to different energies, the overlap is also referred to as energy overlap. The low-resolution reflections are especially affected, leaving the low-resolution data incomplete (*low resolution hole*) if the harmonic overlaps are excluded. Using multiple measurements of the same reflection at different crystal orientations and measurements of corresponding symmetry mates, the harmonic overlaps can be resolved into the component reflections after the λ-curve has been determined. The procedure is referred to as harmonic deconvolution.

Laue data reduction is also affected by a substantial scattering background generated by the polychromatic radiation. As a consequence, weak reflections that occur predominantly at higher resolution are more difficult to determine accurately than for the monochromatic data. The background is substantially reduced by use of undulators (*see* **Subheading 3.3.1.**) which improves the overall resolution of the collected data.

3.5.2. Time-Dependent Difference Electron Density Maps

The result of a TR experiment are SF amplitudes of the initial, dark state $|\mathbf{F}^{\mathbf{D}}(hkl)|$ and a corresponding set of time-dependent SF amplitudes $|\mathbf{F}(hkl,t)|$. From these amplitudes, time-dependent difference SF amplitudes $\Delta F(hkl,t) = |\mathbf{F}(hkl,t)| - |\mathbf{F}^{\mathbf{D}}(hkl)|$ are calculated for each time-point t. Phases ϕ^{D}_{hkl} are obtained from the known dark state structural model. If both $|\mathbf{F}(hkl,t)|$ and $|\mathbf{F}^{\mathbf{D}}(hkl)|$ are on the absolute scale, it can be shown ***(65,66)*** that DED maps calculated using

$\Delta F(hkl,t)$ and the phases ϕ^D_{hkl} are accurate maps on roughly half the absolute scale as long as structural differences remain small and noise moderate. This is referred to as the difference approximation. However, significant random noise in the difference SF amplitudes $\Delta F(hkl,t)$ will have a degrading effect on the maps ***(27,66)***. Moreover, as a result of errors in data acquisition and reduction some $\Delta F(hkl,t)$ may be erroneously large. Consequently, both, extremely large $\Delta F(hkl,t)$ and those with a high experimental error should be weighted down in calculations of DED maps. Several weighting schemes have been proposed ***(16,20,27,67)***. The final DED map is calculated as:

$$\Delta\rho(t) = \frac{1}{V_e}\sum_{hkl} w\Delta F(hkl,t)e^{i\phi^D_{hkl}}e^{-2\pi i(hX+kY+lZ)} \tag{2}$$

where X, Y, Z are the components of the position vector in the coordinate system of the unit cell (fractional coordinates), h, k, l are the reflection indices, V_e is the volume of the unit cell and w is the weighting factor of the difference SF amplitude $\Delta F(hkl,t)$.

These maps can be inspected using specialized molecular modeling programs such as XtalView ***(68)*** or O ***(69)***. In order to display the features that are above the noise of the map, the DED maps need to be contoured above the 3 σ level. The σ value is the root mean square (RMS) deviation of the DED from the mean value determined from all grid points in the asymmetric unit. It includes contributions from both signal and noise. Positive and negative DED features are observed. Negative electron density features represent loss of electrons and account for atoms that have moved away. Positive features represent gain of electrons and account for new positions occupied by atoms that moved. If SF amplitudes are represented on the absolute scale and nonoverlapping DED features are considered, the total (integrated) DED count in a particular feature multiplied with the volume of the grid unit is equivalent to about half the number of electrons displaced from or into this volume.

3.5.3. Structures of the Intermediates and Chemical Kinetic Mechanism From Difference Electron Density Maps

The ultimate goal of any TR experiment is the determination of the kinetic mechanism together with the atomic structures of the underlying intermediates (*see* **Subheading 1.**). Chemical kinetics is the key to analyze the experimental DED maps, to separate admixtures of intermediates present in these maps, and to determine the structures of the intermediates.

The kinetic analysis starts with the search for the number of intermediates and determination of relaxation times (*see* **Subheading 1.**) by examining the series of time-dependent, experimental DED maps. How to find the number of relaxation times from the series of DED maps represented by a large number of grid points which can amount to 10^5 even in moderately sized DED maps?

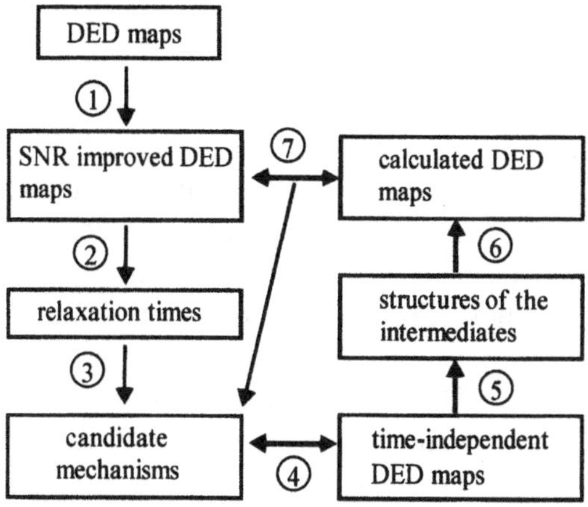

Fig. 7. The seven steps from the time-dependent DED maps to the structures of intermediates and compatible kinetic mechanisms. 1) Noise reduction via SVD-flattening of the DED maps. 2) Fit of the right singular vectors, the main temporal components of the experimental maps derived by SVD, by exponential functions. 3) Fit of candidate reaction mechanisms to the right singular vectors. 4) Construction of the time-independent DED maps corresponding to intermediates from left singular vectors (major spatial components derived by SVD). 5) Extrapolated, conventional electron density maps of the intermediates and modeling of intermediate structures. 6) Calculation of time-dependent DED maps from structures of intermediates for chosen candidate mechanisms. 7) Comparison of calculated and observed time-dependent DED maps on the absolute scale, post-refinement and selection of candidate mechanisms by a *posterior analysis*.

Tools are available from linear algebra, which perform just this task. These tools are all related to a component analysis. One of the tools, the Singular Value Decomposition (SVD), has been shown to work successfully with crystallographic data ***(27–29)***. The SVD separates time and space variables: from a series of time-dependent difference maps it determines only a few main common spatial components and their time variations, which constitute main common temporal components. The temporal components are then used to determine the number of intermediate states and relaxation times.

Figure 7 shows schematically the main steps in the process of deriving the time-independent maps of the intermediate states from the measured, time-dependent DED maps. Before we assess the main features of each step, the application of SVD to the crystallographic data is explained.

3.5.4. Singular Value Decomposition in Time-Resolved Crystallography

To perform an SVD analysis, a data matrix $\boldsymbol{A}$ is prepared where the difference maps from a time-series are entered one by one and in temporal order as column vectors of the matrix. How the grid points of the DED maps are assigned to the elements of the column vectors of ***A*** is irrelevant. However, a chosen order must be consistently used for all maps. If crystallographic symmetry is present, only the asymmetric unit needs to be included. Even more, a mask can be used to consider only the volume occupied by protein atoms. Around $M = 10^5$ grid points per map have to be considered for a protein of 20 kD molecular mass. A further reduction in the number of considered grid points is possible if those that do not contain significant difference electron densities throughout the time-course are disregarded. For this purpose a grid point is included only if DED is above or below a chosen σ level for at least one time point (typically the $\pm 2\ \sigma$ level). Thus, the number of useful grid points can be reduced to around $2 \cdot 10^4$ for the mentioned small protein.

An $M \times N$ data matrix ***A*** composed of DED maps at N time points, each consisting of M grid points, is decomposed into an $M \times N$ dimensional matrix ***U***, which contains the left singular vectors (lSV), an $N \times N$ square diagonal matrix ***S***, diagonal elements of which are the singular values (SV) and the transpose of a $N \times N$ square matrix ***V***:

$$A = \boldsymbol{USV}^T \tag{3}$$

The rows of $\boldsymbol{V}^T$ are the right singular vectors (rSV). The lSV are the main spatial components of the experimental, time-dependent DED maps and are DED maps themselves. Each rSV contains the temporal variation of the corresponding lSV DED map, whereas the SV weights the contribution of the lSV map to the experimental DED maps.

Although N singular vectors result from the SVD decomposition, not all of them contain signal. To those that do, we refer to as significant. The number of significant singular values and vectors is related to the number of intermediates in the reaction. Consider a reaction where molecules initially in the state I_0, when photo-excited, go through two sequential intermediate states, I_1 and I_2, and back to the initial state, $I_1 \rightarrow I_2 \rightarrow I_0$. In this hypothetical experiment, time-dependent DED maps are collected at $N = 10$ time delays. In this case, only two difference maps are present, I_1–I_0 and I_2–I_0, corresponding to two intermediates. All measured DED maps are a mixture of various ratios of these two difference maps. After the SVD, there are only two significant lSV. The first significant lSV_1 represents the average DED map of the relaxation processes and the rSV_1 describes its temporal variation. Consequently, the lSV_1 is a mixture (a linear combination) of the I_1–I_0 and I_2–I_0 difference maps. The second significant lSV_2 is also a linear combination of the I_1–I_0 and I_2–I_0 maps. It has a lower SV and contains the average deviation from the average difference map lSV_1. The second rSV_2 describes its temporal variation. The lSV_1 and lSV_2 maps occupy the first two columns in the ***U*** matrix while the rSV_1 and rSV_2 occupy the first two rows of the matrix $\boldsymbol{V}^T$, since the singular vectors and values are ordered according to the magnitude of the singular values, i.e., according to their significance. The remaining eight vectors at positions 3 to 10 in the matrices ***U*** and ***V*** are insignificant and contain only noise.

It is straightforward to expand these considerations to more intermediate states. In general, if the number of time-points is larger than the number of intermediate states, the data matrix ***A*** can be approximated by a matrix ***A'*** reconstructed from a small number S of significant singular vectors containing signal. There is an inherent SNR improving capability in the SVD analysis as reconstructing the data matrix from the significant singular vectors in effect filters the noise out. When substantial noise is present, the signal will artificially spread to insignificant singular vectors. In this case, the procedure of rotation can be used to recollect the signal ***(28,29,70)***. The SVD noise-filtering is still effective.

3.5.5. ***Step 1:*** *Noise Filtering (Singular Value Decomposition Flattening)*

The experimental noise results in errors in magnitude and, more importantly, sign of the difference SF amplitudes (negative sign instead of positive or vice versa). This gives rise to DED features varying randomly in time ***(28,66)***. The SVD can correct for these errors. However, it is crucial that the number of significant singular vectors, containing signal, is determined correctly. All vectors containing the signal have to be used to reconstruct the approximate data matrix ***A'***. The methods for selecting the significant vectors ***(27–29)*** are based on the magnitude of the singular value and on the autocorrelation of the rSV, but, most importantly, on identifying the regions of the molecule where signal is present by inspecting the DED maps represented by the lSV (**Fig. 8**).

From the DED maps reconstructed using only significant singular vectors and values, new difference SF amplitudes and phases, $\Delta F^{SVD'}$ and $\phi^{SVD'}$, are obtained by an inverse Fourier transform. These difference SF are combined with the SF for the dark (initial) state by a phase recombination scheme ***(27,28)***, to result in improved values ΔF^{SVD} and ϕ^{SVD} and, from these, to DED maps $\Delta\rho(t)^{SVD}$ with improved SNR. It has been shown with the mock data ***(27)*** that the initial dark phases used for the difference SF can be improved by this procedure by 10°–15°, depending on the noise. In other words, the initial dark phases approached by 10°–15° to the true phases of the difference SF. In mock and experimental difference maps the SNR is greatly enhanced ***(27–29)*** and the maps are much better suited for a subsequent kinetic analysis. This new procedure for noise suppression has been named SVD-flattening. **Figure 9** shows an experimental DED map before and after the SVD-flattening is applied to a time series of 15 experimental DED maps (drawing produced by Ribbons *[71]*). It is clearly demonstrated that for the flattened maps (panel B), the signal is enhanced and the noise reduced.

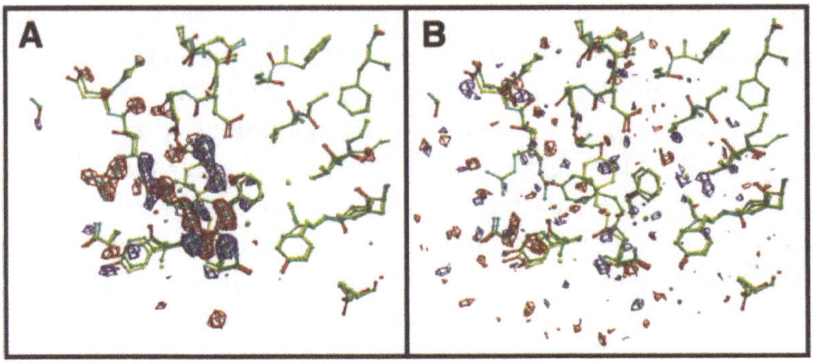

Fig. 8. **(A)** Significant left singular vector **(B)** Insignificant left singular vector. (Both from mock data.)

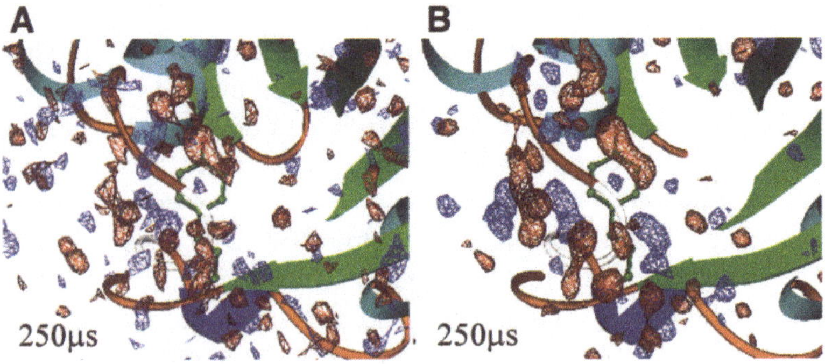

Fig. 9. DED map at 250 μs from a set of 15 experimental DED maps *(28)* **(A)** original DED map **(B)** SVD-flattened DED map, contour levels: red/white: –3 σ/–4 σ, blue/cyan: 3 σ /4 σ.

3.5.6. ***Step 2:*** *Relaxation Times*

The next step in the SVD analysis is to subject the SVD-flattened maps, $\Delta\rho\,(t)^{SVD}$, to the SVD procedure again. The resulting significant rSV are examined as they contain the temporal variations of significant lSV, the main spatial components. The rSV time-traces are linear combinations of the true time-dependent concentrations of intermediates. Hence, the rate coefficients λ_i and the corresponding relaxation times τ_i remain unchanged and can be determined from the rSV. All rSV are fit globally with a sum of exponential terms featuring common relaxation times τ_i (step 2 in **Fig. 7**). **Figure 10** shows the first four rSV from an analysis of real experimental data. Three relaxation times are identified at 170 μs, 620 μs, and 8.5 ms (marked by vertical lines).

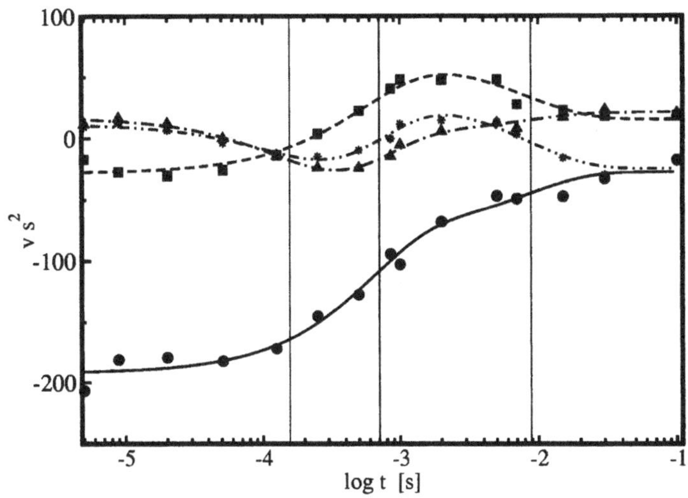

Fig. 10. rSV from a SVD analysis of experimental data *(28)*. Vertical lines mark the identified relaxation times. Data points: ● first rSV, ■ second rSV, ▲ third rSV, * forth rSV. Fitted exponentials are shown for rSV1 to rSV4 by solid, dashed, dashed dotted, and dashed double dotted lines, respectively. Some signal has spread into the fourth rSV owing to noise *(27)*.

3.5.7. Steps 3 *and* 4*: Time-Independent Difference Electron Density Maps*

This part of the SVD analysis is more complex, because here the DED maps corresponding to intermediates must be synthesized from the significant lSV. Although lSV are DED maps, they are only linear combinations of the DED of the intermediates. In order to determine the contribution of each significant lSV to the DED map for each intermediate, time-dependent concentrations of intermediates are needed. This requires the assumption of a mechanism. The mechanism must be compatible with the relaxation times observed in the rSV, i.e., it must generate both the correct number and the correct values of the relaxation times (**step 3** in **Fig. 7**). Because the number of relaxation times is related to the number of states, a general mechanism can be set up that contains all states and all possible rate coefficients between them (as in **Fig. 1**, panel G). For chosen plausible simple candidate mechanisms based on this general mechanism, the time-dependent concentrations of intermediates, I_j, are derived by solving a system of coupled differential equations for each mechanism (**Fig. 2**). The significant rSV are then fit again globally, this time by the sum of concentrations, I_j, with the scale factors E_{nj} as amplitudes for the n-th rSV and the j-th concentration term (*see* **Note 1, Eq. 4**). Concentrations are expressed in terms of rate coefficients (**Fig. 2**). Both the numerical values of the rate coefficients and the scale factors E_{nj} are varied to reproduce (fit) the magnitude of all relaxation times in rSV (*see* **Note 1**, **Eq. 4**).

The resulting scale factors E_{nj} determine the contribution of the *n*-th lSV to the DED map of the *j*-th intermediate. The intermediate DED maps can now be synthesized (*see* **Note 1**, **Eq. 5**, and **step 4** in **Fig. 7**). The mixture of intermediates in the experimental time-dependent DED maps has therefore been separated into the time-independent DED maps of intermediates. An example is shown in **Fig. 11**, panel A *(28)*. Most of the red, negative DED features are located on dark state atoms (green) whereas the blue, positive DED features can be interpreted by an atomic model of a single intermediate.

3.5.8. ***Step 5:*** *Structures of the Intermediates*

DED maps are difficult to interpret. Density is often disconnected, positive and negative densities are not always paired, and positive and negative densities can overlap as molecular rearrangements occur, leaving no density at the overlap locations. The interpretation difficulties can be overcome using conventional electron density maps calculated from the so-called extrapolated structure factors. These SF are obtained by a vector summation of the calculated SF of the dark state and the difference SF resulting from the Fourier transform of the time-independent DED map of an intermediate determined in **step 4** (*see* **Note 2**, **Eq. 6**). The amplitude of the difference SF is extended to correspond to 100% photo-initiation. As a result, each of these conventional electron density maps represents one pure intermediate only (no contribution from the dark state). The structures of intermediates can therefore be modeled and refined using these maps.

However, at this stage of the analysis, several candidate mechanisms can still extract similar, interpretable DED and conventional maps for intermediates, and cannot be distinguished. This is a general difficulty that arises when rSV are analyzed with a kinetic model. In the next paragraph a possible approach, named *posterior analysis*, is proposed to partially solve this problem.

3.5.9. ***Steps 6*** *and* ***7:*** *Chemical Kinetic Mechanism: Calculated Time-Dependent DED Maps and Posterior Analysis*

To compare different candidate mechanisms, one has to consider values for both relaxation rates λ_i and amplitudes P_{ji} in the expressions for concentrations of the intermediates I_j (**Fig. 2**) when fitting the rSV. Even when relaxation rates are very similar for different mechanisms, the amplitudes P_{ji} will be different, and it should be possible to use them to distinguish between these mechanisms. However, the rSV are not on the absolute and common scale, but rather on an arbitrary and unknown scale. As mentioned earlier, the scale factors E_{nj} are needed to bring the concentrations I_j (which are on the absolute scale) to the scale of the rSV. Therefore, the amplitudes P_{ji} cannot be used to distinguish between the candidate mechanisms. However, once additional, chemical information such as the structural and stoichiometric constraints enter the analysis, a further discrimination becomes possible.

Stoichiometric constraints are automatically included when the observed SVD-flattened DED maps, $\Delta\rho(t)^{SVD}$, are presented on the absolute scale: DED is expressed in the absolute units of e/Å^3 and fractional contributions of atoms (occupancies) can be determined. The $\Delta\rho(t)^{SVD}$ can be used to distinguish between mechanisms and, even more importantly, to estimate the extent of reaction initiation. Given the models of the ground state and the intermediates, and a candidate kinetic mechanism, time and mechanism (*k*) dependent DED maps, $\Delta\rho(t,k)^{calc}$, are calculated (**step 6** in **Fig. 7** and **Note 3**, and **Eq. 7**). Initially, the extent of the reaction initiation is set to 1.0 and reasonable initial conditions are assumed, which determines the coefficients P_{ji}. The $\Delta\rho(t,k)^{calc}$ are compared (fit) at all time-points to the $\Delta\rho(t)^{SVD}$. The difference is initially large for all time-points as the reaction initiation has to be adjusted to the observed level. In subsequent steps the rate coefficients for the candidate mechanism are refined (*see* **Note 3**, **Eq. 8**).

After the fit has converged, the residual maps $\Delta\Delta\rho(t) = \Delta\rho(t)^{SVD} - \Delta\rho(t,k)^{calc}$ are inspected. If they are free of density for all time-points, the mechanism is considered compatible. It has generated concentrations that reproduced the observed difference electron density. However, if there is pronounced residual density $\Delta\Delta\rho(t)$ at some or all of the time-points the mechanism is considered inconsistent and should be disregarded. A new candidate is then tested (**Fig. 7**, **Step 7**). Usually, from a set of simple mechanisms in the general mechanism scheme for the given number of intermediates, only a few mechanisms prove to be incompatible and the remaining ones constitutes a set of possible kinetic mechanisms. The degeneracy of the problem is therefore diminished, but not removed.

3.6. Applications of Time-Resolved Crystallography

Several review articles have summarized the progress in studies of structural intermediates by TRX and by trapping methods on a number of proteins ***(1–3,15,25,72,73)***. For some proteins, a detailed characterization of intermediates on the reaction pathway has been obtained using either one or both approaches. Examples include: isocitrate dehydrogenase ***(42)***, hammerhead ribozyme ***(39)***, bacteriorhodopsin ***(8)***, cytochrome $P450_{cam}$ ***(74)***, horseradish peroxidase ***(75)***, myoglobin ***(13,16–18,76–80)***, and photoactive yellow protein ***(20–22,81)***. A comprehensive table of TR Laue diffraction experiments from 1986 to 1998, with a time resolution from 10 ns to several seconds, is given in **ref. *15***. We will summarize here the results for two proteins that have been studied with the best time resolution to date: myoglobin and photoactive yellow protein.

3.6.1. Myoglobin

This small, oxygen-binding heme protein (18 kD) has been extensively studied by numerous experimental techniques for many decades. It is the first protein for which a three-dimensional structure was determined by X-ray diffraction ***(82)***. As the structure revealed no open pathway for the ligand access to the heme, the importance of protein dynamics for the process of ligand binding became evident. Since then, numerous studies of myoglobin established it as a model system to understand the complex nature of protein dynamics and protein–ligand interactions. The cabonmonoxy complex of myoglobin (MbCO), as very stable and easily photolyzed, is particularly well-suited for TRX studies. The goals of these studies are to elucidate the structural basis of the photolysis-induced protein relaxation processes ***(10,11)*** and the nature of the *proteinquake* ***(83)***, as well as to determine the pathway of the photodissociated ligand through the protein matrix.

The ns and sub-ns TR room temperature studies of myoglobin ***(13,16–18)*** demonstrate the capability of the technique. Myoglobin is a challenging case for TRX studies because structural changes following ligand photodissociation are expected to be small (0.2–0.4Å), based on high-resolution, static structures of MbCO and deoxy forms of myoglobin ***(84)***. Nevertheless, such small changes have been detected and the photodissociated CO molecule (CO*) has been identified on its migration pathway through the protein, even at relatively low occupancy levels of 10 to 20%.

In the first ns TR experiment ***(18)*** data sets to 1.8 Å resolution were collected at six time delays, ranging from 4 ns to 1.9 ms. The DED map (MbCO photoproduct–MbCO) at 4 ns shows a clear loss of the CO ligand caused by the photolysis. A CO* docking site is identified in the distal heme pocket, in the region where CO* has been observed in the low temperature photolysis studies ***(76–78)***. The docking site is absent in the 1 μs map indicating that the ligand has moved out of the distal pocket. The heme iron is displaced out of the heme plane whereas the distal histidine His64 moved inward, towards the location of the bound ligand. Small DED features observed along the E and F helices that surround the heme indicate that a more global structural relaxation also occurred, at least partially, by 4 ns.

The subsequent, more comprehensive study ***(16)*** that included 14 time delays, ranging from 1 ns to 1.9 ms, confirmed the initial findings and revealed more details about the ligand migration pathway, changes at the heme and protein relaxation. The ligand photodissociation was estimated to 40%. The iron motion, heme buckling and rotation, His64 swing, and initial displacement of the F and E helices have already occurred by 1 ns (*see* **Fig. 12**). Half of the photodissociated CO* molecules is detected at the distal docking site at 1 ns. The half-life of this site was estimated to approx 80 ns. A second CO* docking

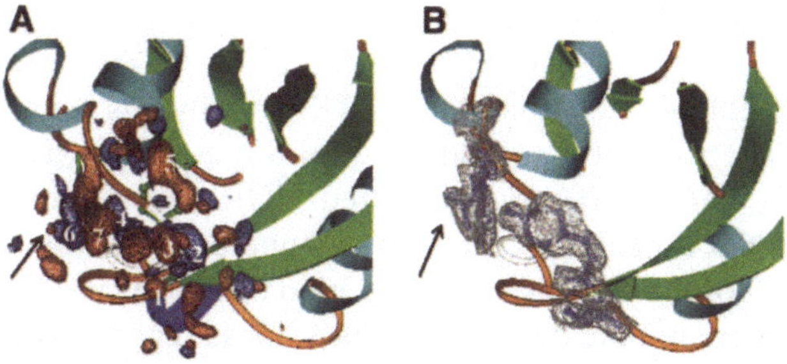

Fig. 11. **(A)** Time-independent difference map of an intermediate extracted from 15 time-points collected on the PYP *(28)*. Features 1,3,5,7: negative DED for the pCA chromophore and Arg52; features 2,4,6: corresponding positive DED features. Dark-state structure of the pCA chromophore shown in green. **(B)** Extrapolated, conventional electron density map (gray) covering the pCA chromophore and Arg52 (blue structures). Arrows in **(A)** and **(B)** extrapolated electron density is present at the position of no or little DED in **(A)**.

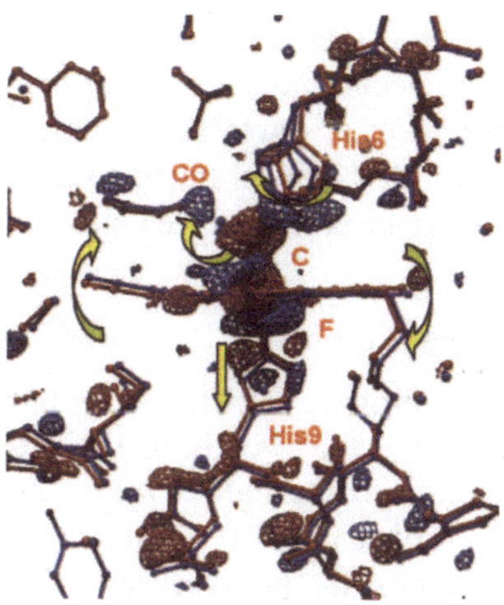

Fig. 12. DED map of the heme region at 1 ns following ligand photodissociation *(16)*. The map is contoured at $\pm$ 3.5 σ and $\pm$ 7 σ. The red features represent negative electron density (loss of electrons) and the blue features positive density (gain of electrons). The arrows indicate observed structural changes: motion of the CO ligand to a docking site CO* in the distal pocket, downward motion of the iron, proximal histidine His93 and the F-helix, and swing of the distal histidine His64. The MbCO model is shown in red, deoxy Mb model in blue.

site was also detected, located on the proximal side of the heme. This is the so-called Xe1 site, one of four hydrophobic cavities where Xe is observed under pressure. The peak occupancy of this site of 20% is reached at about 100 ns and its half-life is estimated to about 10 μs.

The most recent ns and sub-ns MbCO TR experiments were conducted on two mutants. The ns experiment involved a triple mutant (Leu29(B10)Tyr, His64(E7)Gln, Thr67(E10)Arg), denoted YQR, and followed structural changes between 3 ns and 3 ms after the CO photodissociation ***(17)***. The loss of bound CO, tilt of the heme, swinging motion of Tyr-29(B10), and migration of CO* to the more remote docking site Xe4, located on the distal side of the heme, were all detected by 3 ns. By 316 ns, CO* has reached the proximal Xe1 site, similar as in wild-type myoglobin ***(18)***. Structural changes in the distal E-helix and CD-turn regions are found in this case to lag significantly (100–300 ns) behind local changes in the heme region. These results establish a structural basis for the extended time-course of the protein conformational relaxation observed by TR spectroscopy ***(10,11)***.

The first sub-ns TR experiment was conducted recently at the ID09 beamline, ESRF ***(13)***. The L29F mutant of MbCO was chosen because TR IR spectroscopy revealed a short-lived intermediate, with a lifetime of 140 ps. The observed structural changes at 100 ps involve large displacements of the side chains Phe29 and His64 to accommodate CO* at the primary, distal docking site, as well as more subtle correlated rearrangements throughout the entire protein. By 1 ns, CO* has migrated to the more distant Xe4 distal site and by 32 ns it moved to the proximal Xe1 site where it persists for microseconds.

3.6.2. Photoactive Yellow Protein

The photoactive yellow protein (PYP), a blue light photoreceptor with a bright yellow color and a molecular weight of 14.8 kD, was first identified by Meyer ***(85)*** in the halophilic bacterium *Ectothiorodospira halophila* (now called *Halorhodospira halophila*). It is a paradigm for light triggered reactions in living organisms and is most likely involved in the phototactic behavior of this bacterium ***(86)***. Upon illumination with blue light PYP enters a photocycle. The central chromophore, para-coumaric acid (pCA), changes its configuration from *trans* to *cis*. The consequent conformational change of the chromophore causes the protein to relax through numerous intermediates, with lifetimes ranging from sub-ps to 100 ms, to the dark state ***(5)***. The photocycle has been extensively studied by spectroscopy (*see*, e.g., **ref.** ***87*** for a review) and conventional crystallography using trapping methods ***(81)***.

The first TR experiment on PYP followed the relaxation from a photostationary state, produced by a 100 ms exposure to laser light ***(21)***. One Laue data set was collected with a 10 ms X-ray exposure time, 2 ms after the laser was switched off. The major observed structural change involves a swing of the pCA chromophore head towards the surface of the protein and correlated displacement of the adjacent residue Arg52 into the solvent.

The first ns TR crystallographic experiment on PYP by Perman et al. *(22)* revealed the structure of an intermediate populated at about 1 ns. The pattern of positive and negative DED features showed the *trans* to *cis* isomerization of the pCA chromophore already at 1ns, in accordance with results from TR spectroscopy *(7)*. However, refined atomic structures of the early intermediate and those populated in the ns to μs time region are still missing.

Closely-spaced data sets in the time region from early μs to ms, following phototriggering by ns laser pulses, were collected recently at the APS beamline 14-ID *(28)*. At the beginning of the analyzed time range, roughly 15% of the molecules were photo-activated. As demonstrated by mock data *(27)*, this is sufficient for a successful analysis. The structures of two late intermediates in the photocycle were determined using the SVD-driven analysis described in **Subheadings 3.5.3–3.5.9**. The structures of the two late intermediates are similar to each other. They are also similar but not identical to the structure derived from the relaxation of the photostationary state mentioned previously. Detailed differences include the conformation of Tyr42, which has a different orientation in earlier time-points, and Arg52, which shows a double conformation in the pulsed experiment rather than a single conformation evident in the photostationary state. This study shows that an SVD-based analysis can be applied to real data and intermediates can be separated from an admixture. The photocycle of PYP is likely to be characterized completely in the near future on the time-scale from 100 ps to 1 s.

3.7. Conclusions and Future Outlook

With the demonstrated ability to detect small structural changes even at relatively low levels of reaction initiation (15 to 40%), TR crystallography has completed the phase of feasibility studies. Further improvements of the X-ray sources and optics at the third-generation synchrotrons and continuing advances in the development of methods for data analysis, such as the SVD-based analysis, provide us with the opportunity to expand the application of this technique beyond the first test cases, like myoglobin and PYP.

The major challenge in applying the technique to new systems of wider biological interest is to find a suitable and efficient method for the reaction initiation. Whereas the use of pulsed lasers is straightforward for a broad family of proteins containing a chromophore, like heme proteins and photoreceptors, other systems, like enzymes, require more system-specific efforts to determine a suitable caging method. Even irreversible processes can be considered, as improvements in the X-ray flux at the existing third-generation sources and the proposed next-generation, substantially more intense X-ray sources, such as the X-ray free-electron laser (XFEL) *(88,89)*, will minimize the need for signal averaging and therefore the number of crystals required. The increased X-ray flux will also allow the use of smaller crystals, down to μm and possibly sub-μm size. This will help to lower the diffusion barrier and enable diffusion triggering for reactions faster than presently possible given the crystal sizes used today.

Other important frontiers for the ultra-fast TRX involve further improvements in time resolution, the read-out time of the X-ray area detectors, methods for the determination of structures of intermediates and reaction mechanism, and in combining the TR technique with computational approaches. A considerable effort is given to the development of ultra-fast hard X-ray sources, such as the hard X-ray FEL mentioned earlier. In this device, extremely bright, fully coherent X-ray pulses of ~100 fs duration are to be generated, with peak brilliance many orders of magnitude higher than presently available at the third generation synchrotrons. This will provide an opportunity to explore important biological processes that involve events that occur on the fs time scale, such as photosynthesis. The development of large area pixel-array detectors with readout times in the sub-μs time scale ***(90)*** will greatly reduce the need for repeated pump-probe cycles. It will permit following a reaction in real-time by synchronizing a single reaction initiation with multiple cycles of X-ray exposure and detector readout.

The application of the SVD method to extract structures of time-independent intermediates from measured time-dependent data and the development of the *posterior* analysis for identification of possible reaction mechanism, clearly represent significant advances in the analysis of TR data. Further improvements of these methods as well as exploring new avenues in the TR data analysis ***(91)*** are the major goals in TRX. As discussed in the **Subheading 1.**, the ultimate goal of TRX is to provide structures of intermediate states that correspond to the energy minima on the potential energy surface of the system. However, to describe a reaction pathway completely, the characteristics of unstable transition states between the intermediates have to be known, as well as these states actually determine the reaction rates. Combining the experimental results from TRX with computational and theoretical approaches will greatly facilitate achievement of this goal. The high-resolution structures of intermediates from TRX provide solid constraints for methods such as free-energy perturbation simulations ***(92)*** and nudged elastic band calculations ***(93)*** in an effort to completely characterize the reaction pathways.

4. Notes

1. Time-independent DED maps of intermediates are determined in the following way. Preliminary concentrations of each intermediate I_j are calculated from a candidate mechanism by solving the system of coupled differential equations. They are then used to fit the significant rSV globally. In this nonlinear fitting

process the k are varied to match the magnitude of the observed vector elements $v(t)_n^{obs}$:

$$s_n^2 v(t)_n^{obs} \approx v(t)_n^{fit} = s_n^2 \cdot \sum_{j=1}^{J} E_{nj} \cdot I_j(k,t) \tag{4}$$

The fit is weighted by the square of the corresponding singular value s_n^2. The E_{nj} are linear fit parameters that have to be computed for each intermediate j and each singular vector n. The E_{nj} bring the concentrations to the (unknown) scale of the rSV.

Once the scale factors E_{nj} are determined, the time-independent DED for the j_{th} intermediate $\Delta\rho_{Ij}$ is computed using the significant singular values s_n, significant left singular vectors $\boldsymbol{u}_n$ and the scale factors E_{nj} :

$$\Delta\rho_{Ij} = \sum_{j=1}^{S} \boldsymbol{u}_n \cdot s_n \cdot E_{n,j} \tag{5}$$

2. Extrapolated, conventional electron density maps for intermediates are determined in the following way.
   The extracted time-independent DED map $\Delta\rho_{Ij}$ is Fourier transformed and difference SF $\boldsymbol{\Delta F}_j$ are obtained. A multiple f of the difference SF vector is added to the calculated SF vector of the dark state $\boldsymbol{F}_D^{calc}$:

$$\boldsymbol{F}_j^{ext} = \boldsymbol{F}_D^{calc} + f\boldsymbol{\Delta F}_j \tag{6}$$

   The extrapolated map for the j_{th} intermediate is then calculated from the extrapolated structure factors $\boldsymbol{F}_j^{ext}$. The factor f is adjusted, so that the negative density features initially observed on prominent atoms of the dark structure in the extrapolated map just vanish. The atomic model can be built into this extrapolated map and refined by conventional methods against the $/\, \boldsymbol{F}_j^{ext}\, /$. The residual $\boldsymbol{F}_j^{ext} - \boldsymbol{F}_j^{calc}$ difference map (the $\boldsymbol{F}_j^{calc}$ are SF determined from the refined model of the intermediate j), indicates if some electron density features are still not interpreted. This may be the case, if a wrong mechanism is used or if states, which are populated at earlier times where no data are available, are mixed in.
3. Posterior analysis is performed in the following way. SF $\boldsymbol{F}_{Ij}$ are calculated from the structures of the initial dark state and intermediates. Difference SF $\boldsymbol{\Delta F}_{Ij}$ are determined by subtracting the dark state structure factors $\boldsymbol{F}^D$ from those of the intermediates ($\boldsymbol{\Delta F}_{Ij} = \boldsymbol{F}_{Ij} - \boldsymbol{F}^D$). The $\boldsymbol{\Delta F}_{Ij}$ are used to calculate time independent difference maps, $\Delta\rho_{Ij}^{calc}$, for each intermediate. Time-dependent DED maps are then calculated on the absolute scale using the rate k dependent concentrations I_j of the intermediates from **Note 1**:

$$\Delta\rho(k,t)^{calc} = \sum_{j=1}^{J} I_j(k,t) \cdot \Delta\rho_{Ij}^{calc} \tag{7}$$

   The $\Delta\rho(k,t)^{calc}$ must be represented on the same three-dimensional grid as the observed, SVD-flattened DED maps $\Delta\rho(t)^{SVD}$. Hence, the observed DED values

$\Delta\rho_m(t)^{SVD}$ can be compared to the calculated $\Delta\rho_m(k,t)^{calc}$ at every grid point m. **Equation 8** is used as a kernel of a fit routine to refine the rate coefficients k for the mechanism. The inner loop sums over all M grid points in a particular map, whereas the outer loop sums over all T available maps.

$$\sum_{t=1}^{T}\sum_{m=1}^{M}\frac{1}{\left\langle\left|\Delta\rho(t)\right|^{SVD}\right\rangle}\left(\Delta\rho_m(t)^{SVD}-C_{RI}\cdot\Delta\rho_m(k,t)^{calc}\right)^2 \qquad (8)$$

C_{RI} is a linear fit parameter, which scales the calculated DED maps to the observed DED maps. C_{RI} is equivalent to the concentration of activated molecules in the reaction and therefore represents the extent of reaction initiation. The fit is weighted by the average of the observed absolute difference electron density $<|\Delta\rho(t)^{SVD}|>$. This ensures that all DED maps, with weak or strong signal, are considered on an equal footing. To reduce the effect of noise on the fit, only the grid points below or above some σ value should be used. A cutoff of $\pm 2\sigma$ is a reasonable choice.

Acknowledgments

MS was supported by the Deutsche Forschungsgemeinschaft, SFB 533 and HI was supported by the Korea Research Foundation Grant (KRF-2004-003-C00100). RP and VŠ were supported by the NIH/NCRR grant RR07707.

References

1. Moffat, K. and Henderson, R. (1995) Freeze trapping of reaction intermediates. *Curr. Opin. Struct. Biol.* **5,** 656–663.
2. Schlichting, I. and Chu, K. (2000) Trapping intermediates in the crystal: ligand binding to myoglobin. *Curr. Opin. Struct. Biol.* **10,** 744–752.
3. Stoddard, B. L. (2001) Trapping reaction intermediates in macromolecular crystals for structural Analysis. *Methods in Enzymology* **24,** 125–138.
4. Ursby, T., Weik, M., Fioravanti, E., Delarue, M., Goeldner, M., and Bourgeois, D. (2002) Cryophotolysis of caged compounds: a technique for trapping intermediate states in protein crystals. *Acta Cryst.* **D58,** 607–614.
5. Hellingwerf, K., Hendriks, J., and Gensch, T. (2003) On the configurational and conformational changes in photoactive yellow protein that leads to signal generation in *Ectothiorhodospira halophila. J Biol. Phys.* **28,** 295–412.
6. Brudler, R., Rammelsberg, R., Woo, T. T., Getzoff, E. D., and Gerwert, K. (2001) Structure of the I_1 early intermediate of photoactive yellow protein by FTIR spectroscopy. *Nature Struct. Biol.* **8,** 265–270.
7. Ujj, L., Devanathan, S., Meyer, T. E., Cusanovich, M. A., Tollin, G., and Atkinson, G. H. (1998) New photocycle intermediates in the photoactive yellow protein from *Ecothiorhodospira halophila*: picosecond transient absorption spectroscopy. *Biophys. J.* **75,** 406–412.
8. Haupts, U., Tittor, J., and Oesterhelt, D. (1999) Closing in on bacteriorhodopsin: progress in understanding the molecule. *Annu. Rev. Biophys. Biomol. Struct.* **283,** 67–99.

9. Mizutani, Y. and Kitagawa, T. (2001) Ultrafast structural relaxation of myoglobin following photodissociation of carbon monoxide probed by time-resolved resonance Raman cpectroscopy. *J. Phys. Chem. B* **105,** 10,992–10,999.
10. Jackson, T. A., Lim, M., and Anfinrud, P. (1994) Complex nonexponential relaxation in myoglobin after photodissociation of MbCO: measurement and analysis from 2 ps to 56 µs. *Chem. Phys* **180,** 131–140.
11. Ansari, A., Jones, C. M., Henry, E. R., Hofrichter, J., and Eaton, W. A. (1994) Conformational relaxation and ligand binding in myoglobin. *Biochemistry* **33,** 5128–5145.
12. Xie, X. and Simon, J. D. (1991) Protein conformational relaxation following photodissociation of co from carbonmonoxymyoglobin–picosecond circular-dichroism and absorption studies. *Biochemistry* **30,** 3682–3692.
13. Schotte, F., Lim, M., Jackson, T. A., Smirnov, A. V., Soman, J., Olson, J. S., Phillips, G. N. J., Wulff, M., and Anfinrud, P. (2003) Watching a protein as it functions with 150ps time-resolved X-ray crystallography. *Science* **300,** 1944–1947.
14. Krinsky, S., Fundamentals of Hard X-ray Synchrotron Radiation Sources, in *Third-Generation Hard X-ray Synchrotron Radiation Sources*, Mills, D., Ed., John Wiley & Sons, Inc., New York (2002).
15. Ren, Z., Bourgeois, D., Helliwell, J. R., Moffat, K., Šrajer, V., and Stoddard, B. L. (1999) Laue crystallography: coming of age. *J. Synchrotron Rad.* **6,** 891–917.
16. Šrajer, V., Ren, Z., Teng, T.-Y., Schmidt, M., Ursby, T., Bourgeois, D., Pradervand, C., Schildkamp, W., Wulff, M., and Moffat, K. (2001) Protein conformational relaxation and ligand migration in myoglobin: a nanosecond to millisecond molecular movie from time-resolved Laue X-ray diffraction. *Biochemistry* **40,** 13,802–13,815.
17. Bourgeois, D., Vallone, B., Schotte, F., Arcovito, A., Miele, A. E., Sciara, G., Wulff, M., Anfinrud, P., and Brunori, M. (2003) Complex landscape of protein structural dynamics unveiled by nanosecond Laue crystallography. *Proc. Natl. Acad. Sci.* **100,** 8704–8709.
18. Šrajer, V., Teng, T.-Y., Ursby, T., Pradervand, C., Ren, Z., Adachi, S., Schildkamp, W., Bourgeois, D., Wulff, M., and Moffat, K. (1996) Photolysis of the carbon monoxide complex of myoglobin: nanosecond time-resolved crystallography. *Science* **274,** 1726–1729.
19. Neutze, R., Pebay-Peyroula, E., Edman, K., Royant, A., Navarro, J., and Landau, E. M. (2002) Bacteriorhodopsin: a high resolution structural view of vectorial proton transport. *Biochim. Biophys. Acta* **1565**, 144–167.
20. Ren, Z., Perman, B., Šrajer, V., Teng, T.-Y., Pradervand, C., Bourgeois, D., Schotte, F., Ursby, T., Kort, R., Wulff, M., and Moffat, K. (2001) A molecular movie at 1.8 A resolution displays the photocycle of photoactive yellow protein, a eubacterial blue-light receptor, from nanoseconds to seconds. *Biochemistry* **40,** 13,788–13,801.
21. Genick, U. K., Borgstahl, G. E., Ng, K., Ren, Z., Pradervand, C., Burke, P. M., Šrajer, V., Teng, T.-Y., Schildkamp, W., McRee, D. E., Moffat, K., and Getzoff, E. D. (1997) Structure of a protein photocycle intermediate by millisecond time-resolved crystallography. *Science* **275,** 1471–1475.

22. Perman, B., Šrajer, V., Ren, Z., Teng, T-.Y., Pradervand, C., Ursby, T., Bourgeois, D., Schotte, F., Wulff, M., Kort, R., Hellingwerf, K., and Moffat, K. (1998) Energy transduction on the nanosecond time scale: early structural events in a xanthopsin photocycle. *Science* **279,** 1946–1950.
23. Crosson, S. and Moffat, K. (2002) Photoexcited structure of a plant photoreceptor domain reveals a light-driven molecular switch. *Plant Cell* **14,** 1067–1075.
24. Karplus, M. (1999) In *Simplicity and Complexity in Proteins and Nucleic Acids,* (Frauenfelder, H., Deisenhofer, J., and Wolynes, P., eds.). Dahlem University Press, Belin, p. 139.
25. Moffat, K. (2001) Time-resolved biochemical crystallography: a mechanistic perspective. *Chem. Rev.* **101,** 1569–1581.
25a. Steinfeld, J. I., Francisco, J. S., and Haase, W. L. (1989) *Chemical Kinetics and Dynamics.* Prentice Hall, Englewood Cliffs, NJ.
26. Moffat, K. (1989) Time-Resolved Macromolecular Crystallography. *Annu. Rev. Biophys. Biophys. Chem.* **18,** 309–332.
27. Schmidt, M., Rajagopal, S., Ren, Z., and Moffat, K. (2003) Application of singular value decomposition to the analysis of time-resolved macromolecular x-ray data. *Biophys. J.* **84,** 2112–2129.
28. Schmidt, M., Pahl, R., Šrajer, V., Anderson, S., Ren, Z., Ihee, H., and Moffat, K. (2004) Protein kinetics: structures of intermediates and reaction mechanism from time-resolved x-ray data. *Proc. Natl. Acad. Sci.* **101,** 4799–4804.
29. Rajagopal, S., Schmidt, M., Anderson, S., Ihee, H., and Moffat, K. (2004) Analysis of experimental time-resolved crystallographic data by singular value decomposition. *Acta Cryst. D* **60,** 860–871.
30. Abdel-Meguid, S. S., Jeruzalmi, D., and Sanderson, M. R. (1996) Preliminary characterization of crystals. In *Crystallographic methods and Protocols,* Vol. 56 (Jones, C., Mulloy, B., and Sanderson, M. R., eds.). Humana Press, Totowa, NJ, pp. 55–86.
31. Carrell, H. L. and Glusker, J. P. (2001) *Crystallography of Biological Macromolecules,* Kluwer Academic Publishers, Dordrecht, The Netherlands.
32. Ren, Z. and Moffat, K. (1995) Quantitative analysis of synchrotron laue diffraction patterns in macromolecular crystallography. *J. Appl. Cryst.* **10,** 461–481.
33. Ren, Z. and Moffat, K. (1995) Deconvolution of energy overlaps in laue diffraction. *J. Appl. Cryst.* **28,** 482–493.
34. Campbell, J. W. (1995) LAUEGEN, an X-windows-based program for the processing of Laue diffraction data. *J. Appl. Cryst.* **28,** 228–236.
35. Arzt, S., Campbell, J. W., Harding, M. M., Hao, Q., and Helliwell, J. R. (1999) LSCALE - the new normalization, scaling and absorption correction program in the Daresbury Laue software suite. *J. Appl. Cryst.* **32,** 554–562.
36. Wakatsuki, S. (1993) LEAP, Laue evaluation analysis package, for time-resolved protein crystallography, in *CCP4 Study Weekend Proceedings: Data Collection and Processing,* Sawyer, L., Isaacs, N. W., and Bailey, S., eds., CLRC Daresbury, Warrington, UK, pp. 71–79.
37. Bourgeois, D., Nurizzo, D., Kahn, R., and Cambillau, C. (1998) An integration routine based on profile fitting with optimized fitting area for the evaluation of

weak and/or overlapped two-dimensional Laue or monochromatic patterns. *J. Appl. Cryst.* **31,** 22–35.

38. Ren, Z. (2003) Precognition, commercial package; www.renzresearch.com.
39. Scott, W. G., Murray, J. B., Arnold, J. R. P., Stoddard, B. L., and Klug, A. (1996) Capturing the structure of a catalytic RNA intermediate: the hammerhead ribozyme. *Science* **274,** 2065–2069.
40. Singer, P. T., Smalas, A., Carty, R. P., Mangel, W. F., and Sweet, R. M. (1993) The hydrolytic water molecule in trypsin revealed by time-resolved Laue crystallography. *Science* **259,** 669–673.
41. Fulop, V., Phizackerley, R. P., Soltis, S. M., Clifton, I. J., Wakatsuki, S., Erman, J., Hajdu, J., and Edwards, S. L. (1994) Laue diffraction study on the structure of cytochrome c peroxidase compound I. *Structure* **2,** 201–208.
42. Bolduc, J. M., Dyer, D. H., Scott, W. G., Singer, P., Sweet, R. M., Koshland, D. E. J., and Stoddard, B. L. (1995) Mutagenesis and Laue structures of enzyme intermediates: isocitrate dehydrogenase. *Science* **268,** 1312–1318.
43. Gouet, P., Jouve, H. M., Williams, P. A., Andersson, I., Andreoletti, P., Nussaume, L., and Hajdu, J. (1996) Ferryl intermediates of catalase captured by time-resolved Weisenberg crystallography and UV-VIS spectroscopy. *Nature Struct. Biol.* **3,** 951–956.
44. Helliwell, J. R., Nieh, Y. P., Raftery, J., Cassetta, A., Habash, J., Carr, P. D., Ursby, T., Wulff, M., Thompson, A. W., C., N. A., and Hadener, A. (1998) Time-resolved structures of hydroxymethylbilane synthase (Lys59Gln mutant) as it is loaded with substrate determined by Laue diffraction. *J. Chem. Soc. Faraday Trans.* **94,** 2615–2622.
45. Schlichting, I. and Goody, R. S. (1997) Triggering methods in crystallographic enzyme kinetics. *Methods in Enzymology* **277,** 467–490.
46. Chen, Y. (1994) PhD. Thesis. Cornell University, Ithaca, NY.
47. Hori, T., Moriyama, H., Kawaguchi, J., Hayashi-Iwasaki, Y., Oshima, T., and Tanaka, N. (2000) The initial step of the thermal unfolding of 3-isopropylmalate dehydrogenase detected by the temperature-jump Laue method. *Protein Engineering* **13,** 527–533.
48. McCray, J. A. and Trentham, D. R. (1989) Properties and uses of photoreactive caged compounds. *Annu. Rev. Biophys. Biophys. Chem.* **18,** 239–270.
49. Corrie, J. E. T., and Trenhtam, D. R. (1993) *Biological Applications of Photochemical Switches*, Vol. 2, Wiley, New York, NY.
50. Schlichting, I., Almo, S. C., Rupp, G., Wilson, K., Petratos, K., Lentfer, A., Wittinghofer, A., Kabash, W., Pai, E. F., Petsko, G. A., and Goody, R. S. (1990) Time-resolved X-ray crystallographic study of the conformational change in Ha-ras p21 protein on GTP hydrolysis. *Nature* **345,** 309–315.
51. Stoddard, B. L., Koenigs, P., Porter, N., Petratos, K., Petsko, G. A., and Ringe, D. (1991) Observation of the light-triggered binding of pyrone to chymotrypsin by Laue X-ray crystallography. *Proc. Natl. Acad. Sci.* **88,** 5503–5507.
52. Duke, E. M. H., Wakatsuki, W., Hadfield, A., and Johnson, L. N. (1994) Laue and monochromatic diffraction studies on catalysis in phosphorylase b crystals. *Protein Sci.* **3,** 1178–1196.

53. Stoddard, B. L., Cohen, B. E., Brubaker, M., Mesecar, A. D., and Koshland, D. E. J. (1998) Millisecond Laue structures of an enzyme-product complex using photocaged substrate analogs. *Nature Struct. Biol.* **5,** 891–897.
54. Chen, Y., Šrajer, V., Ng, K., LeGrand, A., and Moffat, K. (1994) Optical monitoring of protein crystals in time-resolved x-ray experiments— microspectrophotometer design and performance. *Rev. Sci. Instrum.* **65,** 1506–1511.
55. Hadfield, A. and Hajdu, J. (1993) A fast and portable microspectrophotometer for protein crystallography. *J. Appl. Cryst.* **26,** 839–842.
56. Sakai, K., Matsui, Y., Kouyama, T., Shiro, Y., and Adachi, S.-I. (2002) Optical monitoring of freeze trapped reaction intermediates in protein crystals: A microspectrophotometer for cryogenic protein crystallography. *J. Appl. Cryst.* **35,** 270–273.
57. Bourgeois, D., Vernede, X., Adam, V., Fioravanti, E., and Ursby, T. (2002) A microspectrophotometer for UV-visible absorption and fluorescence studies of protein crystals. *J. Appl. Cryst.* **35,** 319–326.
58. Anderson, S., Šrajer, V., Pahl, R., Rajagopal, S., Schotte, F., Anfinrud, P., Wulff, M., and Moffat, K. (2004) Chromophore conformation and the evolution of tertiary structural changes in photoactive yellow protein. *Structure* **12,** 1039–1045.
59. Bourgeois, D., Ursby, T., Wulff, M., Pradervand, C., LeGrand, A., Schildkamp, W., Laboure, S., Šrajer, V., Teng, T.-Y., Roth, M., and Moffat, K. (1996) Feasibility and realization of single-pulse laue diffraction on macromolecular crystals at ESRF. *J. Synchrotron Rad.* **3,** 65–74.
60. Schotte, F., Techert, S., Anfinrud, P., Šrajer, V., Moffat, K., and Wulff, M. (2002) Picosecond structural studies using pulsed synchrotron radiation. In *Third-Generation Hard X-Ray Synchrotron Radiation Sources.* (Mills, D., ed.). John Wiley & Sons, Inc., New York, NY, pp. 345–401.
61. Knapp, J. E., Šrajer, V., Pahl, R., and Royer, W. E. J. (2004) Immobilization of *Scapharca* HbI crystals improves data quality of a time-resolved crystallographic experiment. *Micron.* **35,** 107–108.
62. Šrajer, V., Crosson, S., Schmidt, M., Key, J., Schotte, F., Anderson, S., Perman, B., Ren, Z., Teng, T.-Y., Bourgeois, D., Wulff, M., and Moffat, K. (2000) Extraction of accurate structure factor amplitudes from Laue data: wavelength normalization with wiggler and undulator X-ray sources. *J. Synchrotron Rad.* **7,** 236–244.
63. Bourgeois, D., Wagner, U., and Wulff, M. (2000) Towards automated Laue data processing: application to the choice of optimal X-ray spectrum. *Acta Cryst.* **D56,** 973–985.
64. Helliwell, J. R. (1992) *Macromolecular Crystallography With Synchrotron Radiation.* Cambridge University Press, Cambridge, UK.
65. Drenth, J. (1999) *Principles of Protein X-Ray Crystallography*, Springer, New York, NY.
66. Henderson, R., and Moffat, J. K. (1971) The difference Fourier technique in protein crystallography: errors and their treatment. *Acta Cryst. B* **27,** 1414–1420.
67. Ursby, T., and Bourgeois, D. (1997) Improved estimation of structure-factor difference amplitudes from poorly accurate data. *Acta Cryst. A* **53,** 564–575.

68. McRee, D. E. (1999) *Practical Protein Crystallography*, Academic Press, San Diego, CA.
69. Jones, T. A., Bergdoll, M., and Kjeldgaard, M., O. (1990) A macromolecular modeling environment. In *Crystallographic and Modeling Methods in Molecular Design.* (Bugg, C. and Ealick, S., eds.). Springer-Verlag Press, New York, NY, pp. 189–195.
70. Henry, E. R. and Hofrichter, J. (1992) Singular value decomposition: Application to analysis of experimental data. *Methods in Enzymology* **210,** 129–192.
71. Carson, M. (1997) Ribbons. In *Methods in Enzymology*, Vol. 277. (Sweet, R. M., and Carter, C. W., eds.) Academic Press, pp. 493–505.
72. Hajdu, J., Neutze, R., Sjogren, T., Edman, K., Szoke, H., Wilmouth, R. C., and Wilmot, C. M. (2000) Analyzing protein functions in four dimensions. *Nature Struct. Biol.* **7,** 1006–1012.
73. Moffat, K. (1998) Time-Resolved Crystallography. *Acta Cryst. A* **54,** 833–841.
74. Schlichting, I., Berendzen, J., Chu, K., Stock, A. M., Maves, S. A., Benson, D. E., Sweet, R. M., Ringe, D., Petsko, G. A., and Sligar, S. G. (2000) The catalytic pathway of cytochrome $P450_{cam}$ at atomic resolution. *Science* **287,** 1615–1622.
75. Berglund, G. I., Carlsson, G. H., Smith, A. T., Szoke, H., Henriksen, A., and Hajdu, J. (2002) The catalysis pathway of horseradish peroxidase at high resolution. *Nature* **417,** 463–468.
76. Schlichting, I., Berendzen, J., Phillips, G. N., and Sweet, R. M. (1994) Crystal structure of photolyzed carbonmonoxy-myoglobin. *Nature* **371,** 808–812.
77. Teng, T. Y., Šrajer, V., and Moffat, K. (1994) Photolysis-induced structural changes in single crystals of carbonmonoxymyoglobin at 40K. *Nature Struct. Biol.* **1,** 701–705.
78. Teng, T. Y., Šrajer, V., and Moffat, K. (1997) Initial trajectory of carbon monoxide after photodissociation from myoglobin at cryogenic temperatures. *Biochemistry* **36,** 12,087–12,100.
79. Chu, K., Vojtechovsky, J., McMahon, B. H., Sweet, R. M., Berendzen, J., and Schlichting, I. (2000) Structure of a ligand-binding intermediate in wild-type carbonmonoxy myoglobin. *Nature* **403,** 921–923.
80. Ostermann, A., Waschipky, R., Parak, F. G., and Nienhaus, G. U. (2000) Ligand binding and conformational motions in myoglobin. *Nature* **404,** 205–208.
81. Genick, U. K., Soltis, S. M., Kuhn, P., Canestrelli, I. L., and Getzoff, E. D. (1998) Structure at 0.85 angstrom resolution of an early protein photocycle intermediate. *Nature* **392,** 206–209.
82. Kendrew, J. C., Dickerson, R. E., Strandberg, B. E., Hart, R. G., Davies, D. R., Phillps, D. C. and Shore, V. C. (1960) Structure of myoglobin. A three-dimensional Fourier synthesis at 2 A resolution. *Nature* **185**.
83. Ansari, A., Berendzen, J., Bowne, S. F., Frauenfelder, H., T., I. I. E., Sauke, T. B., Shyamsunder, E., and Young, R. D. (1985) Protein states and protein quakes. *Proc. Natl. Acad. Sci.* **82,** 5000–5004.
84. Kachalova, G. S., Popov, A. N., and Bartunik, H. D. (1999) A steric mechanism for inhibition of CO binding to heme proteins. *Science* **284,** 473–476.

85. Meyer, T. E. (1985) Isolation and characterization of soluble cytochromes, ferredoxins and other chromophoric proteins from the halophile phototrophic bacterium Ectothiorhodospira halophila. *Biochim. Biophys. Acta.* **806,** 175–183.
86. Sprenger, W. W., Hoff, W. D., Armitage, J. P., and Hellingwerf, K. J. (1993) The eubacterium Ectothiorhodospira halophila is negatively phototactic, with a wavelength dependence that fits the absorption spectrum of the photoactive yellow protein. *J. Bacteriol.* **175,** 3096–3104.
87. Cusanovich, M. A. and Meyer, T. E. (2003) Photactive yellow protein: a prototypic PAS domain sensory protein and development of a common signaling mechanism. *Biochemistry* **42,** 4759–4770.
88. Winick, H. (1995) The linac coherent light source (LCLS): A fourth generation light source using the SLAC linac. *J. Elec. Spec. Rel. Phenom.* **75,** 1–8.
89. Wiik, B. H. (1997) The TESLA project: an accelerator facility for basic science. *Nucl. Inst. Meth. Phys. Res. B* **398,** 1–17.
90. Rossi, G., Renzi, M., Eikenberry, E. F., Tate, M. W., Bilderback, D. H., Fontes, E., Wixted, R., Barna, S., and Gruner, S. M. (1999) Tests of a prototype pixel array detector for microsecond time-resolved X-ray diffraction. *J. Synchrotron Rad.* **6,** 1096–1105.
91. Rajagopal, S., Kostov, K., and Moffat, K. (2004) Analytical trapping: Extraction of time-independent structures from time-dependent crystallographic data. *J. Struct. Biol.*, in press.
92. Schweins, T., Langen, R., and Warshel, A. (1994) Why mutagenesis studies not located the general base in Ras P21. *Nature Struct. Biol.* **1,** 476–484.
93. Henkelman, G., Johannesson, G., and Jonsson, B. (2000) Methods for finding saddle points and minimum energy paths. In *Progress on Theoretical Chemistry and Physics.* (Schwartz, S. D., ed.). Kluwer Academic Publishers, Dordrecht, The Netherlands, p. 269.

8

X-Ray Crystallography of Protein–Ligand Interactions

Ilme Schlichting

Summary

Crystal structures of protein–ligand complexes provide a detailed view of their spatial arrangement and interactions. In the case of stable, unreactive ligands, such as inhibitors or allosteric regulators, the complexes can be generated by cocrystallization or by soaking the ligand into fully grown crystals. In order to obtain highly occupied stochiometric complexes, the concentration and amount of ligand used needs to be considered. Protein complexes with reactive short-lived species that occur in chemical or binding reactions can be determined using monochromatic X-ray diffraction techniques via kinetic trapping approaches. To this end, the kinetics of the reaction has to be determined in the crystalline state and triggering methods to start the reaction need to be established. To facilitate data interpretation, the experimental conditions are usually chosen such that the peak concentration of the reactive species under investigation is maximized.

Key Words: Cocrystallization; concentration; trapping; soaking; diffusion; triggering; cryocrystallography; reaction intermediate; crystallography; monochromatic; kinetic crystallography; diffraction.

1. Introduction

For an in-depth understanding of a macromolecule's interaction with a ligand it is not only necessary to know the energetics and kinetics of their relation, but also their precise arrangement in space. A complete view of the latter can be provided by NMR spectroscopy (Chapter 10) or X-ray crystallography. In this chapter applications of conventional monochromatic X-ray diffraction techniques will be described focusing on sample preparation. This is particularly important for the study of reactive species such as substrate complexes. *Ligand* is defined here as a molecule considerably smaller than the macromolecule under investigation, such as the classic small molecules or peptides. As a

From: *Methods in Molecular Biology, vol. 305: Protein–Ligand Interactions: Methods and Applications*
Edited by: G. U. Nienhaus © Humana Press Inc., Totowa, NJ

result of experimental considerations, ligands are divided here into two categories, namely those that are 1) chemically stable such as inhibitors, substrate analogues, products, and allosteric ligands, or 2) unstable such as substrates or reaction intermediates which will be referred to as *reactive ligand complexes* in the following. Although proteins, and specifically mostly enzymes will be treated explicitly, the considerations are general and thus also applicable to nucleic acids.

2. Structure Determination of Stable Ligand Complexes

Crystals of macromolecules are characterized by a high content of solvent (typically 30 to 80%), which is arranged in large, interstitial channels spanning the crystals. The molecules are arranged in the crystal lattice by relatively few weak interactions that allow for some mobility and often for catalytic activity ***(1,2)***. This allows one to study reactions in crystals ***(3–10)*** as described later and in Chapter 7 and to generate meaningful ligand complexes *in situ* by diffusing a ligand in a crystal of a somewhat plastic and thus functional macromolecule.

Thus, from a practical point of view, the first choice to be made when studying protein–ligand complexes is whether to cocrystallize the complex or to diffuse (*soak*) the compound into a fully-grown crystal. In either case, care has to be taken that sufficient amounts of the ligand are present to ensure the formation of a fully occupied stoichiometric complex. Important parameters are the concentration of the ligand, and in the case of soaking experiments, the volume of the solution. As a rule of thumb, to ensure very high occupancy at the binding site of a protein or enzme **E**, the concentration of the ligand **L** should be about ten times higher than the dissociation constant K_d. This is rooted in the law of mass action $K_d = [E] \cdot [L]/[EL]$ or $[L] = K_d \cdot \{[EL]/[E]\}$.

Obviously, affinities may change under crystallization conditions which can differ significantly from the standard biochemical assay conditions, e.g., in ionic strength or pH. Thus, affinities should be determined under near crystallization conditions ***(11)*** (provided that no protein aggregation occurs, which can be verified by, e.g., dynamic light scattering).

Once an estimate is obtained for the protein–ligand affinity under crystallization conditions, the protein–ligand complex can be formed in solution and set up for crystallization. Usually, one screens around the conditions where the native protein crystallizes, possibly using seeding to help with nucleation. If the underlying assumption is not correct that ligand binding does not induce larger structural changes or conformational changes of residues involved in crystal contacts, crystals may not form. In that case, new crystallization conditions have to be identified from scratch by testing widely varying conditions.

In order to generate stoichiometric crystalline protein–ligand complexes by soaking native crystals in ligand containing mother liquor, one needs to know not only their affinity but also the number of protein molecules in the crystal. These can be calculated from the number of molecules **N** in the asymmetric unit of the crystal lattice. Its Laue group (symmetry) determines the number n of asymmetric units per unit cell. Thus, a crystal with the macroscopic volume **V** [in μm^3] und unit cell volume **v** [in $Å^3$] contains $(V/v) \cdot n \cdot N \cdot 10^6$ molecules, their molar concentration [Mol/liter] in the crystal is given by $n \cdot N \cdot 623 / v$. This allows one to obtain an estimate of the volume of the ligand containing mother liquor required for soaking the crystal, which is a useful number to know in case of expensive ligands.

Soaking can be done most easily by placing the solution in sitting drop bridges or glass depression slides before adding the crystal with a loop or a capillary. Because crystals may crack upon exposure to new solutions it is advisable to test this with crystals of lesser quality. Variables that one should consider checking are the time-frame of the addition of the new mother liquor (step-wise vs immediate complete change) or the composition of the mother liquor (particularly the concentration of the precipitant which could be increased by a couple of percent).

If the protein–ligand affinity is not very high (lower than micromolar) the ligand should be added with the appropriate concentration to all solutions to which the crystal is exposed, particularly the cryoprotectant. Otherwise, one risks lower occupancy of the ligand as a result of loss by diffusion. Diffraction data are usually collected at cryogenic temperatures to reduce the effects of radiation damage *(12)*. In case of unexpected observations that are not in line with the biochemical data one should keep in mind a possible influence of the cryoprotectant and/or cryogenic temperatures on ligand affinities, or of photoelectrons produced via X-ray radiolysis during data collection.

3. Structure Determination of Unstable Ligand Complexes

A very large fraction of the functionally important protein–ligand complexes, such as the ones with substrates or reaction intermediates, are short-lived with typical life-times ranging from ns to s. Therefore, the approaches described above cannot be applied directly.

The first thing to consider is whether one will actually be able *to see* the reactive complex of interest *(13)*. Obviously, its absolute occupancy depends on the occupancy of the starting complex. Therefore, the same considerations in respect to the required concentration and amount of ligand as described for stable ligand complexes have to be applied. The relative occupancy of an intermediate state compared to the initial state is determined by the kinetics of the system. As with any direct method, intermediates can only be detected crystallographically if they are sufficiently occupied (detection limit typically 30%). This translates into the necessary but not sufficient requirement that the apparent rate constant for generation of the intermediate is higher than that of its disappearance. Thus, the characteristics of the system, especially the kinetics of the reaction to be studied, govern the experimental strategy. However, occupancy is not the only issue for deciding almost beforehand whether one can actually detect the intermediate–resolution of the diffraction data is equally critical. For instance, many reactions depend on protonation or ionization states, but in most cases (near) atomic resolution data are required to decide these issues.

Next, one has to make sure that the reaction does take place in the crystalline state ***(14)***. As mentioned previously, crystallization conditions (i.e., ionic strength, pH, solvents, etc.) or steric effects (such as blocked active sites or conformational restraints caused by crystal contacts) may affect the kinetics of the system. This may be in terms of binding constants, rates (often slowed), the equilibrium distribution of catalytic intermediates, or even the catalytic mechanism itself. The effect of crystallization conditions (e.g., ionic strength, pH) on affinity and apparent k_{cat} can be analyzed by studying the kinetics in solution under near crystallization conditions (provided that no aggregation occurs, which can be verified by, e.g., dynamic light scattering). In this case, diffusion and steric hindrance are not an issue. Their effect can be studied by using a microcrystalline slurry, the dependence of the apparent k_{cat} on crystal size indicating the limiting effect of diffusion. Thus, a detailed knowledge of the kinetics is not only required for the reaction as it takes place in solution, but also in the crystal. Obviously, in the ideal case the crystal lattice should not affect the reaction, but the inverse is also true because meaningful measurements cannot be carried out otherwise.

Activity monitoring within crystals is done most conveniently using a microspectrophotometer (www.4dx.se) ***(15)***. The advantage is that it is a noninvasive method and can be done *in situ* while actually collecting the data. This minimizes unpleasant surprises such as low occupancy of the species in question as a result of incomplete reaction initiation (e.g., partial photolysis), or insufficient turnover when collecting the data. In addition, heating effects or photoreduction via X-ray radiolysis can be detected while the experiment is being performed, and countermeasures can be taken.

In general, reactive ligand complexes cannot be obtained simply by cocrystallization due to the lengthy crystallization process (> days) product would have formed long before crystal formation even starts. It is therefore necessary to crystallize stable, biochemically inert complexes that can be activated immediately before data collection, a process often termed *triggering* ***(16)***. This has to be done rapidly, gently (i.e., without affecting the activity of the protein or quality of the crystal lattice), with high yield, and uniformly.

Concentration jumps of substrates, cofactors, protons, etc., to start reactions can be achieved by diffusion, an experimentally straightforward approach. Because of the intrinsic generation of gradients and the competing effects of diffusion and catalysis, reaction initiation by diffusion is suitable only for very slow processes. Typical diffusion times across 200 μm thick crystals are seconds to minutes depending on the size of the compound, the solvent channels, and the viscosity of the mother liquor ***(17,18)***. Changes in pH—if tolerated by the crystal lattice—cannot only be used to chemically trap intermediate states ***(19)*** but also to initiate a reaction whose time course will be followed by time-resolved crystallography ***(20)*** using a flow cell ***(21)***. Depending on the solvent (i.e., water v, e.g., 70 % methanol) this set-up can be used both at ambient and at cryogenic temperatures ***(22–24)***. Reaction initiation by diffusion of substrate was used in time-resolved studies on cytochrome c peroxidase ***(25)*** and catalase ***(26,27)***, and hydroxymehtylbilane synthase.

Because triggering methods in general have been reviewed extensively elsewhere ***(16)*** and photo-initiation in particular in Chapter 7, they will not be addressed here in more detail.

Once one has identified conditions of generating and analyzing reactive ligand complexes in the crystal one needs to think about how to structurally capture them, i.e., acquire the diffraction data. In principle, one has to balance the inherent time-scales of the reaction itself, the triggering of the reaction, and data collection. The latter two have to be significantly faster, otherwise one cannot generate or temporarily resolve the reactive species under investigation.

Suppose one wants to characterize the intermediate *B* in the reaction $A \xrightarrow{k_1} B \xrightarrow{k_2} C$, which is assumed to be nonreversible for simplicity. There are three special cases to consider for the ratio of the two rate coefficients k_1 and k_2: $k_1 \ll k_2$, $k_1 = k_2$, and $k_1 \gg k_2$. In the first case, the intermediate B will not accumulate and thus be nonobservable, because the detection limit is usually around 30% occupancy in crystallographic experiments. In the second case, the intermediate will accumulate to some 37% after time $1/k_1$, and all three states will be occupied during the reaction. In the third case, the intermediate B will build up to high occupancy. Thus, in general, the ratio of the rate coefficients will be somewhere between the $k_1 = k_2$, and $k_1 \gg k_2$ limits. There are three principal (nonexclusive) possibilities for conducting the experiment. The first is to collect the data very rapidly on a time-scale faster than the one given by the reaction under *standard conditions*, an approach named time-resolved crystallography (in its strict definition, sometimes also called *true* time-resolved crystallography to distinguish it from the other two, (*see* Chapter 7). The second approach is to alter the ratio of the rate coefficients k_1 and k_2 so that $k_1 \gg k_2$. Usually, this is done by decreasing k_2 (*see* below). This is referred to as *trapping* because the reaction is effectively halted in B. A third possibility arises if the rate coefficient k_3 which characterizes the reaction $C \xrightarrow{k_3} A$ is significantly higher than k_2: $k_1, k_3 \gg k_2$. In this case, the reaction can be studied under multiturnover conditions (steady-state) because the intermediate will accumulate. One is back to a (quasi)stable situation, provided there is enough substrate available. This can be provided by mounting the crystal in a *flow cell* ***(21)***, a capillary filled with pipe cleaner fibers or sephadex to prevent crystal slippage while substrate containing mother liquor is flowed over the crystal. This is made possible by attaching tubing to either end of the capillary, one serving as a supply line the other as drain. Diffraction data are usually collected at ambient temperature, but cryogenic temperatures are accessible when using a high percentage of organic solvent ***(28)*** in the mother liquor (e.g., 70% methanol in the case of ribonuclease A) ***(29)*** to prevent freezing. An experimentally easier approach for data collection of a steady-state accumulated reactive species is to trap the complex by cryocooling ***(30)***. In order to keep the steady-state condition, the compounds required for generating the complex (e.g., substrate, metal ions, etc.) need to be added to the cryoprotectant solution.

In the case of reactive complexes that can only be studied under single turnover conditions because the rate limiting step of the reaction is not the one through which the species under investigation decays (e.g., substrate binding or product release), one is left with two options, either to reduce data collection times to significantly shorter time scales than the ones inherent to the reaction (*see* Chapter 7) or to prolong the life-time of the reactive species. The latter approach is often called kinetic crystallography or trapping. Trapping can be done chemically (e.g., by modification of the macromolecule, for example by, mutation, the substrate, cofactor, solvent, pH) ***(31–33)*** or physically (commonly by temperature) ***(34–37)***. Because crystals are usually flash-cooled anyway to slow radiation damage during data collection ***(12)***, trapping intermediates by freeze quenching a reaction has become very popular ***(38–47)***. In addition to slowing kinetics, temperature can be used to separate different reaction steps if the respective rate coefficients have different temperature dependencies (e.g., different activation enthalpies). One way of exploiting this is to either initiate a reaction by electromagnetic radiation or pressure in a cryo-cooled crystal or to trap a reaction by freeze-quenching and to let it proceed at low temperature to a *temperature-limited* intermediate ***(22,48–54)***. Often intermediates accumulate because a barrier of the reaction cannot be overcome at temperatures below the glass transition temperature (around 180–200 K), at which nonharmonic collective motions are *frozen out*. This trapping approach has been reviewed recently (*see* **refs. *55–57***).

Generally, the cryoprotectant needed for preventing ice formation in crystals upon freeze trapping may change the reaction characteristics as may temperature-induced changes in pH, dielectric constant, proton activity. etc. Moreover, the equilibrium distribution of the protein structure or spin distribution in heme proteins may change with temperature, and may also depend critically on cooling rates. Changes with temperature observed specifically are slightly differing orientations of α-helices in the carbonmonoxy complex of myogobin, slightly changed positions of catalytic water molecules in Ras ***(58)***, and somewhat different kinetics in the case of the photoactive yellow protein ***(59,60)***. Thus, one has to make sure that the trap does not affect the results of the experiment.

The structure determination of reaction intermediates requires intimate knowledge of the biochemistry and kinetics of the system under investigation. This means that the experimental approach to be taken is intimately related to the reaction itself, and depends critically on its biochemical and kinetic boundary conditions. Therefore, there are no general *cooking recipes* for conducting trapping experiments, and the same applies for the structure determination of unstable species by time-resolved crystallography (*see* Chapter 7). Although the experiments are usually not straightforward, the structures provide unique insights at atomic resolution about the system during biochemical function. Moreover, they provide experimental boundary conditions for theoretical studies and molecular dynamics simulations. Maximizing the peak concentration of an intermediate and stabilizing it long enough for data collection by physical or chemical trapping usually gives better data than the fast-data collection approach, in particular the refinement of the structures is much more straightforward. However, one has to make sure that the reaction (mechanism) is not affected by the trap.

4. Materials

1. Crystals.
2. Mother liquor, cryoprotectant.
3. Ligand (in appropriate concentration and amount).
4. Capillaries and/or loops for manipulating crystals and for data collection.
5. For flowcells: quartz capillaries, diameter >0.5 mm, pipe cleaner fibers, 5 min epoxy, PE50 tubing.
6. Microbridges or glass depression slides for soaking experiments.

References

1. Hajdu, J., Acharya, K. R., Stuart, D. I., Barford, D., and Johnson, L. N. (1988) Catalysis in enzyme crystals. *Trends Biochem. Sci.* **13,** 104–109.
2. Mozzarelli, A. and Rossi, G. L. (1996) Protein function in the crystal. *Annu. Rev. Biophys. Biomol. Struct.* **25,** 343–365.
3. Schlichting, I. (2000) Crystallographic structure determination of unstable species. *Acc. Chem Res.* **33,** 532–538.

4. Schlichting, I. and Chu, K. (2000) Trapping intermediates in the crystal: ligand binding to myoglobin. *Curr. Opin. Struct. Biol.* **10,** 744–752.
5. Schmidt, K. and Henderson, R. (1995) Freeze trapping of reaction intermediates. *Curr. Opin. Struct. Biol.* **5,** 656–663.
6. Schmidt, K. (2001) Time-resolved biochemical crystallography: a mechanistic perspective. *Chem. Rev.* **101,** 1569–1581.
7. Schmidt, K. (1989) Time-resolved macromolecular crystallography. *Annu. Rev. Biophys. Biophys. Chem.* **18,** 309–332.
8. Petsko, G. A. and Ringe, D. (2000) Observation of unstable species in enzyme-catalyzed transformations using protein crystallography. *Curr. Opin. Chem Biol.* **4,** 89–94.
9. Hajdu, J., Neutze, R., Sjogren, T., Edman, K., Szoke, A., Wilmouth, R. C., and Wilmot, C. M. (2000) Analyzing protein functions in four dimensions. *Nat. Struct. Biol.* **7,** 1006–1012.
10. Hajdu, J. and Andersson, I. (1993) Fast crystallography and time-resolved structures. *Annu. Rev. Biophys. Biomol. Struct.* **22,** 467–498.
11. Stoddard, B. L. and Farber, G. K. (1995) Direct measurement of reactivity in the protein crystal by steady-state kinetic studies. *Structure* **3,** 991–996.
12. Garman, E. F. and Schneider, T. R. (1997) Macromolecular Cryocrystallography. *J. Appl. Cryst.* **30,** 211–237.
13. Schlichting, I. and Chu, K. (2000) Trapping intermediates in the crystal: ligand binding to myoglobin. *Curr. Opin. Struct. Biol.* **10,** 744–752.
14. Stoddard, B. L. and Farber, G. K. (1995) Direct measurement of reactivity in the protein crystal by steady-state kinetic studies. *Structure* **3,** 991–996.
15. Hadfield, A. and Hajdu, J. (1994) On the photochemical release of phosphate from 3,5–dinitrophenyl phosphate in a protein crystal. *J. Mol. Biol.* **236,** 995–1000.
16. Schlichting, I. and Goody, R. (1997) Triggering methods in kinetic crystallography. *Methods in Enzymology* **277,** 467–490.
17. Ohara, P., Goodwin, P., and Stoddard, B. L. (1995) Direct measurement of diffusion rates in enzyme crystals by video absorbance spectroscopy. *J. Appl. Cryst.* **28,** 829–834.
18. Stoddard, B. L. and Farber, G. K. (1995) Direct measurement of reactivity in the protein crystal by steady-state kinetic studies. *Structure* **3,** 991–996.
19. Verschueren, K. H., Seljee, F., Rozeboom, H. J., Kalk, K. H., and Dijkstra, B. W. (1993) Crystallographic analysis of the catalytic mechanism of haloalkane dehalogenase. *Nature* 3**63,** 693–698.
20. Singer, P. T., Smalas, A., Carty, R. P., Mangel, W. F., and Sweet, R. M. (1993) The hydrolytic water molecule in trypsin, revealed by time-resolved Laue crystallography. *Science* **259,** 669–673.
21. Petsko, G. A. (1985) Diffraction methods for biological macromolecules. Flow cell construction and use. *Methods Enzymol.* **114,** 141–146.
22. Douzou, P. and Petsko, G. A. (1984) Proteins at work: "stop-action" pictures at subzero temperatures. *Adv. Protein Chem* **36,** 245–361.

23. Douzou, P. (1980) Cryoenzymology in aqueous media. *Adv. Enzymol. Relat Areas Mol. Biol.* **51,** 1–74.
24. Douzou, P. (1983) Cryoenzymology. *Cryobiology* **20,** 625–635.
25. Fulop, V., Phizackerley, R. P., Soltis, S. M., Clifton, I. J., Wakatsuki, S., Erman, J., Hajdu, J., and Edwards, S. L. (1994) Laue diffraction study on the structure of cytochrome c peroxidase compound I. *Structure* **2,** 201–208.
26. Gouet, P., Jouve, H. M., Williams, P. A., Andersson, I., Andreoletti, P., Nussaume, L., and Hajdu, J. (1996) Ferryl intermediates of catalase captured by time-resolved Weissenberg crystallography and UV-VIS spectroscopy. *Nat. Struct. Biol.* **3,** 951–956.
27. Jouve, H. M., Andreoletti, P., Gouet, P., Hajdu, J., and Gagnon, J. (1997) Structural analysis of compound I in hemoproteins: study on Proteus mirabilis catalase. *Biochimie* **79,** 667–671.
28. Douzou, P. and Balny, C. (1977) Cryoenzymology in mixed solvents without cosolvent effects on enzyme specific activity. *Proc. Natl. Acad. Sci. USA* **74,** 2297–2300.
29. Rasmussen, B. F., Stock, A. M., Ringe, D., and Petsko, G. A. (1992) Crystalline ribonuclease A loses function below the dynamical transition at 220 K. *Nature* **357,** 423–424.
30. Weyand, M. and Schlichting, I. (1999) Crystal structure of wild-type tryptophan synthase complexed with the natural substrate indole-3-glycerol phosphate. *Biochemistry* **38,** 16,469–16,480.
31. Stoddard, B. L. (1996) Caught in a chemical trap. *Nat. Struct. Biol.* **3,** 907–909.
32. Stoddard, B. L. (2001) Trapping reaction intermediates in macromolecular crystals for structural analyses. *Methods* **24,** 125–138.
33. Stoddard, B. L. (1996) Intermediate trapping and laue X-ray diffraction: potential for enzyme mechanism, dynamics, and inhibitor screening. *Pharmacol. Ther.* **70,** 215–256.
34. Douzou, P. (1979) The study of enzyme mechanisms by a combination of cosolvent, low-temperature and high-pressure techniques. *Q. Rev. Biophys.* **12,** 78.
35. Schmidt, K. and Henderson, R. (1995) Freeze trapping of reaction intermediates. *Curr. Opin. Struct. Biol.* **5,** 656–663.
36. Schmidt, K. (1995) X-ray crystallography at extremely low temperatures. *Biotechnology (N. Y.)* **13,** 133.
37. Schlichting, I. and Chu, K. (2000) Trapping intermediates in the crystal: ligand binding to myoglobin. *Curr. Opin. Struct. Biol.* **10,** 744–752.
38. Burzlaff, N. I., Rutledge, P. J., Clifton, I. J., Hensgens, C. M., Pickford, M., Adlington, R. M., Roach, P. L., and Baldwin, J. E. (1999) The reaction cycle of isopenicillin N synthase observed by X-ray diffraction. *Nature* **401,** 721–724.
39. Wilmot, C. M., Hajdu, J., McPherson, M. J., Knowles, P. F., and Phillips, S. E. (1999) Visualization of dioxygen bound to copper during enzyme catalysis. *Science* **286,** 1724–1728.
40. Murray, J. B., Szoke, H., Szoke, A., and Scott, W. G. (2000) Capture and visualization of a catalytic RNA enzyme-product complex using crystal lattice trapping and X-ray holographic reconstruction. *Mol. Cell* **5,** 279–287.

41. Luecke, H., Schobert, B., Richter, H. T., Cartailler, J. P., and Lanyi, J. K. (1999) Structural changes in bacteriorhodopsin during ion transport at 2 angstrom resolution. *Science* **286,** 255–261.
42. Edman, K., Nollert, P., Royant, A., Belrhali, H., Pebay-Peyroula, E., Hajdu, J., Neutze, R., and Landau, E. M. (1999) High-resolution X-ray structure of an early intermediate in the bacteriorhodopsin photocycle. *Nature* **401,** 822–826.
43. Royant, A., Edman, K., Ursby, T., Pebay-Peyroula, E., Landau, E. M., and Neutze, R. (2000) Helix deformation is coupled to vectorial proton transport in the photocycle of bacteriorhodopsin. *Nature* **406,** 645–648.
44. Sass, H. J., Buldt, G., Gessenich, R., Hehn, D., Neff, D., Schlesinger, R., Berendzen, J., and Ormos, P. (2000) Structural alterations for proton translocation in the M state of wild-type bacteriorhodopsin. *Nature* **406,** 649–653.
45. Schlichting, I., Berendzen, J., Chu, K., Stock, A. M., Maves, S. A., Benson, D. E., Sweet, R. M., Ringe, D., Petsko, G. A., and Sligar, S. G. (2000) The catalytic pathway of cytochrome p450cam at atomic resolution. *Science* **287,** 1615–1622.
46. Chu, K., Vojtchovsky, J., McMahon, B. H., Sweet, R. M., Berendzen, J., and Schlichting, I. (2000) Structure of a ligand-binding intermediate in wild-type carbonmonoxy myoglobin. *Nature* **403,** 921–923.
47. Genick, U. K., Borgstahl, G. E., Ng, K., Ren, Z., Pradervand, C., Burke, P. M., Srajer, V., Teng, T. Y., Schildkamp, W., McRee, D. E., Schmidt, K., and Getzoff, E. D. (1997) Structure of a protein photocycle intermediate by millisecond time-resolved crystallography. *Science* **275,** 1471–1475.
48. Vitkup, D., Ringe, D., Petsko, G. A., and Karplus, M. (2000) Solvent mobility and the protein 'glass' transition. *Nat. Struct. Biol.* **7,** 34–38.
49. Specht, A., Ursby, T., Weik, M., Peng, L., Kroon, J., Bourgeois, D., and Goeldner, M. (2001) Cryophotolysis of ortho-nitrobenzyl derivatives of enzyme ligands for the potential kinetic crystallography of macromolecules. *Chembiochem.* **2,** 845–848.
50. Ursby, T., Weik, M., Fioravanti, E., Delarue, M., Goeldner, M., and Bourgeois, D. (2002) Cryophotolysis of caged compounds: a technique for trapping intermediate states in protein crystals. *Acta Crystallogr. D. Biol. Crystallogr.* **58,** 607–614.
51. Edman, K., Royant, A., Nollert, P., Maxwell, C. A., Pebay-Peyroula, E., Navarro, J., Neutze, R., and Landau, E. M. (2002) Early structural rearrangements in the photocycle of an integral membrane sensory receptor. *Structure (Camb.)* **10,** 473–482.
52. Royant, A., Edman, K., Ursby, T., Pebay-Peyroula, E., Landau, E. M., and Neutze, R. (2000) Helix deformation is coupled to vectorial proton transport in the photocycle of bacteriorhodopsin. *Nature* **406,** 645–648.
53. Chu, K., Vojtchovsky, J., McMahon, B. H., Sweet, R. M., Berendzen, J., and Schlichting, I. (2000) Structure of a ligand-binding intermediate in wild-type carbonmonoxy myoglobin. *Nature* **403,** 921–923.
54. Ostermann, A., Waschipky, R., Parak, F. G., and Nienhaus, G. U. (2000) Ligand binding and conformational motions in myoglobin. *Nature* **404,** 205–208.

55. Petsko, G. A. and Ringe, D. (2000) Observation of unstable species in enzyme-catalyzed transformations using protein crystallography. *Curr. Opin. Chem Biol.* **4,** 89–94.
56. Schlichting, I. (2000) Crystallographic structure determination of unstable species. *Acc. Chem Res.* **33,** 532–538.
57. Ringe, D. and Petsko, G. A. (2003) The 'glass transition' in protein dynamics: what it is, why it occurs, and how to exploit it. *Biophys. Chem* **105,** 667–680.
58. Scheidig, A. J., Burmester, C., and Goody, R. S. (1999) The pre-hydrolysis state of p21(ras) in complex with GTP: new insights into the role of water molecules in the GTP hydrolysis reaction of ras-like proteins. *Structure Fold. Des* **7,** 1311–1324.
59. Perman, B., Srajer, V., Ren, Z., Teng, T., Pradervand, C., Ursby, T., Bourgeois, D., Schotte, F., Wulff, M., Kort, R., Hellingwerf, K., and Schmidt, K. (1998) Energy transduction on the nanosecond time scale: early structural events in a xanthopsin photocycle. *Science* **279,** 1946–1950.
60. Kort, R., Ravelli, R. B., Schotte, F., Bourgeois, D., Crielaard, W., Hellingwerf, K. J., and Wulff, M. (2003) Characterization of photocycle intermediates in crystalline photoactive yellow protein. *Photochem. Photobiol.* **78,** 131–137.

9

Combined Use of XAFS and Crystallography for Studying Protein–Ligand Interactions in Metalloproteins

Richard W. Strange and S. Samar Hasnain

Summary

This chapter describes the method of X-ray absorption spectroscopy when applied to the study of metal sites in proteins. The method requires the intense X-rays found only at synchrotron radiation sources, and is equally applicable to metalloproteins in dilute solutions, in fibers, films, and in crystalline states. In each case, structural changes occurring at metal sites during catalysis or ligand-binding are revealed with an accuracy and precision equivalent to atomic resolution crystallography. When combined with crystallographic data, of any resolution, X-ray absorption spectroscopy can yield atomic resolution three-dimensional structural models of the metal sites, thus providing the level of structural detail necessary for understanding the chemical mechanisms involved in the active states of metalloproteins.

Key Words: X-ray absorption spectroscopy; XAFS; XANES; metalloproteins; crystallography; 3D-EXAFS.

1. Introduction

Metalloproteins belong to a class of proteins that use the chemical properties of metals to perform a wide range of biological processes essential for sustaining life. They make up at least 30% of all proteins in currently known genomes. The in vivo concentrations and distributions of metals in the cellular environment are highly regulated and controlled, functions that are often performed by specific metal-chaperone proteins. The biological importance of metalloproteins and their chaperones is highlighted by the disease-causing consequences of genetic mutation that may lead to catalytic misfunction or incorrect metallation. For example, these factors are both relevant to understanding the role that Cu_2Zn_2 superoxide dismutase plays in motor neuron

From: *Methods in Molecular Biology, vol. 305: Protein–Ligand Interactions: Methods and Applications*
Edited by: G. U. Nienhaus © Humana Press Inc., Totowa, NJ

disease. X-ray protein crystallography and high-field nuclear magnetic resonance (NMR) are the main techniques for providing three-dimensional structures of proteins and their complexes, often giving unique insights into their workings. For metalloproteins, a key functional feature is the role, catalytic or structural, performed by the metal. X-ray absorption fine structure (XAFS) spectroscopy is an ideal technique for obtaining accurate and precise structural data, comparable to that of small molecule crystallography, that are specific to the metal site and its local environment. These data are complementary to those provided by protein crystallography, which are normally measured at less than atomic resolution.

This chapter outlines how the techniques of X-ray absorption fine structure XAFS spectroscopy and protein crystallography (PX) can work together to provide detailed three-dimensional structural information about protein–ligand interactions in metalloproteins. Because Chapters 7 and 8 in this volume deal comprehensively with PX methods *per se*, the main focus of this chapter is on XAFS. The basic methods used in XAFS, both experimentally and analytically, follow the same course whether dealing with general applications of XAFS or specific questions related to protein–ligand interactions. We give a brief outline of the general principles of the XAFS method, both theory and practice, and then present a more detailed and, we hope, practicable guide to performing experiments and analyzing the XAFS data. Finally, we show how to combine XAFS with PX to yield highly accurate and precise metrical information of metal-site–ligand complexes in proteins. Several examples involving ligand binding to metalloproteins will be used to illustrate the main points.

The information available from XAFS data on a metalloprotein includes the oxidation state of the metal, bonding and orbital occupancies, the metal–ligand distances, the number and type of coordinated ligands, their static or thermal disorder with respect to the metal, and, when combined with PX data, a detailed and highly accurate three-dimensional model of the metal site. The basic physics underlying XAFS involves initial absorption of X-rays by core (1s or K, 2p or L; *see* **Note 1**) atomic electrons via the photoelectric effect. At sufficient X-ray energies the excited electron is ejected from the absorbing atom, any excess energy being taken up by the electron as kinetic energy. The scattering of this X-ray generated photoelectron by the neighboring atoms of the metal gives rise to the XAFS data (**Fig. 1**). Because of the short mean free path of the photoelectron, information is limited to the environment local to the metal site (approx 6 Å radius). As a result, the structure of the metal site can be probed with an accuracy often approaching that routinely achieved in *small molecule* crystallography. XAFS data can be collected on aqueous and crystalline protein samples with the same resolution, so there is no necessity for growing single crystals. In a typical application, XAFS is used to accu-

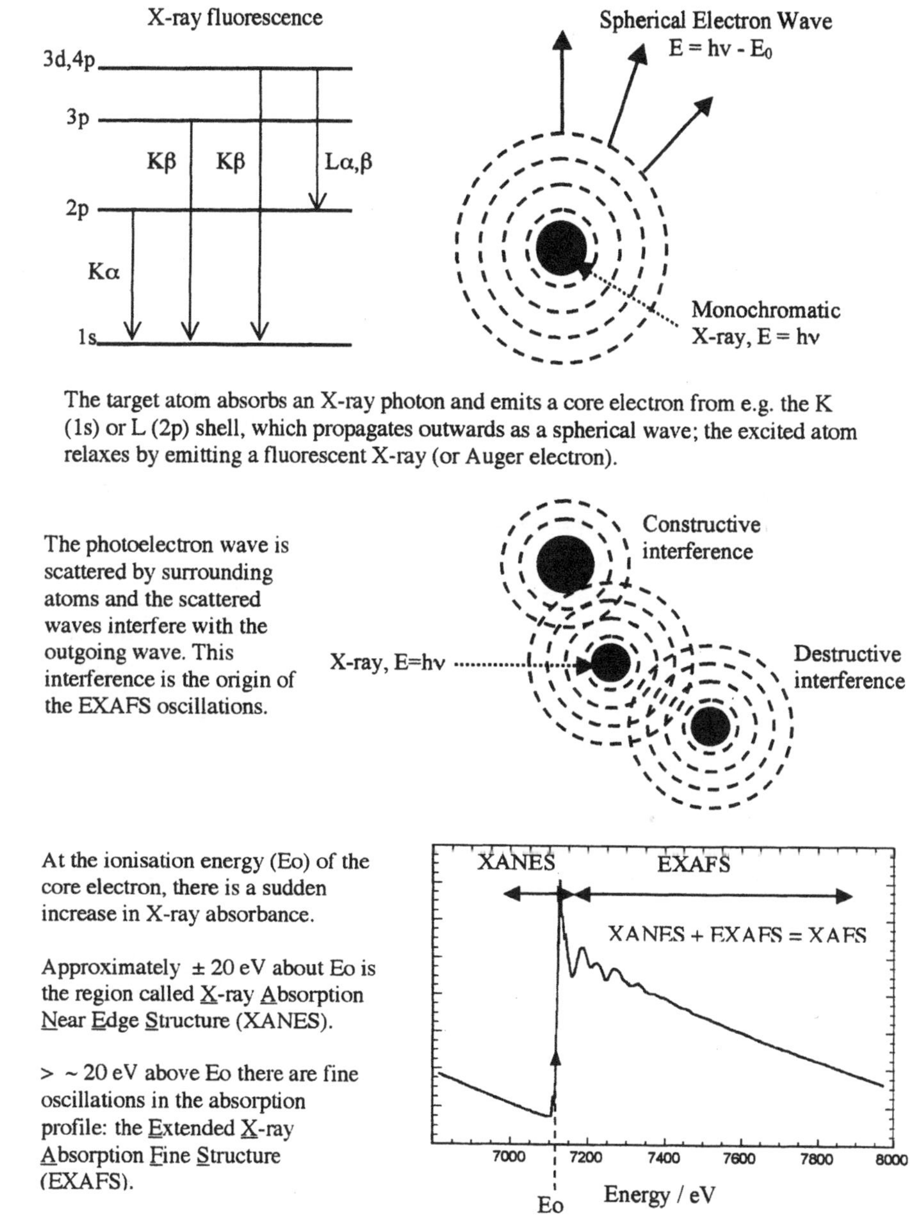

Fig. 1. Heuristic view of the XAFS phenomenon, involving absorption of X-rays by a core electron, emission of the excited photoelectron, and relaxation of the excited atom by emission of fluorescent X-rays. The photoelectron, treated as a wave, is scattered by neighboring atoms giving rise to interference effects that are the extended EXAFS. Formally, the absorption of X-rays is treated using quantum theory and the interference effects are seen as modulations of the absorption cross-section of the target atom.

rately define the metal site of a native *resting* protein, and is then applied to study changes in the local stereochemistry produced by a biochemical reaction such as that involved in ligand binding. These changes are often quite subtle (approx 0.1 Å) and well within the error limits of any but atomic resolution crystallographic structure determinations (for further discussion of these points *see* **Subheading 2.4.**).

Almost all atoms in the periodic table are accessible but for biological systems the natural target atoms are principally the first row transition metals (V, Mn, Fe, Co, Ni, Cu, Zn), plus a handful of other metals including P, S, K, Ca, Se, Cd, W, and Mo.[1] These metals cover a wide range of X-ray absorption edge energies; for example, an experiment that involves changing from Fe to Mo K-edges (e.g., XAFS of the nitrogenase MoFe protein) requires a change in X-ray energy of 13,000 eV (**Table 1**). This is one of the main reasons why XAFS, unlike PX, can only be performed at synchrotron radiation[2] laboratories, where a tuneable source of monochromatic X-rays is available. Additional advantages of using synchrotron radiation (SR) are the high intensity (more than 10^5 times that of a rotating anode), high resolution ($\Delta\lambda/\lambda$ approx 10^{-4}) and extremely good inherent collimation of the X-ray beam. These parameters are shaped and optimized by a combination of optical devices (e.g., collimating mirrors, monochromators, and focusing mirrors) placed in the path of the *white light* emitted by the synchrotron. Under some circumstances, the intrinsic polarization of the SR X-rays can also be an advantage, for example in measuring XAFS of protein single crystals. A detailed description of SR physics and of the optical elements making up a XAFS optimized *beamline* and *experimental station* is beyond the scope of this chapter. The interested reader is referred to **refs. *1*** and ***2***.

The basic ingredients needed to understand XAFS may be understood by considering the high energy (plane wave) approximation (*see* **Note 2**) for the oscillatory extended X-ray absorption fine structure (EXAFS) function $\chi(k)$, which may be written for the K-edge:

$$\chi(k) = \sum_j -\frac{N_j}{kR_j^{\,2}} \left| f_j(k,\pi) \right| \sin(2kR_j + 2\delta_1 + \psi_j) e^{-2R_j/\lambda} e^{-2\sigma_j^2 k^2} \tag{1}$$

Equation 1 shows that $\chi(k)$ is dependent on the number of scattering atoms N_j, the distance of the scattering atoms R_j from the absorber, and on the type of

[1] Other elements of biological importance, like Na and Mg are not readily studied using XAFS because of their low absorption edge energies

[2] Synchroton radiation is emitted by charged particles accelerated to relativistic speeds. Because the radiated power S approx (energy/particle-mass)^4, only electron or positron beams provide useful radiation.

Table 1
X-Ray Absorption K- and L_{III}-Edge Energies and Wavelengths for Metals of Biological Interest, Including Exogenous Metals

Z	Element	K-edge energy eV	(Å)
15	P	2142	(5.787)
16	S	2470	(5.018)
20	Ca	4038	(3.070)
23	V	5464	(2.269)
25	Mn	6539	(1.896)
26	Fe	7113	(1.743)
28	Ni	8332	(1.488)
29	Cu	8984	(1.380)
30	Zn	9663	(1.283)
34	Se	12654	(0.978)
42	Mo	19996	(0.620)
48	Cd	26720	(0.464)
55	Cs	35936	(0.345)
56	Ba	37119	(0.334)

Z	Element	L_{III}-edge energy eV	(Å)
74	W	10204	(2.215)
78	Pt	11586	(1.07)
79	Au	11921	(1.04)
80	Hg	12275	(1.01)
82	Pb	13050	(0.95)

Most of the naturally occurring target metals found in metalloproteins are from the first row transition series.

scattering atom through the characteristic energy dependence of its backscattering amplitude $|f_j(k,\pi)|$ and phaseshift ψ_j. The effect of the electronic potential as a result of the absorbing atom is measured by the phaseshift $2\delta_i$. The mean square variation in the interatomic distance between the absorbing atom and a scattering atom is related by the Debye–Waller factor, σ^2_j, which assumes a harmonic distribution. λ is the elastic mean-free path of the photoelectron. The damping term $\exp^{-2R_j/\lambda}$ and the $1/R^2$ dependence of **Eq. 1** limit the backscattering contribution to approx 6 Å from the absorbing atom. The X-ray energy is given by $k = \sqrt{2m_e(E - E_0)/\hbar^2}$, where m_e is the electron mass, E is the photoelectron energy, E_0 is the electron binding energy, and $\hbar$ is Planck's constant. Because k is in units of Å^{-1} the Fourier transform of $\chi(k)$ gives the radial distribution, in Å, of the scattering atoms from the absorbing atom. The effects of several of the factors in **Eq. 1** on EXAFS data are illustrated by the simple model calculations in **Fig. 2**. More rigorous formulations of XAFS theory *(3–7)*, taking into account electron-wave curvature and complications resulting from photoelectron multiple-scattering events, yield more accurate results than the high-energy approximation, and are more likely to be used in practical data analysis. **Equation 1**—or its more accurate curved wave version—is used to extract R_j, N_j, and $2\sigma_j^2$ given that accurate scattering amplitude and phase functions are known.

In concrete terms, the average XAFS experimenter will be looking for an answer to a specific question, such as the effect on the active site of a metalloprotein during an oxidation-reduction cycle, or the details of a metal-substrate interaction (*see* **Note 3**). **Figure 3** shows an example where the XAFS data for the active site of oxidized and reduced bovine Cu_2Zn_2 superoxide dismutase (SOD) unequivocally demonstrates a change from 5-coordinate to

Fig. 2. (*opposite page*) Theoretical spectra, using Cu as the absorbing atom, illustrating some of the consequences of the single scattering plane wave approximation summarized by Eq. 1. EXAFS data are shown in the left-hand panels with corresponding Fourier transforms shown on the right. **(A)** The data with 4N or 4O backscattering atoms are nearly identical and in practice cannot be distinguished. This is because the photoelectron scattering is dependent upon the number and distribution of electrons belonging to the scattering atom and these are different by only one electron for N and O. In fact, because the *core electrons* dominate the scattering process, EXAFS is generally able only to discriminate between atoms within $Z \pm 2$ in the periodic table (this depends to some extent on the experimental data range, the wider the better). **(B)** The effect of the Debye–Waller (DW or σ^2) factor on the EXAFS for a shell of 4N atoms, showing the fall-off of the EXAFS signal with the increase of DW. **(C)** A comparison between N and S backscattering atoms is shown. The Fourier transform peaks are identical because the Fe – N and Fe – S distances are equal and the coordination numbers and DW factors have been adjusted to match the peak height. However, the EXAFS data are out of phase by π. In practice, the choice of N or S depends on which one of these curves fits the experimental data. **(D)** A typical example of a *first shell* made up of more than one type of backscattering atom contributing. In this case, a first shell similar to that found in the type 1 copper proteins, with 2N atoms at 2 Å and 1S atom at 2.2 Å, is shown by the solid line to give a single Fourier transform peak. This is obtained by summing the contributions from the individual components, shown by the dashed lines.

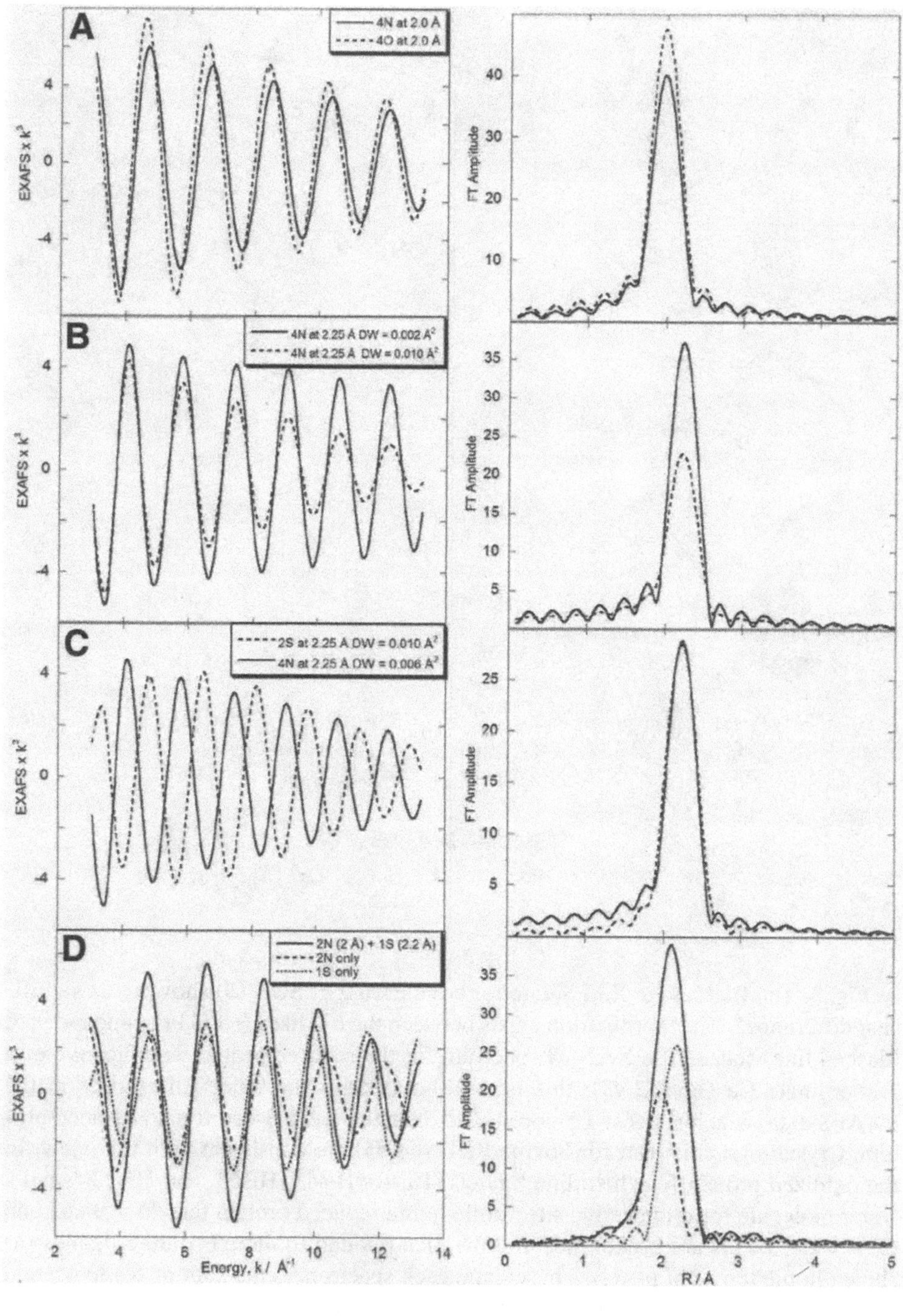
A
4N at 2.0 Å
4O at 2.0 Å
B
4N at 2.25 Å DW = 0.002 Å²
4N at 2.25 Å DW = 0.010 Å²
C
2S at 2.25 Å DW = 0.010 Å²
4N at 2.25 Å DW = 0.006 Å²
D
2N (2 Å) + 1S (2.2 Å)
2N only
1S only
EXAFS x k³
FT Amplitude
Energy, k / Å⁻¹
R / Å

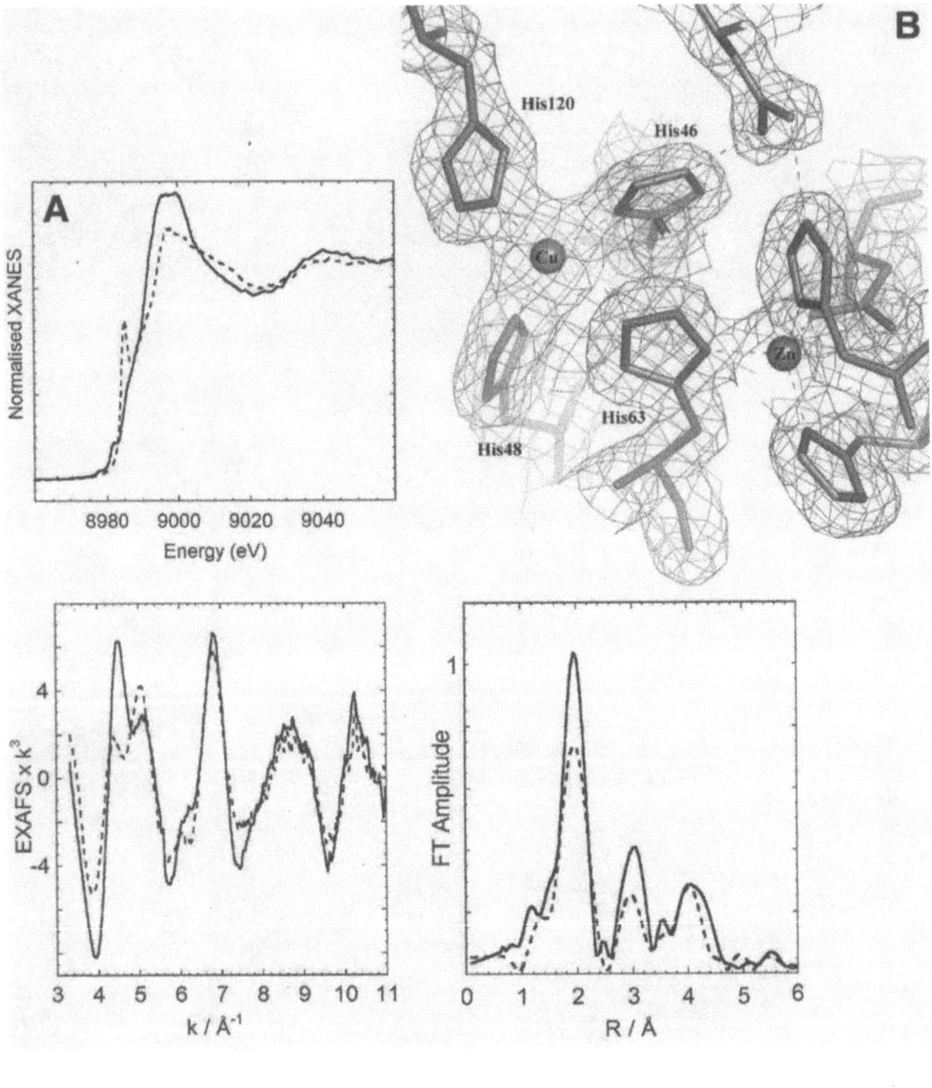

Fig. 3. The Cu K-edge XAFS data for bovine Cu_2Zn_2 SOD *(8)* show that a significant difference in Cu coordination exists between the oxidized (solid line) and reduced (dashed line) states. The XANES spectrum for the reduced protein is indicative of a 3-coordinate Cu site *(42,43)*; this is corroborated by amplitude differences in the EXAFS data, which suggest a 5-coordinate oxidized and a 3-coordinate reduced protein. Crystallographic data for bovine SOD *(44,45)* are consistent with this view. In the oxidized protein four histidine ligands (His46, His48, His63, and His120) and a water molecule form the active site, while in the reduced protein the Cu – His63 and Cu – water bonds are broken, leaving the metal bound to three histidine ligands (as shown in the top right panel). These data, each spectrum is the sum of seven 45-min scans, were collected using EXAFS station 9.2 at Daresbury Laboratory, UK.

3-coordinate Cu *(8)*. This *answer* is revealed with hardly any need for detailed data analysis, but to reach it involves several steps from protein preparation to structural interpretation, via a trip to a synchrotron XAFS beamline for many hours of data collection. These steps and how to perform them are discussed in detail in the following section.

2. Methods

This section gives an account of the requirements and procedures to 1) perform a XAFS experiment, 2) extract normalized XANES and EXAFS spectra from the raw experimental data, 3) analyze and interpret these spectra to obtain meaningful structural information, and 4) combine PX and XAFS data to extract the maximum amount of biologically relevant information, including an *atomic resolution* three-dimensional model of the metal site. Several examples involving protein–ligand interactions will be used to illustrate these tasks. The limitations as well as the advantages of the XAFS method will also be discussed.

2.1. Experimental Aspects

The sample. One of the most important considerations is the concentration of the sample, or more specifically its metal concentration. The typical protein sample will be dilute in metal (e.g., there are two Mo atoms per functional 240 kD dimer in MoFe nitrogenase) and to obtain a usable EXAFS signal (minimum signal-to-noise [S/N] = 2) over a significant energy range (typically approx 10 13 Å^{-1}) several hours of data collection time are required, even using the most intense SR X-ray source and the best fluorescence detector. In practice, a number of *scans* (n) of a sample of 15–60 min duration are recorded and summed together to improve the S/N ratio, which is proportional to $\sqrt{n}$. In terms of the time taken to perform the experiments and the quality of the data obtained, it is current practice to use a metal concentration in the 0.5–5 m*M* range, although various properties of the sample itself may limit the final concentration (e.g., solubility, precipitation effects).

Whereas it is desirable to be able to perform experiments with *physiological* concentrations of protein, in most cases these concentrations are too low for EXAFS measurements to be made at all, or to be done within a reasonable time-scale at existing synchrotron beamlines (e.g., within 1–2 d per sample).[3] For example, the cellular concentration of human superoxide dismutase (SOD) amounts to approx 20 μ*M* in Cu, whereas a practical concentration limit for

[3] XANES data can be more easily obtained using samples at physiological concentrations because the signal close to the absorption edge is sufficiently intense.

EXAFS data collection time is at least tenfold higher (0.2 m*M*) on present day SR sources (*see* **Note 4**). At the second generation Daresbury Laboratory SRS, a typical time scale would be about 12 h for a 1 m*M* Cu sample. In a recent experiment carried out at the third generation European Synchrotron Radiation Facility (ESRF) on the Zn K-edge of adenosine monophosphate deaminase, about 40 h was needed to obtain data to k = 12 $Å^{-1}$ for a sample containing approx 0.2 m*M* Zn (***9***). Significantly longer data collection times will be needed at submillimolar concentrations when using samples containing lighter metals like Ca or Fe (*see* **Note 5**).

The volume of protein solution used is normally less than 200 µL and depends partly on the optimal *shape* of the incident X-ray beam. For example, typical sample dimensions for using wiggler station 9.2 at the SRS are 3 × 20 × 1.5 mm (vertical × horizontal × depth), whereas for wiggler station 16.5, where the X-rays are focused at the sample position, a 3 mm spot size is more appropriate. On the new multiple wavelength anomalous diffraction (MAD) instrument at the SRS on the multipole wiggler, 10, where XAFS fluorescence has been incorporated, a sample size of 1 mm^2 is optimum. The protein solution is normally injected into a perspex, or aluminium sample holder, or other material that is easily handled and is suitable for experiments at temperatures as low as liquid He. The sample space is formed by windows made of mylar or kapton that are attached to the sample holder by strong adhesive. The adhesive must also be suitable for low temperature work. Care must be taken when injecting the protein solution to ensure that bubbles are not formed (*see* **Note 6**). The sample is normally then frozen and stored in dry ice or liquid nitrogen for transport to the synchrotron beamline. In some cases, electron paramagnetic resonance (EPR) compatible Lucite cells are used.

The XAFS instruments. At first impression, a synchrotron facility and XAFS beamlines and their associated experimental stations tend to present the average *visiting* scientist with a bewildering array of optical instruments, electronics, and control systems. Synchrotron centers have worked hard to ensure that expert help in the form of *station scientists* is available with the proposed experiments. In general, the beamline and station instrumentation is set up so that *user* exchange can take place smoothly. For protein XAFS, the situation is not as good as is it for PX as in general XAFS lines are shared among different science communities, e.g., material scientists, biologists, and environmental scientists imposing different and sometimes conflicting requirements at the end-stage of the experimental stations. Some of the most productive XAFS lines for biological work are dedicated for protein work. In these cases the whole beamline including the end station, cryostat, detector, etc., is well optimized for biological experiments and the situation is then very similar to the PX beamlines from a user's perspective. Thus, during a typical XAFS

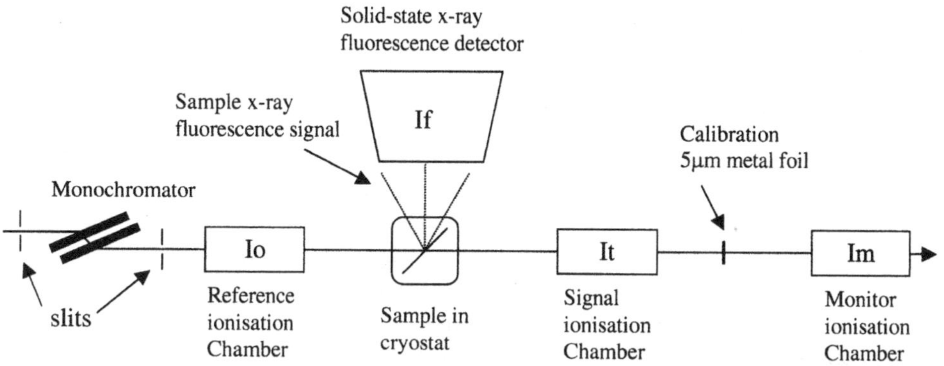

Fig. 4. The basic instrumentation for recording *transmission* XAFS data consists of gas filled ionization chambers to record the incident (*Io*) and transmitted (*It*) monochromatic X-ray intensities. For biological samples, a solid-state (or ionization) detector (*If*) is used, rather than *It*, to record the absorbance of the sample. The sample is oriented at 45º to the incident X-ray beam, while *If* is oriented at 90º to the incident beam; this geometry is used to maximize the fluorescence signal and minimize the background X-ray scatter reaching the SSD. Both methods measure the energy dependence of the X-ray absorption coefficient $\mu(E)$ before and after the absorption edge of the target metal atom. In transmission mode, $\mu(E)$ is given by $\mu(E)x = -\ln(It/Io)$, where x is the sample thickness. In fluorescence mode, $\mu(E) = If/Io$. For many experiments, a third ionization chamber (*Im*) is used to simultaneously record the absorption spectrum of a standard, for example a 5 mm thick metal foil. This gives a calibration of the edge energy for each scan recorded and makes it easy to accurately compare XANES data on samples measured at different times and places.

beamtime the only features of the beamline or station that need concern the experimenter in any detail are (*see* **Fig. 4**):

- *The monochromator*. This is normally a Si or Ge crystal (or double-crystal) that is used to select the X-ray energy with high resolution (approx 10^{-4} Å) according to the Bragg relation $\lambda = 2dSin\theta$, where d is a constant characteristic of the crystal spacing (e.g., for a Si(220) crystal d = 1.92 Å) and 2θ, in millidegrees, is the scattering angle. The monochromator angle and scan range (*see* **Table 1**) are chosen by the user to match the element of interest. In the double-crystal configuration, the first crystal determines λ and the second crystal redirects the monochromatic radiation parallel to the incident beam. An important practical consideration is the fact that the X-ray radiation from a storage ring contains shorter wavelengths ($\lambda/2$, $\lambda/3$, etc.) that also fulfil the diffraction conditions. These unwanted higher orders (harmonics) are therefore diffracted at the same angle as the primary beam and must be removed or minimised. Harmonic rejection is achieved by *detuning* one of the two crystals from parallelism such that the overlap of their *rocking curves* is maximized for λ and minimized for the harmonics.

In some cases, a mirror is used to minimize the harmonics' contribution; also, sometimes monochromator crystal planes are chosen such that higher order is minimized, e.g., for Si(111) second order is absent. The user must accurately set the harmonic rejection before data collection begins.

- *The Slits*. There may be several sets of horizontal and vertical slits present on the beamline and these are positioned before and after the monochromator and mirror vessels. They are used to define the shape and size of the X-ray beam incident on the optical elements and the sample. By varying the vertical slit separation the resolution of the data obtained can be optimized. For example, for XANES data collection the slit width should be as small as possible to maximize resolution; for EXAFS data collection the slits can be opened up to increase the X-ray flux on the sample, improving S/N. On some beamlines, for example, where a collimating mirror and sagittal focusing is employed, the use of slits is not necessary except when optimizing the optical setting by expert users.
- *The X-ray fluorescence detector*. The XAFS of metalloprotein samples is normally obtained by recording its X-ray fluorescence signal. In most cases, multi-element Ge (or Si) solid-state detectors (SSDs) are employed. These consist of an array of high-purity Ge (or Si) diodes mounted in a cryostat and provided with high count-rate signal processing electronics. The number of elements used is variable and SSDs with between 4 and 30 elements are currently in use at XAFS beamlines. Recently, a monolithic SSD has been developed successfully for XAFS allowing the count rate limit to be increased as a result of a significant decrease in the capacitance of the material ***(10)***. The monolithic design also minimizes the nonactive area of the detector, which had been as much as 70% of the active surface area of the first generation of multi-element solid state detectors ***(11)***. Count rates of up to 1 MHz per channel are now available without loss of linearity caused by pileup. One important property of SSDs is their excellent energy resolution, currently approx 170 eV at 5.9 KeV, which permits electronic discrimination between the X-ray fluorescence emission by the target atom and the background noise from X-rays scattered by the sample (*see* **Note 7**). Each element of the SSD is separately optimized before data collection using a *standard* sample as absorber. Further optimization of the signal from very dilute samples is possible by placing a thin (5 μm) Z-1 filter between the sample and the SSD.
- *The ionization chambers*. The role of the ion chambers is to record the intensity of the monochromatic X-ray beam, both before and after absorption by a sample. In a typical set-up, the ion chambers are filled with inert gases that are ionized by the X-ray photons, producing an electric current between two metal plates that have a potential difference between them of approx 800 V. This current is amplified and the signal is fed to a voltage-to-frequency converter. The digital output is then sent to a scalar counter and recorded by the data collection control software. The gases used to fill the ion chambers are chosen so that the first *reference* ion chamber absorbs approx 10 to 20%, and the second *signal* ion chamber absorbs approx 80%, of the incident X-rays. The optimal gas mixture used

depends upon the absorption edge energy and these details are normally available at XAFS beamlines. In most biological XAFS experiments, sample absorption is monitored by an X-ray fluorescence SSD, as described previously, rather than by the signal ion chamber. The signal ion chamber in these cases is often used to simultaneously collect absorption data from a standard sample, such as a 5 μm thick metal foil. These data provide an accurate calibration of the absorption edge energy for each scan, an important prerequisite for analyzing and comparing the XANES of samples measured at different times and places.

2.2. Data Processing

An example X-ray absorption spectrum of a metalloprotein, recorded in fluorescence mode at the Fe K-edge, is shown in **Fig. 5A**. The absorption, $\mu(E)$, has energy dependent fine structure oscillations (EXAFS) that result from backscattering of the excited photoelectron by the neighboring atoms to the Fe. In the absence of these neighboring atoms, the isolated (*atomic*) Fe absorption spectrum is a smoothly varying energy dependent curve, $\mu_0(E)$. The EXAFS signal is then $\chi(E) = (\mu(E) - \mu_0(E))/\Delta\mu_0(E)$, where $\Delta\mu_0(E_0)$ is the change in the atomic absorption, or edge-step. Division by $\mu_0(E_0)$ normalizes the data to unit atom absorption (**Fig. 5B**). E_0 is the energy at which the onset of absorption occurs. In practice, this threshold energy is difficult to set accurately, and E_0 is therefore selected at an energy corresponding to the half-height position of the absorption spectrum as shown in **Fig. 5A**, and is then refined in subsequent data analysis. The procedure required to extract the EXAFS signal from the raw absorption data consists of the following steps:

- The background absorption is removed by (least squares) fitting a polynomial curve to the pre-edge region of the data and extrapolating the curve to the end data point (curve P1, **Fig. 5A**). This curve should be parallel to the post-edge absorption.
- The extrapolated pre-edge curve is subtracted from the data.
- The data is normalized to unit atom absorption (**Fig. 5B**)
- The post-edge background of the normalized data is fitted by one or more polynomials (curve P2, **Fig. 5A**). Care should be taken in choosing the number and order of the polynomials used—too many high order polynomials will fit the EXAFS instead of the background. A maximum of two or three polynomials is sufficient for most cases.
- The post-edge curve is subtracted from the normalized data, yielding the required EXAFS signal (**Fig. 5C**; the spectrum is weighted by k^2 to compensate for the decrease of EXAFS amplitude with increasing energy).

A successful background subtraction should show an approximately even distribution of amplitudes about $\chi(k) = 0$ and the Fourier transform of the data should not have any large amplitude low R peaks. An example of a poor background subtraction of the data is shown in **Fig. 5D**. Choosing the correct set of

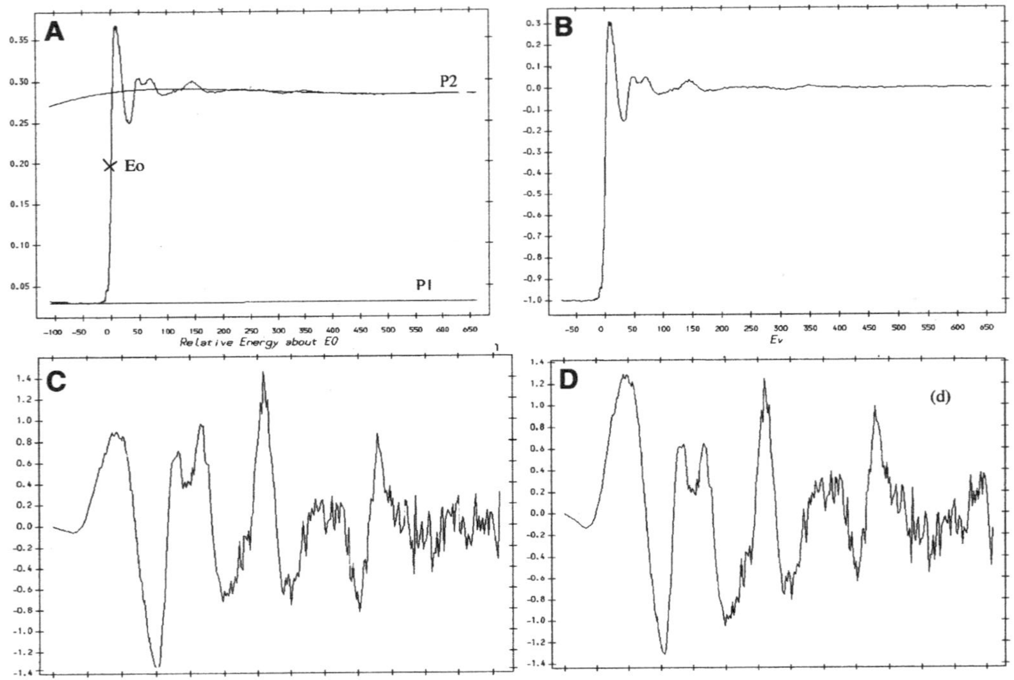

Fig. 5. Background subtraction of raw experimental XAFS data. In **(A),** energy E_0 is chosen at the half-height on the absorption edge, to represent the approximate threshold energy for photoelectron excitation. E_0 is sometimes taken as the maximum of the first derivative of the absorption curve. During subsequent curve fitting, E_0 is refined. A polynomial curve, P1, is fitted to the pre-edge part of the absorption data and extrapolated to the end-energy point. A second polynomial curve, P2, is fitted to the post-edge part of the absorption, and the EXAFS oscillations shown in **(C)** are extracted as described in the text. Note that the spectrum is normalized to unit amplitude as shown in **(B)** before the EXAFS is extracted. In **(D),** a poorer background subtraction of the same raw data is shown. In this example, the EXAFS oscillations are superimposed on a low frequency background oscillation (most noticeable at low k). The Fourier transform (not shown) of this rejected spectrum gives a *false* metal-ligand peak at approx 1 Å.

polynomials is usually a matter of trial and error; although computer programs for automated background subtraction do exist, care still needs to be exercised in their use (*see* **Note 8**).

2.3. Data Analysis and Interpretation

The calibrated, normalized, and background subtracted data can be analyzed in a number of ways to extract useful information. The XANES and EXAFS regions are generally studied separately, largely for theoretical reasons, although some steps have been made towards *whole spectrum* analysis ***(12)***.

The XANES Region. The theory associated with biological applications of XANES is well advanced (*see* **refs. *13*** and ***14***), though accurately simulating this region of the XAFS spectrum remains one of the outstanding challenges for further development of the method. The theoretical difficulties arise largely because of the complexity and richness of the various physical events that occur within the first approx 30–50 eV of the onset of the absorption edge. These events include bound-state photoelectron transitions (e.g., for the K-edge there are 1s → 3d quadrupole, 1s → np dipole transitions, 1s → np + md mixed orbital states, etc.), multiple electron excitations, long range MS pathways resulting from the long mean free path of the photoelectron at low energies, etc. Aside from specialist studies of these processes, the main use of XANES in a typical investigation is confined to qualitative and comparative studies of spectral data and the use of model compounds.[4] These studies can reveal important information related to metal valence and the coordination environment. The Cu K-edge XANES of oxidized and reduced bovine SOD (**Fig. 3A**) has already been mentioned in this respect.

The EXAFS Region. Several ways to treat EXAFS spectra will be outlined in this section.

2.3.1. Comparative Analysis and Model Compounds

The standard method is to compare spectra with data from other, known or closely related, chemical or biological systems. In fact, this essential step can be used during the data collection itself, to confirm, for example, that metal–ligand binding has occurred or that the metal site has been perturbed. A typical approach here is to collect and sum together a number[5] of scans over the complete range of the putative ligand-bound sample, S_L, perform background subtraction and compare the EXAFS (and XANES) with that of the native sample, S. If S and S_L are identical, the experiment can be terminated at this

---

[4] i.e., a crystallographically characterized series of small molecule complexes relevant to the protein under investigation, e.g., for analyzing the Mn K-edge XAFS on Mn-catalase or photosystem II, Mn(II), Mn(III), and Mn(IV) complexes are useful models.

[5] Several scans are needed to obtain a sufficient S/N.

early stage: either ligand-binding has not occurred or the ligand is present but does not perturb the metal site. If S and S_L are different, data collection continues and beamtime will not be wasted. It should be noted that, unlike powder diffraction, there is no *fingerprint* or reference database of XAFS data for identification purposes. The nature of the phenomenon and the complex phase relationships that may exist between different atom types distributed about the metal means that for part of the *k*-range, similar looking spectra may arise from metal environments that are quite different in composition and detail. Collecting structurally relevant *model compound* data can prove quite important in cases where little prior structural information may be available for system under study. Data from a range of chemical systems may prove very helpful for comparative purposes and for generating initial models for the system. The model systems are also important in validating a reliable set of theoretical parameters (e.g., phaseshifts) for more quantitative data analysis.

2.3.2. Shell-by-Shell Curve Fitting

This involves fitting *shells* of atoms to the experimental EXAFS starting with the strongest contribution to the spectrum and adding successively weaker contributions. For biological samples, the strongest backscattering shell is invariably the innermost (first) coordination sphere, consisting of either N, O, or S atoms, or a combination of these, directly bound to the metal atom. Knowledge of the system under study will provide clues about the likely identity of the atoms involved, but the distinction between N and O is generally not possible by this method of analysis, in which one is assuming averaged distances for the coordinating atoms. Nevertheless, the different contributions of N or O atoms to the data and their different bond lengths can sometimes be separated. As an example, consider the Fe^{2+} binding site in photosynthetic reaction center proteins from *Rhodobacter sphaeroides* R26, which crystal structures show is ligated with four N(His) atoms and two O(Glu) atoms. A recent XAFS study ***(15)*** of temperature and light-induced structural changes at the Fe site showed that the best first shell model consisted of 4N atoms at 2.09 Å and 2O atoms at 1.88 Å at 290 K, with an expansion of the coordination sphere to 2.14 Å and 1.91 Å, respectively at 15 K. When the crystal structure is not known, other chemical knowledge should be used to make reasonable assumptions about the likely coordination environment. In such cases, a number of different coordination models may be tried and least squares curve-fitting analysis used to yield the most probable or best model. The best model is decided by a combination of several factors:

- The least squares fit index (FI), which is defined as $FI=\Sigma_i|k^3[\chi^{exp}(k_i)-\chi^{th}(k_i)]|^2/n^2$, where $n = \Sigma_j k^3|\chi^{exp}(k_j)|$ and $\chi^{exp}(k_i)$ and $\chi^{th}(k_i)$ are the experimental and theoretical EXAFS.

- An *R*-factor (R_ε), which is given by $R_\varepsilon = \Sigma_i k^3 |[\chi^{exp}(k_i) - \chi^{th}(k_i)]|/n \times 100\%$. For metalloprotein data, an R_ε of ~ 20 % indicates a good fit, while $R_\varepsilon > 40$ % is poor. *R*-factors for fits to model compound data (or Fourier filtered metalloprotein data) are around 10 % or lower.
- The number of independent parameters (N_p) used in the least squares refinement must be less than the number of independent data points, or $N_{ind} / N_p > 1$, where N_{ind} is given by $2(R_{max} - R_{min})(k_{max} - k_{min})/\pi$. For metalloprotein data, N_p approx 20 is typical. This imposes a clear restriction on the number of different shells of atoms that may be included in a simulation before the refinement is statistically *overdetermined*. Use of chemical knowledge, e.g., the makeup of amino acids, is therefore important in the same way as is done in most protein structure refinements.

2.3.3. Ligand Fitting

This method of refinement makes use of chemical knowledge previously obtained from small molecule crystallography to fit EXAFS data using ligands rather than individual shells of atoms *(16)* (*see* **Note 9**). A good example to illustrate the advantages of this approach is provided by the structure of the haem ring. This may be generated by a fourfold rotation of a pyrrole ring, the latter serving as the basic scattering unit with a coordination number of four (**Fig. 6**). In *constrained refinement*, the pyrrole ring is treated as a rigid body. Fitting the metal – $N_{pyrrole}$ bond length and the metal – $N_{pyrrole}$ – $C_{pyrrole}$ bond angle automatically fixes the positions of all the $C_{pyrrole}$ atoms. Thus, N_{ind}/N_p is significantly increased by minimizing the number of parameters being refined. In many cases, constrained refinement is sufficient to provide good simulations. However, there is some variability in interatomic bond-lengths (e.g., $C_{pyrrole}$ – $C_{pyrrole}$ bonds) and it can be useful to lift the constraints to mimic this when simulating data. This approach is known as *restrained refinement* and is similar to that implemented by Konnert and Hendrickson *(17)* more than 20 yr ago for protein structure refinement. It effectively increases N_{ind} (and therefore N_{ind}/N_p). These aspects of data refinement are discussed in detail by Binsted et al. *(18)*. An example of the practical application of the ligand-fitting method is shown by the EXAFS of the Fe site of the truncated soluble haem domain of the oxygen sensor protein FixL involved in a two component system of a symbiotic bacterium. This EXAFS study *(19)* was carried out before the related crystal structure was solved *(20)*. The interatomic distances and angles of the Fe-ligand bond and the Fe displacement from the haem plane were obtained from the EXAFS of the Fe^{2+}, Fe^{2+}–O2, Fe^{2+}–CO, Fe^{3+}, Fe^{3+}–F, and Fe^{3+}–CN states of FixL. A correlation between the haem-N(His) distance of the haem domain and the phosphorylation activity of the histidine kinase domain of FixL was revealed. The Fe^{2+}–ligand coordination geometry determined by the EXAFS analysis also suggested that the kinase domain directly or indirectly

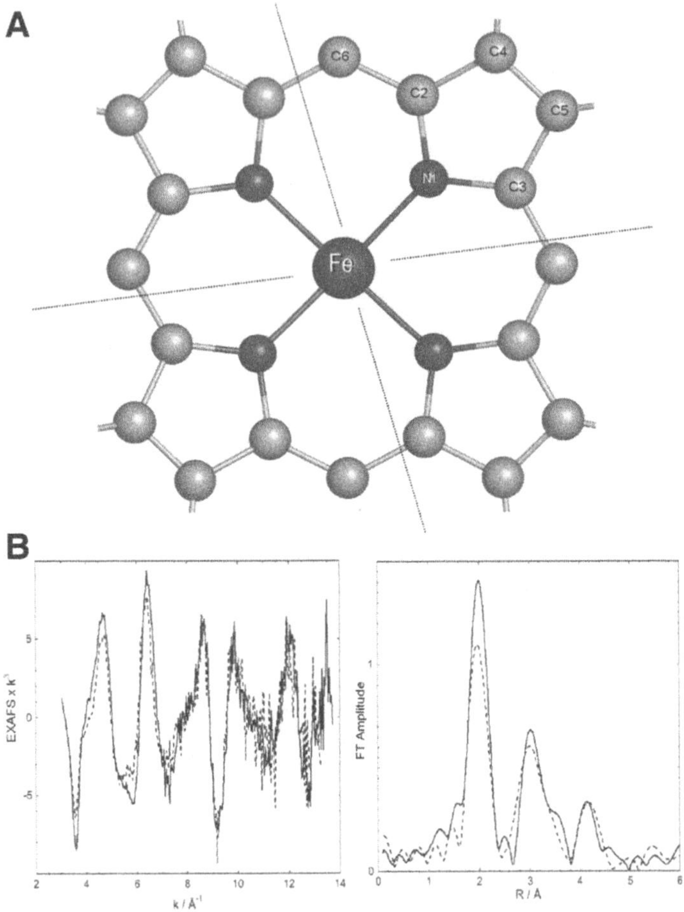

Fig. 6. (**A**) Ball and stick figure of the Fe-haem site found in hemoglobins, myoglobins, and the other haem proteins such as the oxygen sensor protein FixL (*see* text). In the ligand-fitting method, the basic EXAFS scattering unit consists of the five C atoms (C2–C6) and the nitrogen N1 atom: to obtain the total scattering signal from the haem ring, the theoretical spectrum calculated from this basic scattering unit is multiplied by four. An alternative is to use the entire haem ring as the scattering unit. This is a particularly useful approach when coordinates from a metalloprotein crystal structure are directly used to simulate EXAFS data, as used in 3D-XAFS refinements. (**B**) the k^3-weighted experimental EXAFS spectra and Fourier transforms for oxy (solid line) and carbonmonoxy (dashed line) forms of FixL. The differences observed in the EXAFS are a result of the differences in the coordination of the fifth (histidine) and sixth (O_2 or CO) ligands: for oxy-FixL, the Fe – N(His) distance is 2.01 Å and the Fe – O – O angle is 142°; for CO-FixL, they are 2.08 Å and 157°, respectively. Subtle changes like these in the Fe-haem site structure, not normally observable by PX, were important in explaining the functional mechanism of FixL ***(19)***.

influences steric interaction between the Fe-bound ligand and the haem pocket. The subsequent 1.4 Å resolution crystal structure together with the geometric knowledge obtained by the EXAFS analysis, suggested a novel oxygen sensing mechanism for FixL. This example shows one way in which crystal structure data and XAFS can be combined for understanding mechanistic features of enzymatic activity.

Both the *shell-by-shell* and *ligand fitting* methods assume average values for the distances of the coordinating atoms. For example, the four $Cu - N_{imidazole}$ bond lengths of the coordinating histidine ligands in SOD would be set to a single average distance during refinement. In **Subheading 2.4.**, we discuss the most recent and possibly most effective way of integrating PX and XAFS to obtain atomic resolution structural information on metal centers in metalloproteins. In this approach, we no longer use an averaged metal–ligand environment to fit the EXAFS data, but treat each coordinating ligand separately in a three-dimensional model of the metal site that is based directly upon structurally relevant crystallographic data.

2.4. 3D-XAFS: Combining XAFS and PX

Information from PX may be used with XAFS in a number of different ways. One typical application is to extract the metal site coordination from a *wild-type* crystal structure and use it as the starting point for analysis of the XAFS data of wild-type and derivative (e.g., ligand-bound) forms of the protein. A powerful version of this approach that exploits both the inherent MS information present in XAFS and the intrinsic resolution independence of the technique uses PX data, measured at any resolution, to build a highly accurate three-dimensional picture of metal site coordination ***(21–24)*** (*see* **Note 10**). The accuracy and precision of the metrical information determined by PX is strongly dependent upon the resolution of the diffraction data used to build the structural model.[6] Structural changes at a metal site resulting from ligand binding are generally approx 0.1 Å and require atomic resolution (< 1.2 Å or better) PX data to reveal them. Unfortunately, obtaining crystals with such diffraction properties is not achievable on demand. Very few atomic resolution structures have been solved to date and there are no obvious grounds to suppose that the future will bring a major increase in output. The diffraction range of protein

---

[6] Other factors impinging upon the accuracty and precision of the model are the completeness, redundancy, and merging statistics of the diffraction data. For data of equal quality in these respects, but of variable resolution, the higher resolution data will yield the superior metrical model. It is important to note that the completeness and redundancy of data are factors completely under the control of the experimenter (good datasets should be complete, etc.), whereas growing crystals that diffract to high resolution is often a matter of luck.

crystals is primarily limited by their inherent crystalline disorder—a difficult problem to overcome, although methods using cryogenic annealing or rehydration of crystals look promising *(25–28)*. In any case, the current fashion of high-throughput PX structure determination may even set a trend against the pursuit of atomic resolution data collection. In these circumstances, the finer details of metal–ligand interactions cannot be revealed and sufficiently understood by PX methods alone. However, by using the three-dimensional atomic coordinates of low-to-high resolution (approx 3–1.5 Å) protein crystal structures as initial input for analysis of XAFS spectra, the resolution at the metal site can be improved to atomic resolution (*see* **Note 11**). Ideally, the crystal structure of the system being studied would be appropriate, but a closely related structure can also be expedient. For example, the EXAFS of the different liganded states of the Fe-haem site of FixL, analyzed by the ligand-fitting method in the previous section, could instead have been modeled using the atomic coordinates of myoglobin or haemoglobin. The general benefits of combining PX and XAFS data (3D-XAFS) over the shell-by-shell or ligand fitting methods of analysis described in **Subheading 2.3.** are:

- Known and relevant protein structures are used as starting structures.
- The three-dimensional geometry of the metal site is retained (no averaging!).
- Crystallographically derived restraints on ligand geometry can be applied so that chemically and physically realistic results are guaranteed.
- The XAFS results are directly comparable to the input PX model and three-dimensional models of the metal site derived by XAFS can be compared (e.g., before and after ligand binding).
- The XAFS output may be used as restraints in the final phase of crystallographic analysis of a metal center when the resolution of the crystallographic structure is not close to atomic *(22)*
- In the future, protein single crystal XAFS and PX experiments can be combined and the polarized nature of the X-rays from SR exploited to pick out specific features of the metal binding environment (e.g., in catalytically active crystals). Time-resolved *movies* of the metal site in crystals might be possible using this combination. In principal, the *steady-state* crystallographic data could be used for 3D-XAFS refinements of *frozen* reaction intermediates, using freeze-quench, or stopped flow methods.

The three-dimensional approach is possible because inter- and intra-ligand multiple-scattering pathways are used in the theoretical calculation of the XAFS *(21)*, allowing determination of the entire X-ray absorption spectrum, including the near-edge region *(12)*. The input model from the PDB is constrained or restrained during refinement. Because no averaging is used in the 3D-XAFS method, care must be taken to ensure that the $N_{ind} / N_p > 1$.

A typical example of the application of 3D-XAFS goes something like this: the crystal structure of a metalloprotein in the *resting* or native state is known.

Although substantial spectroscopic, kinetic, etc., data exist for other states of the protein, there is no structural information available (e.g., growing crystals suitable for PX is proving difficult). If the targets of interest are states of the protein with altered metal environments, caused by ligand binding/ligand loss, redox reactions, etc., then 3D-XAFS can step in to provide the missing structural details. The data analysis proceeds like this:

- The first step is to make a 3D-XAFS model of the native state of the protein using the crystallographic coordinates. If the structure is known to atomic resolution, the PX coordinates generally provide an acceptable fit to the EXAFS requiring minor adjustments; for low-to-high resolution structures, constrained refinement should be performed to improve the model.
- In the next step, the refined coordinates are used as the starting set for analyzing the EXAFS of the derivative form of the protein. This starting model is modified in a number of ways, depending upon the nature of the derivative being studied. Possibilities include: (i) ligands near or bound to the metal atom being substituted by other ligands (e.g., a substrate molecule replacing a water ligand, as in nitrite binding to the type 2 Cu site of nitrite reductase *[29,30]*); (ii) ligands being removed from the coordination sphere altogether (e.g., following catalytic reduction, a water and a histidine ligand are lost from the Cu site of SOD *(see* **Fig. 3** and **refs.** *8* and *31*); and (iii) addition of ligands to the metal site (e.g., ligand binding to the sixth coordination position in Fe-haem proteins). The easiest challenge for analysis is highlighted in (ii). The loss of ligand(s) in such cases normally means a shortening of the metal–ligand distances, a straightforward task accomplished by constrained refinement. The most challenging situation is often (iii) because an addition of one ligand to an already crowded coordination sphere makes a comparatively small change to the EXAFS signal. There is even more reason to tackle this kind of problem using 3D-XAFS than by other methods.

The important point to emphasize in all these cases is not just the successful fitting of EXAFS data (this can also be accomplished using the other fitting methods), but that the three-dimensional structure of the metal site is retained. As a practical example, consider the binding of small molecules to the haem site of carbonmonoxy myoglobin (MbCO). This metalloprotein, from different sources, has been extensively studied using a number of spectroscopic, structural, and theoretical methods and over the last 20 yr the resolution of the crystal structures has improved from medium to high to atomic. **Figure 7A,B** show 3D-XAFS simulations for the EXAFS spectrum of sperm whale MbCO, using crystal structures measured at two different resolutions, 1.5 Å (*32*) and 1.15 Å *(33)*. The fit using the atomic resolution data is clearly superior. The simulation is further improved by performing a restrained least squares refinement of the coordinates (**Fig. 7C** and **Table 2**). These data illustrate two important points previously discussed: first, the higher resolution PX data provides the

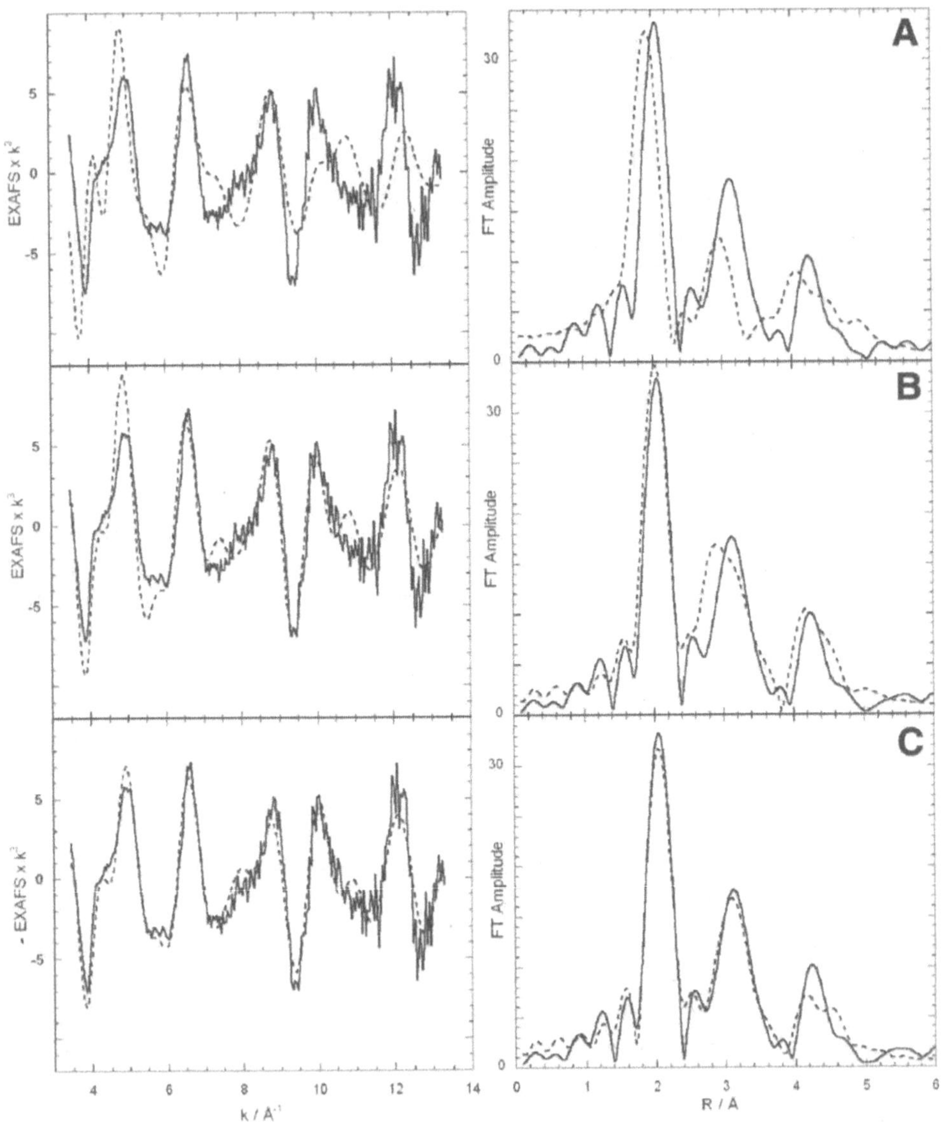

Fig. 7. 3D-XAFS simulations of the Fe K-edge k^3-weighted EXAFS of sperm whale carbonmonoxy myoglobin using crystal structure coordinates measured at **(A)** 1.5 Å resolution (*32*) [PDB: 1MBC] and **(B)** 1.15 Å resolution (*33*) [PDB: 1A6G]; **(C)** restrained refinements, starting from **(A)** or **(B)** converge to the same fit. The difference between theory and experiment seen in the Fourier transform in **(B),** at approx 3 Å, is removed by adopting a smaller (151°) Fe – C – O bond angle compared to that of the atomic resolution data (171°) (*see* **Table 2** for fit parameters).

Table 2
Principal Fe–Ligand Distances (in Å) for 3D-XAFS Simulations of MbCO Using Crystal Structure Coordinates Obtained at High (1.5 Å) and Atomic (1.15 Å) Resolutions and Following Restrained Refinement

Fe–ligand	Crystal structure resolution (Å) 1.5	1.15	Restrained fit
1 (haem)	1.90	1.94	1.89
N2 (haem)	1.99	1.98	2.02
N3 (haem)	1.99	1.99	2.01
N4 (haem)	2.00	2.01	2.02
N5 (His)	2.19	2.06	2.02
C	1.92	1.82	1.78
O	2.93	2.91	2.75[b]
Fe – C – O angle (°)	120 / 141[a]	171	151
Fit index	3.68	1.40	0.80

[a] Two different angles for the Fe-C-O bond angle were found in the crystal structure.
[b] The 3D-XAFS derived Fe – C - O bond angle is significantly different from that of the 1.15 Å resolution crystal structure but in agreement with single crystal polarized XANES of MbCO *(34)*.

better initial fit to the experimental EXAFS; second, an EXAFS simulation beginning with either the 1.5 Å or 1.15 Å resolution parameters converges to the same fit, shown in **Fig. 7C**. In other words, there is no requirement for atomic resolution PX data to use the 3D-XAFS method; lower resolutions—which make up 95% of the PX structures reported—also provide good starting points for the analysis. The combined use of XAFS and PX in this example also confirms the results of a single crystal polarised XANES study of MbCO made nearly 20 yr ago *(34)*, where the Fe – C – O bond angle was predicted to be approx 150°, some 20° smaller than seen in the current 1.15 Å resolution structure. 3D-XAFS refinements even of atomic resolution crystallographic structures can therefore yield significant structural information. The overall conclusion is that low-, medium-, or high-resolution PX data are capable of yielding atomic resolution quality 3D structural information when they are analyzed with 3D-XAFS methods.

2.5. XAS and Genomics (Metallogenomics)

From the genomics viewpoint *(35)*, it is important to note that at least 30% of the proteins encoded by genomes contain metals. The identification, through bioinformatics analysis, of potential metal sites in the genetic information; the expression of the potential target metalloproteins; the incorporation of the bio-

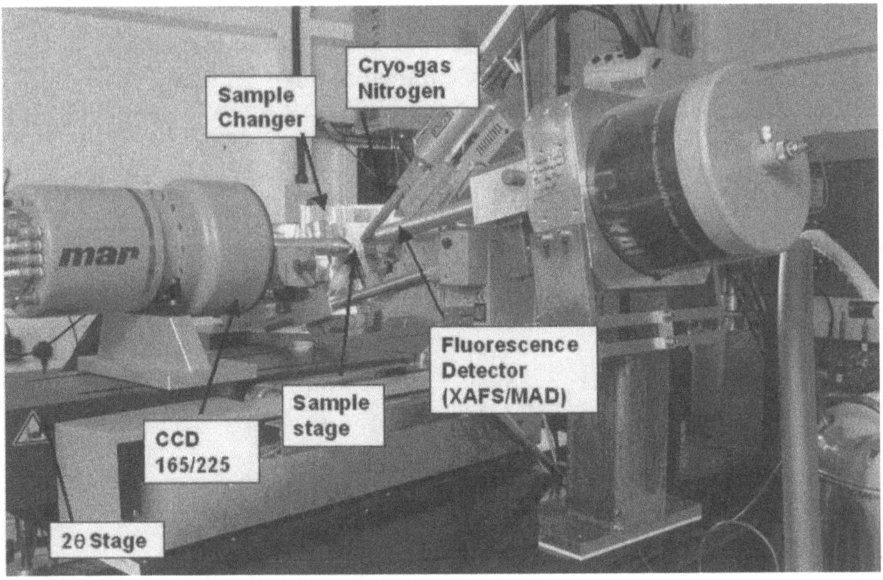

Fig. 8. The sagittally focused MAD PX facility on beamline 10 at the Daresbury SRS is designed for performing single- and multiple-wavelength crystallographic data collections on crystals as small as 50 microns. It is also fully fitted for XAFS data collection, especially for use with protein single crystals. The station is equipped with a monolithic 9-element SSD, for X-ray fluorescence measurements during MAD optimization or XAFS work. The SSD can also be usefully employed to monitor the oxidation states of functional metals while PX data collection is in progress.

logically relevant metal (an important post-transcriptional event); and structural information obtained by model chemistry. These are the essential steps needed to build one, or a few, structural models to be used as starting points for analysis of metal-site ligand interactions, using XAS data. The challenge of high-throughput studies in post-genome programs has led to a reconsideration of the traditional approaches to structural biology. Accordingly, PX and NMR are now frequently allied to other structural techniques, including neutron and X-ray solution scattering, cryoelectron microscopy, near-field microscopy, circular dichroism, and mass spectrometry. Biological XAFS has its own unique contribution to bring to the list of methods used by structural biologists, and is available for studies of solution-state or crystalline samples. A recent advance is the development of a synchrotron-based facility at the SRS for performing PX and XAFS at a single station using the same beamline optical elements (*see* **Fig. 8**). Crystallographic data collection and single crystal XAFS on the same metalloprotein crystal are possible using this set-up

2.6. The Future of Biological XAFS

This chapter has outlined some of the methods and applications of XAFS for structural biology research. At the same time, some of the complexities of the method have been mentioned. Probably the main difficulty facing scientists wishing to use XAFS in their research programs is that they are likely to be unacquainted with the method, and have no certain idea of how to proceed. XAFS is often seen as a specialist technique, carried out by a handful of experts, involving sophisticated instruments and yielding abstruse data. The contrast with protein crystallography is worth making here: PX is possible at the home laboratory as well as at the synchrotron. Protein crystallographers get their training at home and by the time they reach the synchrotron already understand what is required to collect good data, process it, and analyze it; the basic instrumentation particularly near the sample end is also already familiar. Advances in hardware and software automation mean that data collection is often a routine matter. Furthermore, the role played by CCP4[7] and other such efforts in bringing together crystallographic expertise from around the world, for software development and distribution, as well as training, has had a significant impact over the last decade in opening up the field of protein crystallography to scientists from a wide variety of subject areas: biologists can become proficient as well as physicists. There is no version of this *in-house* training and no CCP for scientists wishing to do XAFS. The practical aspects of data collection and analysis are usually only learned at a synchrotron beamline. The lack of training (and of training opportunities) for candidate practitioners and the *specialist* reputation of XAFS are a current prohibition on the development of a flourishing international biological XAFS community. In our view, correcting this problem is the most important task for the future of biological applications of XAFS and is the area where further development of XAFS as a method in molecular biology is most needed. In this respect, recent efforts made via the BioXAS study weekends (*see* **ref. *35*** and other papers in this issue of *J. Synchr. Rad.*) are highly promising.

3. Notes

1. The notation of atomic physics is often encountered in XAFS literature, where references to K-edges (equivalent to the 1s shell) and L-edges (i.e., 2s, 2p shells) of metal atoms will be found. Measurements at L-edges become necessary for high Z elements as a result of the large lifetime broadening that results from the short electron—core hole lifetimes at high excitation energies. For example, iodine has its K-edge at 0.37 Å (33.5 keV) with a lifetime broadening >7 eV. Eventually, lifetime broadening *smears out* the XAFS signal making K-edge

[7]Computational Computer Project 4.

measurements impossible. Ba (Z = 56) is probably the heaviest element that may be reasonably studied at the K-edge (0.33 Å).

2. More sophisticated and accurate versions of theory are available (e.g., *see* **ref. *6***) and implemented in several computer programs. Lee and Pendry ***(36)*** and Ashley and Doniach ***(37)*** showed that when the energy of the photoelectron is sufficiently high the curvature of the electron wave can be neglected, and thus the theory can be greatly simplified into what became known as the *plane-wave approximation*. However, at lower electron energies, the plane wave approximation breaks down and leads to errors in the calculated phase, which in turn can result in incorrect determination of the interatomic distances. The low energy part of the EXAFS spectrum is best treated by use of the exact theory given by Lee and Pendry ***(36)*** which takes account of the curvature of the electron wave and thus has been named *spherical (or curved) wave theory*. It has not been used in a majority of studies because of its mathematical complexity and requirement for large computational time. However, by averaging over the angular positions of the scattering atoms relative to the X-ray beam direction—a simplification that does not compromise the exact nature of the theory for polycrystalline, solution or amorphous samples—a more practical version of the theory has been implemented ***(3)***, for example, in the EXCURVE program ***(12,18)***. For anisotropic samples, oriented membranes or single crystals, for example, the exact theory must be used.
3. The questions posed in this example might be addressed solely by a PX study, provided appropriate crystals are available and diffract to sufficiently high resolution to reveal the structural differences at the metal site. These conditions are not always met and even when they are the crystallographic evidence may be conflicting (*see* **refs. *8*** and ***21*** for examples that illustrate these points). Problems where structural information is already available from other techniques form the bulk of XAFS applications in biology, but work on completely *unknown* systems is also possible.
4. The ability to measure proteins under *physiological* concentrations will be eventually realized with the next generation SR machines and advances in X-ray instrumentation, though other problems for the experimenter will then arise. For example, even with the third generation SR machines already in operation the problems of (normally unwanted) X-ray induced photochemical change (e.g., of metal oxidation states) and radiation damage to samples have been observed, even at low temperatures.
5. This is because the probability of core-hole relaxation via X-ray fluorescence (*see* **Fig. 1**) increases with atomic number. For the first row transition metals the X-ray fluorescence yields range from 16% for Ca to 48% for Zn, a threefold inherent decrease in detectable signal. Measurements of very dilute biological samples at the Ca K-edge are challenging but it has been done, notably on photosystem II. Relaxation by Auger electron emission dominates at low X-ray energies and the Auger electrons can be detected to provide a XAFS signal. This is a commonly used method in material science experiments especially for looking at surface structures, but it is not generally suitable for biological specimens.

6. Proteins with low metal content may have a *treacle-like* consistency when highly concentrated. In such cases, pasting the sample into a sample holder with only one window attached and slowly freezing it can work. The same approach can be used for samples prepared from crushed crystalline material or *crystal slurries*.
7. The energy resolution and energy discrimination properties of SSDs also allow an energy-resolved X-ray fluorescence spectrum to be recorded at once, enabling elemental composition of the sample to be determined. Relative proportions of different metals present in a protein can be rapidly determined down to μM concentrations whereas quantitative measurements are possible with the aid of standards.
8. Computer programs for performing background subtraction and data analysis of XAFS spectra are available at many synchrotron radiation laboratories. Please see the following links: http://srs.dl.ac.uk/XRS/index.html; http://cars9.uchicago.edu/~newville/autobk/; http://www.esrf.fr/computing/scientific/exafs/links.html. Most of the software is free for academic users. There is also an International XAFS Society: http://ixs.csrri.iit.edu/IXS/.
9. The ligands can be manually constructed using bond-length and bond-angle information available in databases (*see* http://cds.dl.ac.uk/cds/datasets/dbases.html). Some of these data are conveniently tabulated by Orpen (*38*). Another useful web-site is the Metalloprotein Database and Browser at http://metallo.scripps.edu, which provides quantitative information on geometrical parameters of metal-binding sites in proteins ***(39)***.
10. Conventionally, PX data is divided into low (>3 Å), medium (3.0–1.5 Å), high (1.5–1.2 Å), and atomic (<1.2 Å) resolutions, the atomic resolution limit being taken as the distance at which individual atoms and peptide carbon bonds may be resolved (***40***). There are relatively few atomic resolution datasets in the Protein Data Bank (PDB) ***(41)*** accounting, for example, for only 2.5% of copper protein structures submitted. Medium resolution structures make up the majority of entries in the PDB.
11. At least one data analysis program (EXCURVE) currently allows PDB files to be read in by a simple command, centered on the metal coordinates, and used in refinement. PDB format files can also be written-out when the analysis is complete, so that the three dimensional EXAFS model can be manipulated in molecular graphics programs. If necessary, the atomic Cartesian coordinates can be entered manually.

References

1. Koningsberger, D. C. and Prins, R. (eds.) (1988) X-ray *Absorption: Principles, Applications, Techniques of EXAFS, SEXAFS and XANES*. Wiley, New York.
2. Duke, P. (2000) *Synchrotron Radiation—Production and Properties*, Oxford University Press, Oxforf, UK.
3. Gurman, S. J., Binsted, N., and Ross, I. (1984) A rapid, exact curved-wave theory for EXAFS calculations. *J. Phys. C.* **C17,** 143–151.

4. Gurman, S. J., Binsted, N., and Ross, I. (1986) A rapid, exact curved-wave theory for EXAFS calculations: II. The multiple scattering contributions. *J. Phys. C.* **C19,** 1845–1861.
5. Vvedensky, D. D. (1992) Theory of X-ray absorption fine structure. *Topics in Applied Physics* **69,** 139–176.
6. Rehr, J. J. and Albers, R. C. (2000) Theoretical approaches to X-ray absorption fine structure. *Rev. Mod. Physics* **72,** 621–654.
7. Rehr, J. J. and Ankudinov, A. L. (2001) Progress and challenges in the theory and interpretation of X-ray spectra. *J. Synchrotron Rad.* **8,** 61–65.
8. Murphy, L. M., Strange, R. W., and Hasnain, S. S. (1997) A critical assessment of the evidence from XAFS and crystallography for the breakage of the imidazolate bridge furing vatalysis in CuZn superoxide dismutase. *Structure* **5,** 371–379.
9. Ranieri-Raggi, M., Raggi, A., Martini, D., Benvenuti, M., and Mangani, S. (2003) XAFS of dilute biological samples. *J. Synchrotron Rad.* **10,** 69–70.
10. Derbyshire, G., Cheung, K. C., Sangsingkeow, P., and Hasnain, S. S. (1999) A low profile monolithic multi-element Ge X-ray detector for fluorescence applications. *J. Synchrotron Rad.* **6,** 62–62.
11. Cramer, S. P., Tench, O., Yocum, M., and George, G. N. (1988) A 13 element Ge detector for fluorescence EXAFS. *Nucl. Inst. A* **266,** 586–591.
12. Binsted, N. and Hasnain, S. S. (1996) State of the art analysis of whole X-ray absorption spectra. *J. Synchrotron Rad.* **3,** 185–196.
13. Joly, Y. (2003) Calculating X-ray absorption near-edge structure at very low energy. *J. Synchrotron Rad.* **10,** 58–63.
14. Benfatto, M., Della Long, S., and Natoli, R. C. (2003) The MXAN procedure: a new method for analyzing the XANES spectra of metalloproteins to obtain structural quantitative information. *J. Synchrotron Rad.* **10,** 51–57.
15. Chen, L. X., Utschig, L. M., Schlesselman, S. L., and Tiede, D. M. (2004) Temperature and light-induced structural changes in photosynthetic reaction center proteins probed by X-ray absorption fine structure. *J. Phys. Chem.* **B108,** 3912–3924.
16. Hasnain, S. S. and Strange, R. W. (1990) In *Biophysics and Synchrotron Radiation.* (Hasnain, S. S., ed.), Ellis Horwood Ltd, Chichester, UK, pp. 104–122.
17. Konnert, J. H. and Hendrickson, W. A. (1980) A restrained parameter thermal-factor refinement procedure. *Acta Cryst.* **A36,** 344–350.
18. Binsted, N., Strange, R. W., and Hasnain, S. S. (1992) Constrained and restrained refinement in EXAFS data analysis with curved wave theory. *Biochemistry* **31,** 12,117–12,125.
19. Miyatake, H., Mukai, M., Adachi, S., Nakamura, H., Tamura, K., Iizuka, T., Shiro, Y., Strange, R. W., and Hasnain, S. S. (1999) Iron coordination structures of oxygen sensor FixL characterized by FeK-edge extended X-ray absorption fine structure and resonance Raman spectroscopy. *J. Biol. Chem.* **274,** 23,176–23,184.
20. Miyatake, H., Mukai, M., Park, S. Y., Adachi, S., Tamura, K., Nakamura, H., Nakamura, T., Tschuya, T., Iizuka, T., and Shiro, Y. (2000) Sensory mechanism

of oxygen sensor FixL from Rhizobium meliloti: crystallographic, mutagenesis and resonance Raman studies. *J. Mol. Biol.* **301,** 415–431.
21. Cheung, K. C., Strange, R. W., and Hasnain, S. S. (2000) 3D EXAFS refinement of the Cu site of azurin sheds light on the nature of the structural change at the metal centre in an oxidation-reduction process: an integrated approach combining EXAFS and crystallography. *Acta Cryst. D.* **D56,** 697–704.
22. Strange, R. W., Eady, R. R., Lawson, D., and Hasnain, S. S. (2003) XAFS studies of nitrogenase: the MoFe and VFe proteins and the use of crystallographic coordinates in three-dimensional EXAFS data analysis. *J. Synchrotron Rad.* **10,** 71–75.
23. Hasnain, S. S. and Strange, R. W. (2003) Marriage of XAFS and crystallography for structure-function studies of metalloproteins. *J. Synchrotron Rad.* **10,** 9–15.
24. Strange, R. W., Ellis, M. J., and Hasnain, S. S. (2004) Atomic resolution crystallography and XAFS. *Coord. Chem. Rev.,* in press.
25. Samygina, V. R., Antonyuk, S. V., Lamzin, V. S., and Popov, A. N. (2000) Improving the X-ray resolution by reversible flash-cooling combined with concentration screening, as exemplified with PPase. *Acta Cryst. D.* **D56,** 595–603.
26. Stevenson, C. E. M., Mayer, S. M., Delarbre, L., and Lawson, D. M. (2001) Crystal annealing—nothing to lose. *Journal of Crystal Growth* **232,** 629–637.
27. Kriminski, S., Caylor, C. L., Nonato, M. C., Finkelstein, K. D., and Thorne, R. E. (2002) Flash-cooling and annealing of protein crystals. *Acta Cryst. D* **58,** 459–471.
28. Ellis, M. J., Antonyuk, S., and Hasnain, S. S. (2002) Resolution improvement from "in situ annealing'' of copper nitrite reductase crystals. *Acta Cryst. D* **58,** 456–458.
29. Adman, E. T., Godden, J. E., and Turley, S. (1995) The structure of copper nitrite reductase from achromobactor cycloclastes at 5 pH values, with NO_2^- bound, and with type-2 Cu(II) depleted. *J. Biol. Chem.* **270,** 27,458–27,474.
30. Dodd, F. E., Hasnain, S. S., Abraham, Z. H. L., Eady, R. R., and Smith, B. E. (1997) Structures of a blue-copper nitrite reductase and its substrate-bound complex. *Acta Cryst. D.* **53,** 406–418.
31. Tainer, J. A., Getzoff, E. D., Richardson, J. S., and Richardson, D. C. (1983) Structure and mechanism of copper, zinc superoxide dismutase. *Nature* **306,** 284–287.
32. Kuriyan, J., Wilz, S., Karpus, M., and Pesko, G. A. (1986) X-ray structure and refinement of carbonmonoxy Fe(II) myoglobin at 1.5Å resolution. *J. Mol. Biol.* **192,** 133–154.
33. Vojtechovsky, J., Chu, K., Berendzen, J., Sweet, R. M., and Schlichting, I. (1999) Crystal structures of myoglobin-ligand complexes at near-atomic resolution. *Biophys. J.* **77,** 2153–2174.
34. Bianconi, A., Congio-Castellano, A., Durham, P. J., Hasnain, S. S., and Phillips, S. (1985) The CO bond angle of carboxymyoglobin determined by angular resolved XANES spectroscopy. *Nature* **318,** 685–687.
35. Ascone, I., Fourme, R., and Hasnain, S. S. (2003) Introductory overview: X-ray absorption spectroscopy and structural genomics. *J. Synchrotron Rad.* **10,** 1–3.

36. Lee, P. A., and Pendry, J. B. (1975) Theory of the extended X-ray absorption fine structure. *Phys. Rev. B* **11,** 2795–2811.
37. Ashley, C. A., and Doniach, S. (1975) Theory of extended X-ray absorption spectroscopy (EXAFS) in crystalline solids. *Phys. Rev. B.* **B11,** 1279–1288.
38. Orpen, A. G., Brammer, L., Allen, F. H., Kennard, O., Watson, D. G., and Taylor, R. (1989) Tables of bond lengths determined by X-ray and neutron diffraction. Part 2. Organometallic compounds and coordination complexes of the d- and f-block metals. *J. Chem. Soc. Dalton Trans.* **Supplement,** S1–S83.
39. Castagnetto, J. M., Hennessy, S. W., Roberts, V. A., Getzoff, E. D., Tainer, J. A., and Pique, M. E. (2002) MDB: the Metalloproetin Database and Browser at The Scripps Research Institute. *Nucl. Acids Res.* **30,** 379–382.
40. Sheldrick, G. M. and Schneider, T. R. (1997) SHELX: High-resolution refinement. *Meth. Enzymol.* **227,** 319–343.
41. Abola, A., Bernstein, F. C., Bryant, S. H., Koetzle, T. F., and Weng, J. (1987) In *Crystallographic Databases—Information Content, Software Systems, Scientific Applications.* (Allen, F. H., Bergerhoff, G., and Sievers, R., eds.), pp. 107–132, Data Commision of the International Union of Crystallography, Bonn/Cambridge/Chester.
42. Blackburn, N. J., Strange, R. W., Reedijk, J., Volbeda, A., Farooq, A., Karlin, K. D., and Zubieta, J. (1989) X-ray absorption edge spectroscopy of copper(I) complexes. Coordination geometry of copper(I) in reduced forms of copper proteins and their derivatives with carbon monoxide. *Inorg. Chem.* **28,** 1349–1357.
43. Kau, L., Spira-Solomon, D., Penner-Hahn, J. E., Hodgson, K. O., and Solomon, E. I. (1987) X-ray Absorption edge determination of the oxidation state and coordination number: application to the type 3 Cu site in rhus vernicefera laccase and its reaction with oxygen. *J. Am. Chem. Soc.* **109,** 6433–6442.
44. Hough, M. and Hasnain, S. S. (1999) Crystallographic structures of bovine copper-zinc superoxide dismutase reveal asymmetry in two subunits: functionally important three and five coordinate copper sites captured in the same crystal. *J. Mol. Biol.* **287,** 579–592.
45. Hough, M. A., Strange, R. W., and Hasnain, S. S. (2000) Conformational variability of the Cu site in one subunit of bovine CuZn superoxide dismutase: the importance of mobility in the Glu119-Leu142 loop region for catalytic function. *J. Mol. Biol.* **304,** 231–241.

10

NMR Studies of Protein–Ligand Interactions

Till Maurer

Summary

Interaction between biological macromolecules or of macromolecules with low-molecular-weight ligands is a central paradigm in the understanding of function in biological systems. It is also the major goal in pharmaceutical research to find and optimize ligands that modulate the function of biological macromolecules. Both technological advances and new methods in the field of nuclear magnetic resonance (NMR) have led to the development of several tools by which the interaction of proteins or DNA and low molecular weight-ligands can be characterized at an atomic level.

Information can be gained quickly and easily with ligand-based techniques. These need only small amounts of nonisotope labeled, and thus readily available target macromolecules. As the focus is on the signals stemming only from the ligand, no further NMR information regarding the target is needed.

Techniques based on the observation of isotopically labeled biological macromolecules open the possibility to observe interactions of proteins with low-molecular-weight ligands, DNA or other proteins . With these techniques, the structure of high-molecular-weight complexes can be determined. Here, the resonance signals of the macromolecule must be identified beforehand, which can be time consuming but with the benefit of obtaining more information with respect to the target ligand complex.

Key Words: STD NMR; WaterLOGSY; transferred NOE; $T_1\rho$; heteronuclear NMR; protein assignment; chemical shifts.

1. Introduction

1.1. Outline of the Theory

Using nuclear magnetic resonance (NMR) methods, it is possible to observe and characterize the interaction between biological macromolecules and their natural or synthetic ligands. The basis of NMR spectroscopy is the property of

From: *Methods in Molecular Biology, vol. 305: Protein–Ligand Interactions: Methods and Applications*
Edited by: G. U. Nienhaus © Humana Press Inc., Totowa, NJ

many elements to have a nuclear magnetic moment—those stable isotopes of particular importance in biological macromolecules being ^{1}H, ^{13}C, or ^{15}N. When placed into a static magnetic field B, the different nuclear spin states of these nuclei become quantized with energies proportional to their projection onto B (the so-called Zeeman Splitting). The energy difference depends on the type of nucleus, is proportional to field strength of the static magnet, and is dependant on the chemical environment of the nucleus.

This energy difference corresponds to electromagnetic radiation in the megahertz range. The transition between these states can be induced by irradiation with a radio-frequency field with characteristic frequencies for each type of nucleus and its chemical environment. The frequency of the NMR signal is extremely sensitive towards changes in covalent bonds such as neighboring groups, but also to noncovalent bonding often found in the interaction of biological macromolecules. Furthermore, transfer of magnetization through bonds or through space results in a characteristic change of the shape and size of the NMR signal and reflects, for example, the bond angle in the case of scalar coupling or spatial distance in the case of dipolar coupling.

All these phenomena can be observed using the corresponding NMR experiments and give rise to spectra that when correspondingly evaluated provide an insight into the interaction between a target macromolecule and a ligand.

1.2. Techniques

The NMR-based procedures described here can be roughly subdivided into two groups:

1. Observation of the NMR signals of the usually low molecular weight-ligand and its behavior upon binding to the target.
2. Focus on the signals of the usually much higher molecular weight protein or DNA target and the effect of the binding ligand.

The former relies on the transfer of magnetization between target and bound ligand giving rise to ligand signals, whereas the latter observes the effect of ligand binding on the chemical shift of the target resonances, thus changing the position of the target NMR signals (the so-called shift map). The NMR measurements described here are performed in aqueous (mostly buffered) solution, measuring conditions that are fairly close to those found in biological systems. These NMR measuring conditions make it very easy to add ligands and observe the effects that these ligands have on the target molecule.

Two ligand-based methods commonly are used in pharmaceutical research. They are called saturation transfer difference (STD) NMR ***(1,2)*** and water-ligand optimized gradient spectroscopy (WaterLOGSY) ***(3)***. Both techniques require microgram quantities of target, usually protein available from natural sources or overexpression in prokaryotic cells such as *E. coli* or eukaryotic cells.

1.3. Applications

Using these techniques, it is possible to observe ligand binding directly and without the need for a readout based on a biochemical or biophysical assay. This makes NMR a versatile and robust method to detect interaction in an affinity range between millimolar and nanomolar binding constants, thus spanning six orders of magnitude. When applying ligand techniques to protein or DNA, binding information attainable via NMR can be divided into levels of increasing information content:

1. Does a ligand bind? The basic STD or WaterLOGSY experiment delivers information in the form of ligand signals in the NMR spectrum. Appearance of ligand signals or changes in their phase indicates binding interaction with the target under investigation.
2. Where does a ligand bind? With a ligand of known binding site often referred to as a reporter ligand, addition of a second ligand and observation of the intensity of the signals of the first ligand can allow differentiation of binding to the same or a second binding site. With knowledge of the binding affinity of the first, an approximation of the affinity of the second ligand can be accomplished.
3. In which orientation does the ligand bind? In particular, STD signal intensity relies on the transfer of magnetization through space and thus is dependent on the distance between target and respective ligand proton. Ligand protons with more intense signals can indicate close proximity to the target surface, and thus give information with respect to the binding orientation.

2. Materials

The NMR sample consists of 320–600 uL of a target solution usually in buffer containing as few as possible proton carrying substances. Often phosphate buffer (PBS) is used, but also organic buffers such as TRIS or HEPES and stabilization agents such as EDTA or DTT are commercially available in deuterated form. Measurements in protonated water (H_2O) are possible utilizing water signal suppression routines, but higher sensitivity can be attained by substituting water with deuterium oxide (^2H_2O, D_2O). The (nonlabeled) protein concentration should be in the low micromolar range that, although close to the detection limit of the NMR spectrometer, is enough to observe the ligand effects. The ligand is added to achieve a 20–100-fold excess. Ligand solubility is often a problem when screening for ligands based on chemical libraries. The use of stock solutions in deutero DMSO (d6-DMSO) is a good way of dissolving otherwise reluctantly soluble ligands by addition of this stock solution to a DMSO concentration of 1 to 5% in the final aqueous buffer. Addition of freeze-dried protein or millimolar concentrated protein solution to a previously heated or sonicated and filtered ligand solution is preferred if the ligand precipitates when the d6-DMSO stock is pipeted into the protein solution. Precipitate should always be removed. Of great importance in measurements for ligand-binding detection is the concentration of the ligand, as at concentrations up to the millimolar range one observes signals stemming from interaction with secondary or unspecific sites within the ligand. Using cryogenic probe technology, only small amounts are needed to attain signals of sufficient intensity within a reasonable time. Interestingly, ligands with low nanomolar binding affinity can also be observed in contrast to published data. This is important in pharmaceutical research, as the low nanomolar affinity range is a clear goal in the characterization of candidates for drug development.

3. Methods

3.1. Implementation of Ligand Nuclear Magnetic Resonance Experiments

The NMR spectrometers are coded in pulse programs, which are part of the software that accompanies the spectrometers. These are usually supplied by the NMR vendors. Often, pulse programs are also available from literature. The techniques are described in the references. Parameters of typical experiments are also given in the **Notes** section.

For the ligand detection experiments described in this chapter, a sample consisting of a solution of human serum albumin at a concentration of 100 μM and glucose and benzoic acid at a 10- to 50-fold excess in PBS at a pH of 7.4 in either H_2O or D_2O is well-suited to test the experimental setup.

3.1.1. Saturation Transfer Difference Nuclear Magnetic Resonance Experiments to Detect Ligand Binding

In the STD NMR experiment signal attenuation of ligand signals in the comparison of two separate experiments is observed. In the two experiments, the target is alternately left unsaturated and saturated with magnetization by radiofrequency irradiation (*see* **Note 1**). If interaction of nonlabile protons is the focus, STD NMR experiments can be carried out in 100% D_2O, eliminating problems arising from the dominant water signal. The WATERGATE solvent suppression sequence or the excitation sculpting method can be used to suppress the H_2O signal in samples where labile protons need to be observed. Subsequent acquisition leads to two separate NMR spectra where the ligand signals and in cases where the protein is of low molecular weight also target signals are visible. As a result of line broadening effects, the spurious target signals become less intense with increasing molecular weight. They can be further suppressed by adding filters that remove magnetization of only the high-molecular-weight components (spin lock). This comes at a cost of signal intensity on the whole and should be applied with care. When comparing the ligand signals in the two spectra, target interaction is indicated by a reduction of the ligand signal intensity in the spectrum where the protein is saturated. Alter-

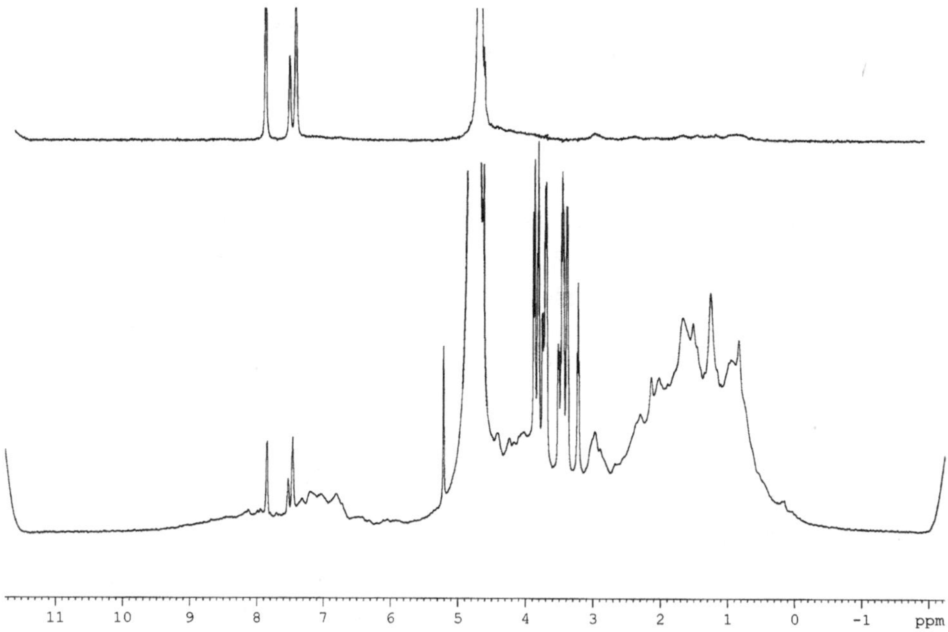

Fig. 1. 1D (bottom trace) and STD (top trace) ^1H NMR spectra measured at 600 MHz and 298K of a mixture of human serum albumin (HSA) (100 μ*M*), glucose (1 m*M*) and benzoic acid (1 m*M*) in PBS in 99% ^2H_2O at pH 7.0. The NMR resonance signals from protein and both ligands can be seen in the 1D spectrum. Glucose signals appear in the range of 3–5.2 ppm and those of benzoic acid at 7.4–7.9 ppm. In the top trace, only signals of benzoic acid remain, indicating binding to HSA.

nately, the two spectra can be subtracted during the measurement by phase cycling. In the resulting difference spectrum, only ligand signals should be visible when interaction takes place. **Figure 1** shows HSA and a mixture of glucose and benzoic acid.

3.1.2. Saturation Transfer Difference Nuclear Magnetic Resonance Experiments to Detect the Ligand-Binding Site

The binding site of a target to which a novel ligand binds can be determined through titration experiments using a reporter ligand with a known binding site. For example, acetaminophen is known to bind to the site II in human serum albumin with an affinity of 690 μ*M (4)*. The STD NMR spectra show the signals of acetaminophen. These can be modulated by addition of ibuprofen, known to bind to the same site but with higher affinity (1.2 μ*M*). Addition of ibuprofen to the mixture of human serum albumin (HSA)/acetaminophen results in a reduction of the STD signal intensity of acetaminophen and a proportional increase of the ibuprofen signals. Here, a competition for the same binding site results in a displacement of acetaminophen, giving the described results. **Figure 2** shows HSA/acetaminophen titrated with ibuprofen.

3.1.3. Saturation Transfer Difference Nuclear Magnetic Resonance Experiments to Detect the Ligand-Binding Orientation

As previously mentioned, the signal intensity in STD NMR experiments can reflect structural aspects such as the orientation of the bound ligand, as STD NMR relies on the transfer of magnetization through space, the dipolar interaction with the effectiveness having an r^{-6} dependence. Signal intensities in 1D NMR spectra normally reflect the number of nuclei giving rise to a certain signal. When comparing normal 1D and STD signal intensities, this does not hold true for STD signals. The STD intensities of individual protons are modulated by several experimental, physicochemical, and molecular parameters such as duration of presaturation, relaxation time, binding constant, and dissociation kinetics. There is no easy way to obtain a distance calibration to extract the target proximity information from STD NMR experiments. On the other hand, quantification of STD effects is achieved by signal integration. Often, these intensity values are divided by the corresponding intensities obtained from a normal 1D spectrum and then expressed as the percentage. This ratio (the relative STD effect) is useful for the comparison of different experimental conditions. The relative STD effect can be used to gain information not only as to whether a ligand binds, but also to some degree which orientation the ligand has in the bound conformation. Here, the signal of the ligand protons should be of different intensity depending on the distance between them and the target surface. An example is the binding of enantiomers, the signals of which appear in exactly the same position in the 1D spectrum but with modulated intensity in the STD NMR spectrum. In the case of the amino acid tryptophane binding to HSA the binding constants the two enantiomeric forms differ by the factor of 100.

Figure 3A shows the build-up experiment of saturation time in a mixture of HSA and D- and L-tryptophane. **Figure 3B** shows the STD spectrum at the saturation time of 2 s. The α-proton in the weaker binding D-tryptophane has a lower signal intensity.

Fig. 2. (*opposite page*) A titration of the mixture of human serum albumin and acetaminophen with ibuprofen. Top and bottom traces show the 1D proton NMR spectra of the ligands, the middle traces show STD NMR spectra of varying ligand ratios. The binding affinity of acetaminophen to HSA is 690 μ*M*, that of ibuprofen 1.2 μ*M* (*4*). As they compete for the same binding site, the STD signals stemming from acetaminophen as the weaker binder are reduced upon addition of the stronger binder ibuprofen.

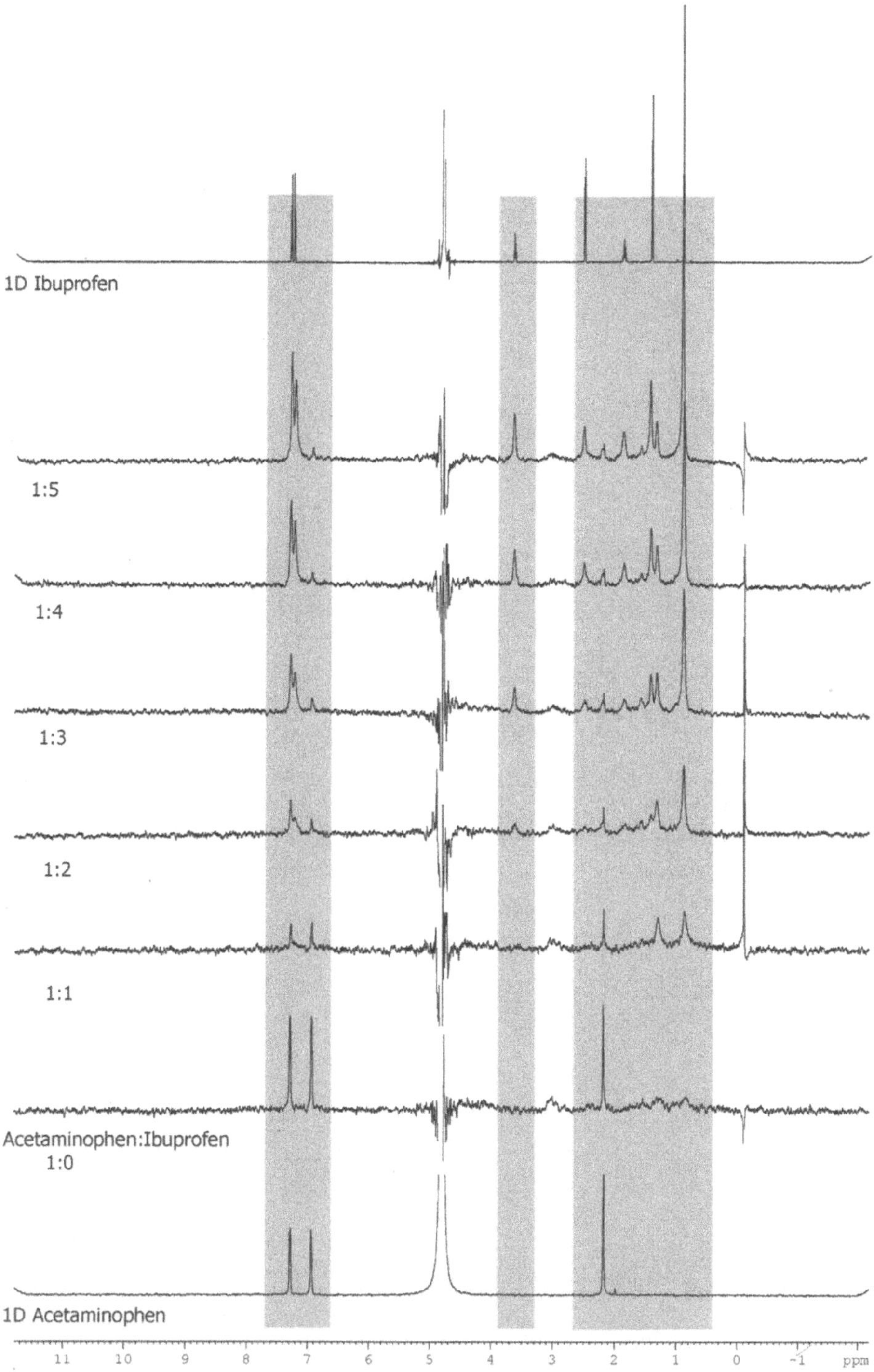
1D Ibuprofen
1:5
1:4
1:3
1:2
1:1
Acetaminophen:Ibuprofen
1:0
1D Acetaminophen
11
10
9
8
7
6
5
4
3
2
1
0
-1
ppm

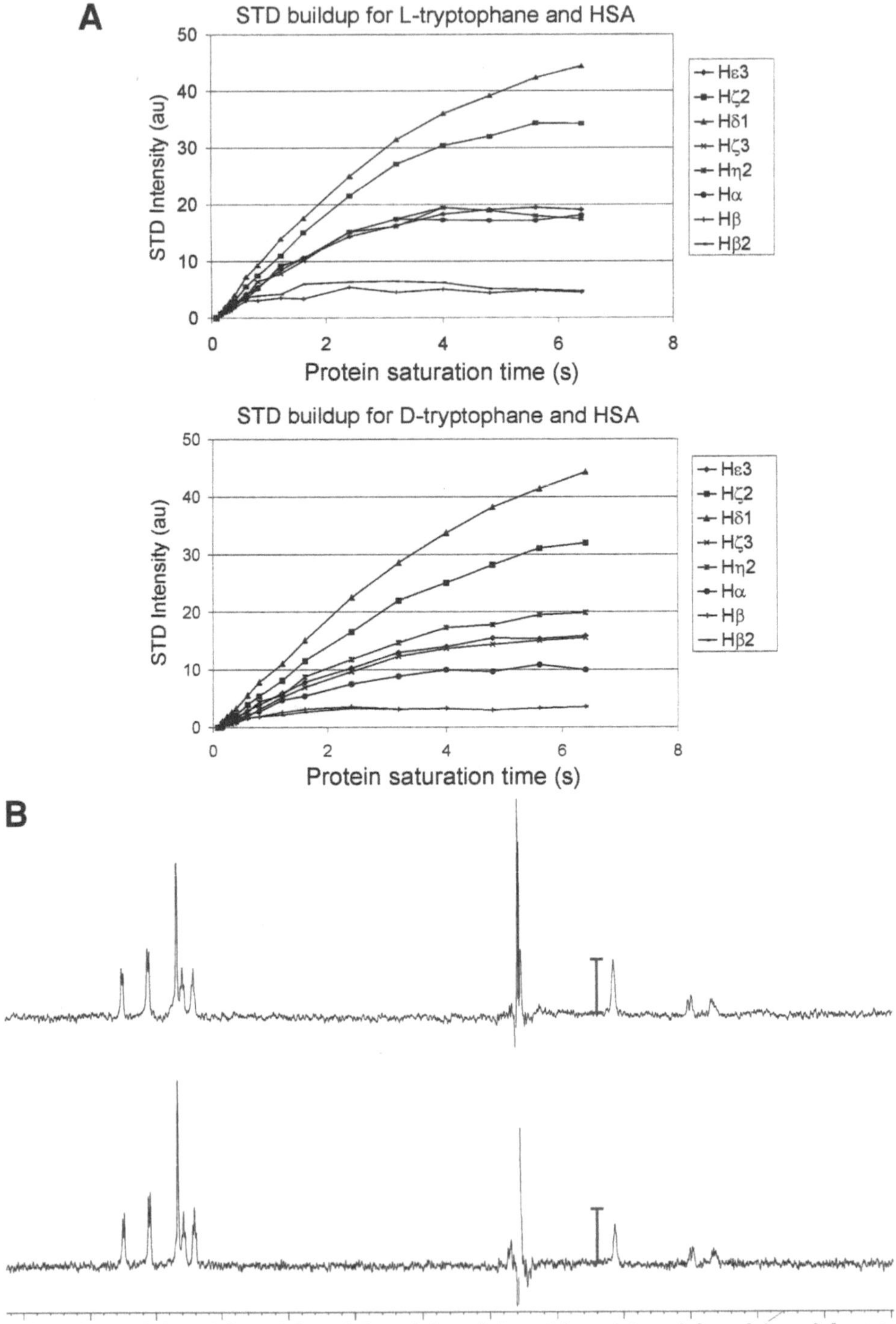
A
STD buildup for L-tryptophane and HSA
STD Intensity (au)
Protein saturation time (s)
Hε3
Hζ2
Hδ1
Hζ3
Hη2
Hα
Hβ
Hβ2
STD buildup for D-tryptophane and HSA
STD Intensity (au)
Protein saturation time (s)
Hε3
Hζ2
Hδ1
Hζ3
Hη2
Hα
Hβ
Hβ2
B
ppm

As a side effect, a differentiation of the two enantiomers is possible using STD NMR. Assuming the different binding constant reflects a difference in the geometry of the bound state, the distance of the α-proton in D-tryptophane to the protein surface must be larger in the L-tryptophane.

3.1.4. Water-Ligand Optimized Gradient Spectroscopy

Similar to STD NMR, WaterLOGSY relies on the transfer of magnetization between ligand and target. The transfer path is not direct (as in STD NMR) but uses H_2O molecules in the binding pocket as an intermediate magnetization pool. Thus, the sample solution must consist predominantly of H_2O with only a small quantity of D_2O for locking. Because the sign of the magnetization transfer signals via the nuclear overhauser effect (NOE) is dependent on the molecular weight of the interacting molecules under investigation, a signal stemming from transfer between the water and the ligand (both of low molecular weight) will have a positive sign in the WaterLOGSY spectrum.

Figure 4 shows HSA with a mixture of glucose and benzoic acid using WaterLOGSY. Upon binding to the (high molecular weight) target the sign reverses and signals of negative sign are observed, indicating interaction with the target. In contrast to STD NMR measurements, both binders and nonbinders can be seen in one experiment.

3.1.5. Differential Line Broadening

Information on the binding orientation of a ligand can be obtained using differential line broadening experiments *(5)*. Dipolar relaxation reduces the lifetime of magnetization and leads to broadening of NMR signals. Here, the effect of faster relaxation in the bound state of a low molecular weight-ligand bound to a high molecular weight-target can be observed. A realistic molecular weight ratio should be 1:500 up to 1:50,000. Especially in a titration experiment the gradual increase of line widths of ligand signals is shown clearly. Starting with a pure ligand solution, protein is added in steps to reach a 10- to 20-fold excess. In the case that some but not all ligand signals become broader

---

Fig. 3. *(opposite page)* (**A**) STD build-up curves of a mixture of human serum albumin with D-tryptophane (bottom trace) and L-tryptophane (top trace). The signal intensity for each proton resonance is plotted against the duration of the protein saturation. The binding affinity of D-tryptophane to HSA is of a factor of 100 lower *(4)*. The most obvious difference can be seen in the curve for the α-proton. (**B**) STD spectra of the same mixture at the saturation time of 2 s. The spectra were adjusted to match the intensities of the signals between 7.2–7.8 ppm. The signal stemming from the proton bound to the α-carbon is clearly less intense in the D-tryptophane spectrum.

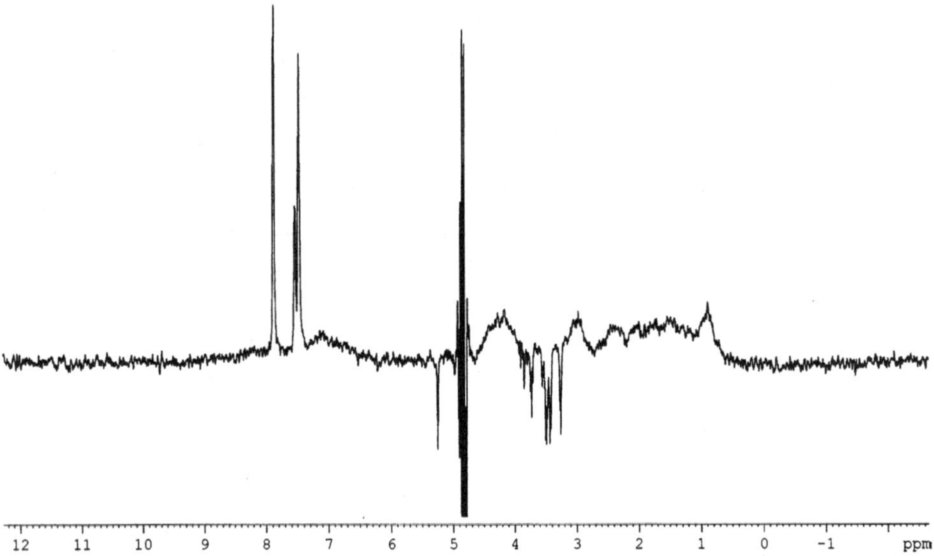

Fig. 4. Water-LOGSY spectrum of the same mixture used in example 1 (human serum albumin with glucose/benzoic acid). Positive signals stem from benzoic acid, negative ones from glucose. As HSA is not of very high molecular weight (66 kDa) the protein signals are also visible.

earlier than others, one speaks of a differential effect. Ligand protons closer to the protein surface interact via dipolar coupling more efficiently than those more distant and their signals thus grow broader at lower ligand protein ratios. This effect obviously reflects the binding orientation. Differential line broadening has been investigated in numerous publications and is successfully applied in pharmaceutical research. This technique is better applicable to ligands with a large number of proton signals and a binding mode that covers a large area of the target-binding pocket. Many successful experiments have been performed with peptide-based ligands. In combination with transferred NOE measurements to gain information on the conformation of the bound state, a model of the bound conformation is possible. The technique is sensitive towards the binding kinetics, ligands with very slow off kinetics can be difficult to detect.

3.1.6. $T_1\rho$ Measurements

Similar to the effect in differential line broadening experiments, relaxation in the rotating frame, called $T_1\rho$ *(6)* relies on the observation of relaxation effects. Because of the use of a $T_1\rho$ relaxation filter, the signals of free ligands can be effectively separated from those belonging to bound ligands. The observation of signals in regions with spectral overlap becomes easier because the resonances tend to be less overlapped than in the case of broadened signals in the differential line broadening experiment.

3.1.7. Transferred Nuclear Overhauser Effect

Intermolecular dipolar interaction of ligand protons while the ligand is bound to the target can be measured in transferred NOE experiments *(7,8)*. As in the WaterLOGSY experiment, the sign of the NOE signal is used to differentiate between the bound conformation (negative signal in the spectrum) and the free state (positive signal in the spectrum). Often, a ligand does not adopt a stable conformation while moving freely in solution, so that few NOE signals with positive sign are observed in the spectrum in the first place. The appearance of negative NOE cross peaks indicate proximity of the involved protons within a stable bound conformation. This data can be used to calculate the bound conformation. Here, again the effect of binding kinetics plays a role towards the sensitivity and thus the quality of the attained spectra.

3.1.8. Diffusion Edited Techniques

Pulsed field gradients make experiments possible where signal attenuation in the diffusion NMR spectrum corresponds to the translational motion of a molecule in the NMR sample volume. Spatial discrimination is achieved by application of a magnetic field gradient in addition to the static homogenous field of the NMR magnet. The diffusion experiment is set up as a series of 1D spectra with a fixed delay to allow for diffusion and variation of the strength of the gradient. This results in an attenuation of the NMR signals with increasing gradient strength. Small molecules move fast and their signals decay faster than large slowly diffusion molecules. The movement of small ligands (less than 1 kDa) is attenuated when they interact with a large target (more than 10 kDa). This results in the faster decay of signals belonging to the ligands that bind to the target in the diffusion NMR experiment (**Fig. 5**).

Using this technique, a ligand that binds can be detected out of mixtures of potential binders. As the resulting mean diffusion of a ligand detectable by NMR is the sum of free and bound diffusion constant, the method works best if the difference in molecular weight of ligand and target is large. Here, similar to all ligand-based NMR methods, the kinetics of binding influence the applicability and size of the observed effect.

3.2. Techniques Utilizing Isotopically Labeled Biological Macromolecules

NMR methods with a focus on the target *(9)* rely on its production using isotopic labeling, except in the case where the target is very small (less than

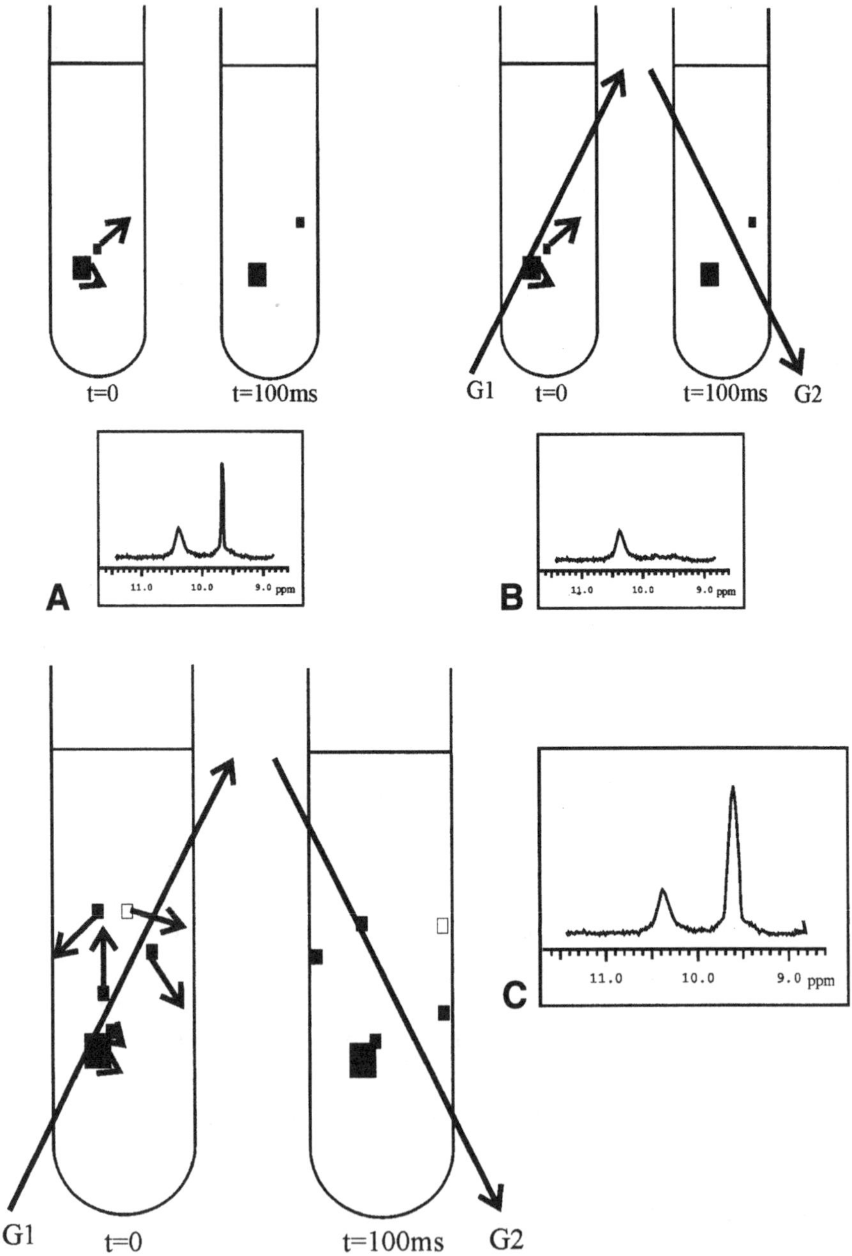

Fig. 5. Detecting ligand binding by diffusion NMR. Pulsed field gradients are employed to achieve spatial resolution. NMR signals of molecules that diffuse during the mixing period are attenuated most when translational movement is fast. Upon binding to a slowly diffusing target, the signal attenuation is reduced. Binding ligands can be identified from a mixture of candidates.

10 kDa). In this case, an assignment of the target using only proton–proton correlations is feasible. For large systems, labeled proteins are routinely expressed in *E. coli* using isotopically labeled nutrients such as ^{15}N ammonium chloride or ^{13}C glucose and isotopically labeled full media are commercially available. Yeast expression is also possible. Expression of isotopically labeled protein in mammalian cells has recently been accomplished and promises to become a standard method in the future ***(10)***. To set up the experiments, isotopically labeled standard samples of, e.g., ubiquitin can be obtained from commercial vendors. The chemical shift information for this protein can be obtained from the biological magnetic resonance database (BMRB).

With ^{13}C and ^{15}N labeled protein, the identification of all backbone and most side chain atom chemical shifts is feasible but still a time-consuming task. Several multidimensional spectra must be recorded whose description is beyond the scope of this chapter. As a rule, experiments that correlate the backbone amide proton and nitrogen with the backbone and side chain carbons, such as HNCA, CBCA(CO)NH, CBCANH, HNCO, and experiments that yield distance information separated by a ^{15}N or ^{13}C dimension such as the ^{15}N separated NOESY spectra are used in data evaluation ***(11)***. With semi-automated routines, the time needed for a backbone assignment is dependent on the molecular weight of the target in question. We have recently assigned a 30 kDa protein in 4 mo, but the time needed will become shorter in semi and full automation with the advent of robust algorithms and larger databases for chemical shifts. With this data in hand the determination of the three dimensional structure can be accomplished.

With shift data from NMR and a three-dimensional structure information from another source, i.e., X-ray crystallography, the binding interaction between the protein and a ligand can be monitored by observation of the protein resonance signals. More common is the comparison of the ^{15}N HSQC or ^{15}N-HSQC-TROSY ***(12)*** spectra of apo protein and complexed protein that should give rise to changes in the position of signals in the binding site stemming from backbone amide groups. The changes are often confined to those amino acids that reside in the binding pocket of the protein and thus, with knowledge of the three-dimensional structure, allow mapping of these chemical shift changes to the protein surface. The resulting picture indicates clearly where binding takes place. **Figure 6** shows the shift map of human macrophage elastase. In contrast to ligand based methods, the binding affinity and the binding kinetics of a ligand under investigation using isotopically labeled target does not play such a dominant role in the applicability of this method. This is because the observed chemical shift change is directly caused by the number of ligand-bound protein molecules, and is caused by the inherent sensitivity of chemical shift changes towards changes in the chemical environment.

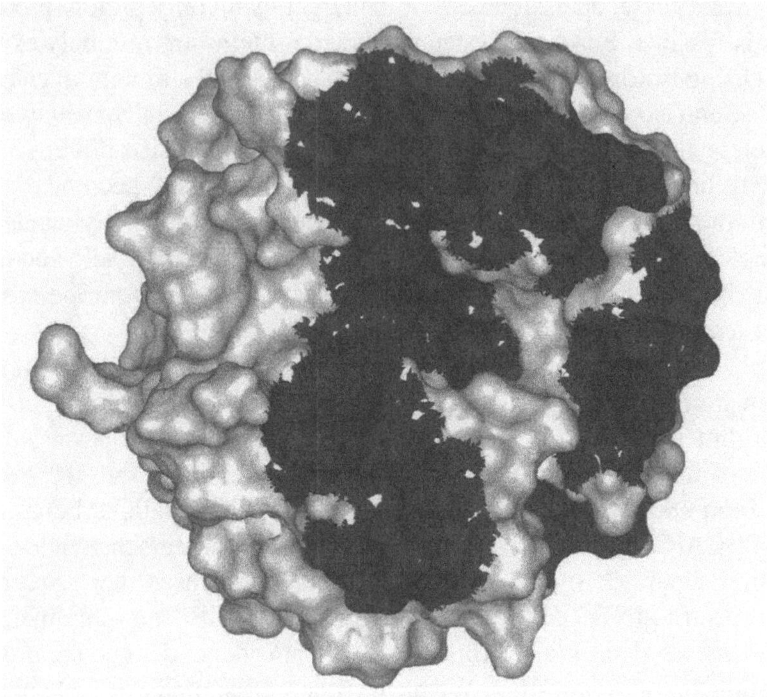

Fig. 6. Chemical shift map of ligand binding to human macrophage elastase. The solvent accessible protein surface is depicted in light gray, the regions where shift changes take place are black. The proposed ligand-binding site is in the centre of the protein molecule. Clearly visible are additional regions where chemical shift changes take place.

Fairly often, the changes observed in a ligand-binding study are not confined to the binding site. Interpretation of the data becomes more difficult in these cases, but the amount of information that can be gained increases also. **Figure 7** shows the human macrophage elastase dimer.

Measurement of the intermolecular dipolar interaction between target and ligand is possible ***(13)***. When one of the two binding partners is isotopically labeled, the use of heteronuclear filtering techniques allows identification of NOE signals that indicate proximity between labeled target and nonlabeled ligand. With this technique, the through space interaction between a labeled target and a nonlabeled ligand can be filtered from the intramolecular interactions within one component. This gives rise to distance information that allows the calculation of the complex to a resolution level that can be compared with the corresponding co-crystal structure.

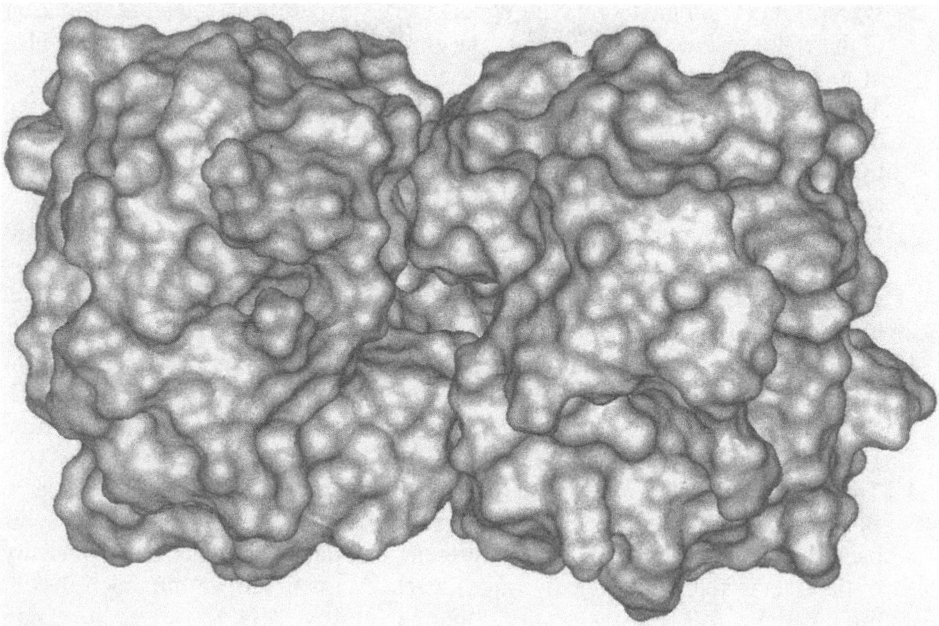

Fig. 7. Possible explanation of the effects observed during ligand titration of human macrophage elastase: the protein is a dimer in the ligand bound state. The observed shift changes are all on one face to the protein molecule. In the dimer, the ligand-binding domains face one another.

In particular, a great advantage of NMR as a method is the possible observation of the dynamics of a target. The future will show whether this will become an important aspect in the description of biological macromolecule interaction.

4. Notes

1. Parameters for STD NMR experiments. The protein presaturation in an STD NMR experiment is accomplished by a pulse train of selective shaped pulses (e.g., Gaussian) of 30 ms duration. Corresponding repetition of the shaped pulse in a loop results in a presaturation time of 2–5 s. The offset is set to –1.0 ppm of the on-resonance and 40 ppm for the off-resonance experiment. A test experiment with only the ligand present should be performed to make sure that none of the ligand resonances are hit by the presaturation cascade. This can happen if the ligand aggregates to micelles or particles in equilibrium with the free form. If ligand resonance signals are near the presaturation offset, it can be set to values up to –10 ppm for the on-resonance case. An alternative is to set the on-resonance offset to lower frequencies (down field, e.g., 12 ppm), which saturates resonances arising from the aromatic protons of the target.

2. WaterLOGSY parameters. In the WaterLOGSY experiment, selective excitation of the water resonance is pertinent. The experiment uses a selective proton pulse of 8–25 ms duration. Gradient pulses of 1–2 ms duration and 2–6 G/cm then destroy unwanted magnetization. A weak gradient pulse is applied during the mixing time ending with a short gradient recovery delay of 1–2 ms and followed by the detection pulse at high power. Water suppression is vital and achieved by an excitation sculpting sequence with gradients.
3. Differential line broadening. Basically, any 1D proton sequence can be used to observe the occurring differential line broadening in a titration of ligand with target. Several successive NMR spectra are recorded at constant target and increasing ligand excess concentration.
4. $T_1\rho$ measurements. Relaxation edited NMR experiments are based on standard established pulse sequences using a relaxation time of 400 ms. This time can be varied similar to the standard T2 experiments to attain a 2D experiment where 1D spectra are recorded with increasing relaxation time.
5. Transferred NOE. The transferred NOE (tr NOE) experiment is basically the same as the standard NOESY experiment. The tr NOE buildup takes place in the complex of a large target molecule (e.g., protein) and the bound ligand. The intensity of the transferred NOE (negative sign) reaches a maximum within much shorter mixing times (50–200 ms) than the normal positive NOE observed for small molecules. Large ligand excess factors lead to large positive portion in the NOE crosspeaks. To remove the broad target signals from the spectrum, the NOESY experiment can be supplemented with a spin lock pulse.
6. Diffusion based techniques. Diffusion edited NMR experiments are carried out by recording a series of experiments with increasing gradient power resulting in a modulation of signal intensity that is dependant on the diffusion rate. A delay in the pulse sequence allows for diffusion and must be chosen so that the signals of the ligand in question are nearly zero at the highest gradient power. Values of 100–500 ms are usually sufficient. The gradient power is increased in steps of 10%.

Acknowledgments

The author would like to thank Dr. Claudio Dalvit, Dr. Helmut Hanssum, Dr. Wolfgang Jahnke, Prof. Bernd Meier, Dr. Robert Meinecke, Dr. Herbert Nar, Prof. Thomas Peters, and Dr. Rüdiger Weisemann for helpful discussions and support.

References

1. Mayer, M. and Meyer B. (1999) Characterising of ligand binding by daturation transfer difference spectroscopy. *Angew. Chem. Int. Ed.* **38,** 1784–1788.
2. Mayer, M. and Meyer, B. (2001) Group epitope mapping by saturation transfer difference NMR to identify segments of a ligand in direct contact with a protein receptor. *J. Med. Chem.* **123,** 6108–6117.
3. Dalvit, C., Pevarello, P., Tatò, M., Veronesi, M., Vulpetti, A., and Sundström, M. (2000) Identification of compounds with binding affinity to proteins via magnetization transfer from bulk water. *Journal of Biomolecular NMR* **18,** 65–68.

4. Peters, T., Jr. (1996) *All About Albumin*. Academic Press, San Diego, CA.
5. Hammes, G. G., Tallman, and Dennis E. (1971) Nuclear magnetic resonance study of the interaction of L-epinephrine with phospholipid vesicles. *Biochim. Biophys. Acta* **233,** 17–25.
6. Kopple, K. D., Wang, Y.-S., Cheng, A. G., Bhandary, and Krishna K. (1988) Conformations of cyclic octapeptides. 5. Crystal structure of cyclo(Cys-Gly-Pro-Phe)$_2$ and rotating frame relaxation ($T_{1}\rho$) NMR studies of internal mobility in cyclic octapeptides. *J. Am. Chem. Soc.* **110,** 4168–4176.
7. Ni, F. (1994) Recent developments in transferred NOE methods. Prog. *Nucl. Magn. Reson.* Spectrosc. **26,** 517–606.
8. Peters, T., Jr. (2000) Transfer NOE experiments for the study of carbohydrate-protein interactions. In *Carbohydrates in Chemistry and Biology* (Ernst, B., Hart, G. W., and Sinay, P., eds.). Wiley-VCH, Verlag GmbH, Weinheim, Germany, pp. 1003–1023.
9. Roberts, G. C. K. (2000) Applications of NMR in drug discovery. *Drug Discovery Today* **5,** 230–240.
10. Strauss, A., Bitsch, F., Cutting, B., Fendrich, G., Graff, P., Liebetanz, J., Zurini, M., and Jahnke, W. (2003) Amino-acid-type selective isotope labeling of proteins expressed in Baculovirus-infected insect cells useful for NMR studies. *J. Biomol. NMR* **26,** 367–372.
11. Sattler, M., Schleucher, J., and Griesinger, C. (1999) Heteronuclear multidimensional NMR experiments for the structure determination of proteins in solution employing pulsed field gradients. *Prog. NMR Spectrosc.* **34,** 93–158.
12. Pervushin K., Riek R., Wider G., and Wüthrich K. (1997) Attenuated T2 relaxation by mutual cancellation of dipole-dipole coupling and chemical shift anisotropy indicates an avenue to NMR structures of very large biological macromolecules in solution. *Proc. Natl. Acad. Sci. USA* **94,** 12,366–12,371.
13. Breeze, A. L. (2000) Isotope-filtered NMR methods for the study of biomolecular structure and interactions. *Prog. NMR Spectrosc.* **36,** 323–372.

11

Probing Heme Protein–Ligand Interactions by UV/Visible Absorption Spectroscopy

Karin Nienhaus and G. Ulrich Nienhaus

Summary

Ultraviolet/visible (UV/vis) absorption spectroscopy is a powerful tool for steady-state and time-resolved studies of protein–ligand interactions. Prosthetic groups in proteins frequently have strong electronic absorbance bands that depend on the oxidation, ligation, and conformation states of the chromophores. They are also sensitive to conformational changes of the polypeptide chain into which they are embedded. Steady-state absorption spectroscopy provides information on ligand binding equilibria, from which the Gibbs free energy differences between the ligated and unligated states can be computed. Time-resolved absorption spectroscopy allows one to detect short-lived intermediate states that may not get populated significantly under equilibrium conditions, but may nevertheless be of crucial importance for biological function. Moreover, the energy barriers that have to be surmounted in the reaction can be determined. In this chapter, we present a number of typical applications of steady-state and ns time-resolved UV/vis absorption spectroscopy in the study of ligand binding to the central iron in heme proteins.

Key Words: Ultraviolet/visible absorption spectroscopy; flash photolysis; ligand binding; time-resolved spectroscopy; geminate recombination; heme proteins; hemoglobin; myoglobin; neuroglobin.

1. Introduction

A myriad of chemical reactions take place in a living cell at any given point in time. These processes are arranged in a delicate network of interdependent, coupled reactions that are mutually regulated and coordinated. The fundamental process underlying the proper functioning of such a network is molecular recognition, the specific interaction between biomolecules. Of particular importance, are interactions of proteins with DNA/RNA, lipids, sugar moi-

From: *Methods in Molecular Biology, vol. 305: Protein–Ligand Interactions: Methods and Applications*
Edited by: G. U. Nienhaus © Humana Press Inc., Totowa, NJ

eties, and ligand molecules. The latter can be as small as diatomic molecules, such as nitric oxide (NO), which has been recognized as an important signaling molecule. Protein–ligand interactions play a key role in metabolic processes. Ligand binding can strongly modulate the behavior of a protein in its interactions with other proteins. Therefore, it is of utmost importance to elucidate the physicochemical basis of these interactions to understand biological function.

Whereas structural studies have always been appreciated, recent years have witnessed a growing awareness among researchers that explorations of the energetics and dynamics are essential for understanding protein–ligand interactions. The study of ligand binding equilibria gives insights into the free energy differences between the stable, bound, and unbound states, whereas kinetic measurements yield information on the rates of interconversion between different species and the energy barriers separating the states. Moreover, they allow detection of short-lived intermediates that may be only weakly populated in equilibrium so that they escape detection. Nevertheless, such transient species may be key factors for biological function.

Absorption spectroscopy in the visible and ultraviolet spectral regions is a powerful technique by which ligand binding equilibria can be studied. The technique also offers time resolution over the entire range of time scales relevant for biomolecular dynamics. A crucial prerequisite is that a suitable chromophore exists that is sensitive to ligation changes. The strongly absorbing heme prosthetic group has made heme proteins particularly attractive as model systems for investigating the physicochemical basis of ligand binding reactions *(1–3)*. Its electronic absorption spectrum depends on the charge of the central heme iron, binding of ligands, and even protonation events and structural changes in its vicinity *(4)*. The latter effects are more subtle and, therefore, require a careful spectral analysis. Myoglobin (Mb) and hemoglobin (Hb), the best-studied heme proteins, bind small diatomic ligands such as carbon monoxide (CO), dioxygen (O_2), and NO. In recent years, various novel ligand-binding heme proteins have been discovered. Neuroglobin (Ngb), for example, is a globin expressed in the vertebrate brain with an as yet unknown biological function. In the following, we present selected examples from our ongoing research on ligand binding to Mb, Hb, and Ngb as applications of steady-state, ns time- and temperature-resolved ultraviolet/visible (UV/Vis) absorption spectroscopy.

1.1. Background

1.1.1. Absorption of Light by Proteins

In proteins, we distinguish three different internal chromophores that give rise to electronic absorption bands: 1) The peptide bond linking the amino acids absorbs weakly around 220 nm; 2) the aromatic amino acids phenylalanine, tyrosine, tryptophan, and histidine contribute with bands in the range of 230–300 nm; and 3) most importantly, many biological molecules show strong absorbance in the visible region of the spectrum as a result of the presence of metal ions and prosthetic groups with extended π-electron systems, such as chlorophyll, carotinoid, flavin, heme, and retinal. These bands are sensitive to the surrounding polypeptide environment and reflect structural changes, oxidation states, and the binding of ligands, making UV/Vis spectroscopy an excellent tool for physical studies of protein function.

In absorption spectroscopy, the loss of light from an incident beam is measured as a function of the frequency (or wavelength) of the radiation. The absorbance or optical density, $A(\lambda)$, of a given sample is described by Beer's Law:

$$A(\lambda) = \log\left(\frac{I_0}{I}\right) = \varepsilon(\lambda)cl, \tag{1}$$

with incident light intensity, I_0, light intensity, I, after passing through the sample, extinction coefficient $\varepsilon(\lambda)$, concentration c and path length l. With the concentration in [M] and the path length in cm, the unit of the extinction coefficient ε is M^{-1} cm^{-1}. Note, that A is a logarithmic quantity and therefore unitless. Frequently, absorbances are quoted with the unnecessary pseudo-unit *optical density* (OD) attached. As is evident from **Eq. 1**, UV/Vis absorption by biomolecules is a very convenient method to measure the concentration of a protein solution.

Room temperature absorbance spectra of protein molecules are typically measured in aqueous solvent, buffered at approx pH 7.0, to mimic physiological conditions. For work at cryogenic temperatures, cryosolvents are employed that turn into a glass at low temperature in which the proteins are embedded. Frequently, glycerol and buffer are mixed at a volume ratio of 3:1, but other cryosolvents have also been explored *(5)*. With strongly absorbing chromophores such as the heme group ($\varepsilon \approx 100{,}000\ M^{-1}$ cm^{-1} for the most prominent Soret band), typical protein concentrations are in the range of 10–100 µ*M* when using 1 cm pathlength cuvets.

1.1.2. Steady-State Spectroscopy

We consider a simple, bimolecular binding reaction of a ligand L with a protein P,

$$P + L \leftrightarrow PL\,. \tag{2}$$

The equilibrium between the free species and the protein–ligand complex is characterized by the dissociation coefficient, K_d, or the association coefficient, K_a:

$$K_d = \frac{1}{K_a} = \frac{[L][P]}{[LP]} \tag{3}$$

Here, the brackets denote the concentrations of the respective species. The degree of saturation, or fraction of proteins with a bound ligand, Y, is given by

$$Y = \frac{[LP]}{[P]+[LP]} \tag{4}$$

By combining **Eqs. 3** and **4**, we obtain the saturation as a function of the (free) ligand concentration,

$$Y = \frac{[L]}{[L]+K_d} \tag{5}$$

From this hyperbolic dependence, it is also apparent that K_d is the ligand concentration at which half-saturation occurs. Equilibrium studies provide K_d, thus enabling the calculation of the standard free energy change upon ligand dissociation, ΔG,

$$K_d = \exp\left[-\frac{\Delta G}{RT}\right], \tag{6}$$

with gas constant R and absolute temperature T. **Equation 5** describes the simplest case of a single binding site per protein molecule. Many sophisticated schemes can be found in the literature that were developed to treat proteins with multiple binding sites and different ligands, and to also introduce energetic coupling between the sites ***(1–3)***.

1.1.3. Time-Resolved Spectroscopy

The measurement of ligand binding equilibria merely yields the free energy change associated with the reaction, but does not provide information on the time it takes for the reaction to occur. Therefore, time-resolved spectroscopic experiments are carried out to complement equilibrium studies. Frequently, rate coefficients governing the reaction are determined by exploiting the fact that the equilibrium coefficient of a reaction, K, and thus the Gibbs free energy difference, ΔG (**Eq. 6**), can be changed by applying a *sudden* perturbation, i.e., a perturbation on a time scale faster than the time response of the reaction itself. This can be achieved, for example, by changing the sample pressure, temperature, or the concentrations of reaction components in fast mixing experiments, or by photodissociating ligands from proteins using short laser pulses. Another elegant method to quickly change the concentration of reaction components is based on the use of so-called *caged* compounds. In this approach, the reactants are mixed but cannot react because the ligand is deactivated (caged) by a protecting group. It is uncaged by a short UV flash after which the reaction proceeds (***6***).

The sample is in a nonequilibrium state immediately after the external perturbation; and the ensuing relaxation towards the new equilibrium can be fol-

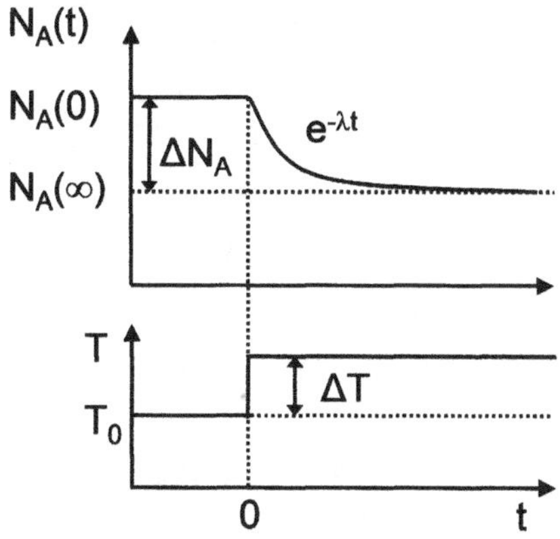

Fig. 1. Schematic representation of a temperature jump experiment, where the sample temperature is suddenly raised by ΔT. ΔN_A is the total extent of the population change in state A upon the temperature increase. $N_A(0)$ and $N_A(\infty)$ are the equilibrium populations in state A prior to and after the perturbation, respectively.

lowed by time-resolved spectroscopy, as shown schematically in **Fig. 1** for a temperature jump experiment. We mention in passing that another, fundamentally different approach exists for determining reaction rate coefficients. It involves the observation of number fluctuations in equilibrium. These occur with significant amplitude only in small molecular ensembles; in the extreme case, one observes chemical reactions at the level of single molecules *(7,8)*.

In the following, we briefly sketch the very basics of chemical kinetics that suffices for modeling the examples presented in this chapter. A number of textbooks can be consulted for an in-depth treatment (*see* **refs.** ***9–13***). The simplest reaction mechanism is that of a unimolecular reaction, in which a molecule changes from state A to state B (**Fig. 2A**). The relative fraction of proteins in state A (B) is N_A (N_B), with $N_A + N_B = 1$. The rate coefficient, which represents the probability that a molecule performs a transition from A (B) to B (A) per unit time is k_{AB} (k_{BA}). The Gibbs free energy difference between states A and B is ΔG. In equilibrium, the number of molecules changing from B to A and A to B is identical,

$$N_A \cdot k_{AB} = N_B \cdot k_{BA}, \tag{7}$$

and therefore, the equilibrium coefficient, K, is given by

$$K = \frac{N_A}{N_B} = \frac{k_{BA}}{k_{AB}}. \tag{8}$$

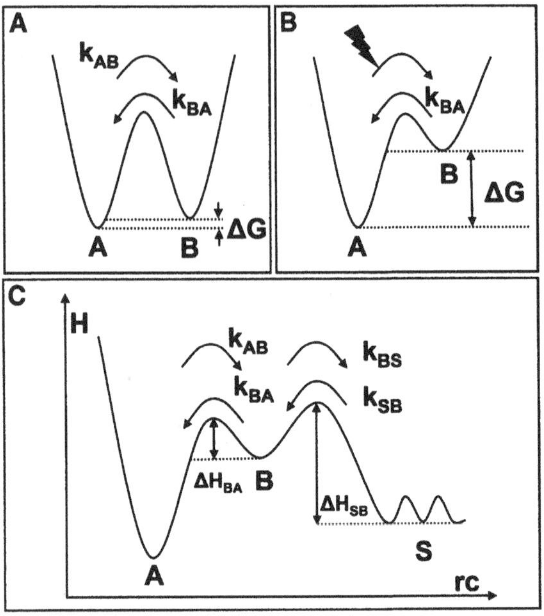

Fig. 2. Schematic depictions of reaction energy surfaces. **(A)** Two-state system with a small Gibbs free energy difference ΔG. States A and B are both significantly populated. State A (B) represents the ligand-bound (unbound) state. **(B)** Two-state system with a large Gibbs free energy difference ΔG. In thermal equilibrium, only state A is populated. **(C)** Three-state system with ligand-bound state A, unbound state B with the ligand still inside the protein and solvent state S, where the photolyzed ligand has escaped from the protein into the solvent.

Its connection to the Gibbs free energy difference ΔG has already been introduced in **Eq. 6. Equation 8** shows that equilibrium studies yield only the ratio between rate coefficients k_{AB} and k_{BA}. To obtain information on the individual rate coefficients, two cases can be distinguished.

1. If the free energy difference ΔG is small (i.e., on the order of RT), $N_A \approx N_B$ (**Fig. 2A**). A small perturbation (e.g., a temperature or pressure jump) quickly changes ΔG and thus K, the population ratio in equilibrium. Subsequently, the system readjusts its populations N_A and N_B to restore equilibrium, as shown schematically in **Fig. 1**. The change in population N_A with time is governed by the differential equation,

$$\frac{dN_A}{dt} = k_{BA} \cdot N_B - k_{AB} \cdot N_A \tag{9}$$

Rearrangement and integration yields

$$N_A(t) = N_A(\infty) + \Delta N_A \exp(-\lambda t) \tag{10}$$

Here, $N_A(\infty)$ is the new equilibrium population in state A after the perturbation, and ΔN_A is the total extent of the change (**Fig. 1**). The time course is governed by the apparent rate coefficient,

$$\lambda = k_{AB} + k_{BA} \tag{11}$$

The microscopic rate coefficients k_{AB} and k_{BA} can be determined separately by also measuring their ratio, the equilibrium coefficient K (**Eq. 8**).

2. If ΔG is much larger than RT, $N_B \approx 0$ and $k_{AB} \approx 0$ (**Fig. 2B**). A major perturbation has to be applied to populate state B, for example, by using stopped flow, continuous flow, pulse radiolysis, or photoexcitation techniques. In this chapter, we focus on flash photolysis experiments with nanosecond laser pulses that photodissociate ligands from proteins. As long as the ligands do not exit the protein, a unimolecular mechanism is appropriate to model this reaction, with states A and B referring to the ligand-bound and photodissociated states, respectively. Because of the large ΔG, $k_{BA} >> k_{AB}$, and we may thus neglect k_{AB} in the calculation of N_A and N_B with **Eqs. 10** and **11**. We usually present flash photolysis data on the basis of the normalized survival probability of the protein in the photolyzed (unbound) state at time t after photolysis,

$$N(t) = \exp(-k_{BA} \cdot t), \tag{12}$$

which equals N_B for a two-state model. Frequently, the microscopic rate coefficients such as k_{BA} are observed to follow the Arrhenius law,

$$k_{BA} = A_{BA} \cdot \frac{T}{T_0} \cdot \exp(-\Delta H_{BA} / RT) \tag{13}$$

with pre-exponential A_{BA}, universal gas constant R, temperature T and reference temperature T_0, set to 100 K. The enthalpy barrier, ΔH_{BA}, is depicted in **Fig. 2C**. At very low temperatures ($T \leq 50$ K), the temperature dependence of ligand binding deviates from Arrhenius behavior because of quantum-mechanical tunneling. Moreover, in cryospectroscopy experiments, nonexponential kinetics is usually observed at temperatures below the glass transition temperature of the cryosolvent. This behavior arises because protein molecules become frozen in many slightly different conformations, each with a different ΔH_{BA}. Consequently, each protein has a different k_{BA}, which results in the nonexponential kinetics. To model the data, a probability density of enthalpy barriers, $g(H_{BA})$, is introduced to account for the structural heterogeneity ***(14)***. The time dependence of recombination is given by integration over all the different barriers,

$$N(t) = \int g(H_{BA}) \exp[-k_{BA} \cdot t] dH_{BA} \tag{14}$$

At ambient temperature, conformational fluctuations are usually faster than rebinding, which causes exponential rebinding with an average rate coefficient, $\langle k_{BA} \rangle$. Moreover, a simple two-well model with bound and unbound state fails at temperatures above the glass transition temperature of the cryosolvent. In that temperature range, both protein and solvent can perform structural fluctuations that enable ligands to exit the protein and migrate into the solvent. Therefore, photodissociated ligands do not only recombine from within the protein (geminately) but also escape into the solvent, and afterwards, a ligand from the solvent rebinds in a bimolecular reaction. To include this additional process in the model, we introduce an additional state S that refers to the protein with the ligand in the solvent. States A, B and S are interconnected by four microscopic rate coefficients, k_{AB}, k_{BA}, k_{BS}, and k_{SB} (**Fig. 2C**). The set of differential equations governing the time evolution of the populations in the three wells can be solved in a straightforward manner ***(9)***. However, the expressions are rather clumsy and provide little physical insight. A much simpler solution can be obtained, which also provides a physical picture of the processes, by introducing the approximations $k_{BA} = 0$ and $k_{SB} >> k_{BA}$, k_{BS}, which are justified under the typical experimental conditions. The latter assumption holds if two temporally well-separated processes are observed in the kinetics. Immediately after photolysis, all ligands are photodissociated and reside in the protein; thus, $N_B = 1$. The temporal development of the photodissociated fraction,

$$N(t) = 1 - N_A(t) = N_B(t) + N_S(t) \tag{15}$$

is given by two sequential exponential processes,

$$N(t) = N_I \exp(\lambda_I t) + N_S \exp(\lambda_S t), \tag{16}$$

with amplitudes

$$N_I = \frac{k_{BA}}{k_{BA} + k_{BS}} \tag{17}$$

and

$$N_S = \frac{k_{BS}}{k_{BA} + k_{BS}}, \tag{18}$$

and apparent rate coefficients

$$\lambda_I = k_{BA} + k_{BS} = \frac{k_{BA}}{N_I}, \tag{19}$$

and

$$\lambda_S = k_{SB} \cdot N_I = k_{SB} \frac{k_{BA}}{k_{BA} + k_{BS}} = k_{BA} \cdot \frac{k_{SB}}{k_{BS}} \cdot \frac{k_{BS}}{k_{BA} + k_{BS}}. \tag{20}$$

N_I denotes the fraction of ligands that rebind internally (geminately) and N_S those that recombine from the solvent. Whereas λ_I describes the time course of depletion of state B, λ_S represents bimolecular rebinding from the solvent. $N_S \approx 1$ implies that essentially all ligands escape into the solvent from state B. If $N_S << 1$, most ligands cannot escape after dissociation and rebind from within the protein. In the geminate process, the same ligand that was initially photolyzed will rebind at the heme iron. Thus, λ_I does not depend on the ligand concentration in the solvent. Rebinding from the solvent, however, is a bimolecular process; its rate depends on both the concentrations of ligand and protein,

$$-\frac{d[P]}{dt} = \lambda'_S [P][L]. \tag{21}$$

The solution of **Eq. 21** is a complex expression that crucially depends on the relative concentrations of protein and ligand. To simplify the problem, ligand binding studies are frequently carried out with a large excess of ligand, $[L] >> [P]$. As a consequence, the ligand concentration can be taken as effectively constant during the course of the reaction. Then, the reaction can be described with a pseudo-first order rate coefficient,

$$\lambda_S = \lambda'_S [L], \tag{22}$$

that depends linearly on the ligand concentration.

2. Materials

2.1. Reagents

1. Gases: N_2, CO, NO, O_2.
2. 0.05–0.1 *M* Sodium phosphate buffer, pH 8.0.
3. 0.05–0.1 *M* Potassium phosphate buffer, pH 8.0.
4. Identical concentrations (0.05–0.1 *M*) of sodium citrate and Na_2HPO_4 are mixed to obtain buffers with final pH between 4.0 and 8.5.
5. Identical concentrations (0.05–0.1 *M*) of Na_2CO_3 and $NaHCO_3$ are mixed for buffers with final pH 8.5–11.0.
6. Glycerol (Sigma).
7. 1 *M* Sodium dithionite ($Na_2S_2O_3$) in H_2O (preparation described below).
8. Saturated solution of potassium hexacyanoferrate [$K_3Fe(CN)_6$] in H_2O.

All gases are commercially available in pressure bottles and purchased at purities of at least 99.97%. Equilibration of a buffer solution with 1 bar of a particular gas will result in a well-defined concentration of dissolved gas (*see* **Note 1**). To obtain lower concentrations, gas mixtures (e.g., 5% CO, 95% N_2) can be ordered.

All buffer salts are of ACS grade and available from chemical companies like Sigma. They are used without further purification. Buffers should be prepared with Millipore water and stored at 4°C. pH measurements are carried out at ambient temperature (*see* **Note 2**). Typical buffer concentrations are 10–100 m*M*. Glycerol is hygroscopic and should thus be acquired in small aliquots. Sodium dithionite oxidizes and should thus not be stored over long periods of time.

2.2. Sample Preparation

In the following, we describe the preparation of heme protein samples with different oxidation and ligation states of the heme iron, starting with freeze-

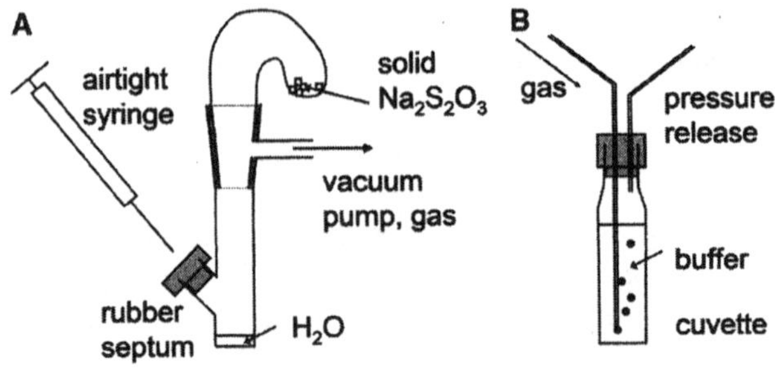

Fig. 3. **(A)** Homebuilt glassware to anaerobically prepare a sodium dithionite solution. **(B)** Experimental set-up to prepare a buffer solution with a well-defined concentration of dissolved gas.

dried protein powder or concentrated stock solutions (*see* **Note 3**). The samples are prepared in commercial quartz or glass cuvets ($1 \times 1 \times 3$ cm^3) with septum inlets to allow sealing with rubber septa.

A: Met sample (ferric heme iron):

Recombinant expression and purification sometimes yields (at least partially) the ferrous form (ligated with O_2 or CO) for a number of heme proteins. To obtain a sample with ferric heme iron only, it has to be oxidized, typically with $K_3Fe(CN)_6$.

1. Dissolve the protein at $\geq$ 100 μ*M* in 80 μL of 0.1 *M* buffer (Eppendorf tube).
2. Add 20 μL of a saturated $K_3Fe(CN)_6$ solution.
3. Wait 5 min.
4. Submit the mixture to a G25 gel filtration column (vol. 5 mL), equilibrated with the same buffer. The protein will be eluted first.

B: Deoxy (ferrous heme iron, no exogenous ligand), CO-ligated and NO-ligated samples:

1. 2 mL of buffer are put in a quartz cuvet ($1 \times 1 \times 3$ cm^3) sealed with a rubber stopper.
2. For a deoxy sample, equilibrate the buffer with N_2 gas for at least 5 min to remove dissolved O_2 (**Fig. 3**). For a CO-ligated sample, equilibrate with CO gas for 15 min. For an NO-ligated sample, equilibrate with N_2 for at least 15 min to remove O_2 and then with NO for another 15 min. Note that NO will react with residual O_2 and water to give NO_2^- and NO_3^-.
3. Add 5 μL of an anaerobically prepared 1 *M* $Na_2S_2O_3$ solution (see paragraph D) with an airtight syringe. For a ferric NO sample, skip this step.
4. Add a few microliters of a concentrated protein stock solution (1–5 m*M*) with an airtight syringe until the absorbance of the Soret band is between 1 and 2 OD. For the ferric NO sample, the stock solution has to be oxidized and deaerated.

C: O_2-ligated heme protein (ferrous heme iron, O_2 ligand):

1. Add a few grains of solid sodium dithionite to the met solution (prepared as described in *A*) to reduce the heme iron. Note, that a large excess of sodium dithionite will acidify the protein solution to the point where the protein denatures.
2. Remove excess reducing agent by passing the solution through a G25 column. In the absence of dithionite, the reduced heme iron will bind O_2 dissolved in the buffer (*see* **Note 1**).
3. If necessary, dilute the protein solution with additional buffer.
4. Seal the cuvet with a rubber septum. Oxygenated samples are usually unstable and autoxidize rapidly.

D: Sodium dithionite solution:

1. Add 200 mg of solid $Na_2S_2O_3$ and 1 mL water to a special glass container (**Fig. 3A**).
2. Evacuate with a vacuum pump to close to the boiling pressure of water.
3. Fill with N_2 (for deoxy, NO) or CO gas to approx 1 atm.
4. Repeat **steps 2** and **3** five times.
5. Dissolve the $Na_2S_2O_3$.
6. Draw a few µL with an airtight syringe.
7. Discard.
8. Repeat **steps 6** and **7** three times.
9. Draw 5 µL and add them to the prepared buffer solution through a rubber septum.

E: Samples for experiments at cryogenic temperatures:

1. Instead of aqueous buffer, use 3:1 glycerol/potassium phosphate buffer solutions. The viscosity of this cryosolvent increases upon lowering the temperature. Below its glass transition (T_g = 175 K), it behaves like a solid in which the proteins are embedded.
2. Transfer the samples from quartz cuvets into home-built $10 \times 10 \times 2.5$ mm^3 polymethyl methacrylate (PMMA) cuvets, using an airtight syringe.
3. To minimize gas exchange with the environment and to facilitate the application of vacuum for thermal insulation in the cryostat, the cuvet is hermetically sealed with a PMMA plug and glue.

F: Affinity determination (described here for CO):

1. Equilibrate buffer with 1 bar CO gas for 15 min at room temperature. This will result in a 1 m*M* aqueous solution of CO (*see* **Note 1**).
2. Add specific amounts of this solution in a step-wise manner to a ferrous deoxy sample prepared as described in B.

3. Methods

3.1. Steady-State Spectroscopy

UV/Vis absorption spectra are usually recorded with dispersive dual-beam spectrometers available from several companies (e.g., Varian, Hewlett Packard,

Perkin Elmer). Monochromatic light is generated from a white light source (tungsten lamp for the visible and hydrogen lamp for the ultraviolet) by means of a monochromator and passed through the sample and/or the reference path in an alternating fashion. The spectral resolution of the instrument can be varied by adjusting the slit width of the monochromator. Absorbance bands of proteins are rather broad, with a full width at half maximum of typically a few tens of nanometers. Therefore, it suffices to record them with approx 1 nm resolution. Note, that for better spectral resolution, the required narrower slit admits less light to the sample and causes noisier spectra.

For routine measurements of absorption spectra at room temperature, the sample solutions are loaded into a commercial $1 \times 1 \times 3\ cm^3$ cuvet (glass or quartz) and placed into the cuvet holder of the instrument. A reference sample with solvent only may be put into the reference beam (*see* **Note 4**). The cuvet has to be filled sufficiently so as to ensure that the entire cross section of the light beam impinges onto the sample solution. For visual inspection, it is helpful to set the monochromator to 550 mm (green light). For smaller sample volumes, cuvets with reduced pathlengths and widths are commercially available. When working with reduced width, special care has to be taken to focus the beam onto the sample, for example, by reducing the slit.

3.2. Nanosecond Time-Resolved Spectroscopy

Figure 4 shows the experimental setup of our homebuilt flash photolysis system. Photoexcitation of the sample is achieved by a 532 nm, 6 ns (full width at half maximum) pulse from a frequency doubled, Q-switched Nd:YAG laser (either model Surelite II or NY 61, Continuum, Santa Clara, CA); typical pulse energies are approx 100 mJ. For the monitoring beam, we use a 150-W tungsten lamp in a parabolic reflector housing (model A 1010, PTI, Brunswick, NJ). The light is passed through a monochromator (mc) and, with the help of lenses, guided through the sample, a notch filter and a second monochromator to strongly suppress scattered light from the 532 nm laser pulse, onto a photomultiplier tube (model R5600U, Hamamatsu Corp., Middlesex, NJ). The photocurrent is converted into a voltage, amplified and recorded with a digital storage oscilloscope from 10 ns to 50 μs (model TDS 520, Tektronix, Wilsonville, OR) and a homebuilt logarithmic time-base digitizer (Wondertoy II) from 2 μs to 1000 s. This setup provides high temporal stability so that absorbance changes as small as 100 μOD can be measured for times up to 1000 s. A sample cryostat for temperature adjustment between 10 and 320 K completes the setup (vide infra). The experimental apparatus is entirely controlled by LabView programs (LabView release 4, National Instruments, Austin, TX). User-defined protocols enable data collection at different temperatures in a fully automated fashion over many hours.

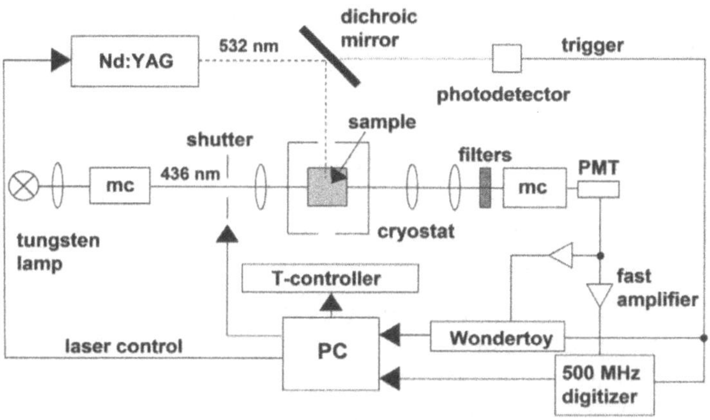

Fig. 4. Schematic of our home-built flash photolysis system. PMT, photomultiplier tube; mc, monochromator; Nd:YAG, ns pulse laser; Wondertoy, homebuilt logarithmic time-base digitizer with microsecond resolution.

In flash photolysis experiments on heme proteins, the bond between heme iron and the ligand is broken by the Nd:YAG laser pulse, thus creating a nonequilibrium state in the ensemble of protein molecules. Subsequently, recombination of ligands is observed by monitoring the light intensity, $I(t)$, at a particular wavelength λ as a function of time. The transmittance, $T(t)$, and absorbance change, $\Delta A(t)$, at a particular wavelength λ are calculated as

$$T(t) = \frac{I(t)}{I(0^-)}, \tag{23}$$

$$\Delta A(t) = -\log[T(t)], \tag{24}$$

where $I(0^-)$ denotes the transmitted light intensity just before the laser flash impinges on the sample. Absorbance changes may not only arise from ligand binding, but can also occur because of spectral shifts resulting from conformational changes in the protein (***15***). Great care has to be taken to disentangle these effects. Neglecting spectral shifts, the fraction of proteins that has not bound a ligand at time t after the photolyzing pulse, $N(t)$, is related to the absorbance change $\Delta A(t)$ by

$$N(t) = \frac{\Delta A(t)}{\Delta A_{max}}. \tag{25}$$

Here, ΔA_{max} denotes the absorbance change upon complete photolysis. In the absence of processes that are faster than the time resolution of the instrument, ΔA_{max} can be determined directly from the experimental data at time

$t = 0^+$ immediately after photolysis. The presence of fast processes can be inferred from a smaller total absorbance change than what one would expect from the steady-state spectra.

For an experiment, the sample (or the entire cryostat with the sample) is carefully positioned with respect to the monitoring light to obtain maximum intensity at the photomultiplier. Moreover, for homogeneous excitation, one has to ensure that the photolysis beam impinges on the whole sample. Data collection then proceeds as follows:

1. Select the monitoring wavelength to maximize the photolysis-induced absorbance difference between the ground state and the photoproduct species. This is usually done based on the steady-state spectra of bound and unbound species.
2. Adjust the power of the monitoring light source (*see* **Note 5**).
3. Adjust the laser power. The power necessary for complete (saturating) photolysis can be determined by test measurements with increasing laser power (*see* **Note 6**).
4. Take an offset with monitoring beam shutter closed (measure the voltage signal corresponding to no light on the detector).
5. Take an offset with the monitoring light passing through the sample (measure the voltage signal corresponding to the light transmitted through the sample without photoexcitation).
6. Fire the laser and record the voltage signal that represents the light intensity transmitted through the sample as a function of time. From this signal and the offset readings, the software calculates the absorbance. Typically, traces should be recorded multiple times and averaged, which is especially important to improve the signal-to-noise ratio in the submicrosecond region.

Commercial instruments are also available, for example the LKS.60 Nanosecond Laser Photolysis Spectrometer by Applied Photophysics Ltd. (Leatherhead, Surrey KT22 7PB, United Kingdom) (*see* **Note 7**).

3.3. Low-Temperature Spectroscopy

For low temperature data collection, samples loaded in our homebuilt PMMA cuvets are mounted in a block of oxygen-free high-conductivity copper. Good thermal contact is ensured by spring loading and the use of indium gaskets. The copper block is attached to the cold finger of a closed cycle helium refrigerator cryostat (for example, model 22, CTI Cryogenics, Mansfield, MA). For insulation, the cold finger with the sample is enclosed by a vacuum shroud. An assembly of quartz windows (diver's helmet) admits the photolysis pulse and the monitoring beam to the sample. To achieve the minimum temperature (approx 10 K for the model 22 refrigerator), a vaccum of at least 10^{-3} mbar is required to minimize heat flow to the cold finger. The temperature is measured with a silicon temperature sensor diode and regulated with a 50-Ω resistive heater. A digital temperature controller (model 330, Lake Shore Cryotronics,

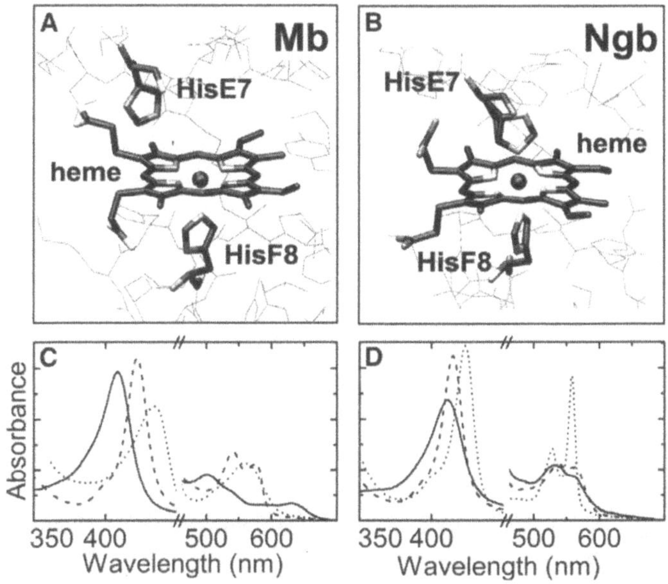

Fig. 5. Active site structures of **(A)** Mb and **(B)** Ngb. Explicitly shown are the heme and the distal and proximal histidines, HisE7 and HisF8. Absorption spectra of **(C)** Mb and **(D)** Ngb in the UV/Vis region of the spectrum. In the met species (solid lines), the ferric heme iron (Fe^{3+}) is coordinated by H_2O/OH in Mb and by the distal histidine HisE7 in Ngb. In the deoxy species (dotted lines), the sixth coordination site on the ferrous heme iron (Fe^{2+}) is vacant in Mb; in Ngb, the iron is still coordinated by the HisE7. In the CO-ligated form (dashed lines), a CO molecule is bound to Fe^{2+} in both proteins.

Westerville, OH) provides an electrical current through the heater to keep the sample at the desired temperature.

3.4. Applications

3.4.1. Steady-State Experiments

3.4.1.1. Determining Oxidation and Ligation States of the Heme Prosthetic Group

The spectra of heme depend sensitively on the oxidation and ligation state of the central iron atom. At physiological pH, Mb has a water molecule at the sixth coordination site in the ferric state (**Fig. 5A**). Ngb, by contrast, is bis-histidine ligated (**Fig. 5B**). The absorbance spectra of the two ferric proteins are shown in **Fig. 5C,D** (solid lines); spectra of the ferrous (dotted) and CO-ligated (dashed) species are also included. Absorbance spectra of Mb and Hb have been discussed in detail by Antonini and Brunori *(2)*.

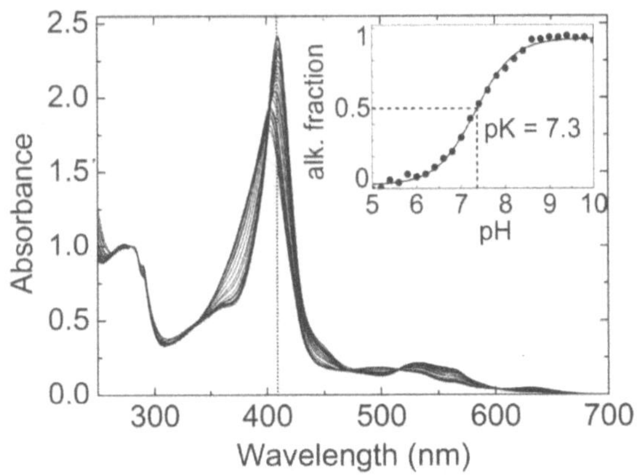

Fig. 6. Absorption spectra of ferric Ngb H64L in the pH range 5.0–11.0. The ligand at the sixth coordination site of the heme iron changes from H_2O to OH^- with increasing pH. Inset: fraction of the alkaline (OH^--bound) species versus pH, calculated from the normalized absorbance change at 409 nm.

3.4.1.2. pH-Dependent Spectra

Many ferric heme proteins bind water at physiological pH, which is replaced by hydroxyl ions at higher pH. To determine the pK of this acidic/alkaline transition, a concentrated (1–10 m*M*) stock solution of the protein is prepared in water or low concentration buffer. Here, we use ferric Ngb, mutant H64L, as an example. Equal aliquots (several μL) are added to buffer solutions (constant volume of 1–3 mL) of different pH, and absorbance spectra are collected, as shown in **Fig. 6**. Isosbestic points at 510 and 600 nm are indicative of a two-state interconversion. To analyze the data in a simple way, the absorbances at a selected wavelength, where large changes are observed (here at 409 nm), are normalized from 0 to 1 by subtracting the lowest value and dividing by the largest value. The normalized absorbances are plotted versus pH in the inset of **Fig. 6**. The pH dependencies of the fractional populations of the protonated (acid) and deprotonated (base) species, denoted by c_+ and c_0, are given by

$$c_+(\mathrm{pH}) = 1 - c_0(\mathrm{pH}) = \frac{1}{1+10^{n(\mathrm{pH}-\mathrm{pK})}}. \qquad (26)$$

For a single protonation, the Henderson–Hasselbalch relation holds, in which the parameter n equals 1. A cooperative transition involves simultaneous protonation at multiple sites and thus $n > 1$. A fit of **Eq. 26** to the data in **Fig. 6** reveals a single protonation with a pK of 7.3 ± 0.1 (*see* **Notes 8** and **9**).

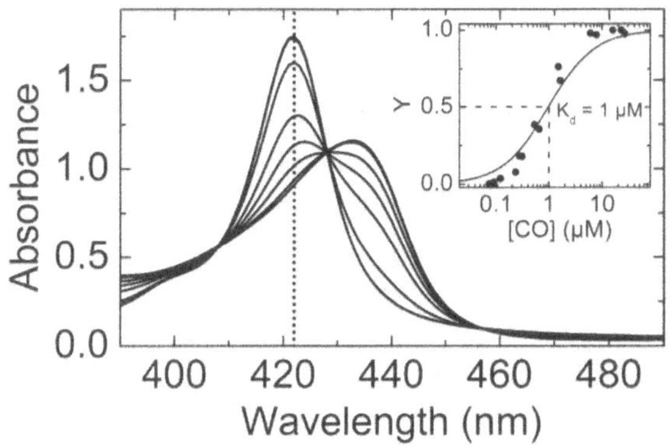

Fig. 7. Absorption spectra of ferrous Mb in the Soret region upon successive, quantitative addition of CO. Inset: Scaled absorbance change in the Soret band of the CO-ligated species at 422 nm, plotted versus the free ligand concentration.

3.4.1.3. Measuring Ligand Affinity

In a typical ligand binding experiment, the concentration of protein (ligand) is kept constant while the concentration of the ligand (protein) is varied. As an example, we present the absorbance spectra of ferrous Mb in the Soret region, recorded upon successive addition of CO (**Fig. 7**). With increasing CO concentration, the deligated species, characterized by a prominent peak at 434 nm, decreases, whereas the CO-bound form, characterized by the Soret at 422 nm, increases (*see* **Note 10**).

In the simplest fashion, the fraction of ligand-bound protein can be determined from the absorbance at a particular wavelength, $A(\lambda_0)$, at which spectral changes are large. In analogy to the treatment of the pH dependence, $A(\lambda_0)$ can be rescaled between 0 and 1. If the absorbance decreases upon ligation, $1-A(\lambda_0)$ is calculated. Finally, $A(\lambda_0)$ is plotted vs the (free!) ligand concentration (*see* inset **Fig. 7**), and the data are fitted with **Eq. 5**.

More elaborate approaches exist to analyze binding equilibria, which make use of the entire spectral information instead of only using data at a particular wavelength, thereby disregarding most of the data. If the spectra of the ligated and deligated forms are known, the spectrum at each ligand concentration can be fitted with a linear combination of the basis spectra of ligand-bound, S_b, and unbound species S_u, with fractions f_b and f_u:

$$S_{rec} = f_b S_b + S_u f_u \tag{27}$$

Singular value decomposition (SVD) is an even more sophisticated technique, with which we carry out a global analysis of a set of spectra taken as a

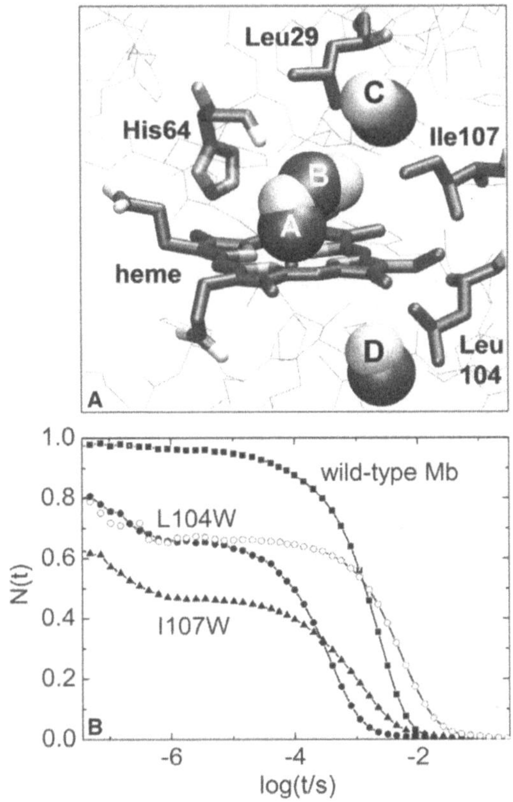

Fig. 8. (**A**) Active site structure of wild-type Mb, including the heme group, amino acid residues 29, 64, 104, and 107, and the CO in the bound state A and in photoproduct states B, C and D. (**B**) Time traces of CO recombination to wild-type Mb and mutants L104W and I107W. Samples were equilibrated with CO partial pressures of 1 bar (solid symbols) and 0.05 bar (open symbols).

function of an external control parameter, for example pH or ligand concentration. The algorithm allows one to determine the minimal set of spectra required for a superposition and the relative weight of each component spectrum ***(16)***.

3.4.2. Nanosecond Time-Resolved Spectroscopy

3.4.2.1. Ligand Binding in Myoglobin Mutants

In some heme proteins, hydrophobic interior cavities exist that are large enough to host xenon atoms or diatomic ligands. In recent years, the functional role of the internal cavities in Mb has been examined in great detail ***(17–23)***. In **Fig. 8A**, structural details near the active site are shown, including the heme group, important residues and the CO ligand in the bound-state position A as well as in photoproduct positions B, C, and D. Two mutants, L104W and I107W, have been constructed so as to prevent access to the internal cavities C and D. In **Fig. 8B**, we have plotted typical kinetic traces of wild-type Mb and mutants L104W and I107W after photolysis of the CO-ligated form at ambient temperature, monitored at 436 nm ***(21)***. The data extend over 7.5 orders of magnitude in time (30 ns to 1 s) and are thus presented on a logarithmic time axis. To present the normalized fraction of proteins without a ligand at time t after photolysis, $N(t)$, the absorbance differences were scaled by the change expected on the basis of the steady-state spectra of the CO-bound and unbound species. Wild-type Mb exhibits a very small internal (geminate) rebinding step of 4% between 100 ns and 1 μs that is barely visible. Consequently, ligands escape efficiently from the protein and subsequently rebind in a bimolecular fashion on the ms time scale. By contrast, L104W and even more marked I107W, display a much more pronounced geminate phase, and a significant fraction of ligands has already rebound at the shortest times that our setup can resolve. These examples show that one has to be exceedingly careful with the normalization of the data. Otherwise, fast processes may escape the attention of the researcher.

To distinguish between geminate and bimolecular rebinding from the solvent, the flash photolysis experiment has to be performed at different ligand concentrations (e.g., 0.1–1 m*M*, *see* **Note 1**) keeping the protein concentration constant (e.g., 10 μ*M*). The rate coefficient of bimolecular recombination scales with the ligand concentration, as shown in **Fig. 8B** for L104W MbCO for 1 m*M* and 0.05 m*M* CO, whereas the kinetics of geminate rebinding is not affected by the amount of CO in the solvent.

The observed kinetic differences result from different CO migration pathways within these proteins. In wild-type Mb, the CO initially settles in site B but can also escape to transient ligand docking sites C and D (**Fig. 8A**). Because rebinding can only occur from state B, the presence of alternative docking sites prevents rebinding on short time scales, and 96% of the ligands stay within the proteins without rebinding until a protein fluctuation opens a transient exit pathway to the solvent. In mutant L104W, site D has been blocked by an indole side chain, and in mutant I107W, pathways to both cavities C and D are occluded. As a consequence, geminate recombination is markedly enhanced, and the fraction of ligands escaping from the protein, N_S, is reduced in the order wild-type > L104W > I107W (**Fig. 8B**). The effect of ligand migration is also reflected in the bimolecular rate coefficients. CO ligands entering the protein from the solvent migrate between the accessible internal docking sites and may eventually escape again. For wild-type Mb, the yield of solvent process of 96% means that, statistically, upon 25 CO entries into the deligated protein,

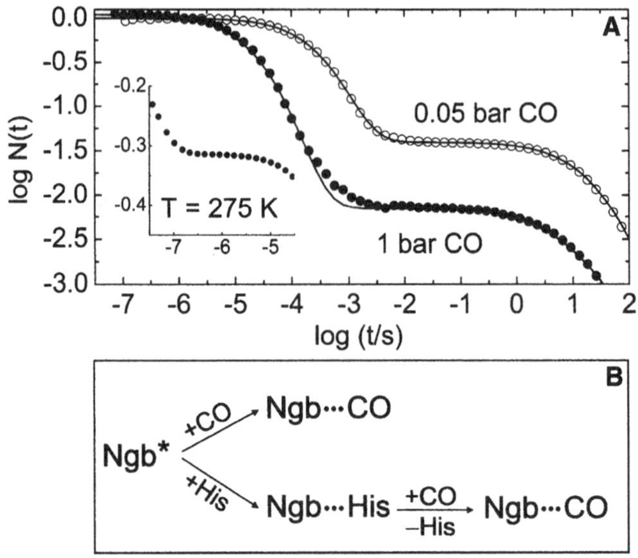

Fig. 9. (**A**) Time traces of CO recombination to murine wild-type Ngb. Sample solutions were equilibrated with 1 bar (closed symbols) and 0.05 bar (open symbols) CO. Inset: At temperatures below 275 K, the geminate process is visible at the earliest times. (**B**) Schematic representation of the individual steps in the recombination process. Ngb* denotes the pentacoordinate species.

only once a covalent bond with the heme iron is formed. Consequently, the bimolecular rate coefficient λ_S increases in the order wild-type < L104W < I107W.

3.4.2.2. Ligand Binding to Neuroglobin

In **Fig. 9A**, we have plotted the time traces of CO recombination to Ngb after photolysis at 293 K for two different CO concentrations. The data were arbitrarily normalized to $N(t) = 1$ the early time points. As for MbCO, the experimental data (symbols) again show two steps that can be fitted by a sum of two Gaussian rate distributions (lines). However, in contrast to MbCO, the first step depends on the CO concentration and thus represents bimolecular recombination, whereas the second step is independent of the CO concentration. A fast geminate process with considerable amplitude can be resolved at 275 K (**Fig. 9A**, inset), but at 293 K, it has become too fast to resolve with the 30 ns time resolution of our instrument.

To correctly interpret the two-step kinetics at 293 K, one has to consider the peculiar structure of Ngb. In the absence of an exogenous ligand, the heme iron is hexacoordinate, with the distal and proximal histidines serving as axial ligands (**Fig. 5B**). After photolysis of NgbCO, the heme is pentacoordinate and extremely reactive. Although geminate rebinding is, therefore, very fast at 293 K, a major fraction of CO ligands is nevertheless able to escape into the solvent. Upon returning to the heme iron, the exogenous CO ligands have to compete with the endogenous HisE7 imidazole side chain for the free binding site at the heme iron. In a certain fraction of molecules, HisE7 will bind at the sixth coordination. Consequently, CO recombination will not be completed, but come to a halt, as seen from the plateau at times longer than approx 1 ms at 293 K (**Fig. 9A**). Note, that CO binding is bimolecular and slows with decreasing CO concentration, and thus, the fraction of endogenous HisE7 binding increases, as inferred from the CO concentration dependence of the amplitude of the plateau in **Fig. 9A**. Only after thermal dissociation of the endogenous ligand, CO can form the thermodynamically more stable complex. This happens at $t > 1$ s in the second kinetic step *(24)*. The resulting reaction scheme is presented in **Fig. 9B**.

3.4.2.3. Flash Photolysis With Spectral Resolution

Single-wavelength kinetic monitoring suffices for analyzing two-state processes. In NgbCO, however, photolysis creates a pentacoordinate species that subsequently converts into a CO-ligated species and a hexacoordinate deoxy species by bond formation between the heme iron and the HisE7 imidazole *(24)*. In this case, we have to consider three species, and great care has to be taken to correctly interpret the kinetics at a single wavelength. Note that the kinetic amplitudes in **Fig. 9A** do not represent the relative fractions of the species because they will in general absorb differently at the monitoring wavelength λ_0. To analyze the process in depth, we examined the time evolution of the entire spectrum by performing ns time-resolved spectroscopy with monitoring at different wavelengths.

Figure 10A shows a contour plot of the spectral changes in the Soret region of NgbCO upon photolysis. Dashed lines represent the disappearance of the CO-ligated form, whereas solid lines represent the increased absorption due to the transient species. At times shorter than approx 1 ms, the absorbance increase peaks at 432 nm because of the pentacoordinate form. At times longer than 10 ms, the maximum has shifted to 425 nm, indicating the transition from a pentacoordinate heme iron to ligation by the distal histidine HisE7 *(25)*. **Figure 10B** shows how the measured absorbance difference signal (open squares) can be analyzed to obtain the spectrum of the pentacoordinate species. The absorbance spectrum of the NgbCO sample (solid spheres) was determined in the ns spectrometer in the absence of a photolysis pulse. From the ligand bound and the difference spectrum, the spectrum of the pentacoordinate species can be reconstructed (solid squares).

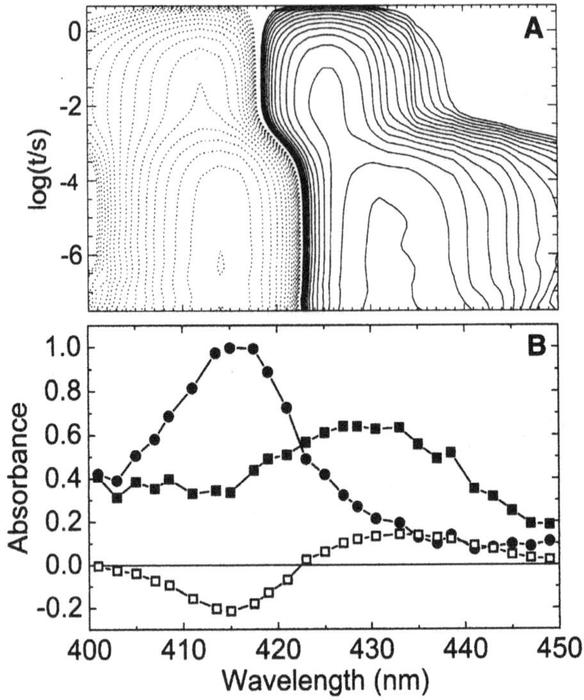

Fig. 10. (**A**) Contour plot of the time dependence of the photolysis difference spectrum of NgbCO. (**B**) Determination of the room-temperature absorption spectrum of pentacoordinate Ngb. The absorption spectrum of NgbCO in the Soret region was determined with the flash photolysis system (solid dots). The absorbance difference spectrum at early times determined from flash photolysis traces (open squares) is used to reconstruct the spectrum of the pentacoordinate Ngb species (closed squares).

3.4.2.4. Flash Photolysis at Cryogenic Temperatures

In cryogenic experiments, ligand recombination occurs exclusively from within the protein below the glass transition of the solvent. Protein dynamics is strongly suppressed in the glass, and consequently, geminate rebinding is nonexponential because of structural heterogeneity giving rise to different activation enthalpy barriers against rebinding. **Figure 11A** shows time traces of CO recombination to mutant ThrE11Val of the monomeric Hb of the nemertean worm *Cerebratulus lacteus* between 60 and 140 K (20 K steps). Only one process is present, thermally activated according to the Arrhenius relation (**Eq. 13**). The lines are the results of a global fit with a single distribution of enthalpy barriers peaking at 5.8 kJ/mol (**Fig. 11A**, inset) *(26)*. At temperatures above the glass transition temperature of the cryosolvent, ligands can escape into the solvent and subsequently rebind in an exponential, bimolecu-

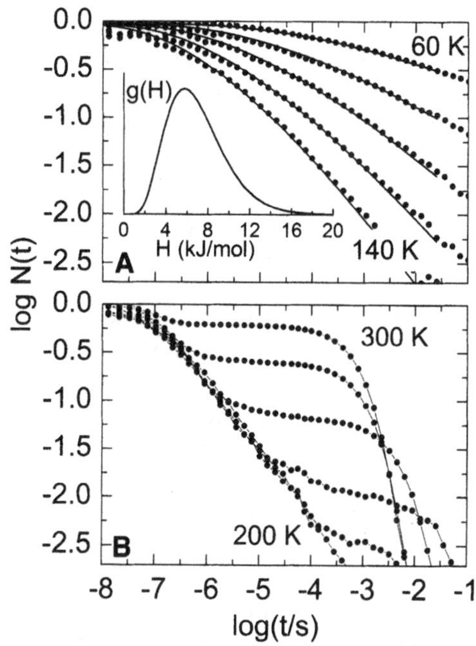

Fig. 11. **(A)** Geminate recombination of CO to the monomeric Hb of *Cerebratulus lacteus* between 60 and 140 K (spheres) after photodissociation. All traces were fitted with an enthalpy barrier distribution (inset) peaking at approx 5.8 kJ/mol (lines). **(B)** Geminate and bimolecular CO recombination between 200 and 300 K.

lar process. **Figure 11B** shows data between 200 and 300 K. Apparently, geminate rebinding remains non-exponential, whereas bimolecular recombination is much slower and essentially exponential. This indicates that, on the millisecond time scale, protein fluctuations cause an averaging of the heterogeneous rebinding barriers. The fraction of solvent process, N_S, continually grows with temperature because the rate coefficient for ligand escape, k_{BS}, is more strongly temperature-activated than the one for rebinding, k_{BA} (**Fig. 2**). A careful analysis of the temperature dependencies of the amplitudes and apparent rate coefficients of the kinetic features provides quantitative information about the detailed mechanism and reaction energy surface governing the ligand binding process.

4. Notes

1. The concentration of dissolved gas in a buffer solution is proportional to the partial pressure of the gas in equilibrium with the solution. A partial pressure of 1 bar above the solution results in the following concentrations of dissolved gas

(at 20° C): 1 bar CO = 1 m*M* dissolved CO; 1 bar O_2 = 1.3 m*M* dissolved O_2; 1 bar NO = 2 m*M* dissolved NO (*27*). Note, that air contains approx 20% of O_2, resulting in a concentration of approx 260 μ*M* dissolved O_2 in air-equilibrated buffer. It takes about 15 min of bubbling gas through the buffer to equilibrate the solution with the gas.

2. Some buffer solutions, for example TRIS (tris-(hydroxymethyl)-aminomethane), exhibit a significant change of pH with temperature and should therefore be avoided when measuring temperature dependencies of absorbance spectra and kinetics (*5,28*).
3. For absorbance spectroscopy, the protein concentration in the sample should be adjusted such that the maximum absorbance is below 2 OD, which means that at least 1% of the light passes through the sample. At higher absorbances, spectra will suffer from noise unless data collection times are extended. Moreover, the nonlinear response of the instrument gives rise to spectral distortions. For strong chromophores such as the heme group (extinction coefficient approx 100,000 M^{-1} cm^{-1} at the Soret peak maximum), sample concentrations are typically in the micromolar range when using standard, 1 cm path length cuvets. It is also not advisable to record spectra with peak absorbances below 0.01 OD.
   In time-resolved spectroscopy, the sample concentration has to be adjusted so as to (i) achieve a high photolysis yield at the wavelength of the laser pulse (532 nm for frequency-doubled Nd:YAG lasers), (ii) to obtain a significant absorbance difference, and (iii) a minimal background absorbance at the monitoring wavelength λ upon photolysis. In practice, one prefers to monitor at a wavelength at which laser excitation causes a bleaching signal so that the absorbance is small at short times, where the data quality is limited by shot noise (photon quantization noise). Typically, the absolute absorbance is kept between 0.1 and 1 OD at all times.
4. The instrument also allows to first taking a baseline that will automatically be subtracted from the subsequent data. Here, the same cuvet can be used for baseline and sample. Therefore, baseline shifts as a result of the cuvet itself are excluded.
5. Maximum power of the monitoring light source yields the best signal-to-noise ratio at short times. However, if slow processes are present in the kinetics, photoexcitation by the monitoring light source can cause a significant buildup of photoproducts on time scales longer than approx 100 ms, thus reducing the detected absorbance difference. Moreover, slow kinetic features can be significantly distorted due to this effect. To avoid this problem, the light source is adjusted to lower power for the study of slower processes.
6. When photodissociating with linearly polarized light, special attention has to be paid to avoiding artifacts as a result of photoselection, which refers to the preferential photoexcitation of molecules that have their transition dipole oriented parallel to the electric field vector of the excitation pulse (*29,30*). After partial photolysis, rotational diffusion of the protein molecules creates additional kinetic features that are superimposed on those due to the photo-induced ligand binding reaction. This problem can be avoided by applying flashes that

ensure photoexcitation yields close to 1, or by monitoring at the magic angle of 54.7° with respect to the excitation polarization, or by separately measuring with the monitoring light polarized parallel or perpendicular to the excitation pulse. The photoexcitation yield can be determined by measuring the power dependence of the absorbance difference signal. Note, that even with saturating flashes, the signal may be smaller than expected from the steady-state spectra of the ligand-bound and ligand-free species because of rebinding on time scales below approx 10 ns, the shortest time that can be measured with ns pulsed laser excitation.

7. Details of the commercial instrument by Applied Photophysics can be found on the web at http://www.photophysics.com/.
8. It is advisable to check the pH of the final protein solution because it might differ from the buffer pH.
9. In the approach described here, each data point requires preparation of a new sample. An elegant way to change the pH in a given sample is the use of *caged* protons that can be successively *uncaged* by near-UV laser flashes ***(31)***.
10. The ligand concentration should not be changed linearly, but in logarithmic steps, for example, by diluting the stock solution sequentially by a constant factor. Frequently, the factor 2 is chosen, yielding 3.3 data points per order of magnitude. In a logarithmic plot, data points will be evenly spaced along the x-axis.

Acknowledgments

This work was supported by the Deutsche Forschungsgemeinschaft (DFG, grant Ni291/3) and the Fonds der Chemischen Industrie.

References

1. Weber, G. (1992) *Protein Interactions*, Chapman & Hall, New York.
2. Antonini, E. and Brunori, M. (1971) *Hemoglobin and Myoglobin in Their Reactions with Ligands*, North-Holland, Amsterdam.
3. Cantor, C. R. and Schimmel, P. R. (1980) *Biophysical Chemistry*, III, W. H. Freeman and Company, New York.
4. Nienhaus, K., Lamb, D. C., Deng, P., and Nienhaus, G. U. (2002) The effect of ligand dynamics on heme electronic transition band III in myoglobin. *Biophys. J.* **82,** 1059–1067.
5. Douzou, P. (1977) *Cryobiochemistry*, Academic Press, London.
6. Marriot, G. (1998) *Caged Compounds. Methods Enzymology* **291**.
7. Leuba, S. H. and Zlatanova, J. (2001) *Biology at the Single Molecule Level*, Pergamon, Oxford, UK.
8. Rigler, R., Orrit, M., and Basche, T. (2002) *Single Molecule Spectroscopy*, Springer-Verlag, New York.
9. Moore, J. W. and Pearson, R. G. (1981) *Kinetics and Mechanism*, John Wiley & Sons, New York.
10. Hammes, G. G. (2000) *Thermodynamics and Kinetics for the Biological Sciences*, Wiley-Interscience, New York.

11. Hammes, G. G. (1978) *Principles of chemical kinetics*, Academic Press, London.
12. Hammes, G. G. (1974) *Techniques of Chemistry*, Wiley-Interscience, New York.
13. Gutfreund, H. (1995) *Kinetics for the life sciences*, University Press, Cambridge.
14. Steinbach, P. J., Chu, K., Frauenfelder, H., Johnson, J. B., Lamb, D. C., Nienhaus, G. U., Sauke, T. B., and Young, R. D. (1992) Determination of rate distributions from kinetic experiments. *Biophys. J.* **61,** 235–245.
15. Ormos, P., Szaraz, S., Cupane, A., and Nienhaus, G. U. (1998) Structural factors controlling ligand binding to myoglobin: a kinetic hole-burning study. *Proc. Natl. Acad. Sci. USA* **95,** 6762–6727.
16. Müller, J. D., McMahon, B. H., Chien, E. Y., Sligar, S. G., and Nienhaus, G. U. (1999) Connection between the taxonomic substates and protonation of histidines 64 and 97 in carbonmonoxy myoglobin. *Biophys. J.* **77,** 1036–1051.
17. Ostermann, A., Waschipky, R., Parak, F. G., and Nienhaus, G. U. (2000) Ligand binding and conformational motions in myoglobin. *Nature* **404,** 205–208.
18. Bourgeois, D., Vallone, B., Schotte, F., Arcovito, A., Miele, A. E., Sciara, G., Wulff, M., Anfinrud, P., and Brunori, M. (2003) Complex landscape of protein structural dynamics unveiled by nanosecond Laue crystallography: Watching a protein as it functions with 150-ps time-resolved X-ray crystallography. *Proc. Natl. Acad. Sci. USA* **100,** 8704–8709.
19. Brunori, M., Cutruzzola, F., Savino, C., Travaglini-Allocatelli, C., Vallone, B., and Gibson, Q. H. (1999) Structural dynamics of ligand diffusion in the protein matrix: A study on a new myoglobin mutant Y(B10) Q(E7) R(E10). *Biophys. J.* **76,** 1259–1269.
20. Chu, K., Vojtchovsky, J., McMahon, B. H., Sweet, R. M., Berendzen, J., and Schlichting, I. (2000) Structure of a ligand-binding intermediate in wild-type carbonmonoxy myoglobin. *Nature* **403,** 921–923.
21. Nienhaus, K., Deng, P., Kriegl, J. M., and Nienhaus, G. U. (2003) Structural Dynamics of Myoglobin: The Effect of Internal Cavities on Ligand Migration and Binding. *Biochemistry* **42,** 9647–9658.
22. Schotte, F., Lim, M., Jackson, T. A., Smirnov, A. V., Soman, J., Olson, J. S., Phillips, G. N., Jr., Wulff, M., and Anfinrud, P. A. (2003) Watching a protein as it functions with 150-ps time-resolved X-ray crystallography. *Science* **300,** 1944–1947.
23. Scott, E. E., Gibson, Q. H., and Olson, J. S. (2001) Mapping the pathways for O_2 entry into and exit from myoglobin. *J. Biol. Chem.* **276,** 5177–5188.
24. Kriegl, J. M., Bhattacharyya, A. J., Nienhaus, K., Deng, P., Minkow, O., and Nienhaus, G. U. (2002) Ligand binding and protein dynamics in neuroglobin. *Proc. Natl. Acad. Sci. USA* **99,** 7992–7997.
25. Nienhaus, K., Kriegl, J. M., and Nienhaus, G. U. (2004) Structural dynamics in the active site of murine neuroglobin and its effects on ligand binding. *J. Biol. Chem.* **279,** 22,944–22,952.
26. Pesce, A., Nardini, M., Ascenzi, P., Geuens, E., Dewilde, S., Moens, L., Bolognesi, M., Riggs, A. F., Hale, A., Deng, P., Nienhaus, G. U., Olson, J. S., and Nienhaus, K. (2004) ThrE11 regulates O_2 affinity in cerebratulus lacteus mini-hemoglobin. *J. Biol. Chem.* **279,** 33,662–33,672.

27. Lide, D. R. (1994) *Handbook of Chemistry and Physics*, CRC Press, Boca Raton, FL.
28. Orii, Y. and Morita, M. (1977) Measurement of the pH of frozen buffer solutions by using pH indicators. *J. Biochem.* **81,** 163–168.
29. Ansari, A., Jones, C. M., Henry, E. R., Hofrichter, J., and Eaton, W. A. (1993) Photoselection in polarized photolysis experiments on heme proteins. *Biophys. J.* **64,** 852–868.
30. Ansari, A. and Szabo, A. (1993) Theory of photoselection by intense light pulses. Influence of reorientational dynamics and chemical kinetics on absorbance measurements. *Biophys. J.* **64,** 838–851.
31. Barth, A. and Corrie, J. E. (2002) Characterization of a new caged proton capable of inducing large pH jumps. *Biophys. J.* **83,** 2864–2871.

12

Ultrafast Time-Resolved IR Studies of Protein–Ligand Interactions

Manho Lim and Philip A. Anfinrud

Summary

Time-resolved mid-IR spectroscopy combines molecular sensitivity with ultrafast capability to incisively probe protein–ligand interactions in model heme proteins. Highly conserved residues near the heme binding site fashion a ligand-docking site that mediates the transport of ligands to and from the binding site. We employ polarization anisotropy measurements to probe the orientation and orientational distribution of CO when bound to and docked near the active binding site, as well as the dynamics of ligand trapping in the primary docking site. In addition, we use more conventional transient absorption methods to probe the dynamics of ligand escape from this site, as well as the ultrafast dynamics of NO geminate recombination with the active binding site. The systems investigated include myoglobin, hemoglobin, and microperoxidase.

Key Words: Time-resolved; femtosecond; mid-IR spectroscopy; photoselection; anisotropy; heme-pocket; myoglobin; hemoglobin; NO; CO.

1. Introduction

An understanding of protein–ligand interactions is crucial to scientists' quest to decipher how enzymes function at an atomistic level of detail. These interactions involve a dynamic interplay between a ligand and its complementary binding site, where functionally significant motion *(1)* can span times ranging from femtoseconds to seconds. Protein–ligand interactions that can be abruptly perturbed with an optical pulse can be studied on the fastest of these time scales. Here, we describe the method of femtosecond time-resolved mid-infrared spectroscopy (mid-IR) spectroscopy and show how this technique has been used to incisively probe a ligand's orientation and dynamics while transiently trapped near the active site of heme proteins. These studies are possible because the mid-IR region of the electromagnetic spectrum excites molecular vibrations

From: *Methods in Molecular Biology, vol. 305: Protein–Ligand Interactions: Methods and Applications*
Edited by: G. U. Nienhaus © Humana Press Inc., Totowa, NJ

whose frequencies can be assigned to a specific ligand. Moreover, vibrational excitation is often polarized along a unique direction in the molecular frame, so polarized absorbance measurements can extract information about the orientation of molecules in the laboratory frame. By studying the orientation and dynamics of ligands with time-resolved mid-IR spectroscopy, we gain deeper insights into the role that nearby protein side chains play in mediating interactions between the active site and a ligand.

2. Materials

Currently, there are no commercially available femtosecond time-resolved mid-IR spectrometers. All spectrometers in existence have been assembled from components acquired from various sources, and no two spectrometers are alike. The cost of the components is in the range of $250–350K, though resourceful individuals with a strong optics background may be able to assemble a functioning spectrometer for less. The key components from which a femtosecond time-resolved mid-IR spectrometer could be assembled are listed below. Web sites of possible suppliers are appended. We elaborate further on these items in **Subheading 3.**

1. Mode-locked femtosecond Ti:sapphire laser operating near 800 nm:
   - www.coherent.com
   - http://www.splasers.com/
   - http://www.cmxr.com/
   - http://kmlabs.com/
2. kHz regenerative amplifier to amplify 800 nm seed pulses to about 1 mJ at 1 kHz:
   - www.coherent.com
   - http://www.splasers.com/
   - http://www.cmxr.com/
   - http://www.poslight.com/
   - http://kmlabs.com/
3. Optical parametric amplifier to generate tunable femtosecond light pulses:
   - http://www.quantron.com/
   - http://www.splasers.com/
   - http://www.poslight.com/
4. Optical parametric amplifier with difference frequency mixing option to generate tunable mid-IR pulses:
   - http://www.quantron.com/
   - http://www.splasers.com/
   - http://www.cmxr.com/
5. IR array detector with electronics suitable to read the array at 1 kHz:
   - http://www.irassociates.com/
   - http://www.infraredsystems.com/
6. Computer controlled translation stage to use as an optical delay:
   - http://www.mellesgriot.com/
   - http://www.newport.com/
7. Mid-IR optics to deliver mid-IR pulses to and from sample:
   - http://www.janostech.com/
   - http://www.layertec.de/
   - http://www.thorlabs.com/
   - http://www.optometrics.com/
   - http://www.eksma.lt/
   - http://www.edmundoptics.com/
   - http://www.cvilaser.com/
8. Rotating sample cell with IR windows.

3. Methods

3.1. Sample Preparation

3.1.1. Preparation of HbCO

A 4 m*M* solution of $Hb^{13}CO$ was prepared from the hemolysate of fresh red blood cells according to the following procedure:

1. Purify hemolysate by chromatography on a DEAE–Sephacel (Pharmacia Biotech) column.
2. Dialyze solution against D_2O buffered with 0.1 *M* potassium phosphate (pD 7.5).
3. Convert HbO_2 to HbCO by stirring under 1 atm of ^{13}CO (Cambridge Isotope Laboratories) for 12 h.
4. Concentrate solution to 4 m*M* (16 m*M* in heme) by centrifugation at 5000*g* in a Centricon-10 concentrator.

3.1.2. Preparation of MbCO

A 15 m*M* solution of $Mb^{13}CO$ was prepared according to the following procedure:

1. Dissolve lyophilized skeletal horse Mb (Sigma) in deoxygenated D_2O buffered with 0.1 *M* potassium phosphate (pD 7.5).
2. Equilibrate solution with 1 atm of ^{13}CO.
3. Reduce with a slight molar excess of freshly prepared sodium dithionate (Aldrich, 85%).
4. Stir reduced Mb solution under 1 atm of ^{13}CO for at least 30 min to insure complete conversion to $Mb^{13}CO$.

3.1.3. Preparation of MbNO and MpNO

A 13 m*M* solution of MbNO (MpNO) was prepared according to the following procedure:

1. Carry out sample preparation in an ice bath to minimize thermal denaturation.
2. Dissolve lyophilized skeletal horse Mb or microperoxidase-11 (Mp) (both from Sigma) in deoxygenated D_2O buffered with 0.1 *M* potassium phosphate (pD 7.5).
3. Reduce solution with a twofold excess of freshly prepared sodium dithionate.
4. Introduce one equivalent of NO by adding a stoichiometric amount of 0.1 *M* degassed sodium nitrite (Aldrich) and 0.1 *M* sodium dithionate *(2)*.

3.1.4. Mid-IR Sample Cell

To characterize a ligand through its mid-IR absorbance spectrum, we must be able to distinguish between its absorbance and that from the solvated protein. The mid-IR absorbance of the protein and the surrounding water is very strong in certain spectral regions, making it impossible to extract spectra of ligands in those regions. For a sample with a path length of 100 μm, even the weak water absorbance band at 2127 cm^{-1} (*see* **Note 1**) has an optical density of 1.9 (transmittance of 1.2%). Keeping the path length of the sample thin (e.g., approx 50 µm) opens up spectroscopically useful *water windows* between 1700–2800 cm^{-1} and 1200–1550 cm^{-1} *(3)*. Owing to the isotope effect, the water windows for D_2O are shifted to lower frequencies and complement those for H_2O. Through judicious selection of the solvent (H_2O or D_2O), mid-IR spectra of the protein and ligand can be recorded over a spectral range spanning from 700 to 4000 cm^{-1} *(3)*.

The material used for the sample cell windows must be transparent over the range of interest. Above 1000 cm^{-1}, fluoride glasses such as CaF_2 and BaF_2 are excellent materials because they are optically transparent and their low index of refraction ($n \approx 1.4$) results in low Fresnel reflection losses (approx 6% through sample cell). However, they are sensitive to thermal and mechanical shock. Window materials that can be used below 1000 cm^{-1} include ZnSe ($n \approx 2.4$, cutoff wavelength, $\lambda_{co} = 20$ µm), ZnS ($n \approx 2.2$, $\lambda_{co} = 14$ µm), CdTe ($n \approx 2.7$, $\lambda_{co} = 31$ µm), KRS-5 ($n \approx 2.4$, $\lambda_{co} = 40$ µm), and Ge ($n \approx 4$, $\lambda_{co} = 17$ µm), Owing to their relatively high refractive indices, they all suffer from a large reflection loss (exceeding 34% through sample cell).

Because the path length of the sample is generally restricted to 10–100 µm, high sample concentrations (e.g., m*M* or higher) are required to record the absorbance spectrum, even for strong IR absorbers. The maximum concentration that can be prepared is limited by the size of the protein. For example, proteins the size of Mb start to experience molecular crowding around 15 m*M*, as evidenced by an increase in the solution viscosity.

To remove light scattering sources such as dust particles and denatured protein aggregates, all samples were filtered through a 0.45 µm membrane filter before loading in a gas-tight rotating sample cell with 2 mm thick CaF_2 windows.

During data collection, the sample cell was rotated sufficiently fast so that each photolyzing laser pulse illuminated a fresh volume of the sample. The temperature of the rotating sample cell was kept at 283 ± 1K (*see* **Note 2**).

3.2. Time-Resolved IR Spectrometer

3.2.1. Pump-Probe Method

A time-resolved spectrometer capable of recording transient IR absorption spectra with approx 100 fs time resolution is shown in **Fig. 1**. The spectrometer provides broad spectral coverage from 290 nm to 18 µm *(4)*. It generates sufficient energy to produce ultraviolet (UV), visible, or IR pump pulses. Here, we will focus on visible pump—IR probe experiments. Visible pulses are used to photolyze the sample, whereas mid-IR pulses are used to measure the transmission of the sample at the selected wavelength and at a delay time established by a computer-controlled optical delay. A synchronous light chopper

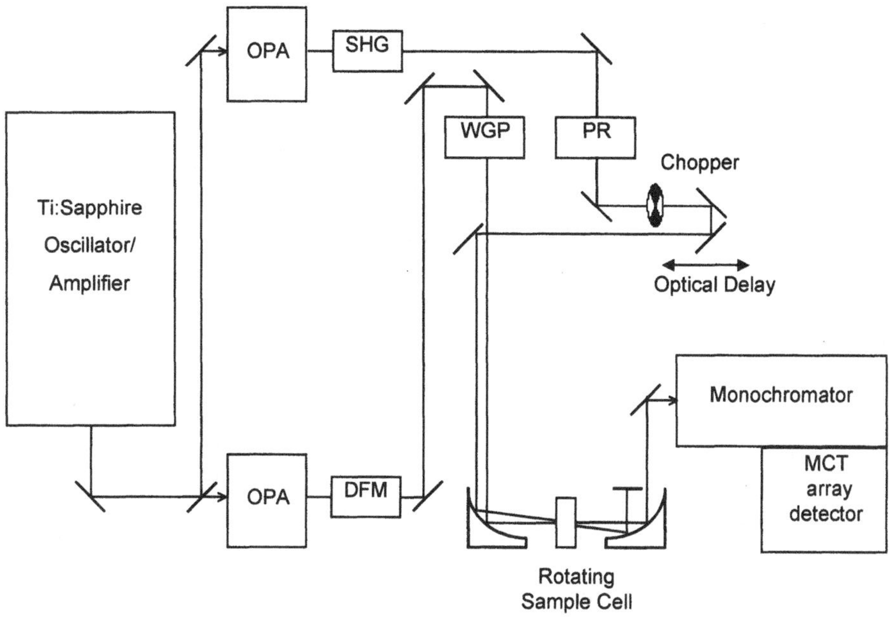

Fig. 1. Schematic diagram of time-resolved mid-IR spectrometer. OPA, optical parametric amplifier; SHG, second harmonic generator; DFM, difference frequency mixer; WGP, wire-grid polarizer; PR, polarization rotator.

blocks every other pump pulse; the transmittances of the pumped and unpumped sample are therefore recorded alternatively. The pump-induced change in the absorbance of the sample $\Delta A(\lambda,t)$, is computed on a shot-to-shot basis and averaged. A liquid crystal polarization rotator controls the polarization direction of the photolysis pulse electronically (*see* **Note 3**).

3.2.2. Generation of Visible Pump Pulses

A commercial Ti:sapphire oscillator/amplifier system (Hurricane, Spectra Physics, CA) is used to generate 100 fs pulses at 800 nm with an energy of 0.7 mJ at a repetition rate of 1 kHz. Two identical homebuilt optical parametric amplifiers (OPA), pumped by the amplified pulse, are used to generate a visible pump pulse *(5)* and a mid-IR probe pulse *(4,6,7)*. In a double-pass OPA setup, a single filament white light continuum produced in a 1 mm thick sapphire window seeds a 4 mm thick type-II BBO crystal, pumped with 3 μJ for the first pass and 200 μJ for the second pass. The OPA produces near transform-limited 70–100 fs pulses, tunable between 1.16 and 2.58 μm, with a total energy of 50 μJ in the signal and idler pulse (*see* **Note 4**). To generate a tunable pump pulse in the visible region, the signal pulse of one OPA is separated from its idler in a dichroic beamsplitter, and frequency doubled in a type-I, 1 mm thick BBO crystal *(5)*. The visible pulses are tunable from 580 nm to 750 nm and contain approx 10 μJ energy.

3.2.3. Generation of Mid-IR Probe Pulses

To generate a tunable mid-IR pulse, the signal and idler pulse of the second OPA are difference frequency mixed in a 1.5 mm thick, type-I $AgGaS_2$ crystal. The IR pulse is tunable from 3.3 μm to 12 μm with a typical duration of 110 fs, a spectral bandwidth of 160 cm^{-1}, and a pulse energy around 1 μJ. The tunability can be extended up to 18 μm by replacing the $AgGaS_2$ with a 1 mm thick GaSe crystal *(8)* (*see* **Note 5**). The generated IR pulse is collimated by a curved mirror and filtered by a long pass filter (FXLP-0300, Janos Technology, VT). Then, a small portion is reflected by a wedged BaF_2 window to serve as an attenuated probe pulse. When performing IR pump/IR probe experiments, the transmitted pulses are used as a mid-IR pump pulse. Two wire-grid polarizers are used in series to set the output polarization to any angle desired, and at the same time, provide variable attenuation. Alternatively, an IR waveplate can be inserted in the beam path to control the polarization direction of the IR pulse.

3.2.4. Detection of Mid-IR Pulses

The broadband transmitted probe pulse is detected with a $N_2(l)$-cooled HgCdTe array detector (Infrared Associates, FL), which consists of 64 elements (0.4 mm high, 0.2 mm wide) whose typical peak detectivity is $D^* = 4 \times 10^{10}$ cm Hz $^{1/2}$ W^{-1} at 10 kHz. The array detector is mounted in the focal plane of a 320 mm IR monochromator (Triax320, Horiba, NJ) with a 150 *l*/mm grating, resulting in a spectral resolution of ca. 1.1 cm^{-1}/pixel at 1600 cm^{-1}. The monochromator can be readily calibrated against a water vapor signal in the spectral region investigated. The signals from each of the detector elements were amplified with a homebuilt 64-channel amplifier and digitized by a 12-bit analog-to-digital converter. Transient spectra spanning 100 cm^{-1} can be constructed as a superposition of two 64-point spectra that overlap by a few elements. Chopping the excitation light pulse affords greater immunity to long-term instrumental drift. Because of the excellent short-term stability of the IR light source (< 0.5% rms), noise of less than 1×10^{-4} rms in absorbance units after 0.5 s of signal averaging is routinely obtained without shot-by-shot referencing against a second detector.

3.3. Determining Ligand Orientation in Myoglobin

3.3.1. Theory of Polarization Anisotropy

To probe the orientation of CO in Mb and Hb, we take advantage of the following facts: 1) the heme in Mb and Hb preferentially absorbs light that is polarized parallel to its plane (preference scales according to $\cos^2\alpha$); 2) photoexcitation triggers release of CO from hemes that are favorably aligned relative to the light polarization. Given these conditions, laser photolysis creates an anisotropic distribution of dissociated CO, whose directional anisotropy can be calculated from the pump-induced change in the CO absorbance spectra measured parallel ($\Delta A^{\parallel}$) and perpendicular ($\Delta A^{\perp}$) to the photolysis polarization direction ***(9,10)***.

$$r(\lambda) = \frac{\Delta A^{\parallel}(\lambda) - \Delta A^{\perp}(\lambda)}{\Delta A^{\parallel}(\lambda) + 2\Delta A^{\perp}(\lambda)} \qquad (1)$$

Because the heme possesses a twofold degenerate transition dipole in its plane (i.e., the heme is a circular absorber) ***(11)***, *r* can theoretically span the range $-0.2 \le r \le 0.1$, and the measured anisotropy can impose stringent constraints on the angle between the CO transition dipole moment and the heme normal. If CO is oriented at a fixed angle θ with respect to the heme normal, then this angle can be determined from the measured anisotropy according to **ref. *9***:

$$\theta = \sin^{-1}\sqrt{(10/3)\,(r + 0.2)} \qquad (2)$$

The orientation of the CO may be narrowly distributed about a central angle, but it is certainly not fixed, so the angle calculated using **Eq. 2** provides an upper limit to the actual angle. When CO is dissociated from the heme, its orientation is much more broadly distributed, so this equation does not apply. However, an electrostatic field around the heme causes a Stark shift of the CO vibrational frequency, and this effect has been used to deduce the orientational distribution of CO when localized in the so-called heme-docking site.

3.3.2. Orientation of CO Bound to Hemoglobin

When HbCO is photolyzed with a green laser pulse, the vibrational feature associated with bound CO (A-state) decreases in amplitude while a new band associated with *free* CO (B-state) appears. The photolysis-induced change in the vibrational absorbance spectrum is shown in **Fig. 2**, where the negative-going feature corresponds to the loss of bound CO while the positive-going feature near 2100 cm^{-1}corresponds to free CO. The polarization anisotropy of the A-state is independent of wavelength. If we assume the bound CO is confined to a fixed angle relative to the heme normal, the angle of its transition moment only 8.3 degrees ***(12)***. This measurement is quite precise provided the anisotropy is measured accurately and systematic errors are avoided (*see* **Note 6**).

3.3.3. Orientation of CO in Heme Pocket Docking Site

The polarization anisotropy of the B-states in **Fig. 2** is strongly wavelength dependent. An enlarged view of its spectra is shown in **Fig. 3**. Because CO

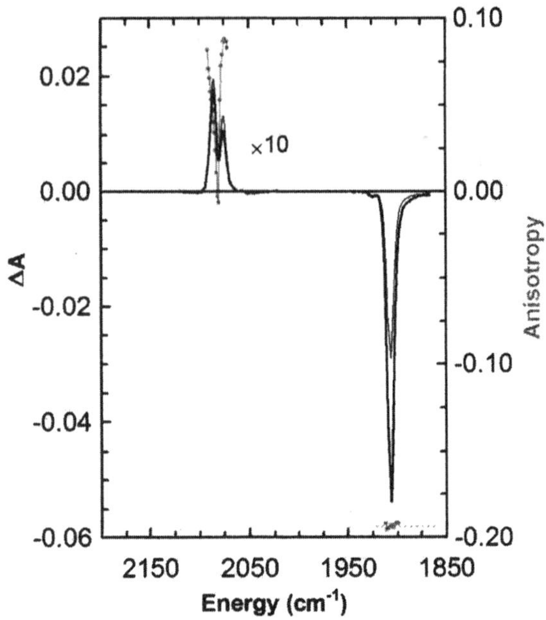

Fig. 2. Time-resolved polarized mid-IR absorbance spectra of photolyzed $Hb^{13}CO:D_2O$ at 10ºC. The A-state (negative-going) and B-state (positive-going) spectra were recorded at ≈10% and ≈20% photolysis, respectively. The spectra were acquired with the pump pulse polarized perpendicular, $\Delta A^{\perp}$ (thick lines) and parallel, $\Delta A^{\parallel}$ (thin lines) to the probe pulse. The left ordinate depicts the pump-induced change in the absorbance of the sample and the right ordinate depicts the polarization anisotropy (filled circles). The B-state spectra have been scaled by × 10 (the absolute intensity of the CO absorbance decreases significantly upon dissociation from the heme). For clarity, the background and hot band contributions to the B-state spectra have been removed.

rattles about in the docking site, its orientation cannot be described by a single angle, and is better defined in terms of its orientational probability distribution function. The shape of this function can be deduced from its polarized CO vibrational absorbance spectra. This determination is possible because *docked* CO is bathed in a static electric field, which leads to an orientation-dependent Stark shift of its vibrational frequency ***(13–15)***. The vibrational Stark shift arises from the change of dipole moment associated with a vibrational transition and is proportional to $\mathbf{E} \cdot \mu$ ***(13,15–17)***, where **E** is the electric field sensed by the CO molecule and $\Delta\mu$ is the Stark tuning rate ***(15–18)***. Consequently, the two features correspond to CO pointing in approximately opposite directions.

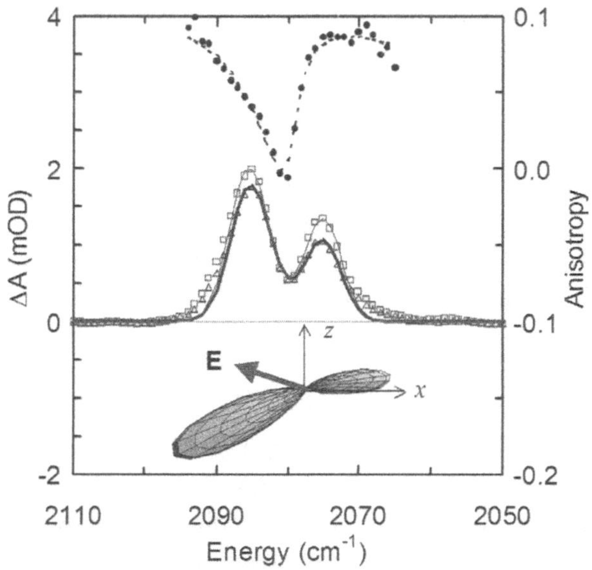

Fig. 3. Time-resolved polarized mid-IR absorbance spectra of photolyzed $Hb^{13}CO$ in D_2O at 10ºC. The B-state spectra were recorded at 20% photolysis with the pump pulse polarized perpendicular, $\Delta A^{\perp}$ (triangles) and parallel, $\Delta A^{\parallel}$ (squares) to the probe pulse. The left ordinate depicts the pump-induced change in the absorbance of the sample and the right ordinate depicts the polarization anisotropy (filled circles). The solid lines correspond to a simple model based on a double well cosine potential for the CO orientation; the least-squares optimization was confined to the region where the absorbance is greater than 30% of its neighboring peak. Except for the wings of the spectra, the absorbance and anisotropy features are reproduced with high fidelity. The best-fit orientational probability distribution function is plotted as a three-dimensional rendered object with the Stark field found in the *xz*-plane. For clarity, the background, and hot band contributions to the B-state spectra have been removed.

The simplest model that can account for the vibrational splitting and the frequency-dependent polarization anisotropy is based on a cylindrically-symmetric double-well potential written in spherical polar coordinates:

$$V(\theta,\phi) = V_0\left(\frac{1}{\sin^2\left\{0.5\times\cos^{-1}\left[\hat{r}(\theta,\phi)\cdot\hat{r}(\theta_1,\phi_1)\right]\right\}}+\frac{1}{\Delta+\sin^2\left\{0.5\times\cos^{-1}\left[\hat{r}(\theta,\phi)\cdot\hat{r}(\theta_2,\phi_2)\right]\right\}}\right)^{-1} \quad (3)$$

where $\hat{r}(\theta_1, \phi_1)$ and $\hat{r}(\theta_2, \phi_2)$ are the unit direction vectors for the equilibrium orientation of CO in the two wells, V_0 is related to the potential barrier separating the two wells, and Δ is the energy difference between the two minima.

Given **Eq. 3** and a Boltzmann distribution of thermal energies, the orientational probability distribution function is computed from

$$P(\theta,\phi) \propto \exp\left[\frac{-V(\theta,\phi)}{RT}\right].$$

This function looks like a bent *peanut* with two lobes of high probability oriented in different directions. Given this orientational probability distribution function, Stark-shifted polarized vibrational spectra and their polarization anisotropy were calculated numerically and plotted in **Fig. 3**.

3.4. Characterizing Ligand Dynamics

3.4.1. Dynamics of Ligand Trapping: MbCO

When CO is photolyzed from MbCO, it quickly becomes trapped within a docking site where it is constrained to lie approximately parallel to the plane of the heme ***(19)***. The dynamics of ligand trapping within the docking site were measured by femtosecond time-resolved polarized IR spectroscopy. As shown in **Fig. 4**, two features (denoted B_1 and B_2) quickly develop, which suggests that CO is funneled into a docking site that is located near the binding site. The time-dependent anisotropies arising from the time-dependent ligand orientation evolve exponentially with time constants of 0.2 ps for the B_1 state and 0.52 ps for the B_2 state ***(19)***. The prompt appearance and independent development of the two B-sates suggest that there are two deterministic CO trajectories. Based on a kinematic argument and isotopic geminate rebinding studies of photolyzed MbCO at 20 K ***(20,21)***, it was rationalized that B_1 (B_2) corresponds to the faster (slower) rotating trajectory where the O (C) end of C–O ends pointing toward the heme iron ***(19)***. The fact that CO becomes trapped in a nearby site (approx 2 Å) so quickly (0.2–0.52 ps) suggests that the structure of the protein in the vicinity of the binding sites provides a channel that facilitates rapid and efficient trapping of CO in the docking site.

3.4.2. Dynamics of Ligand Escape: MbCO

The ns spectral evolution of photolyzed MbCO is shown in **Fig. 5**. The spectrum at 10 ns is virtually identical to that observed at 100 ps, demonstrating that the CO remains trapped in the heme pocket docking site for tens of nanoseconds ***(22)***. At long times, the spectrum reveals a single broad feature corresponding to CO in the solvent (S-state). At intermediate times, an additional

Fig. 5. *(opposite page)* Time-resolved mid-IR spectra of sperm whale $Mb^{13}CO$ in D_2O at 10ºC. All spectra were collected with linearly polarized pump and probe pulses oriented at the magic angle (54.7°) relative to each other.

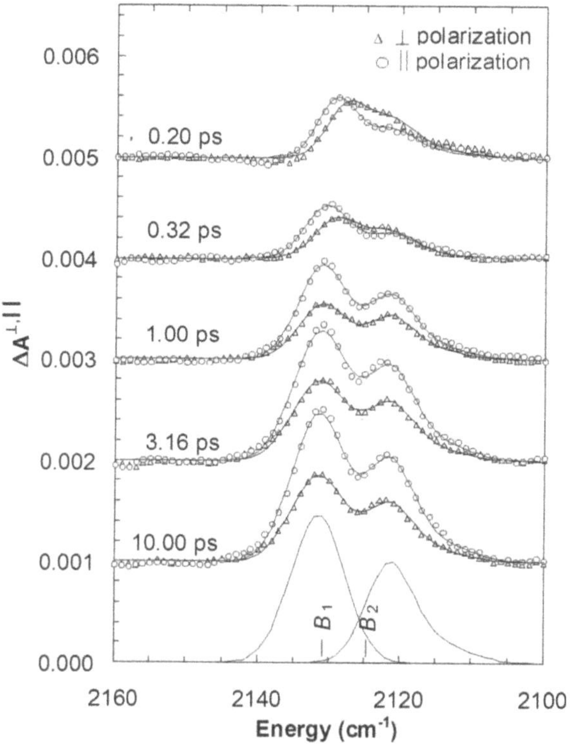

Fig. 4. Time-resolved polarized mid-IR spectra of photolyzed horse $Mb^{13}CO$ at 10°C. The spectra were recorded with the photolysis and probe pulses polarized parallel $\Delta A^{||}$ and perpendicular $\Delta A^{\perp}$ to one another. For clarity, the background and hot band contributions to the time-resolved spectra have been removed and the spectra have been offset from one another.

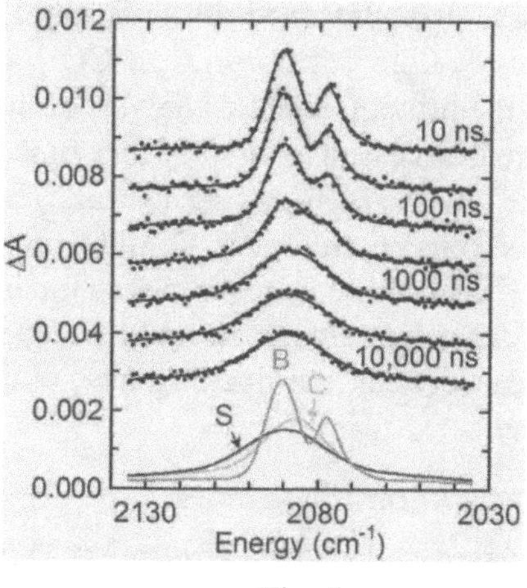

Fig. 5.

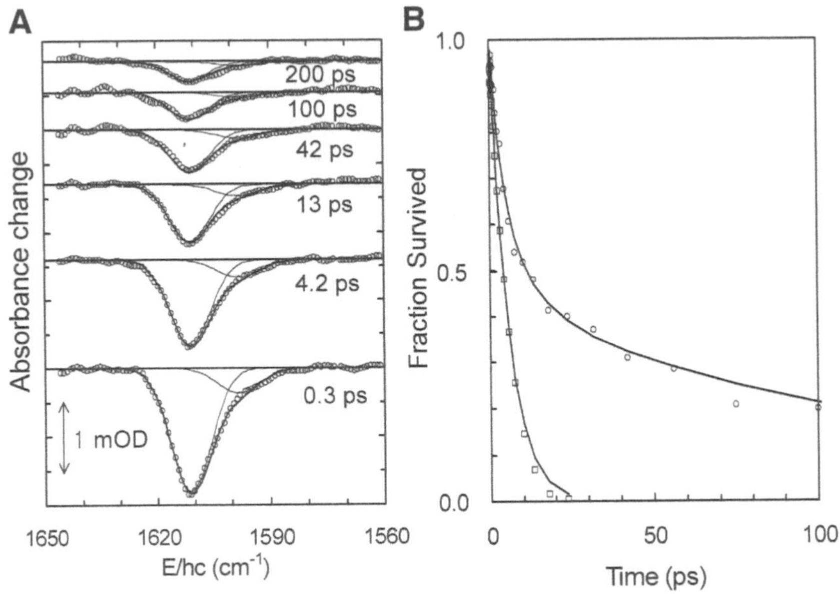

Fig. 6. **(A)** Representative time resolved IR spectra of photolyzed MbNO in D_2O (mOD = 10^{-3} optical density). The spectra are decomposed into two Gaussians, one (83% population) at 1611 cm^{-1} and the other (17%) at 1598 cm^{-1}. The relative intensity of the two Gaussians is time independent. **(B)** Kinetics of GR of NO to Mb (open circles) and Mp (open squares). Whereas rebinding to MpNO is exponential (5.6 ps), that to MbNO is nonexponential, and can be modeled with a time-dependent rebinding rate (the initial and final rebinding time constants are 7.7 ps and 150 ps, respectively).

feature centered between the two B states is observed. This feature (C-state) grows concomitant with the disappearance of the B states, after which it also disappears. Evidently, this feature corresponds to CO transiently trapped in another internal cavity of the protein. Kinetic analysis of the spectral evolution suggests that this new state is accessed reversibly on a time scale similar to that for escape from the B state into the surrounding solvent. The spectral evolution can be modeled by a linear combination of B, C, and S states at all times. This suggests that transport between these states is fast compared to the residence time within each state. Evidently, conformational fluctuations open and close paths for shuttling ligands between the docking site, other internal cavities, and the surrounding solvent.

3.4.3. Dynamics of Ligand Rebinding: MbNO

Figure 6A shows time-resolved mid-IR absorption spectra of MbNO after 580 nm photolysis. The negative going features (bleach) arise from the loss of bound NO and can be modeled with a sum of two Gaussians, which implies that the bound state exists in at least two distinct conformational substates ***(23–25)***. The relative amplitude of the two Gaussians is time independent, which demonstrates that the two conformers have the same geminate recombination (GR) kinetics. As shown in **Fig. 6B**, rebinding of NO to microperoxidase-11 (Mp) is ultrafast (5.6 ps time constant) with near unity geminate yield, suggesting that the rebinding of NO to heme is a nearly barrierless process ***(26)***. The active site of this peptide is exposed to the surrounding water, which forms an ice-like cage that evidently transiently traps dissociated NO at short times and encourages GR. The kinetics of NO rebinding to Mb is nonexponential, but is well described by a time-dependent rebinding rate that presumably arises from a time-dependent barrier to rebinding ***(27, 28)***. The initial rebinding rate, $(7.7 \text{ ps})^{-1}$, is similar to the intrinsic rebinding rate to heme, $(5.6 \text{ ps})^{-1}$, but gradually decreases as the barrier grows toward its equilibrium height, presumably due to conformational relaxation of the protein. The rate constant for the time dependence of the barrier, found to be $(16.5 \text{ ps})^{-1}$, is comparable to the *effective* rate for conformational relaxation of Mb, $(18 \text{ ps})^{-1}$ ***(29)***. These results clearly show that conformational relaxation plays a dominant role in the nonexponential NO rebinding to Mb and that static inhomogeneity (conformational substates) has little influence on the dynamics of NO rebinding to Mb under physiological conditions ***(26)***.

4. Notes

1. H_2O has prominent IR absorption bands ***(30)*** at 3404 cm^{-1} (O–H stretching mode; approx 400 cm^{-1} FWHM; 99.9 $M^{-1}cm^{-1}$ molar absorptivity), 1643.5 cm^{-1} (H–O–H bending mode, approx 100 cm^{-1} FWHM; 21.8 M^{-1} cm^{-1} molar absorptivity), and at 2127 cm^{-1} (combination of bending and liberation mode, approx 300 cm^{-1} FWHM; 3.5 M^{-1} cm^{-1} molar absorptivity).
2. Keeping the temperature at 283 K lengthens the lifetime of the samples. At room temperature, photolyzed heme proteins tend to denature and aggregate more rapidly.
3. A computer-controlled liquid-crystal retarder (Meadowlark) allows us to frequently alternate the polarization of the photolysis pulse between parallel and perpendicular orientations during a time scan. This approach to data collection affords greater immunity to instrumental drift, making for more precise polarization anisotropy measurements.
4. Having stable mid-IR pulses is one of the most important requirements for femtosecond IR spectroscopy. To achieve stable mid-IR pulses, the OPA output must be stable. When optimizing the OPA, it is often useful to monitor the second harmonic of the signal beam and optimize for a uniform spatial profile, rather than for maximum OPA output energy.

5. While GaSe has a broader transmission range and larger nonlinear coefficient (0.62 – 20 μm, 54 pm/V) than $AgGaS_2$ (0.47 – 13 μm, 12 pm/V), it is currently available only with its surface cleaved along the [001] direction, and it cannot be antireflective coated *(8)*. With this cleave, the phase-matching condition for mid-IR generation requires a severe angle of incidence, thereby limiting its usefulness. Nevertheless, research with indium-doped GaSe promises crystals cut at arbitrary angles *(31)*, which may overcome the orientation problem.
6. Although the underlying theory for determining angles with the photoselection technique is rigorous, there are number of experimental pitfalls. Aside from avoiding systematic errors that can be introduced when measuring the polarization anisotropy *(32)*, the following criteria need to be fulfilled to trivially relate the polarization anisotropy to the orientation of the transition dipoles of the sample *(33)*. First, the measurement should be made at a pump-probe delay time that is short compared to the rotational motion of the sample. Since rotational diffusion of the sample randomizes the orientation of the photoselected transition dipoles, the polarization anisotropy must be measured at a pump-probe delay time of that is short compared to rotational diffusion dynamics. Second, the polarization anisotropy should be determined in the limit of zero photolysis. As the level of photolysis increases, the pump-induced change in the absorbance of the sample increases; however, the absolute value of the measured anisotropy decreases. Therefore, the measured polarization anisotropy must be corrected for the effects of fractional photolysis. When the sample is optically thick, the pump pulse intensity decreases as it passes through the sample, so the fraction of photoexcited molecules is depth-dependent, requiring a more sophisticated correction. *(33)*. Third, the angular distribution of the transition dipole in the molecule should be known. Because of thermal motion, the angular distribution of the dipole is not a delta function, but is distributed about some equilibrium orientation. The measured anisotropy needs to be corrected with the order parameter for the distribution. Fourth, the photolyzing beam should be measurably larger than the probe beam to ensure spatial uniformity of photolysis. Only after the above criteria are met can an accurate angle between transition dipoles be obtained from photoselection experiments. Some earlier experiments neglected fractional photolysis effects or corrected for them incompletely, or did not account for the orientational distribution of CO, or were compromised by systematic experimental errors. Except for systematic errors, factors that influence the polarization anisotropy, but are unaccounted for, tend to reduce its absolute value. Special care needs to be taken to ensure that both polarizations are pure (extinction ratios are routinely better than 1000:1) and of equal intensity.

Acknowledgments

M. Lim gratefully acknowledges support from the Korea Science and Engineering Foundation through the Center for Integrated Molecular Systems (POSTECH).

References

1. Ansari, A., Berendzen, J., Bowne, S. F., Frauenfelder, H., Iben, I. E. T., Sauke, T. B., Shyamsunder, E., and Young, R. D. (1985) Protein states and proteinquakes. *Proc. Natl. Acad. Sci. USA* **82,** 5000–5004.
2. Moller, J. K. S. and Skibsted, L. H. (2002) Nitric Oxide and Myoglobins. *Chem. Rev.* **102,** 1167–1178.
3. Maxwell, J. C. and Caughey, W. S. (1978) Infrared spectroscopy of ligands, gases, and other groups in aqueous solutions and tissues. *Methods Enzymol.* **54,** 302–323.
4. Kim, S., Jin, G., and Lim, M. (2003) Structural dynamics of myoglobin probed by femtosecond infrared spectroscopy of the amide band. *Bull. Kor. Chem. Soc.* **24,** 1470–1474.
5. Lim, M., Wolford, M. F., Hamm, P., and Hochstrasser, R. M. (1998) Chirped wavepacket dynamics of HgBr from the photolysis of $HgBr_2$ in solution. *Chem. Phys. Lett.* **290,** 355–362.
6. Hamm, P., Lim, L., and Hochstrasser, R. M. (1998) The Structure of the Amide I Band of Peptides Measured by Femtosecond Nonlinear IR Spectroscopy. *J. Phys. Chem. B.* **102,** 6123–6138.
7. Hamm, P., Kaindl, R. A., and Stenger, J. (2000) Noise suppression in femtosecond mid-infrared light sources. *Opt. Lett.* **25,** 1798–1800.
8. Kaindl, R. A., Wurm, M., Reimann, K., Hamm, P., Weiner, A. M., and Woerner, M. (2000) Generation, shaping, and characterization of intense femtosecond pulses tunable from 3 to 20 μm. *J. Opt. Soc. Am. B.* **17,** 2086–2094.
9. Moore, J. N., Hansen, P. A., and Hochstrasser, R. M. (1988) Iron-carbonyl bond geometries of carboxymyoglobin and carboxyhemoglobin in solution determined by picosecond time-resolved infrared spectroscopy. *Proc. Natl. Acad. Sci. USA* **85,** 5062–5066.
10. Ansari, A. and Szabo, A. (1993) Theory of photoselection by intense light pulses. *Biophys. J.* **64,** 838–851.
11. Eaton, W. A. and Hofrichter, J. (1981) Polarized absorption and linear dichroism spectroscopy of hemoglobin. *Methods Enzymol.* **76,** 175–261.
12. Lim, M., Jackson, T. A., and Anfinrud, P. A. (2004) The orientational distribution of CO before and after photolysis of MbCO and HbCO: A determination using time-resolved polarized mid-IR spectroscopy. *J. Am. Chem. Soc.* **126,** 7946–7957.
13. Augspurger, J. D., Dykstra, C. E., and Oldfield, E. (1991) Correlation of carbon-13 and oxygen-17 chemical shifts and the vibrational frequency of electrically perturbed carbon monoxide: a possible model for distal ligand effects in carbonmonoxyheme proteins. *J. Am. Chem. Soc.* **113,** 2447–2451.
14. Ma, J., Huo, S., and Straub, J. E. (1997) Molecular dynamics simulation study of the B-states of solvated carbon monoxymyoglobin. *J. Am. Chem. Soc.* **119,** 2541–2551.
15. Park, E. S. and Boxer, S. G. (2002) Origins of the sensitivity of molecular vibrations to electric fields: Carbonyl and nitrosyl stretches in model compounds and proteins. *J. Phys. Chem. B.* **106,** 5800–5806.

16. Reimers, J. R. and Hush, N. S. (1999) Vibrational Stark spectroscopy 3. Accurate benchmark ab initio and density functional calculations for CO and CN^-. *J. Phys. Chem. A* **103,** 10,580–10,587.
17. Andrews, S. S. and Boxer, S. G. (2002) Vibrational Stark effects of nitriles II. Physical orgins of Stark effects from experiment and perturbation models. *J. Phys. Chem. A* **10,** 469–477.
18. Park, E. S., Andrews, S. S., Hu, R. B., and Boxer, S. G. (1999) Vibrational Stark Spectroscopy in Proteins: A Probe and Calibration for Electrostatic Fields. *J. Phys. Chem. B.* **103,** 9813–9817.
19. Lim, M., Jackson, T. A., and Anfinrud, P. A. (1997) Ultrafast rotation and trapping of carbon monoxide dissociated from myoglobin. *Nature Struct. Biol.* **4,** 209–214.
20. Alben, J. O., Beece, D., Browne, S. F., Eisenstein, L., Frauenfelder, H., Good, D., Marden, M. C., Moh, P. P., Reinisch, L., Reynolds, A. H., and Yue, K. T. (1980) Isotope effect in molecular tunneling. *Phys. Rev. Lett.* **44,** 1157–1163.
21. Alben, J. O., Beece, D., Bowne, S. F., Doster, W., Eisenstein, L., Frauenfelder, H., Good, D., McDonald, J. D., Marden, M. C., Mo, P. P., Reinisch, L., Reynolds, A. H., Shyamsunder, E., and Yue, K. T. (1982) Infrared spectroscopy of photodissociated carboxymyoglobin at low temperatures. *Proc. Natl. Acad. Sci. USA* **79,** 3744–3748.
22. Schotte, F., Lim, M., Jackson, T. A., Smirnov, A. V., Soman, J., Olson, J. S., Phillips, G. N., Jr., Wulff, M., and Anfinrud, P. A. (2003) Watching a protein as it functions with 150-ps time-resolved x-ray crystallography. *Science* **300,** 1944–1947.
23. Ansari, A., Berendzen, J., Braunstein, D. K., Cowen, B. R., Frauenfelder, H., Hong, M. K., Iben, I. E. T., Johnson, J. B., Ormos, P., Sauke, T. B., Scholl, R., Schulte, A., Steinbach, P. J., Vittitow, J., and Young, R. D. (1987) Rebinding and relaxation in the myoglobin pocket. *Biophys. Chem.* **26,** 337–355.
24. Balasubramanian, S., Lambright, D. G., and Boxer, S. G. (1993) Perturbation of the distal heme pocket in human myoglobin mutants probed by infrared spectroscopy of bound CO: correlation with ligand binding kinetics. *Proc. Natl. Acad. Sci. USA* **90,** 4718–4722.
25. Tian, W. D., Sage, J. T., Srajer, V., and Champion, P. M. (1992) Relaxation dynamics of myoglobin in solution. *Phys. Rev. Lett.* **68,** 408–411.
26. Kim, S., Jin, G., and Lim, M. (2004) Dynamics of geminate recombination of NO with myoglobin in aqueous solution probed by femtosecond mid-IR spectroscopy. *J. Phys. Chem. B.*, in press.
27. Petrich, J. W., Lambry, J. C., Kuczera, K., Karplus, M., Poyart, C., and Martin, J. L. (1991) Ligand binding and protein relaxation in heme proteins: a room temperature analysis of nitric oxide geminate recombination. *Biochemistry* **30,** 3975–3987.
28. Petrich, J. W., Lambry, J.-C., Balasubramanian, S., Lambright, D. G., Boxer, S. G., and Martin, J. L. (1994) Ultrafast measurements of geminate recombination of NO with site-specific mutants of human myoglobin. *J. Mol. Biol.* **238,** 437–444.

29. Lim, M., Jackson, T. A., and Anfinrud, P. A. (1993) Nonexponential protein relaxation: dynamics of conformational change in myoglobin. *Proc. Natl. Acad. Sci. USA* **90,** 5801–5804.
30. Venyaminov, S. Y. and Prendergast, F. G. (1997) Water (H_2O and D_2O) molar absorptivity in the 1000–4000 cm^{-1} range and quantitative infrared spectroscopy of aqueous solutions. *Anal. Biochem.* **248,** 234–245.
31. Suhre, D. R., Singh, N. B., Balakrishna, V., Fernelius, N. C. , and Hopkins, F. K. (1997) Improved crystal quality and harmonic generation in GaSe doped with indium. *Opt. Lett.* **22,** 775–777.
32. Locke, B., Lian, T., and Hochstrasser, R. M. (1995) Erratum of Chemical Physics 158 (1991) 409–419. *Chem. Phys.* **190,** 155.
33. Lim, M. (2002) The orientation of CO in heme proteins determined by time-resolved mid-IR spectroscopy: anisotropy correction for finite photolysis of an optically thick sample. *Bull. Kor. Chem. Soc.* **23,** 865–871.

13

Monitoring Protein–Ligand Interactions by Time-Resolved FTIR Difference Spectroscopy

Carsten Kötting and Klaus Gerwert

Summary

Time-resolved FTIR difference spectroscopy is a valuable tool to monitor the dynamics of protein–ligand interactions, which selects out of the background absorbance of the whole sample the absorbance bands of the protein groups and of the ligands, which are involved in the protein reaction. The absorbance changes can be monitored with time-resolutions down to nanoseconds and followed then over nine orders of time up to seconds even in membrane proteins with the size of 100,000 Dalton. Here, we will discuss the various experimental setups. We will show new developments for sample cells and how to trigger a reaction within these cells. The kinetic analysis of the data will be discussed. A crucial step in the data analysis is the clear-cut band assignment to chemical groups of the protein and the ligand. This is done either by site directed mutagenesis or by isotopically labeling. Examples for band assignments will be presented in this chapter.

Key Words: Infrared; time-resolved; difference spectroscopy; rapid scan; step-scan; bacteriorhodopsin; retinal; reaction center; GTPases; GTP; caged-substances; isotopic labeling; band assignment; flowcell; ATR; global fit.

1. Introduction

The performance of difference spectra between the inactive and active states of proteins selects out of the background absorbance of the whole sample the absorbance bands of the protein groups and of the ligands, which are involved in the protein activation *(1)*. The absorbance changes can be monitored within the same set up with time-resolutions down to nanoseconds and followed than for time periods ranging over nine orders of magnitude up to seconds even in membrane proteins with size of 100,000 Dalton *(2)*. This allows for example to

From: *Methods in Molecular Biology, vol. 305: Protein–Ligand Interactions: Methods and Applications*
Edited by: G. U. Nienhaus © Humana Press Inc., Totowa, NJ

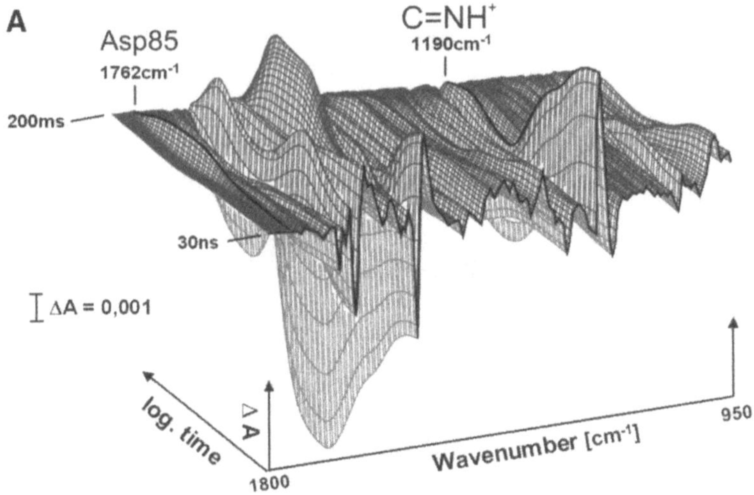

Fig. 1. **(A)** The IR absorbance changes during the photocycle of bR. One band of the protein at 1762 cm^{-1} (protonated Asp85) and one of the protein–ligand interface at 1190 cm^{-1} (protonated Schiff base) are marked. **(B)** (*See* page 263.) Model of the proton pump mechanism of bR, to which many groups have contributed (for references, *see* text). After the light-induced all-trans to 13-cis retinal isomerization in the BR-to-K transition, the Schiff base proton is transferred to Asp 85 in the L-to-M transition. Deprotonation of the protonated Schiff base can be followed at 1190 cm^{-1} in **(A)** and protonation of Asp85 at 1762 cm^{-1} in **(A)**. Simultaneously, an excess proton is released from an ice-like hydrogen-bonded network of internal water molecules to the extracellular site. Glu 204, Glu 194, and Arg 82 control this network. Asp 85 reprotonates the network in the O–BR reaction. The Schiff base is oriented in the M1-to-M2 transition from the proton release site to the proton uptake site, and thereby determines the direction of the proton transfer. A larger backbone movement of the helix F is observed in the M-to-N transition compared with the M1-to-M2 transition. Asp 96 reprotonates the Schiff base in the M-to-N transition also seen at 1190 cm^{-1} in **(A)**. Deprotonation of Asp 96 can also be seen at 1742 cm^{-1} in **Fig. 12**. Asp 96 itself is reprotonated from the cytoplasmic site in the N-to-O transition.

monitor the light induced isomerization reactions of retinal within bacteriorhodopsin *(3,4)* and rhodopsin *(5)*, the succeeding deprotonation of the protonated Schiff base retinal, protonation of carboxylic acids and helix movements in the protein *(6)*. In **Fig. 1** the light-induced absorbance changes during bacteriorhodopsins photocycle **(A)** and the deduced proton pump model **(B)** is shown.

By comparing the FTIR difference spectra with FT-Resonance Raman spectra, the bands of the ligands can be identified. In the FT-Resonance Raman

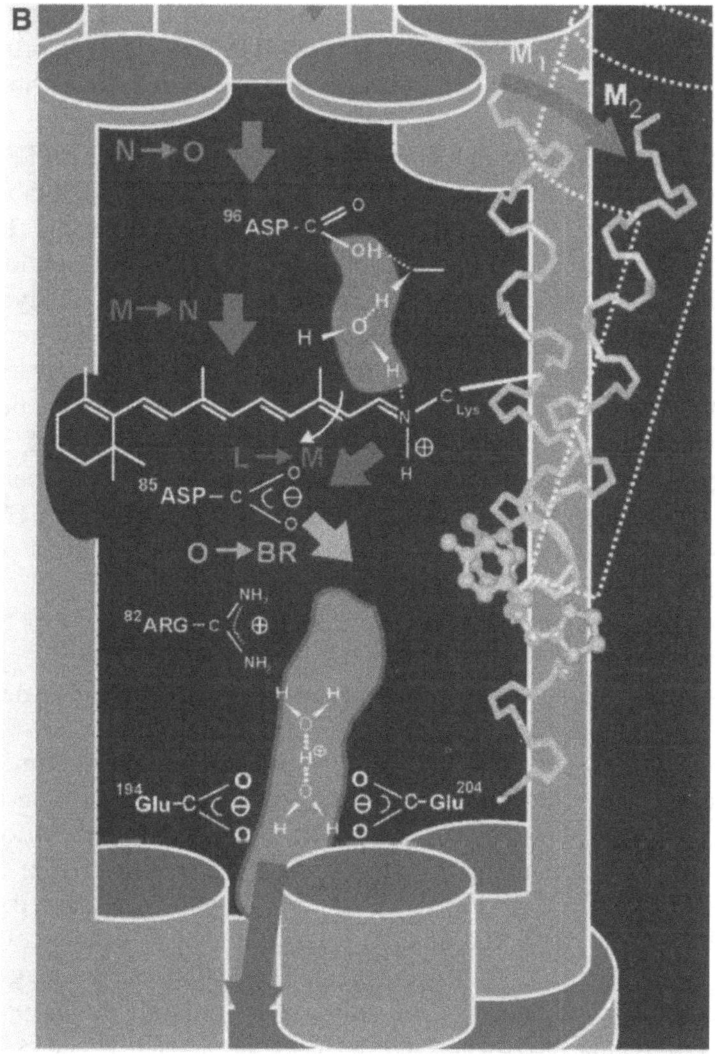

spectra only the chromophore bands appear, whereas in the FTIR difference spectra ligand and protein bands appear. Thereby, for example the chromophore bands in photoactive yellow protein are assigned *(7)*.

Moreover, light-induced redox reactions of ligands are monitored in photosynthetic proteins. The light-induced electron transfer within the prosthetic groups and the proton uptake via amino acid side chains coupled in a ping pong mechanism to electron transfer in photosynthetic reaction centers is elucidated by time resolved FTIR *(2)*.

Using photolabile trigger compounds also protein–ligand interactions can be studied in proteins without chromophores. The redox driven proton transport in cytochrom C oxydase is studied using the electron donor riboflavin as caged electrons *(8)*.

Further, the interactions of ATPases and GTPases with their ATP and GTP ligands can be studied using caged ATP or caged GTP as a photolabile trigger *(9)*. By this approach the interaction of GTP with Ras is studied. It is elucidated that binding of GTP to Ras induces a specific charge distribution in GTP, which reduces the free activation energy *(10,11)*. Thereby, GTP hydrolysis is catalyzed. In addition, protein–protein interactions and their influence on the GTP ligand can be studied. This is elaborated in studies of the activation of Ras by the GAP protein *(12)*. Binding of GAP to Ras catalyses the reaction by five orders of magnitude. In oncogenic Ras this activation is inhibited, and this is thought to be the central event in transformation a cell into a cancer cell. The influence of GAP on the conformation and charge distribution of GTP bound to Ras is studied in detail *(12)*.

In summary, time-resolved FTIR difference spectroscopy is a powerful tool to monitor protein–ligand interaction. Complementary to X-ray or NMR providing three-dimensional structural models of proteins FTIR delivers information on H-bonding, protonation state, charge distribution, and time dependence of the protein–ligand interaction.

Here, we will discuss the various experimental setups for time-resolved FTIR studies. The rapid scan technique allows a time resolution in the ms regime *(3)*, whereas the step-scan technique allows ns time resolution *(6)*. We will show new developments for sample cells and how to trigger a reaction within these cells. The kinetic analysis of the data will be discussed. A crucial step in the data analysis is the clear-cut band assignment to chemical groups of the protein and the ligand. This is done either by site directed mutagenesis *(4)*, where the absorbance bands of the exchanged amino acids disappear or by isotopically labeling *(13)*, where the band of the labeled group is frequency shifted. Examples for band assignments will be presented.

2. Materials

2.1. Spectrometer

1. IFS 66v/s (Bruker Optics, Karlsruhe, Germany) with KBr-Beamsplitter CaF_2 windows (Korth, Kiel, Germany).
2. MCT-detector KMPV11-1-J1 (Kolmar, Newburyport, MA).
3. Dry-Air purge gas generator Balston 75–62 (ParkerBalston, NJ).

2.2. Laser

1. Excimer-Laser LPX 240i (Lambda Physik, Göttingen, Germany) with XeCl (308 nm).

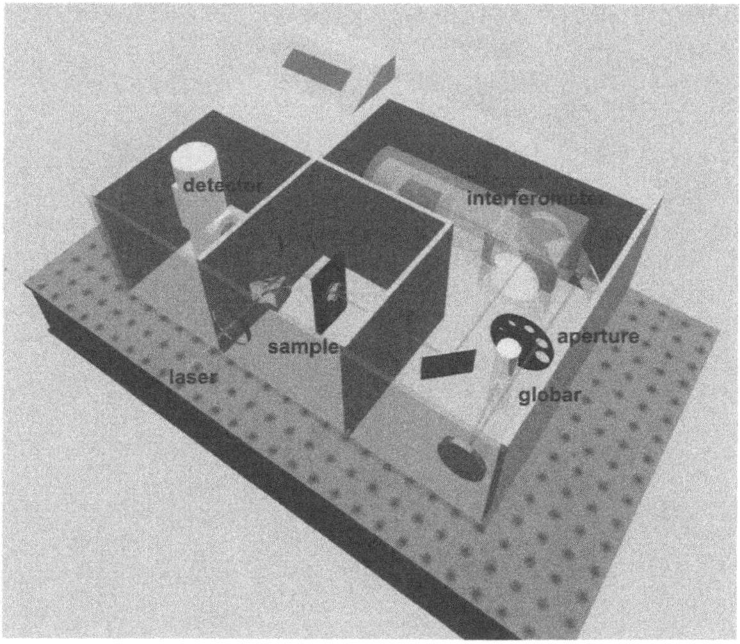

Fig. 2. Scheme of an FTIR spectrometer (from **ref. *37***).

2. Dye-laser FL 105 (Lambda Physik) with Coumarin 153 (540 nm).
3. Nd:YAG-Laser Quanta-Ray GCR-170 (Spectra Physics, Mt. View, CA).

2.3. Step-Scan

1. MCT-detector KV100-1-B-7/190, cutoff 850cm^{-1} (Kolmar, Newburyport, MA).
2. Keithley Adwin-Gold 12bit/400kHz transient recorder (Keithley Instruments, OH).
3. Spektrum Pad82 8bit/200MHz transient recorder (Microelectronic GmbH, Siek, Germany)

2.4. Special Equipment

1. IR-Scope (Bruker Optics, Karlsruhe, Germany).
2. CaF_2 BaF_2 windows (Korth, Kiel, Germany).
3. Diamond-μ-ATR-cell (Resultec, Garbsen, Germany).
4. Flow Cell (Biolytics, Freiburg, Germany).

3. Methods

3.1. General Setup

A typical setup for a time-resolved FTIR experiment is shown in **Fig. 2**. The light source is a globar (SiC heated at 1500 K), which is a black body radiator.

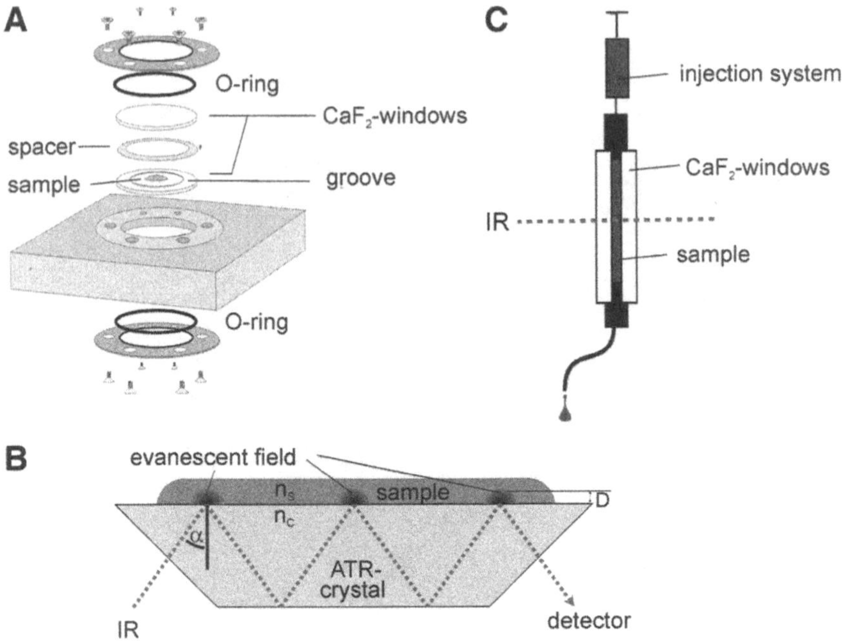

Fig. 3. Various sample cells for the investigation of protein–ligand interaction: **(A)** Transmission-cell; **(B)** ATR-cell; **(C)** flowcell.

Its infrared light passes an aperture (0.25 –12 mm) before entering a Michelson interferometer, consisting of a beamsplitter (KBr for the mid-infrared), a fixed and a movable mirror. Subsequently, the light passes the sample chamber equipped, e.g., with a thermostatic transmission cell. Here, this window can additionally be irradiated by a laser. Finally, the infrared light reaches the liquid nitrogen cooled mercury-cadmium-tellurium (MCT)-detector.

In **Fig. 3** some sample cells are shown. The most common cell is a simple transmission cell with IR-transparent windows (e.g., CaF_2, *see* **Note 1**) as shown in **Fig. 3A**. Because of the high absorptivity of water in the mid-infrared spectral region, meaningful spectra of hydrated proteins are obtained by transmission measurements only through very thin (2–10 µm) films. This involves placing a drop of a protein suspension or solution onto an IR transparent window and then carefully concentrating it under a nitrogen stream or under vacuum. Alternatively, a protein suspension of a membrane bound protein is centrifuged and the pellet is squeezed between two IR transparent windows. A typical measurement requires about 100–200 µg protein. The concentration of the protein in the film is 6–10 m*M*. The sample chamber is closed by a second

IR-window, which is separated from the first by a mylar-spacer of a few μm thickness.

Instead of transmission cells, attenuated total reflection (ATR)-cells (**Fig. 3B**) can be used ***(14,15)***. The IR light is reflected at the interface of a crystal (most common are diamond, ZnSe, and Ge) and the sample. In this process an evanescent wave penetrates into the sample. The depth D, where the intensity of the wave is reduced to (for example) 13.5%, can be calculated by

$$D = \frac{10^4}{2\pi \nu n_c \sqrt{\sin 2\alpha - \frac{n_s}{n_c}}}, \quad (1)$$

where *n* is the wavenumber, n_S is the refractive index of the sample, n_C the refractive index of the crystal and *a* is the angle of incidence.

In most cases *D* is in the low μm regime and the first layers are participating stronger to the absorption. This effect can be enhanced by the surface enhanced infrared absorption (SEIRA)-technique, where the ATR-cell is coated by thin film of gold or silver ***(16,17)***. Thus, thin layers (e.g., of lipids) can be investigated. One can increase the absorption further by using multiple reflections. ATR-cells with 1 to 25 reflections are available.

Both transmission cells and ATR-cells can be used as flowcells (**Fig. 3C**). Here, the sample can be easily exchanged by means of a tubing system. This can increase the quality of difference spectra enormously, because the whole setup (sample thickness, window position, etc.) can be maintained exactly the same.

Figure 4A shows the absorbance spectrum of a protein. Even a smaller protein of 20 kD has about 10^4 absorptions in the infrared. Thus, from the absorption spectrum alone one cannot obtain information on individual bands, but on global features of the protein. The spectrum is dominated by the amide I (C=O stretch) and amide II (NH bend coupled with C–N stretch) bands, where every amino acid contributes. From this backbone absorption, e.g., information on the secondary structure can be gained ***(17a)***. Water absorptions (O–H bend) are found in the same region as the amide I.

For a FTIR difference spectrum (**Fig. 4B**) of a reaction A → B one calculates the absorbance spectrum of B minus the absorbance spectrum of A. Thus, the vibrations from groups that are not changed during the reaction annihilate each other and only the changes during the reaction are seen. Now individual absorptions can be resolved. It is important, to maintain accurately the same conditions during the reactions, otherwise the background, with a 10^3 stronger absorbance, will obscure the difference spectrum. To monitor such small changes highly sensitive instrumentation is required. FT–IR spectroscopy is able to reliably detect such small changes due to the multiplex and the Jacquinot advantages which lead to the crucial increase in the signal to noise ratio.

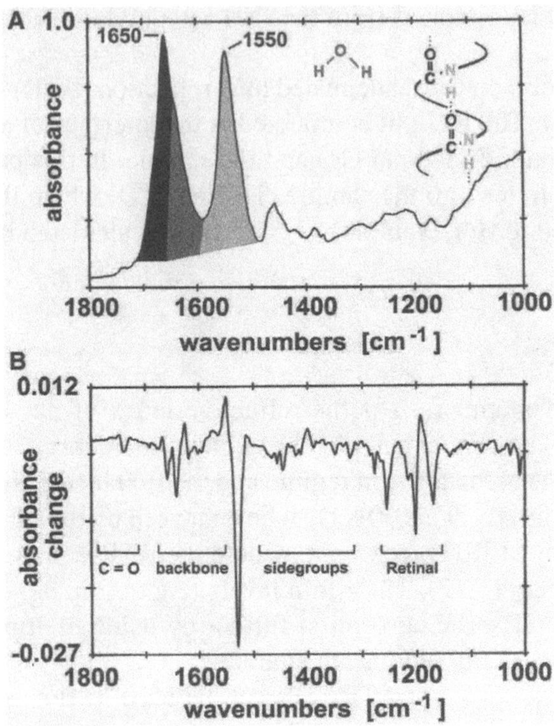

Fig. 4. (**A**) Typical IR absorbance spectrum of a hydrated protein film (BR), showing all absorptions of the protein. (**B**) Typical difference spectrum (BR-L–BR), showing only absorptions which change during the reaction.

3.2. Trigger Techniques

Because the reaction induced absorbance changes are several orders smaller than the background absorbance of the protein, the sample has to be activated without removing it from the sample chamber. The difference technique requires a sharp initiation (triggering) of the protein reaction. This can be achieved by photoexcitation or by fast mixing.

Photobiological systems. Ideally suited to time-resolved study are light-induced reactions in photobiological systems like bacteriorhodopsin (bR) ***(3,4)*** and the photosynthetic reaction centers (RCs) ***(2)*** that carry intrinsic chromophors. In these systems, the chromophore can be directly activated by a laser flash, which induces isomerization or redox reactions of the prosthetic groups.

Caged compounds. A much broader range of applications can be achieved using caged compounds, in which biologically active molecules are released from inactive photolabile precursors. They allow the initiation of a protein reaction with a nanosecond UV laser flash. Caged phosphate, caged GTP, caged ATP, and caged calcium have been established as particularly suitable trigger compounds ***(18,19)***. The most popular caging groups are *ortho*-alkylated nitrophenyl compounds. However, its photoreactions involve several intermediates that limit the time resolution for the reaction. Faster is the *para*-hydroxyphenacyl-cage, where the reaction proceeds on the excited state in the subnanosecond time regime ***(20)***.

Caged GTP: Caged GTP will be discussed as an example of a time-resolved FT–IR study of caged protein–ligands. The 1-(2-nitrophenyl) ethyl (NPE) moiety is frequently used to protect phosphate, nucleotides, and nucleotide analogs. The application of UV flashes leads to the release of the desired phosphate compound. The mechanism of photolysis of compounds containing the 2-nitrobenzyl group has been the topic of several investigations ***(21,22)***. After photolysis, the caged compound decays to GTP and the by-product 2-nitrosoacetophenone. Typical FT–IR difference spectra of this photolysis reaction are shown in **Fig. 5A** ***(9)***. A spectrum of the caged phosphate is measured prior to the photolysis as reference, then after photolysis further spectra are recorded and the absorbance difference spectra, ΔA, are calculated. Only those vibrational modes that have undergone reaction-induced absorbance changes give rise to bands in the difference spectra. Negative bands in the difference spectrum are as a result of the caged GTP, whereas positive bands are caused by the photolysis products. One can resolve an intermediate, the aci-nitro anion. At pH 6.0 and 260 K the formation of GTP is complete after 3000 ms. A faster reaction is observed at pH 7.6 with 300 ms. However, the best time resolution can be obtained with the para-Hydroxyphenyl (pHP) caged group (**Fig. 5B**). Here, the cleavage takes place on an excited state and is complete within 0.5 ps ***(20)***. Already the first spectrum after the laser flash shows that product formation is complete. We cannot resolve any intermediate. The photolysis reaction of caged GTP has been investigated in detail by FT–IR spectroscopy. Bands were assigned by the use of ^{18}O phosphate labeling by Cepus et al. ***(9)***. In addition, FT–Raman spectra and photolysis spectra of caged Ca^{2+} are reported in Cepus et al. ***(9)***.

Micro-mixing-cells. Because silicon is transparent in the mid-infrared, micromachined silicon components offer great potential for establishing FT–IR spectroscopy as a new method for studying microsecond mixing experiments. When the dimensions of liquid channels in silicon devices are reduced to a few micrometers, the flow will be laminar at almost all velocities. Therefore, there are no turbulent vortices to enhance mixing. In micrometer-size dimension channels diffusion is the only way reactants can be brought together, and the flow pattern has to be designed such that diffusion length scales are kept small. It has been shown that microscale-mixing devices can decrease

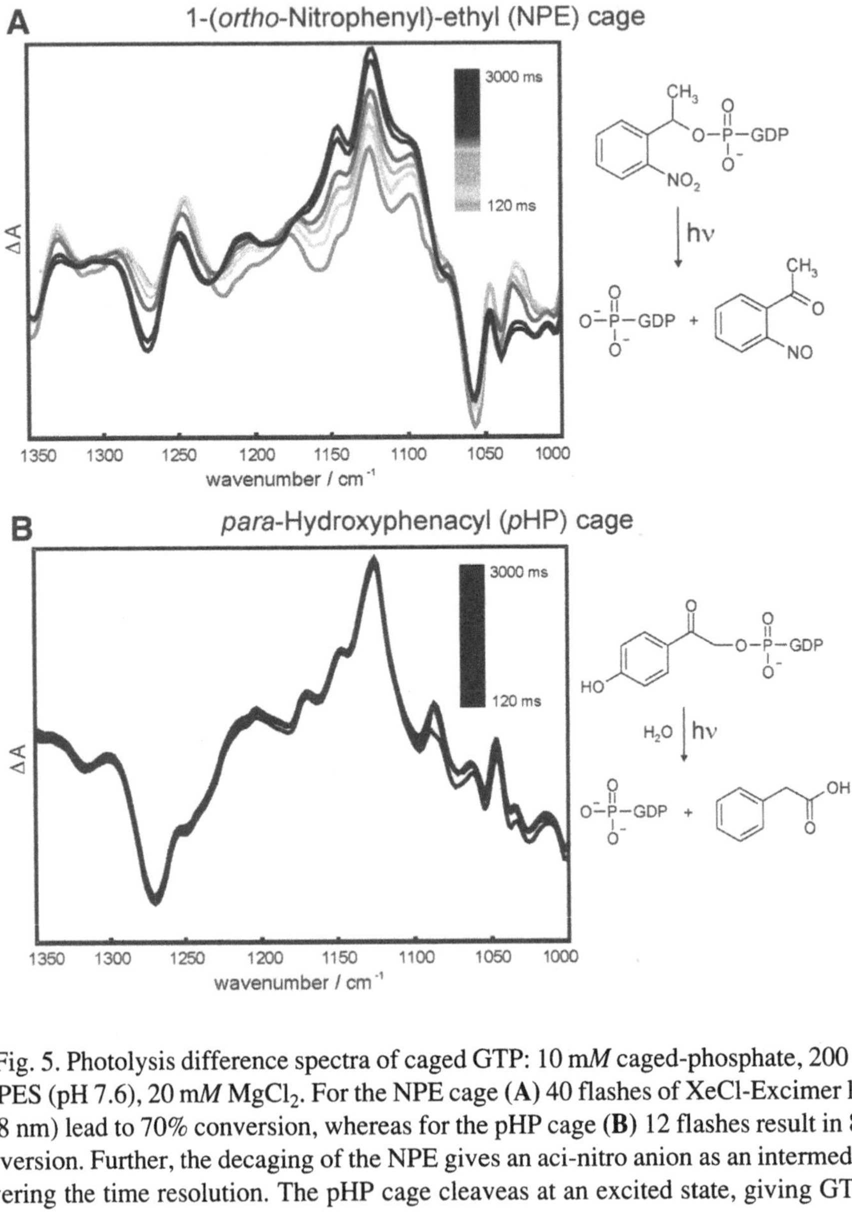

Fig. 5. Photolysis difference spectra of caged GTP: 10 m*M* caged-phosphate, 200 m*M* HEPES (pH 7.6), 20 m*M* $MgCl_2$. For the NPE cage **(A)** 40 flashes of XeCl-Excimer laser (308 nm) lead to 70% conversion, whereas for the pHP cage **(B)** 12 flashes result in 80% conversion. Further, the decaging of the NPE gives an aci-nitro anion as an intermediate, lowering the time resolution. The pHP cage cleaveas at an excited state, giving GTP in 0.5 ns.

the characteristic mixing time from milliseconds down to 10 μs. A new continuous-flow-mixing chip has been designed for FT-IR microscopy **(Fig. 6)**. The protein solution in the center and two streams of mixing buffer enter through 80 μm deep inlet channels, which intersect with the 8 μm deep observation

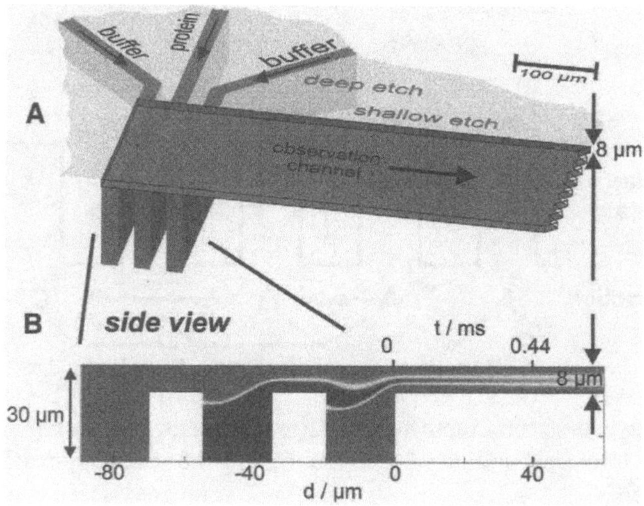

Fig. 6. Scheme of the microfabricated flow-cell. A color figure and more details are given in **ref. *23***.

channel. Because the three inlet channels are a factor of ten deeper than the observation channel, there is almost equal pressure over the whole width of the observation channel. Because of the viscosity determined laminar flow, no turbulence is induced when the buffer channels merge in, and a layer of the center (protein) solution between two buffer layers over the whole width of the observation channel results. A fluid dynamics simulation was performed to verify the desired flow pattern. The resulting flow pattern shows the formation of a protein layer between two layers of buffer solution. Because the protein layer is thin, diffusing reactant molecules (e.g., ligands) stream from the buffer solution into the protein solution, and thus mixing, is fast. The time resolution is achieved by scanning along the observation channel with the focused beam of a FT–IR microscope.

The continuous-flow mixing chips, reported recently ***(23)***, open up new dimensions for FT–IR spectroscopy of protein reactions. The time resolution of 400 µs is about 1000 times faster than recent IR stopped-flow set-ups. Additionally, the miniaturization reduces the sample consumption by an even greater magnitude. For a reactant with a higher diffusion constant than trifluoroethanol (TFE), the time resolution can be further improved by using a smaller IR focus spot and a higher flow velocity. With the current design and the ca. 10 µm diameter focused spot of a synchrotron IR source, a time resolution on the order of 50 µs is feasible.

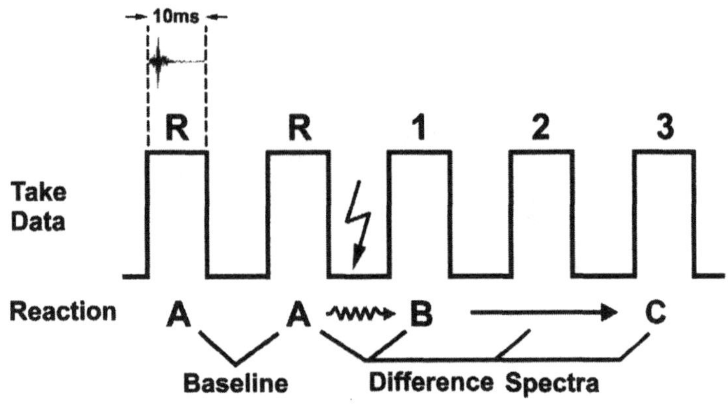

Fig. 7. Time course for a rapid scan FTIR experiment. First, reference spectra (R) are taken. The laser flash initiates a reaction. During the reaction towards C interferograms are taken.

3.3. Rapid Scan

The principle of the rapid scan FT–IR mode of time-resolved spectroscopy is simple: after taking a reference spectrum of the protein in its ground state, one activates the protein (e.g., by a laser fash) and records interferograms in much shorter times than the half-lives of the reactions *(4)*. The time pattern of such an experiment is shown schematically in **Fig. 7**. During each high-state period interferograms are taken. The first two reference (R) interferograms represent the ground state (A), while the following three interferograms are taken during the reaction pathway (B to C). Thus the first interferogram will mainly represent the spectrum of B and in the following the ratio of C will increase according to the kinetics of the observe reaction. The Fourier transformation technique allows observation of processes whose half-life are on the order of the scan time for first-order reactions or even below (*see* **Note 2**). If the half-life of the observed process is shorter than the duration of the scan, the intensity of the interferogram is convolved with the absorption change of the sample. In case of first-order reactions the interferogram is convolved with an exponential function, which results only in Lorentzian line shape broadening in the spectrum after Fourier transformation.

3.4. Step-Scan

In the step-scan mode, the interferometer moving mirror may be visualized as being held stationary at the interferogram data position x_n **(Fig. 8)**; the protein activity is initiated, for example, by a laser flash and the time dependence of the intensity change at this interferogram position x_n is measured. Then, the

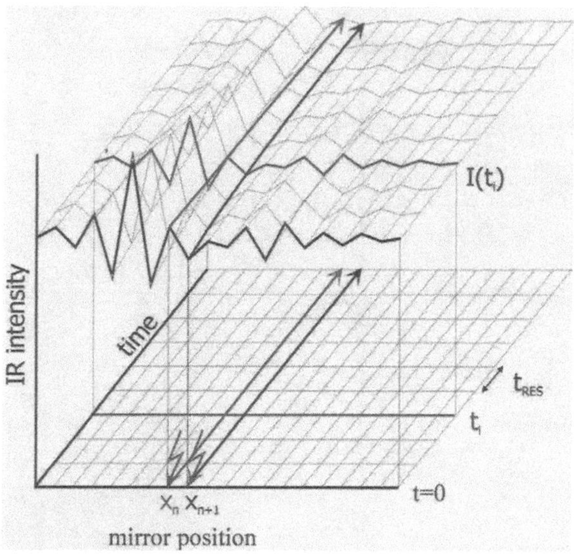

Fig. 8. Principle of the step-scan technique. The interferometer mirror is stepped to a sampling position x_n, the reaction is then initiated and the time dependence of the IR intensity is measured (dark grey arrow). The detector limits the time resolution. After relaxation the interferometer is stepped to the next position, x_{n+1}, and the data recording process is repeated (medium gray arrow). After the measurement at all interferometer positions the data is rearranged to yield time-dependent interferograms $I(t_i)$ (light grey), which after Fourier transformation yield the time-dependent spectra.

interferometer *steps* to the next interferogram data position x_{n+1} and the reaction is repeated and measured again. This process is continued at each sampling position of the interferogram. The position of the interferometer mirror must be kept accurate to about 1–2 nm at x_n while the intensity change of the interferogram during the time of the reaction is measured. Therefore, the method is very sensitive to external disturbances (e.g., noise). The time resolution is usually determined by the response time of the detector, which is about a few ns. After the measurement the data is rearranged to yield time-dependent interferograms $I(t_i)$. Using pulsed IR sources instead of the conventional globar the time resolution is determined by the time duration of the probe pulse. This can give in principle femtosecond resolution with broad-band femtosecond lasers or synchrotron radiation. For more details on the step-scan technique see the literature ***(24–28)***.

A typical step-scan experimental setup is shown schematically in **Fig. 9** ***(28)***. Except for the homebuilt sample chamber, the FT–IR apparatus is evacuated to

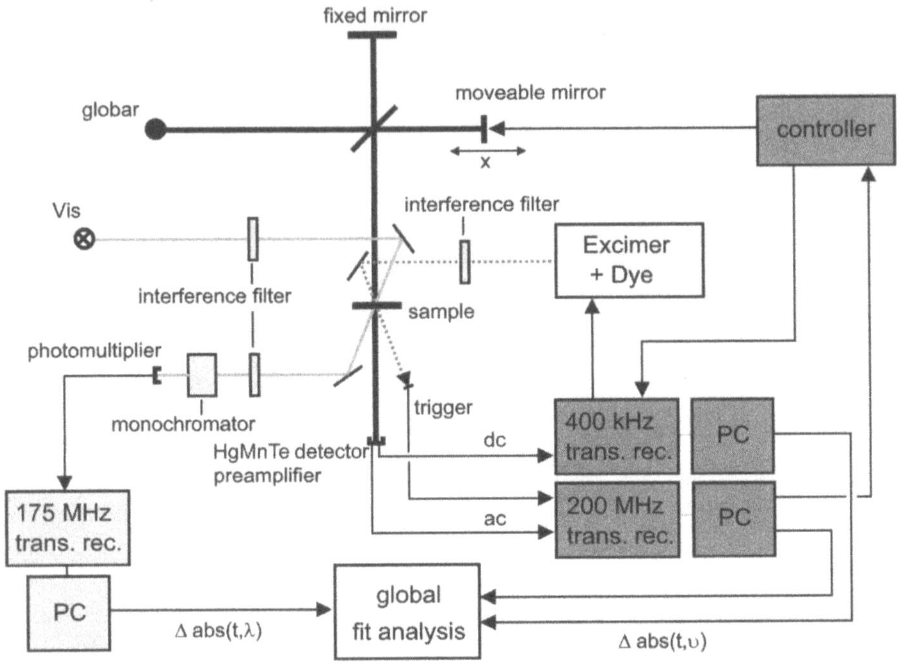

Fig. 9. A typical step-scan setup, consisting of an FTIR instrument and pre-amplifier connected to a 400 kHz and a 200 MHz transient recorder, and a photolysis setup to activate the sample.

3 mbar during the measurement. This increases the stability and reduces the sound-sensitivity of the movable mirror. Furthermore, the set-up is positioned on a vibration isolation table within a temperaturecontrolled laboratory. With this set-up, we have determined the residual spatial fluctuations of the movable mirror to be approx 0.5 nm. The sample chamber is purged with dry air (dew point 70°C). The IR absorbance changes are detected by a photovoltaic HgMnTe detector. The detector's signal is amplified in a homebuilt two-stage preamplifier with an AC- and DC-coupled output. The bandwidth of the DC part is limited to 400 kHz, whereas the bandwidth of the AC part is 200 MHz. Controls ensure that the output-signal of the preamplifier depends linearly on the IR intensity. The DC-coupled output of the preamplifier is digitized by a 12 bit, 400 kHz transient recorder connected to a PC. The offset of the input signal can be compensated to zero. This allows subsequent amplification of the signal and use of the full dynamic range of the transient recorder. In order to reduce the huge amount of data from the ns to the ms time domain, the 400 kHz acquisition

rate is slowed to 50 kHz at 200 μs. This allows time-averaging. The AC-coupled output of the preamplifier is recorded by a 200 MHz 8 bit transient recorder.

At every sampling position of the interferogram the correct positioning of the movable mirror is checked before data acquisition starts. A transistor-to-transistor (TTL) output-signal then triggers the excimer-laser to initiate the reaction. A fast photodiode (rise-time 10 ns) is triggered by the dye-laser flash and starts data acquisition of the 200 MHz transient recorder.

The spectral range is limited below the Nyquist-wavenumber 1975 cm^{-1} by an interference filter to reduce the number of sampling points of the interferogram. This filter also shields the IR detector from both scattered light of the dye laser and the heat emitted by the sample. (The dye laser's pulse causes a small warming of the sample.) The resulting interferogram contains 780 data points. It is multiplied by the Norton–Beer-weak apodization function and then zero-filled by a factor of 2. The phase-spectrum $\varphi(\nu)$ is calculated with a spectral resolution of 50 cm^{-1}, whereas the difference spectra are recorded with a spectral resolution of approx 3 cm^{-1}. Details on the course of a time-resolved measurement is given in **Note 3**.

Step-scan FT-IR of noncyclic reactions. The step-scan technique cannot be applied directly to noncyclic reactions, because the investigated process has to be repeatedly initiated typically at about 1000 sampling positions of the interferogram. Consequently, to investigate irreversible systems the sample has to be renewed at every sampling position. In conventional flow-cells the optical pathlength is too large to perform difference spectroscopy of hydrated biological samples. We need to use 4 μm thin films to depress the water background absorption of biological samples.

In a novel approach, the IR beam and the excitation laser-beam are focused to a very small diameter of 200 μm (**Fig. 10**). Thereby, only a small segment of the sample that has an overall diameter of 15 mm is excited and probed. By moving the sample, which is mounted on a movable *x–y* stage, to different nonexcited segments the reaction can be repeated until a complete interferogram data set has been recorded.

The technique was successfully applied to the noncyclic reaction of the photolysis of caged ATP *(29)*. By this technique the transiently formed aci-nitro anion complex is also measured. This successful demonstration of a study of an irreversible reaction with 20 μs time-resolution now opens the way for many new applications of step-scan FT–IR measurements to noncyclic reactions.

3.5. Global Fit

For analyzing time-resolved data adequate kinetic analysis is important. A so-called global-fit analysis yields the apparent rate constants of the analyzed processes *(6)*. The global-fit analysis not only includes fitting the absorbance

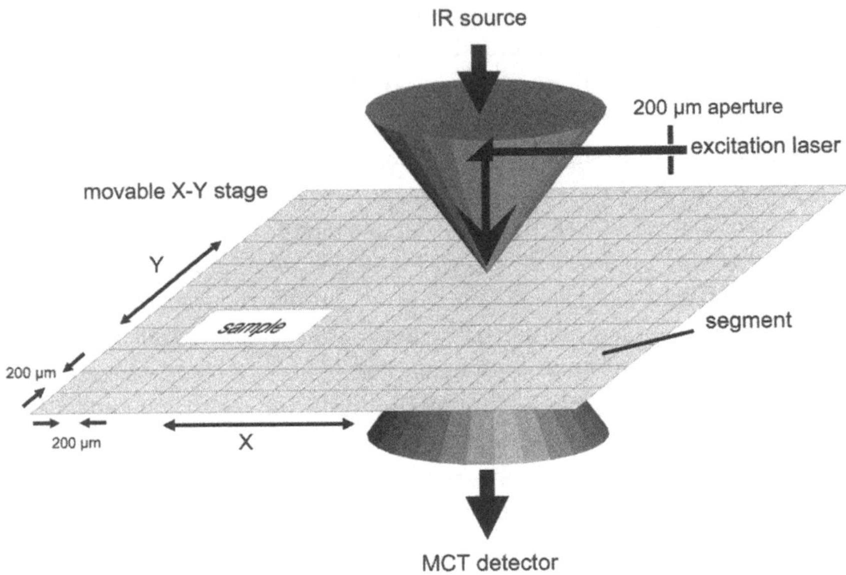

Fig.10. Representation of a step-scan setup for noncyclic reactions. Both IR-beam and excitation laser are focused to a diameter of 200 µm. The sample is mounted on a motorized *x–y* stage divided into several thousand 200 µm × 200 µm segments. The sample is renewed by consecutively moving the next segment into the focused spot.

change at a specific wavenumber, but also changes at up to 800 points in the spectrum simultaneously. All reactions are assumed to be first order, and can therefore be described by a sum of exponentials. The fit procedure minimizes the difference between the measured data $\Delta A_{measured}$ and the theoretical description ΔA, weighted according to the noise w_{ij} at the respective wavenumbers, and summarized not only over the time (t_j) but also over the wavenumbers (r_i). In global-fit analysis the absorbance changes ΔA in the visible and IR are analyzed with sums of n_r exponentials with apparent rate constants k_l and amplitudes a_l:

$$\Delta A(\nu,t) = \sum_{l=1}^{n_r} a_l(\nu)\, e^{-k_l t} + a_0(\nu) \qquad (2)$$

In this analysis, the weighted sum of squared differences f between the fit with n_r apparent rate constants k_l and data points at n_w measured wavenumbers n_i and n_t times t_j is minimized:

$$f = \sum_{i=1}^{n_w} \sum_{j=1}^{n_t} \left(w_{ij}\right) \left[\Delta A_{gemessen}\left(\nu_i, t_j\right) - \Delta A\left(\nu_i, t_j\right)\right]^2 \qquad (3)$$

For unidirectional forward reactions, the determined apparent rate constants are directly related to the respective intrinsic rate constants describing the respective reaction steps *(30,31)*. If in addition significant back-reactions occur, the analysis becomes more complicated *(32)*. Then, the reaction has to be modeled until the guessed intrinsic rate constants fulfil the experimentally observed time-course described by the apparent rate constants. Because the number of the intrinsic rate constants in the model is larger than the number of the experimentally observed apparent rate constants, the problem is experimentally underdetermined and the solution is not unequivocal.

An alternative method is the singular value decomposition or principal component analysis. Thereby, the basis difference spectra are calculated from all difference spectra measured *(6)*. This procedure allows the determination of transient spectra independent of specific kinetic models and independent of the temporal overlap *(6)*.

3.6. Band Assignments

A large number of bands can be seen in the difference spectra in **Fig. 4B**. The frequency range in which a band appears allows a rough tentative assignment of the bands. For example, the retinal vibrations are expected in the fingerprint region between 1300 cm^{-1} and 1100 cm^{-1}, and the carbonyl vibrations of aspartic or glutamic acids between 1700 cm^{-1} and 1770 cm^{-1}. In order to draw detailed conclusions from the spectra, the bands have to be assigned to individual groups. Unambiguous band assignment can be performed by using isotopically labeled proteins or by amino acid exchange via site directed mutagenesis.

Isotopic labeling shifts the stretching frequency, ν, of the labeled group as a result of the increased reduced mass *m*:

$$\nu = \frac{1}{2\pi}\sqrt{\frac{k}{\mu}} \tag{4}$$

where *k* is the force constant. Isotopic labeling can be performed on prosthetic groups like retinal *(33)* or nucleotides like GTP *(9,11,12)*. These compounds can be chemically synthesized. As an example, site-specific isotopic labeled caged GTP is presented below.

Isotopic labeling of all amino acids of one kind can be achieved by biosynthetic incorporation of isotopically-labeled amino acids, e.g., aspartic acid *(13)*. This is relatively easy to perform. Further, site-directed exchange of an amino acid by mutagenesis eliminates the absorption band of the exchanged group.

The principle of band assignment by site-directed mutagenesis is demonstrated in **Fig. 11**, and as an example the bR-N difference spectra of the wild-type (WT) and the Asp-96-Asn mutant are shown in **Fig. 12** *(4)*. Absorbance

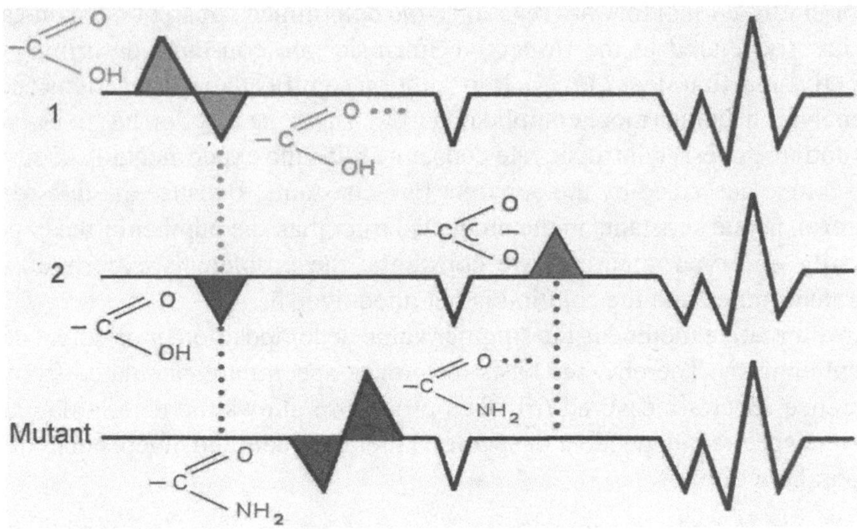

Fig. 11. Schematic showing the expected absorbance changes in the IR difference spectra. If in the transition of BR to an intermediate (case 1), a hydrogen bond of a protonated carboxylic acid in wild-type (WT) is broken, a difference band is expected as a result of the frequency upshift of the carbonyl vibration. For a deprotonation (case 2), a negative carbonyl band should disappear and a carboxylate band appear. If the amino acid giving rise to the absorbance changes is exchanged and the mutant protein is measured, the carbonyl band (in cases 1 and 2) and the carboxylate band (case 2) should disappear as compared with the WT difference spectrum. (These are marked by the dotted lines.) In addition, in the mutant difference spectrum, a new carbonyl band might appear, e.g., for Asp (Glu) to Asn (Gln) mutation. It is important to notice that all other bands remain the same (as indicated), showing that most of the structure is not affected and that the mutation is noninvasive.

changes in the spectral range between 1500 cm^{-1} and 1000 cm^{-1} are highly reproducible, indicating that this specific mutagenesis is noninvasive and does not disturb the protein structure. Only the carbonyl band shift at 1742 cm^{-1} is absent in the spectrum of the mutant. This is the 4-carbonyl vibration of the exchanged Asp 96 *(4)*, which demonstrates that this group deprotonates in this step.

Mutation of an amino acid changes the structure of the protein to a greater or lesser degree, but is easy to achieve by site-directed mutagenesis, a standard molecular biology method ***(34)***. On the other hand, isotopic labeling has the advantage of marking the molecular group noninvasively. An example for band assignment by isotopic labeling of a ligand is given in **Fig. 13**. Here, $\gamma^{18}O_4$-

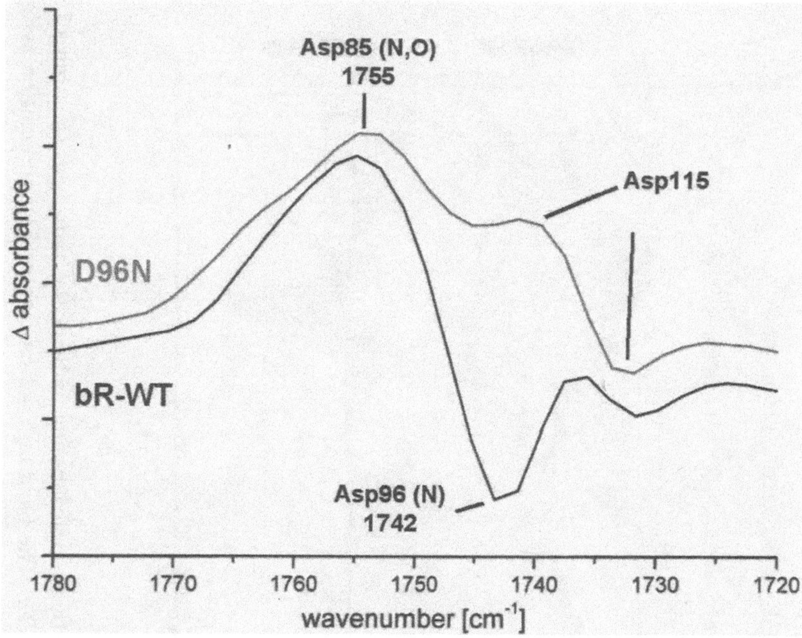

Fig. 12. Spectra, demonstrating the principle of band assignment by site-directed mutagenesis. Difference spectra of bR-N are shown. The spectrum of the bR-WT showing a negative band of Asp96 (deprotonation, case 2 in **Fig. 11**). In the corresponding spectrum of the mutant Asp 96 Asn, this band is absent. Additionally, the difference bands of Asp115 are now seen more clearly.

labeled GTP is used in the Ras catalyzed hydrolysis reaction of GTP *(11)*. In the difference spectrum (Δ) of the hydrolysis the negative bands correspond to the Ras·GTP state. The band at 1143 cm^{-1} is shifted by the γ-label, and can thus be assigned to an absorption of the γ-GTP group. On the other hand, the α-band at 1263 cm^{-1} (assigned by α-labeled GTP, not shown) remains in the same position after γ-labeling. Similarly, one can assign positive bands (product state) at 1078 cm^{-1} and 992 cm^{-1} as absorptions of the appearing P_i after hydrolysis. Often, a double difference spectrum (ΔΔ) and the difference of the labeled and the unlabeled difference spectra gives distinct band positions, because only absorptions which shift as a result of the labeling appear.

However, the most suitable method, site-directed isotopic labeling of an amino acid is an expensive molecular biology method and only very rarely successfully applied. Today, chemical synthesis of a protein allows site-directed isotopic labeling and may become the method of the future *(35,36)*.

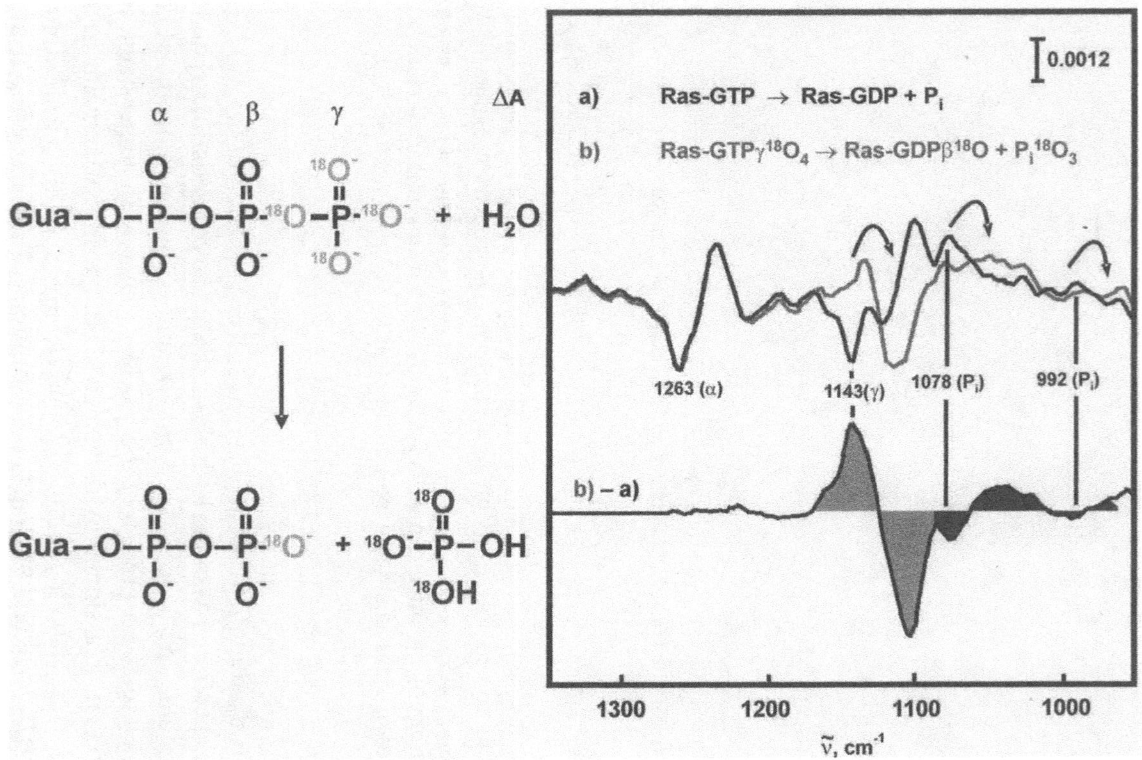

Fig. 13. Spectra, demonstrating the principle of band assignment by isotopic labeling. In black the hydrolysis spectrum of Ras·GTP toward Ras·GDP is shown, in gray the same with $\gamma^{18}O$-labeled GTP. The band at 1143 cm^{-1} is shifted and can be assigned to a γ-vibration, whereas the band at 1263 cm^{-1} (the α-band) is not affected.

Table 1
Frequently Used IR-Window Materials

Material	Transmission Range (cm^{-1})
Silicon (Si)	100–10000
Calcium fluoride (CaF_2)	950–66000
Barium fluoride (BaF_2)	890–50000
Zinc selenide (ZnSe)	450–20000
Silver chloride (AgCl)	400–23000

4. Notes

1. For FTIR measurements numerous types of window material available. **Table 1** gives a short list of frequently used materials for FTIR measurements of hydrated protein samples. Most frequently used is CaF_2, because it is completely water insoluble and transparent in the mid-IR and UV region. However, for investigation at lower wavelength than 950 cm^{-1} other materials are necessary.
2. The velocity of the interferometer moving mirror, V_{max}, and the desired spectral resolution Δv determines the scan duration Δt and thereby the time resolution: Today, the fastest commercially available FT–IR spectrometers are capable of yielding a time resolution of 10 ms at 4 cm^{-1} spectral resolution. Improvements in current interferometer designs that yield a significant increase in mirror scan velocity are unlikely to occur for interferometers with a reciprocating motion. There are practical reasons for this, e.g., the extreme acceleration of the scanner at the turning points. If a rapidly rotating, rather than a reciprocating, mirror is used, the scan-speed can be increased to the point that the time resolution for a 4 cm^{-1} resolution can be decreased to 1 ms. Without a radical change of design, however, the time resolution for the rapid scan is still likely to be limited to the ms range. A typical experimental set-up for rapid scan measurements is illustrated in **Fig. 2**. A laser flash activates the sample. Simultaneously, a conventional photolysis set-up measures the absorbance change in the visible region.
3. A Scheme of the course of a step-scan experiment is shown in **Fig. 14.**
   a. The mirror moves to the first acquisition point of the interferogram. The DC value of the IR intensity is measured. Afterwards, the offset of the DC signal is set to zero.
   b. A laser flash starts the reaction. Both transient recorders simultaneously measure the time-dependent IR intensity changes at the sampling position. The 200 MHz transient recorder measures the time domain from 30 ns to 20 μs, whereas the 400 kHz transient recorder monitors from 5 μs until the end of the reaction in the ms time range. At each sampling position of the interferogram, the reaction is repeated several times to improve the S/N. After measuring the time traces of all interferogram sampling positions the data are rearranged to yield time dependent difference interferograms $\Delta I(t_i)$ (*see* **Fig. 8**). Because these difference interferograms contain positive as well as

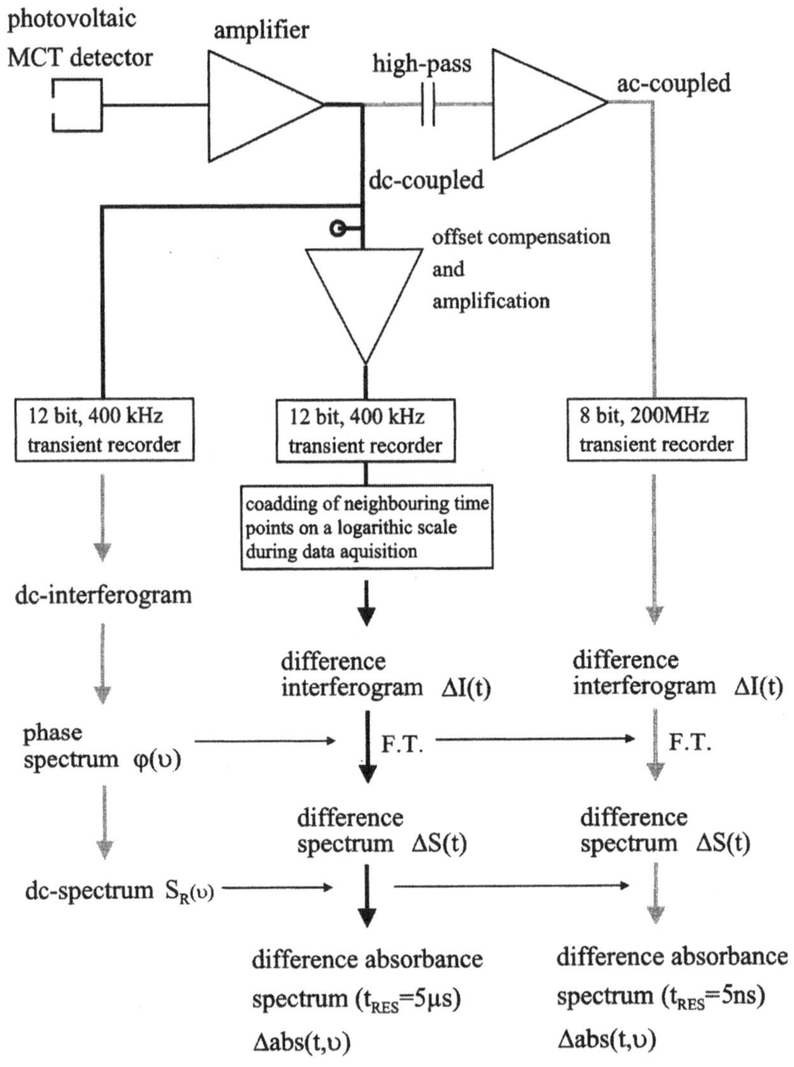

Fig. 14. Schematic representation of the IR signal preamplification, the data recording and the calculation of the difference spectra in the DC- and AC-coupled modes.

negative spectral features, usual Mertz phase correction cannot be directly applied. Therefore, the stored phase $\varphi(\nu)$ from the first measurement in **step a** is used. The phase does not change between both measurements, because the movable mirror stops exactly at the same sampling points and only small absorbance changes take place. Possible errors caused by the transient heating of the sample by the actinic laser flash, baseline distortions, and nonlinearity of the IR detectors are discussed in detail in Rammelsberg et al. *(28)*.

References

1. Gerwert, K. (1993) Molecular reaction mechanisms of proteins as monitored by time-resolved FTIR spectroscopy. *Curr. Opin. Struct. Biol.* **3,** 769–773.
2. Remy, A. and Gerwert, K. (2003) Coupling of light-induced electron transfer to proton uptake in photosynthesis. *Nat. Struct. Biol.* **10,** 637–644.
3. Gerwert, K., Souvignier, G., and Hess, B. (1990) Simultaneous monitoring of light-induced changes in protein side-group protonation, chromophore isomerization, and backbone motion of bacteriorhodopsin by time-resolved Fourier-transform infrared spectroscopy. *Proc. Natl. Acad. Sci. USA* **87,** 9774–9778.
4. Gerwert, K., Hess, B., Soppa, J., and Oesterhelt, D. (1989) Role of aspartate-96 in proton translocation by bacteriorhodopsin. *Proc. Natl. Acad. Sci. USA* **86,** 4943–4947.
5. Siebert, F., Maentele, W., and Gerwert, K. (1983) Fourier-transform infrared spectroscopy applied to rhodopsin. The problem of the protonation state of the retinylidene Schiff base reinvestigated. *Euro. J. Biochem.* **136,** 119–127.
6. Hessling, B., Souvignier, G., and Gerwert, K. (1993) A model-independent approach to assigning bacteriorhodopsin's intramolecular reactions to photocycle intermediates. *Biophys. J.* **65,** 1929–1941.
7. Brudler, R., Rammelsberg, R., Woo, T. T., Getzoff, E. D., and Gerwert, K. (2001) Structure of the I1 early intermediate of photoactive yellow protein by FTIR spectroscopy. *Nat. Struct. Biol.* **8,** 265–270.
8. Luebben, M. and Gerwert, K. (1996) Redox FTIR difference spectroscopy using caged electrons reveals contributions of carboxyl groups to the catalytic mechanism of heme-copper oxidases. *FEBS Lett.* **397,** 303–307.
9. Cepus, V., Ulbrich, C., Allin, C., Troullier, A., and Gerwert, K. (1998) Fourier transform infrared photolysis studies of caged compounds. *Methods Enzymol.* **291,** 223–245.
10. Cepus, V., Scheidig, A. J., Goody, R. S., and Gerwert, K. (1998) Time-resolved FTIR studies of the GTPase reaction of H-ras p21 reveal a key role for the beta-phosphate. *Biochemistry* **37,** 10,263–10,271.
11. Allin, C. and Gerwert, K. (2001) Ras catalyzes CTP hydrolysis by shifting negative charges from gamma- to beta-phosphate as revealed by time-resolved FTIR difference spectroscopy. *Biochemistry* **40,** 3037–3046.
12. Allin, C., Ahmadian, M. R., Wittinghofer, A., and Gerwert, K. (2001) Monitoring the GAP catalyzed H-Ras GTPase reaction at atomic resolution in real time. *Proc. Natl. Acad. Sci. USA* **98,** 7754–7759.
13. Engelhard, M., Gerwert, K., Hess, B., and Siebert, F. (1985) Light-driven protonation changes of internal aspartic acids of bacteriorhodopsin: an investigation of static and time-resolved infrared difference spectroscopy using [4–13C]aspartic acid labeled purple membrane. *Biochemistry* **24,** 400–407.
14. Harrick, N. J. (1987) Nanosampling via internal reflection spectroscopy. *Appl. Spectrosc.* **41,** 1–2.
15. Fringeli, U. P., Baurecht, D., Siam, M., Reiter, G., Schwarzott, M., Burgi, T., and Bruesch, P. (2002) ATR spectroscopy of thin films. *Handbook Thin Film Mat.* **2,** 191–229.

16. Osawa, M. (2001) Surface-enhanced infrared absorption. *Topics in Appl. Phys.* **81,** 163–187.
17. Ataka, K. and Heberle, J. (2003) Electrochemically induced surface-enhanced infrared difference absorption (SEIDA) spectroscopy of a protein monolayer. *J. Amer. Chem. Soc.* **125,** 4986–4987.
17a. Oberg, K. A., Ruysschaert, J. M., and Goormaghtigh E. (2003) Rationally selected basis proteins: a new approach to selecting proteins for spectroscopic secondary structure analysis. *Protein Sci.* **12,** 2015–2031.
18. Pelliccioli Anna, P. and Wirz, J. (2002) Photoremovable protecting groups: reaction mechanisms and applications. *Photochemical & Photobiological Sciences* **1,** 441–458.
19. McCray, J. A. and Trentham, D. R. (1989) Properties and uses of photoreactive caged compounds. *Ann. Rev. Biophys. Biophys. Chem.* **18,** 239–270.
20. Park, C.-H. and Givens, R. S. (1997) New photoactivated protecting groups. 6. p-hydroxyphenacyl: a phototrigger for chemical and biochemical probes. *J. Amer. Chemical Society* **119,** 2453–2463.
21. Corrie, J. E. T. and Trentham, D. R. (1993) Caged nucleotides and neurotransmitters. *Bioorg. Photochem.* **2,** 243–305.
22. Walker, J. W., Reid, G. P., McCray, J. A., and Trentham, D. R. (1988) Photolabile 1–(2-nitrophenyl)ethyl phosphate esters of adenine nucleotide analogs. Synthesis and mechanism of photolysis. *J. Amer. Chem. Soc.* **110,** 7170–7177.
23. Kauffmann, E., Darnton, N. C., Austin, R. H., Batt, C., and Gerwert, K. (2001) Lifetimes of intermediates in the beta-sheet to alpha-helix transition of beta-lactoglobulin by using a diffusional IR mixer. *Proc. Natl. Acad. Sci. USA* **98,** 6646–6649.
24. Palmer, R. A., Chao, J. L., Dittmar, R. M., Gregoriou, V. G., and Plunkett, S. E. (1993) Investigation of time-dependent phenomena by use of step-scan FT-IR. *Appl. Spectrosc.* **47,** 1297–1310.
25. Palmer, R. A., Manning, C. J., Chao, J. L., Noda, I., Dowrey, A. E., and Marcott, C. (1991) Application of step-scan interferometry to two-dimensional Fourier transform infrared (2D FT-IR) correlation spectroscopy. *Appl. Spectrosc.* **45,** 12–17.
26. Weidlich, O. and Siebert, F. (1993) Time-resolved step-scan FT-IR investigations of the transition from Kl to L in the bacteriorhodopsin photocycle-identification of chromophore twists by assigning hydrogen-out-of-plane (Hoop) bending vibrations. *Appl. Spectrosc.* **47,** 1394–1400.
27. Uhmann, W., Becker, A., Taran, C., and Siebert, F. (1991) Time-resolved FT-IR absorption spectroscopy using a step-scan interferometer. *Appl. Spectrosc.* **45,** 390–397.
28. Rammelsberg, R., Hessling, B., Chorongiewski, H., and Gerwert, K. (1997) Molecular reaction mechanisms of proteins monitored by nanosecond step-scan FT-IR difference spectroscopy. *Appl. Spectrosc.* **51,** 558–562.
29. Rammelsberg, R., Huhn, G., Lubben, M., and Gerwert, K. (1998) Bacteriorhodopsin's intramolecular proton-release pathway consists of a hydrogen-bonded network. *Biochemistry* **37,** 5001–5009.

30. Cantor, C. R. and Schimmel, P. R. (1980) *Biophysical Chemistry, Pt. 1: The Conformation of Biological Macromolecules*, Freeman, San Francisco, CA.
31. Fersht, A. (1999) *Enzymes: Structures and Reaction Mechanisms*, Freeman, San Francisco, CA
32. Steinfeld, J. I., Francisco, J. S., and Hase, W. L. (1999) *Chemical Kinetics Dynamics, Second Edition*, Prentice Hall, Englewood Cliffs, NJ.
33. Lugtenburg, J., Mathies, R. A., Griffin, R. G., and Herzfeld, J. (1988) Structure and function of rhodopsins from solid state NMR and resonance Raman spectroscopy of isotopic retinal derivatives. *Trends Biochem. Sci.* **13,** 388–393.
34. Stryer, L. and Editor. (1996) *Biochemistry, 4th Revised Edition*, Freeman, New York, NY.
35. Fischer, W. B., Sonar, S., Marti, T., Khorana, H. G., and Rothschild, K. J. (1994) Detection of a water molecule in the active-site of bacteriorhodopsin—hydrogen-bonding changes during the primary photoreaction. *Biochemistry* **33,** 12,757–12,762.
36. Becker, C. F. W., Hunter, C. L., Seidel, R., Kent, S. B. H., Goody, R. S., and Engelhard, M. (2003) Total chemical synthesis of a functional interacting protein pair: The protooncogene H-Ras and the Ras-binding domain of its effector c-Raf1. *Proc. Natl. Acad. Sci. USA* **100,** 5075–5080.
37. Kolano, C. (2003) PhD thesis, Ruhr-Universität.

14

Proteins in Motion

Resonance Raman Spectroscopy as a Probe of Functional Intermediates

Uri Samuni and Joel M. Friedman

Summary

Elucidating proteins function at a level that allows for intelligent design and manipulation is essential in realization of their potential role in biomedical and industrial applications. It has become increasingly apparent though, that probing structures and functionalities under equilibrium conditions is not sufficient. Rather, many aspects of protein behavior and reactivity are rooted in protein dynamics. Thus, there is a growing effort to probe intermediate structures that occur transiently during the course of a proteins function in particular linked to the binding or release of a ligand or substrate. However, studies following the sequence of conformational changes triggered by the binding of substrate/ligand and the concomitant change in functional properties are inherently difficult because often the diffusion times are of the order of conformational relaxation times. This chapter describes methodologies for generating resonance Raman spectra from transient forms of hemoglobin under conditions that allow for the systematic exploration of conformational relaxation and functionality. Special consideration is given to Raman compatible protocols based on sol-gel encapsulation that allow for the preparation, trapping and temporal tuning of nonequilibrium population generated from either the addition or the removal of ligands/substrates.

Key Words: Resonance Raman; protein dynamics; protein conformation; ligand binding; sol-gel; hemoglobin.

1. Introduction

Proteins are an extraordinary class of materials whose diverse and extensive range of functionalities has great, but as yet underutilized potential for bio-

From: *Methods in Molecular Biology, vol. 305: Protein–Ligand Interactions: Methods and Applications*
Edited by: G. U. Nienhaus © Humana Press Inc., Totowa, NJ

medical and industrial applications. Thus, it is an important objective to understand the biophysical basis of protein function at a level that allows for the design and manipulation of both functional properties and stability. Two major foci of molecular biophysics and physical biochemistry is to expose the three-dimensional structure of proteins and to establish and characterize functional properties. It has become increasingly apparent that probing structures and functionalities under equilibrium conditions do not adequately provide all the needed elements to understand the workings of proteins. Missing in this approach are aspects of protein behavior and reactivity that stem from protein dynamics. In much the same way that a static picture of a car provides little in the way of directly assessing how energy is stored, transduced, and utilized for its actual function, so too equilibrium pictures of proteins have limitations. A time-ordered sequence of events within either the car or the protein is much more likely to provide the causal relationships between structure and function. Thus, there has been a growing effort to probe protein structures of intermediates that occur transiently during the course of a functional cycle. In the generic case where protein function is linked to the binding or release of a ligand or substrate, this effort would consist of following both the sequence of conformational changes triggered by the binding or release of substrate/ligand and the change in functional properties as the conformation is changing.

The process of exposing the role of relaxation in workings of proteins consists of several steps. Ideally, the initial step consists of establishing the overall three dimensional structure of the intermediate species. More detailed dissection of the structure of the intermediates is typically oriented towards finding both the reaction coordinates for the conformational transitions, the conformational coordinates that control functional properties, and finally the time evolution for the functionally important conformational coordinates. Once the functionally important coordinates are established and their relaxation properties exposed, the final step would be following the changes in the functional properties as the appropriate degrees of freedom are changing. This last step provides not only a stringent test of proposed relationships between conformation and function but also a means of revealing the precise nature of the relationship between a conformational degree of freedom and functional parameters.

The previous steps present considerable experimental challenges. To start with, one needs a system that can be triggered on a sufficiently fast time scale and with sufficient amplitude to rapidly create a large nonequilibrium population of structures that evolves in synchronized fashion. Laser-triggered relaxations, such as those initiated by photodissociation, are suitable for those limited systems that have ligands or substrates that are vulnerable to photolysis. Much more difficult is developing protocols to follow conformational evolution subsequent to ligand or substrate binding. Unfortunately, rapid mix technology is limited by diffusion times thus precluding the possibility of probing many initial intermediates. Once the transient population is generated, the next challenge is how to probe the evolution of appropriate conformational degrees of freedom and concomitantly the evolution of functional parameters.

At this time, hemeproteins such as myoglobin and hemoglobin are among the very few protein systems that are suitable for the systematic exploration of relaxation and reactivity that we have described earlier. The database of functional and conformational properties for these proteins under equilibrium conditions are extensive when compared with most other proteins. The reactive site is the heme chromophore whose optical properties allow for in depth probing of both ligand binding reactivity, and the functionally important degrees of freedom that are modulated by the protein. These features coupled with the ease with which several heme-binding ligands can be photodissociated from the heme-iron have made both hemoglobin and myoglobin the foremost molecular laboratories for protein relaxation studies.

This chapter describes methodologies for generating resonance Raman spectra from transient forms of hemoglobin, and myoglobin under conditions that allow for the systematic exploration of relaxation and functionality as described previously. Raman spectroscopy, like FTIR, provides vibrational frequencies derived from the relatively sharp spectral bands. From the absolute frequencies and the shifts in these frequencies as a function of conformational change, very small but functionally significant changes in bond strength and length can be obtained. Resonance Raman spectroscopy allows for the selective and often dramatic enhancement of Raman bands based on the choice of excitation wavelength. For example, the use of laser excitations that overlap the strong absorption of the heme chromophore result in resonance enhancement of heme-associated Raman bands ***(1–5)***. In contrast, excitations in the 210–240 nm region result in the selective enhancement of Raman bands associated with the aromatic amino acids ***(6,7)***. Here, we focus primarily upon resonance enhancement using excitations that enhance several functionally and conformationally important heme vibrations including ν(Fe-His), the iron-proximal histidine stretching mode ***(5,8–12)***.

Typically, Raman would be used to probe the heme degrees of freedom as the overall protein conformation evolves subsequent to a perturbation that initiates the relaxation process. Rapid photodissociation of the heme-ligand complex (e.g., heme-CO or heme-O_2) using a short laser pulse (ns or less) is a standard approach for initiating the relaxation process associated with the relaxation of liganded conformation towards the equilibrium population associated with the ligand-free species (also referred to as deoxy). In these photodissociation experiments, an initial pulse usually at 532 nm photodissociates the samples, and much less intense second pulse in the blue (resonant with the intense heme Soret band) that is variably delayed with respect to the first pulse generates the resonance Raman spectrum of the photoproduct ***(12)***. The photoproduct spectrum is compared to that of the ligand-free species under equilibrium conditions. The time-dependent differences between the photoproduct and equilibrium populations of ligand-free protein expose both how the heme relaxes subsequent to loss of the ligand and how the heme degrees of freedom follow the relaxation of the global protein conformation. In general, the time scales for the two relaxations are well separated with the former occurring within a few picoseconds or faster, and the later on the tens of picosenconds to ms time scale.

The pulse-probe technique for generating the resonance Raman spectra of intermediates as the photodissociated population relaxed has limitations. It can clearly expose how heme degrees of freedom respond to tertiary and quaternary relaxation as in the case of hemoglobin, but it is difficult to evaluate the functional properties of unrelaxed and partially relaxed populations. Furthermore, and perhaps most significantly the two pulse approach is only suited to populations that can be triggered with the first pulse. To overcome these challenges we have pursued Raman compatible protocols based on sol-gel encapsulation ***(13–22)*** that allow for the preparation, trapping, and temporal tuning of nonequilibrium population generated from either the addition or the removal of ligands/substrates. These protocols have the advantage of allowing for not only the spectroscopic probing of conformation, but also the detection of functional properties of the trapped intermediates. **Subheading 3.** will focus upon use of resonance Raman spectroscopy as a probe of intermediates generated both in solution and in sol-gel matrices that were designed to trap and modulate populations prepared under nonequilibrium conditions.

2. Materials

All chemicals were commercially obtained at the highest purity available and used without further purification.

1. N_2, and O_2 (Tech Air, White Plains, NY).
2. CO (Matheson Tri-Gas, Parsippany, NJ).
3. 50 m*M* Bis TrisAcetate pH 6.5.
4. 50 m*M* Bis TrisAcetate + 25% glycerol (by vol.) pH 6.5.
5. 50 m*M* Bis TrisAcetate + 75% glycerol (by vol.).
6. TMOS (Tetramethylorthosilicate) (Aldrich Chemical Co., Milwaukee, WI).
7. Glycerol.
8. NMR tubes: 7 in. long, flat bottom (*see* **Note 1**), 10 mm outer diameter, thin walled. (Wilmad, Buena, NJ or New Era Enterprises, Vineland, NJ).

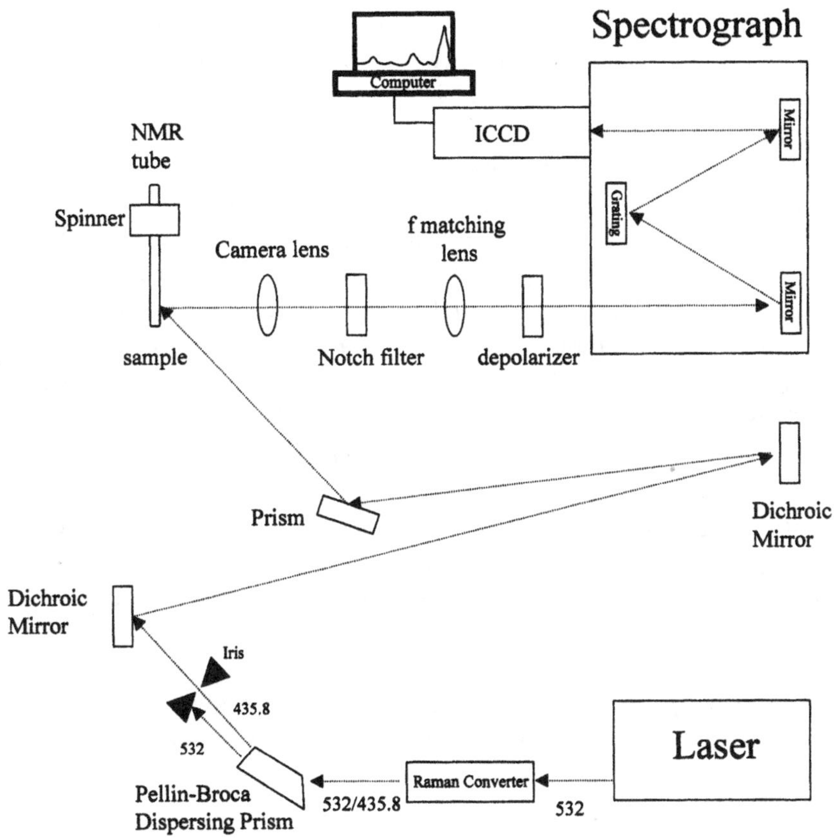

Fig. 1. Schematic description of the time-resolved visible resonance Raman system.

3. Methods

In this section we shall describe 1) the time-resolved visible resonance Raman system as well as data collection, 2) sol-gel encapsulation of proteins, specifically of HbA, and 3) procedures for generating and following the time evolution of nonequilibrium conformations of the protein.

3.1. Time-Resolved Resonance Raman System

Visible resonance Raman (VRR) spectra are generated using an 8 ns pulse at 435.8 nm (Nd:YAG, Continuum, Santa Clara, CA) working at 20 Hz using the second harmonic output at 532 nm. About 4 watts of the 532 nm beam is passed through a Raman converter or shifter (a long stainless steel cell pressurized with hydrogen gas, Light Age, Somerset, NJ) to generate an anti-Stokes up-shifted wavelength of 435.8 nm (*see* **Fig. 1** and **Note 2**). The output of the Raman shifter is passed through a Pellin Broca prism that disperses the different wavelengths emerging from the Raman shifter. The 435.8 nm light is separated spatially from the rest of the beam components (Stokes and anti-Stokes shifted light as well as the 532 nm light) using a combination of long path length, pin holes, and two dichroic mirrors. The beam is focused with a plano-convex lens on the sample at an incidence angle of 45°, approaching from the top hitting the center (sidewise) of the bottom portion of the NMR tube (*see* **Note 3**). The laser power was kept low at the sample (< 3 mw) to avoid sample bleaching. Note, that for samples of ligated HbA, a single pulse was used to both photodissociate the ligand from the heme and to generate the VRR spectra of the protein's photoproduct. The Raman scattered light was collected at normal incidence to the sample (135° to the laser) with a 50 mm Nikon F/1.4 camera lens. A holographic notch filter (Kaiser, Ann Arbor, MI) centered at 442 nm is used to eliminate Rayleigh scattering. An f-matching lens is used to focus the output of the notch filter through a depolarizer (to scramble the polarization and eliminate intensity artifacts caused by polarization dependent grating reflectivity, CVI, Putnam, CT) onto the vertical entrance slit of the monochrometer. The 0.27 m single monochrometer (Spex, Metuchen, NJ) with a 2400 grooves/mm grating is used as a spectrograph with an intensified CCD detector (Princeton Instruments, Trenton, NJ, currently Roper scientific) placed instead of an exit slit. The frequency shifts in the Raman spectra were calibrated using the known Raman spectra of solvents (carbon tetrachloride and toluene). The accuracy of the Raman shifts is about ± 1 cm^{-1} for absolute shifts and about ± 0.5 cm^{-1} for relative shifts

3.2. Time-Resolved Visible Resonance Raman of Solution Samples

About 200 µL of 0.5 m*M* HbA in heme are placed in a 10 mm outer diameter, flat bottom, thin walled, glass NMR tube. The NMR tube is spun to ensure that each laser pulse would sample a fresh HbA population. The spinner (Princeton Photonics, Princeton, NJ) speed was regulated with a flow regulator. As the spinning speed increases the solution rises and forms a ring of thin film on the inner wall of the NMR tube that is sufficient to yield high quality Raman signal. Use of this thin film approach serves to conserve sample volume. This approach works because the Raman scattering is essentially from the samples surface with the laser hardly penetrating the sample (caused by the samples high absorbance). Thus, the required Raman signal collection geometry is in a back scattering configuration. Throughout the measurements the sample was kept at 4°C by blowing a cold nitrogen stream onto the bottom part of the NMR tube (where the sample is).

The time-resolved resonance Raman, pump-probe experiment can be done with two lasers ***(12)***, or as in the case at hand with a single laser. In the latter configuration the same ns pulse both photodissociates the sample and then gen-

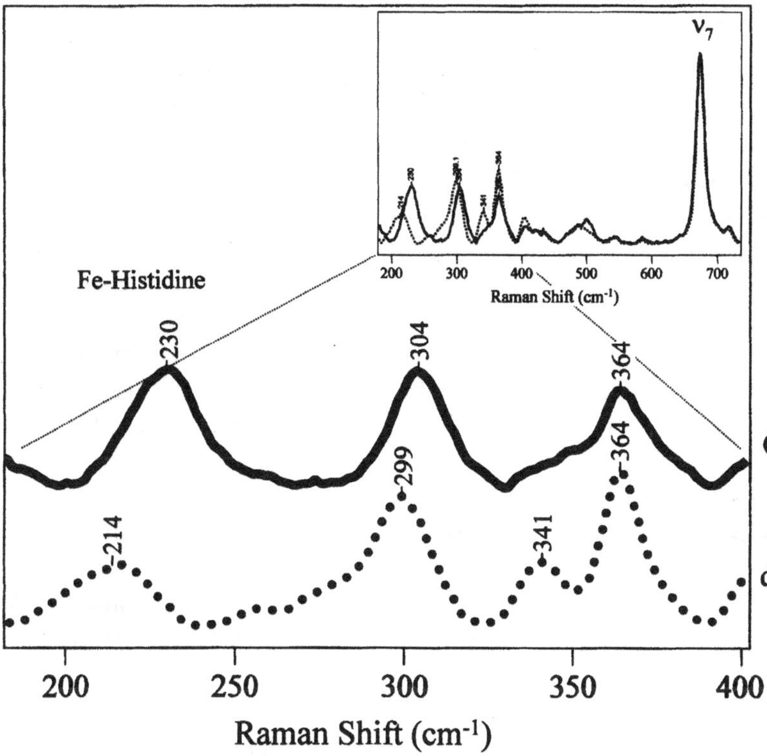

Fig. 2. Resonance Raman of deoxyHbA (dotted trace) vs that of the 8 ns CO photoproduct, COHbA* (solid trace) in the 180–400 cm^{-1} range. The insert shows the 180–735 cm^{-1} spectral range for the same two samples. Excitation wavelength was 435.8 nm, buffer was 50 m*M* BisTris Acetate pH6.5, HbA concentration was 0.5 m*M* in heme and the traces were normalized to the ν_7 band.

erates the resonance Raman spectra of the 8 ns photoproduct. **Figure 2** shows the resonance Raman spectra of deoxy HbA (unliganded HbA) and contrasts it with COHbA* (the 8 ns photoproduct of COHbA). Although the CO–Fe bond has been ruptured within 8 ns, which is the duration of the laser pulse and therefore the duration of the measurement, the protein has undergone only limited conformational relaxation. Thus, the differences in the two resonance Raman traces, most notably in the low frequency region (214 cm^{-1} vs 230 cm^{-1}), reveal a large difference in the frequency of the Fe-Histidine stretching mode when comparing deoxyHbA and the 8 ns photoproduct of COHbA ***(5,10,11)***. The laser's wavelength, 435.8 nm is in resonance with the Soret absorbance band of unliganded HbA (typical of the heme chromophore) which results in the enhancement of heme modes associated with the photoproduct vis a vis any unphotolyzed material.

3.3. Sol-Gel Encapsulation

A great variety of preparative protocols exist in the literature offering a range of sol-gel properties for different applications. Many of the methods used to encapsulate biological molecules are based on the Zink procedure *(18)*, which incorporates sonication of the TMOS mixture. Contrary, in this chapter we shall describe *a no sonication* method that was developed in our laboratory *(20,21)*, and has the advantage of generating sol-gels with improved *locking in* ability with respect to protein conformation.

Sol-gel preparation as well as all solution handling is done in a controlled atmosphere. For example, when making an unliganded HbA (deoxy HbA) all the buffers and solutions were deoxygenated prior to gelation by nitrogen purging, and reduction with dithionite and thereafter maintaining anaerobic conditions until an addition of a ligand is sought. UV/Vis measurements were taken before and after gelation to verify that the samples are deoxy and that no oxy or met was generated during the procedure. All the samples and solutions are kept in an ice bath whenever possible.

The gels are cast on the bottom inner surface of a 5 or 10 mm diameter NMR tube. A thin film was obtained by rapidly spinning the NMR tube using a tube spinner.

The gel preparation steps are:

1. 100 μL of TMOS was added to the NMR tube (*see* **Note 4**).
2. Add 100 μL of buffer (50 m*M* Bis TrisAcetate pH 6.5 + 25% glycerol [in vol.]).
3. Gently vortex the mixture for 5 s without generating any bubbles.
4. Add 100 μL (*see* **Note 5**) of the stock (the solution contain the molecule to be encapsulated for example: 1.5 m*M* HbA (measured in heme units) in a 50 m*M* Bis Tris pH 6.5 buffer).
5. Gently vortex the mixture for 20 s again without generating any bubbles.
   Start spinning the NMR tube. Increase the speed until the solution rises along the inner walls of the NMR tube and forms a thin ring. Continue spinning until the mixture has gelled (*see* **Note 6**).

After gelation is complete, the gels are washed several times and then filled with excess buffer (*see* **Note 7**) and left to age at 4° C for a minimum of 2 d. The final protein concentration in the sol-gel samples is typically in the range of 0.5 m*M* in heme.

3.4. Sol-Gel Encapsulation as a Mean to Study Nonequilibrium Species and Reaction Intermediates

As mentioned in **Subheading 1.**, study of transient species is a significant challenge because often the relaxation time of the nonequilibrium structures is faster than the mixing times. The novel approach of sol-gel encapsulation overcomes this *diffusion barrier* because the gel environment slows down the large-

Protocol 1, trapping the high affinity, R structure

$$\text{COHbA} \xrightarrow{\text{gelatin \& aging}} [\text{COHb}]$$

Protocol 2, trapping the low affinity, T structure

$$\text{deoxyHbA} \xrightarrow{\text{gelaton \& aging}} [\text{deoxyHb}] \xrightarrow{\text{CO exposure}} [\text{deoxyHb}]\text{+CO}$$

Protocol 3, trapping the high affinity, R structure

$$\text{oxyHbA} \xrightarrow{\text{gelaton \& aging}} [\text{oxyHb}] \xrightarrow{\text{+dithionite}} [\text{oxyHb}]\text{–oxygen} \xrightarrow{\text{CO exposure}} [\text{oxyHb}]\text{–oxygen+CO}$$

Protocol 4, trapping the high affinity R structure

$$\text{MetHbA} \xrightarrow{\text{gelaton \& aging}} [\text{MetHb}] \xrightarrow{\text{reduction}} [\text{MetHb}]\text{ to deoxy} \xrightarrow{\text{CO exposure}} [\text{MetHb}]\text{ to deoxy+CO}$$

Scheme 1.

scale conformational changes of the protein, and yet still allows relatively fast diffusion of small substrate or ligand molecules as well as water. The sol-gel encapsulated biomolecules are fully hydrated allowing the results of sol-gel studies to be compared with those from solution-based studies. Indeed, previous studies have demonstrated that encapsulation of myoglobin and hemoglobin does not perturb structural and functional properties of individual tertiary and quaternary states, but does under the appropriate conditions slow down or stop conformational changes.

Thus, sol-gel encapsulation opens the door for a range of otherwise difficult experiments that seek to expose both the structure and function of non-equilibrium species. A useful general approach is to use encapsulation to trap the protein in a specific initial conformation, by controlling its conditions at the time of gelation. Subsequent to gelation, a rapid change (relatively rapid) in solution conditions or components is used to initiate conformational relaxation of the protein under conditions where the full complement of conformational changes is slow enough to be monitored by resonance Raman spectroscopy. **Scheme 1** shows several such experimental protocols that we have developed for the study of the reaction pathway of the T state to R state transition (and vice versa) in HbA. Note, that (as seen in **Scheme 1**) when referring to the samples we use a notation that indicates the history of the samples. Square brackets are used to indicate the species and conditions present during the sol-gel encapsulation and aging. Any changes to the samples subsequent to gelation and aging (e.g., the addition of a ligand or a change in bathing buffer) appear outside and to the right of the square brackets, where each additional change is added on the right in a sequential manner.

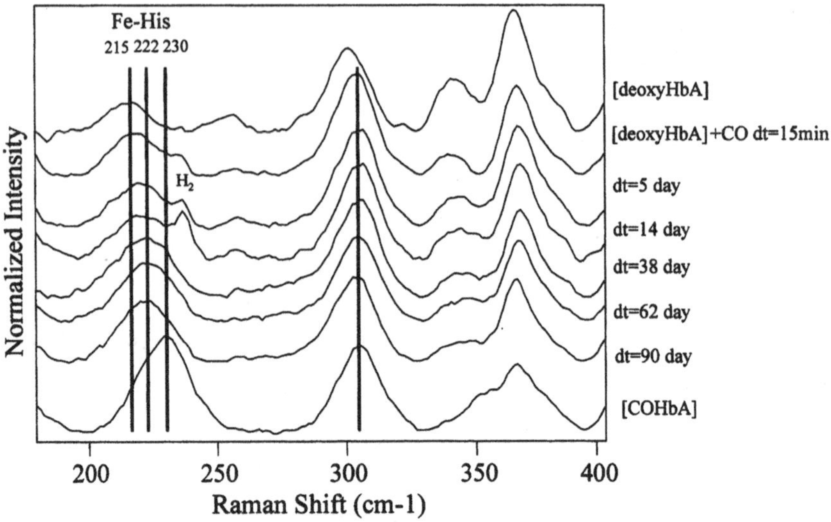

Fig. 3. Time-resolved visible resonance Raman spectra of sol-gel encapsulated deoxyHbA to which CO was added, all in the presence of 75% glycerol. Excitation wavelength was 435.8 nm, the bathing buffer was 50 m*M* BisTris Acetate+75% glycerol in volume, HbA nominal concentration in the gel was 0.5 m*M* in heme and again all the spectra was normalized to the ν_7 band. H_2 marks a spike from the Raman converter.

Protocol 2 shows the initial encapsulation of deoxyHbA, once the gel has solidified and aged, the hemoglobin will be locked in its quaternary T structure, characterized by a lower affinity to ligands. On the other hand, Protocols 1, 3, and 4 show the encapsulation of various HbA species all in the R state. Again, the quaternary structure during gelation and aging is *locked in* by the gel.

3.5. Time-Resolved Visible Resonance Raman of Sol-Gel Encapsulated Samples

To improve *locking in* and minimize the relaxation of nonequilibrium populations trapped within the sol-gel, the samples are always kept at approx 4°C. The *clock* is started by a rapid change in conditions. For example, for encapsulated deoxy HbA, such as in the case of Protocol 2, a rapid addition of CO to the sol-gel encapsulated deoxy sample is the trigger. The process of CO ligation is monitored using UV/VIS spectra. Usually within 5 min the sample turns fully CO. Then the samples are measured on the time-resolved visible resonance Raman (TRVRR) setup and the evolution of the samples with time can be followed by measuring changes in the VRR spectra. **Figure 3** shows the time evolution of [deoxyHbA] samples prepared via Protocol 2. However, in this case an additional slowing down of the conformational relaxation was achieved by changing the bathing buffer after gelation and aging to one that contains 75% glycerol (a high viscosity buffer). Thus, **Fig. 3** displays the time evolution of [deoxyHbA]+75% glycerol+CO (note, that the recorded spectra is still that of the photoproduct of the latter species).

Figure 3 clearly demonstrates that contrary to the results in solution where the addition of CO to a deoxy solution results in the the Fe-His stretching frequency achieving its endpoint value of 230 cm^{-1} within the mixing time (*see* **Fig. 2**), for the sol-gel encapsulated gels the conformational relaxation of a T state deoxy HbA, towards the R state is orders of magnitude slower.

In a similar manner to the previous example, Protocols 3 and 4 were utilized to follow the conformational relaxation along the R to T reaction coordinate, however, this time starting from the R quaternary structure. In Protocol 3 the trigger is the addition of dithionite that reacts with the oxygen to generate a deoxy species still trapped in the oxyHbA conformational structure. Whereas, via protocol 4, reduction with dithionite, reduces the iron from ferric to ferrous, again rapidly generating the nonequilibrium specie of a deoxyHbA with an R quaternary structure. An additional protocol that can be combined with the protocols for generation of the evolving deoxy species is also depicted in **Scheme 1**. This add-on is the addition of CO subsequent to the evolution of the deoxy sample for a specific period of time. The addition of CO triggers a reversal of the conformational relaxation back towards the liganded R state. Once the CO is added both the Raman of the photoproduct as well as the rebinding kinetics of the photodissociated species can be followed. The slow evolution of the populations facilitates the use of the CO addition as a probe of the functional and conformational status of the evolving deoxy population.

4. Notes

1. It is important that the NMR tubes will be flat bottomed, for two reasons. First, with a flat bottom NMR tube, less sample volume is needed to perform the measurement. Second, it reduces scattered light (Rayleigh scattering) and eliminates other artifacts that originate from the optical properties of a curved bottom.
2. The Raman converter is based on Raman shifting the input beam using the vibrational levels of the filling gaseous molecule (H_2 in our system). The 435.8 nm is generated by anti-Stokes upshift. Optimizing the intensity of the 435.8 nm is empirical and involves changes in the laser input power (going into the Raman converter), the filling gas pressure, and the filling gas temperature. A known difficulty with Raman converters is the possible appearance of additional spikes in the Raman spectra. These are often the result of Raman scattering that involves the vibrational and the rotational levels of the filling gas molecule. Our experience is that a trial and error search where both the input laser intensity and the filling gas pressure are modified can eliminate those spikes. The reason being

that within the Raman converter cell several processes can take place simultaneously, however, fine adjustment of the conditions will enhance some on the expense of others.

3. In setting up the Raman optical system, it is important for the polarization of laser beam to approach the sample horizontally (and not vertically), because it increases the amount of scattered Raman light that reaches the vertical slit of the monochrometer.
4. Caution when working with TMOS. We work in the hood. Avoid contact with skin, eyes, or breathing vapor or mist. Discarding of TMOS can be easier by first turning the left over TMOS into a gel. Once a gel has formed and aged it is inert.
5. When casting the sol-gels, the added volume can be changed, (we tried between 10 µL to 400 µL) however, the volume ratio of the three aliquots remains 1:1:1. Furthermore, changing the volume will effect the resultant gel thickness as well as gelation time. When adding the TMOS and the rest of the aliquant, we add them onto the side of the NMR tube so that they will slide down in a vertical path to reduce loss of volume due to adherence to the tube's wall.
6. When making sol-gels, a recommended method to verify that gelation is complete is to slightly reduce the spinning speed and then inspect the gel. If there is no change in the appearance of the gel and no accumulation of liquid at the bottom of the NMR tube then the gelation is complete. Stopping the spinning of the NMR tube completely before the gel has solidified can be problematic if the gel has not completely solidified and yet is already very viscous. Often, in a situation like that the gel will loose it's ring-like shape and would not return to that shape even with vigorous spinning because of its high viscosity.
7. The sol-gel's bathing buffer level has to be higher than the height of the gel to prevent drying of the gels. Further excess of buffer is often also advantageous as it can assist in preventing any changes in pH, or other parameters, as the gel ages.

References

1. Spiro, T. G. (1978) Resonance Raman spectra of hemoproteins. *Methods Enzymol* **54,** 233–249.
2. Spiro, T. G. (1985) Resonance Raman spectroscopy as a probe of heme protein structure and dynamics. *Adv. Protein Chem.* **37,** 111–159.
3. Spiro, T. G., Smulevich, G., and Su, C. (1990) Probing protein structure and dynamics with resonance Raman spectroscopy: cytochrome c peroxidase and hemoglobin. *Biochemistry* **29,** 4497–4508.
4. Spiro, T. G. and Czernuszewicz, R. S. (1995) Resonance Raman spectroscopy of metalloproteins. *Methods Enzymol.* **246,** 416–460.
5. Friedman, J. M. (1994) Time-resolved resonance Raman spectroscopy as probe of structure, dynamics, and reactivity in hemoglobin. *Methods Enzymol.* **232,** 205–231.
6. Asher, S. (1993) UV resonance Raman spectroscopy for analytical, physical and biophysical chemistry. Part 1. *Anal. Chem.* **65,** 59A–66A.

7. Austin, J., Jordan, T., and Spiro, T. (1993) Ultraviolet resonance Raman studies of proteins and related model compounds. In *Biomolecular Spectroscopy Part A* (Clark, R. J. H. and Hester, R. E., eds.), John Wiley and Sons, New York, pp. 55–127.
8. Kitagawa, T. (1988) The heme protein structure and the iron histidine stretching mode. In *Biological Application of Raman Spectroscopy,* Vol. III (Spiro, T. G., ed.), John Wiley & Sons, New York, pp. 97–131.
9. Rousseau, D. L. and Friedman, J. M. (1988) Transient and cryogenic studies of photodissociated hemoglobin and myoglobin. In *Biological Applications of Raman Spectroscopy,* Vol. III (Spiro, T. G., ed.), John Wiley & Sons, New York, pp. 133–215.
10. Friedman, J. M., Scott, T. W., Stepnoski, R. A., Ikeda-Saito, M., and Yonetani, T. (1983) The iron-proximal histidine linkage and protein control of oxygen binding in hemoglobin. A transient Raman study. *J. Biol. Chem.* **258,** 10,564–10,572.
11. Friedman, J. M. (1985) Structure, dynamics, and reactivity in hemoglobin. *Science* **228,** 1273–1280.
12. Scott, T. W. and Friedman, J. M. (1984) Tertiary-structure relaxation in hemoglobin: a transient Raman study. *J. Am. Chem. Soc.* **106,** 5677–5687.
13. Avnir, D., Braun, S., Lev, O., and Ottolenghi, M. (1994) Enzymes and other proteins entrapped in sol-gel materials. *Chem. Mater.* **6,** 1605–1614.
14. Bettati, S. and Mozzarelli, A. (1997) T state hemoglobin binds oxygen noncooperatively with allosteric effects of protons, inositol hexaphosphate, and chloride. *J. Biol. Chem.* **272,** 32,050–32,055.
15. Bruno, S., Bonaccio, M., Bettati, S., Rivetti, C., Viappiani, C., Abbruzzetti, S., and Mozzarelli, A. (2001) High and low oxygen affinity conformations of T state hemoglobin. *Protein Sci.* **10,** 2401–2407.
16. Dave, B. C., Miller, J. M., Dunn, B., Valentine, J. S., and Zink, J. I. (1997) Encapsulation of proteins in bulk and thin film sol-gel matrices. *J. Sol Gel Sci. Technol.* **8,** 629–634.
17. Das, T. K., Khan, I., Rousseau, D. L., and Friedman, J. M. (1999) Temperature dependent quaternary state relaxation in sol-gel encapsulated hemoglobin. *Biospectroscopy* **5,** S64–S70.
18. Ellerby, L. M., Nishida, C. R., Nishida, F., Yamanaka, S. A., Dunn, B., Valentine, J. S., and Zink, J. I. (1992) Encapsulation of proteins in transparent porous silicate glasses prepared by the sol-gel method. *Science* **255,** 1113–1115.
19. Juszczak, L. J. and Friedman, J. M. (1999) UV resonance Raman spectra of ligand binding intermediates of sol-gel encapsulated hemoglobin. *J. Biol. Chem.* **274,** 30,357–30,360.
20. Khan, I., Shannon, C. F., Dantsker, D., Friedman, A. J., Perez-Gonzalez-de-Apodaca, J., and Friedman, J. M. (2000) Sol-gel trapping of functional intermediates of hemoglobin: geminate and bimolecular recombination studies. *Biochemistry* **39,** 16,099–16,109.
21. Samuni, U., Dantsker, D., Khan, I., Friedman, A. J., Peterson, E., and Friedman, J. M. (2002) Spectroscopically and kinetically distinct conformational popula-

tions of sol-gel encapsulated carbonmonoxy myoglobin: a comparison with hemoglobin. *J. Biol. Chem.* **25,** 25.
22. Shibayama, N. and Saigo, S. (1995) Fixation of the quaternary structures of human adult haemoglobin by encapsulation in transparent porous silica gels. *J. Mol. Biol.* **251,** 203–209.

15

Fluorescence Polarization/Anisotropy Approaches to Study Protein–Ligand Interactions

Effects of Errors and Uncertainties

David M. Jameson and Gabor Mocz

Summary

Fluorescence techniques are widely used in the study of protein–ligand interactions because of their inherent sensitivity, and the fact that they can be implemented at true equilibrium conditions. Fluorescence polarization/anisotropy methodologies, in particular, are now extensively utilized in biotechnology and clinical chemistry. In this chapter, we shall discuss both theoretical and practical aspects of polarization/anisotropy methods. We shall also focus attention on considerations of errors and uncertainties in such measurements, and how these uncertainties affect the ultimate estimation of ligand–protein dissociation constants.

Key Words: Fluorescence; anisotropy; polarization; ligand binding; dissociation constants; error propagation.

1. Introduction

The interactions of proteins with other molecules are responsible for the majority of the molecular complexes which make life possible. Hence, these types of interactions have been the subject of a great many theoretical and experimental studies. The biotechnology and pharmaceutical industries, in particular, have a significant interest in quantifying the number and strengths of ligand–protein interactions. Such quantification requires knowledge of the number of interacting molecular species (stoichiometry) and the strengths of the binding interactions (free energies). The Gibbs free energy associated with a particular binding interaction is related to the dissociation constant by **Eq. 1**:

$$\Delta G = -RT \ln K_d \tag{1}$$

From: *Methods in Molecular Biology, vol. 305: Protein–Ligand Interactions: Methods and Applications*
Edited by: G. U. Nienhaus © Humana Press Inc., Totowa, NJ

where R is the universal gas constant, T is the temperature, and K_d is the dissociation constant (which is the reciprocal of the association constant). The dissociation constant that characterizes a reversible equilibrium between a protein and a ligand ($PL \leftrightarrow P + L$) is given by:

$$K_d = \frac{[P][L]}{[PL]} \tag{2}$$

To determine K_d, one must thus be able to determine the concentrations of free protein, free ligand, and protein–ligand complex. Because the total protein (P_T) and total ligand (L_T) concentrations are known, the problem reduces to determination of the concentration of the protein–ligand complex ***(1,2)***. For the case of one binding site, the concentration of protein–ligand complex (PL) is related to K_d, P_T, and L_T via the quadratic equation:

$$[PL] = \frac{(K_d + L_T + P_T) - \sqrt{\left[(K_d + L_T + P_T)^2 - 4P_T L_T\right]}}{2} \tag{3}$$

To determine $[PL]$ in the presence of uncomplexed protein and ligand, optical spectroscopy methods are very popular because they avoid the use of radioactive materials (and hence the attendant health and waste management concerns), and they are generally inexpensive, rapid, and simple. Fluorescence methodologies in particular are extremely sensitive, allowing measurements in the nanomolar to picomolar levels, and given the current commercial availability of literally thousands of fluorescence probes can be applied to almost any protein–ligand system (*see* **ref. *3***). Fluorescence methods may also allow quantification of the complex in the presence of free ligand, i.e., without the need for a separation step such as filtration or chromatography. Such homogeneous methods allow one to study the system at a true equilibrium over a wide range of concentrations. Ideally, in the protein–ligand system being studied, only one component undergoes alterations in its fluorescence signal upon ligand binding. If more than one fluorescence parameter is affected by the binding process, than one unique signal must be isolated by the judicious choice of the excitation and emission wavelengths (for a discussion of optical components such as monochromators and filters *see* **ref. *3***). In many cases, protein–ligand interactions can be studied by following changes in fluorescence intensities of a probe or in the intrinsic fluorescence of the protein subsequent to ligand binding. However, one sometimes encounters systems in which intensity changes do not occur or are minimal (*see* **ref. *4***). The absence of intensity changes upon binding is more common when fluorescence probes with absorption maxima at visible wavelengths are used, which is often necessary to reduce background fluorescence. In such cases, polarization or anisotropy determinations are extremely useful since they rely upon changes in rotational mobility between free and bound probes. We shall focus our attention exclusively on these techniques.

2. Overview of Polarization/Anisotropy

2.1. Theory

The principles underlying polarization/anisotropy measurements have been described numerous times (*see* **refs.** ***3,5–9***), and we shall present only a very brief overview here. Basically, light can be considered as oscillations of an electromagnetic field perpendicular to the direction of propagation - we shall be concerned only with the electric component. Polarizers are optically active devices that can isolate one direction of the electric vector. The most common polarizers used today are 1) dichroic devices, which operate by effectively absorbing one plane of polarization (e.g., Polaroid type-H sheets based on stretched polyvinyl alcohol impregnated with iodine), and 2) double refracting calcite ($CaCO_3$) crystal polarizers—which differentially disperse the two planes of polarization (examples of this class of polarizers are Nicol polarizers, Wollaston prisms, and Glan-type polarizers such as the Glan-Foucault, Glan-Thompson, and Glan-Taylor polarizers). We initially consider that the fluorescence emitted by a sample excited by polarized light can have any direction of polarization. The actual direction of the electric vector of the emission can be determined by viewing the emission through a polarizer, which can be oriented alternatively in the parallel or perpendicular direction relative to the Z axis or laboratory vertical direction. Polarization is then defined as a function of the observed parallel ($I_{\|}$) and perpendicular intensities ($I_{\perp}$):

$$P = \frac{I_{\|} - I_{\perp}}{I_{\|} + I_{\perp}} \tag{4}$$

When plane polarized light of the appropriate wavelength (i.e., a wavelength absorbed by the fluorophore) impinges on a solution of randomly oriented fluorophores a photoselection process occurs. Specifically, the probability (α) of a fluorophore absorbing the incident light (and thus being excited to an upper electronic level) is proportional to the cosine squared of the angle (θ) between the polarization direction of the exciting light and the absorption transition dipole.

$$\alpha = \cos^2\theta \tag{5}$$

If the exciting light is plane polarized parallel to the vertical laboratory axis, then the number of dipoles oriented at an angle, θ, with this vertical axis will be proportional to $\sin\theta$. These two considerations account for the *photoselection* process that occurs when polarized light excites a population of

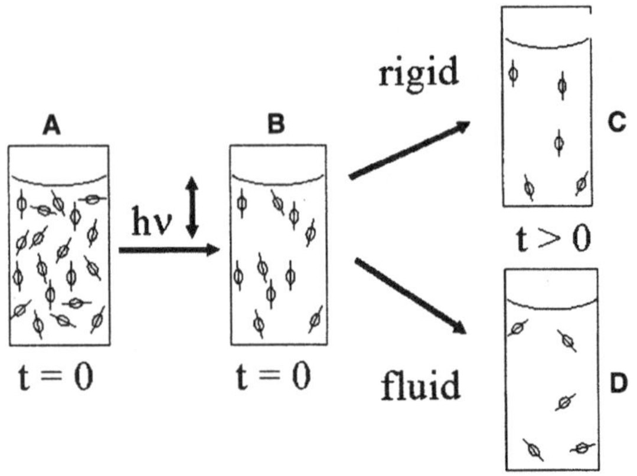

Fig. 1. Depiction of photoselection and rotation of fluorophores. **(A)** depicts a population of fluorophores with their transition dipoles randomly oriented in solution. In **(B)** vertically polarized light impinges on the population and only fluorophores with transition dipoles oriented approximately parallel to the electric vector of the excitation are excited. In **(C)** the solution is considered to be a highly viscous or rigid environment, and after some time has passed the number of excited fluorophores has decreased as a result of the emission process but the orientation of the excited dipoles is unchanged. In **(D)** the solution is considered to be fluid and the dipoles, which remain after the emission process have rotated relative to their initial photoselected orientation.

fluorophores. They also determine the maximum polarization that can be observed from a randomly oriented population of fluorophores that are not free to rotate, e.g., which are immobilized in a very viscous environment. This limiting polarization (in the case where the emission dipole is colinear with the excitation dipole) is + 1/2 (*7*; but *see* **Note 1**).If the fluorophore is excited, it will emit light (fluorescence) after a short duration (known as the fluorescence lifetime) typically on the order of nanoseconds. In aqueous solution at normal temperatures the fluorophore will usually be able to rotate during this time and the direction of the electric vector corresponding to the emitted light will thus be different than that of the exciting light. The direction of this electric vector —usually monitored at right angles to the excitation direction— can then be analyzed. The photoselection and rotation processes are roughly depicted in **Fig. 1**.

Another term frequently used in the context of polarized emission is anisotropy (usually designated as either A or r) which is defined as:

$$r = \frac{I_{\parallel} - I_{\perp}}{I_{\parallel} + 2I_{\perp}} \tag{6}$$

Given the definition of polarization and anisotropy, one can show that:

$$r = \frac{2}{3}\left(\frac{1}{P} - \frac{1}{3}\right)^{-1} \tag{7}$$

or

$$r = \frac{2P}{3 - P} \tag{8}$$

Clearly, the information content in the polarization function and the anisotropy function is essentially identical, and the use of one term or the other is dictated by practical considerations. We also note that the limiting anisotropy, in the case of colinear emission and excitation dipoles, will be + 2/5.

In 1920, F. Weigert was the first to discover that the fluorescence from solutions of dyes was polarized. Specifically, he looked at solutions of fluorescein, eosin, rhodamine, and other dyes and noted the effect of temperature and viscosity on the observed polarization ***(10)***. In Weigert's words "Der Polarisationsgrad des Fluorezenzlichtes nimmt mit wachsender Molekulargröße, mit zunehmender Viskosität des Mediums und mit abnehmender Temperatur, also mit Verringerung der Beweglichkeit der Einzelteilchen zu" ("*The degree of the polarization increases with increasing molecular size, with increasing viscosity of the medium and with decreasing temperature, that is with the reduction of the mobility of the single particles*"). Weigert recognized that all of these considerations meant that fluorescence polarization increased as the mobility of the emitting species decreased.

In 1925–1926, Francis Perrin (son of the famous French physicist, Jean Perrin) published several important papers describing a quantitative theory of fluorescence polarization, including what is now considered his classic paper containing most of the essential information that we use to this day ***(11)***. Specifically, Perrin related the observed polarization to the excited state lifetime and the rotational diffusion of a fluorophore:

$$\frac{1}{P} - \frac{1}{3} = \left(\frac{1}{P_0} - \frac{1}{3}\right)\left(1 + \frac{RT}{\eta V}\tau\right) \tag{9}$$

In this equation, P is the observed polarization, P_0 is the limiting polarization (the polarization observed in the absence of rotation), R is the universal gas constant, T is the absolute temperature, V is the molar volume of the rotating unit, η is the solvent viscosity, and τ is the excited state lifetime. This expression is often further simplified to:

$$\frac{1}{P} - \frac{1}{3} = \left(\frac{1}{P_0} - \frac{1}{3}\right)\left(1 + \frac{3\tau}{\rho}\right) \tag{10}$$

where ρ is the Debye rotational relaxation time which for a sphere is given as:

$$\rho_0 = \frac{3\eta V}{RT} \tag{11}$$

In the case of a protein wherein the partial specific volume (v) and hydration (h) are known one can then write:

$$\rho_0 = \frac{3\eta M(v+h)}{RT} \tag{12}$$

where M is the molecular weight, v the partial specific volume, and h the degree of hydration (*see* **Note 2**). The relationship between observed polarization and the rotational mobility of the fluorophore is what makes fluorescence polarization so useful for studies of protein–ligand interactions. The basic idea is simply that the polarization/anisotropy of a fluorescent ligand, free in solution, is low (assuming that the fluorescent lifetime is not extremely short; *see* **Note 2**) but upon binding to a macromolecule, such as a protein, the observed polarization/anisotropy of the fluorescent ligand will increase as a result of its slower rotational mobility. This fact was realized by Laurence who was the first to apply the method to study the interaction of ligands with proteins ***(12)***. Specifically, Laurence studied the binding of numerous dyes, including fluorescein, eosin, acridine, and others, to bovine serum albumin and used the polarization data to estimate the binding constants. Dandliker and his coworkers later applied these principles explicitly to study antibody–antigen ***(13)*** and hormone-binding site interactions ***(14)***. One of the first commercial instrument designed for clinical chemistry applications of fluorescence polarization was the TDx instrument from Abbott Laboratories ***(15)***.

2.2. Additivity of Polarization/Anisotropy

In order to convert the observed polarization/anisotropy from a mixture of free and bound fluorophore into the fraction of bound ligand we must understand the additivity properties of the relevant functions. The Perrin relationship was extended by Gregorio Weber to consider ellipsoids of revolution with fluorophores attached in random orientations ***(16)***. In this study, Weber also explicitly derived the relationship governing additivity of polarizations from different species, namely:

$$\left(\frac{1}{\langle P\rangle} - \frac{1}{3}\right)^{-1} = \sum f_i \left(\frac{1}{P_i} - \frac{1}{3}\right)^{-1} \tag{13}$$

where P is the actual polarization observed arising from i components, f_i represents the fractional contribution of the ith component to the total emission intensity, and P_i is the polarization of the ith component (*see* **Note 3**). This additivity principle was later expressed in terms of anisotropy (r) by Jablonski ***(17)*** as:

$$r_0 = \sum f_i r_i \tag{14}$$

Although the anisotropy formulation is simpler in appearance, the information content of the two approaches is identical, and given present day computer-assisted data analysis the difference is moot. Clearly, if the quantum yield of the fluorophore changes upon binding, the fractional intensity terms in Weber's additivity equation (**Eq. 13**) will alter. Although many probes (such as fluorescein) do not significantly alter their quantum yield upon interaction with proteins, one should not take this fact for granted and would be well advised to check. If the quantum yield does in fact change, one can readily correct the fitting equation to take the yield change into account. In terms of anisotropy the correct expression relating observed anisotropy (r) to fraction of bound ligand (x), bound anisotropy (r_b), free anisotropy (r_f), and the quantum yield enhancement factor (g) is ***(18)***:

$$x = \frac{r - r_f}{r_b - r_f + (g-1)(r_b - r)} \tag{15}$$

An analogous expression can be derived for polarization measurements:

$$x = \frac{(3 - P_b)(P - P_f)}{(3 - P)(P_b - P_f) + (g-1)(3 - P_f)(P_b - P)} \tag{16}$$

where P_f is the polarization of the ligand free in solution, P_b is the polarization of the bound ligand and P is the observed polarization. We note that although this equation is slightly more complex than the corresponding anisotropy equation, given the fact that such data is universally analyzed using computer programs this apparent complexity is moot.

2.3. Practical Considerations

Clearly, every researcher will have his or her unique system to investigate and it would be impossible to give a particular *recipe* for applying polarization/anisotropy methodologies, which would be appropriate for all cases. Nonetheless, certain general considerations apply to all such studies and can be enumerated.The most critical consideration is probably the choice of fluorophore. In applications to ligand–proteins interactions, one will ideally have a fluorescent ligand (although one could conceivably use the intrinsic fluorescence of the protein if that is appreciably altered by the binding of a nonfluorescent ligand) ***(19)***. Because most ligand systems are nonfluorescent one must find a way to introduce fluorescence into the system. This problem is usually solved by reacting a fluorophore, which has an appropriately reactive chemical group, with the ligand. It is now possible to find fluorophores with the appropriate reactivity for almost any ligand functional group (*see* **Note 5**). Jameson and Eccleston, to give but one example, have reviewed the methodologies for attaching fluorescent probes to the sugar moieties of nucleotides *(20)*. Another critical consideration (alluded to earlier) is the fluorescent lifetime of the fluorophore. As evident from Perrin's equation (**Eq. 9**), the observed polarization will depend not only on the rotational rates of the system but also on the excited state lifetimes. Put simply, if the lifetime is too short (relative to the rotational relaxation times involved) the change in polarization or anisotropy may be negligible (*see* **Note 2**). Ideally, one should choose a lifetime such that the polarization/anisotropy of the free ligand is low (typically a lifetime of several nanoseconds will be sufficient) whereas the polarization/anisotropy of the bound ligand is as large as possible, i.e., near the limiting values (P_0 or r_0). Generally, the lifetimes of the most common probes, e.g., in the fluorescein or rhodamine families, will achieve these goals. However, some popular cyanine and Alexa probes may have very short lifetimes. The reader is urged to either find appropriate lifetime values in the primary literature or to contact the company providing the probe for lifetime information. Aspects of time-resolved fluorescence that impact on studies of ligand–protein interactions, were discussed by Jameson and Sawyer *(20)*. Once the fluorescent ligand is in hand, it then remains to simply measure the polarization/anisotropy function at varying ligand–protein ratios. To obtain the complete ligand dissociation curve, an appropriate method is to start with a solution of the fluorescent ligand in the presence of an excess of protein. One then removes part of the ligand–protein solution and replaces it with an equivalent volume of the ligand solution at the same concentration. Hence, the ligand concentration is kept constant while the protein concentration is decreased. In this manner, one can obtain a complete binding curve as shown in the simulations in **Fig. 2**. **Figure 2A** shows typical curves for polarization or anisotropy data as a function of total protein concentration, whereas **Fig. 2B** demonstrates the effect of quantum yield changes (i.e., enhancement values) on the observed data.

3. Sources of Errors

3.1. Overview

The most fundamental source of error in studies of ligand–protein interactions using fluorescence polarization/anisotropy determinations would be actual errors in the polarization measurements themselves. In the days before the commercial availability of fluorescence polarization instruments one had to be particularly aware of such considerations. Researchers now usually trust the manufacturer to provide an accurate instrument. However, knowledge of the potential sources of errors in the apparatus is still useful, and can certainly guide the design of the next generation of instruments.In his classic report of the first photoelectric polarization instrument *(21)*, Gregorio Weber presented a careful and detailed consideration of potential sources of errors in the polar-

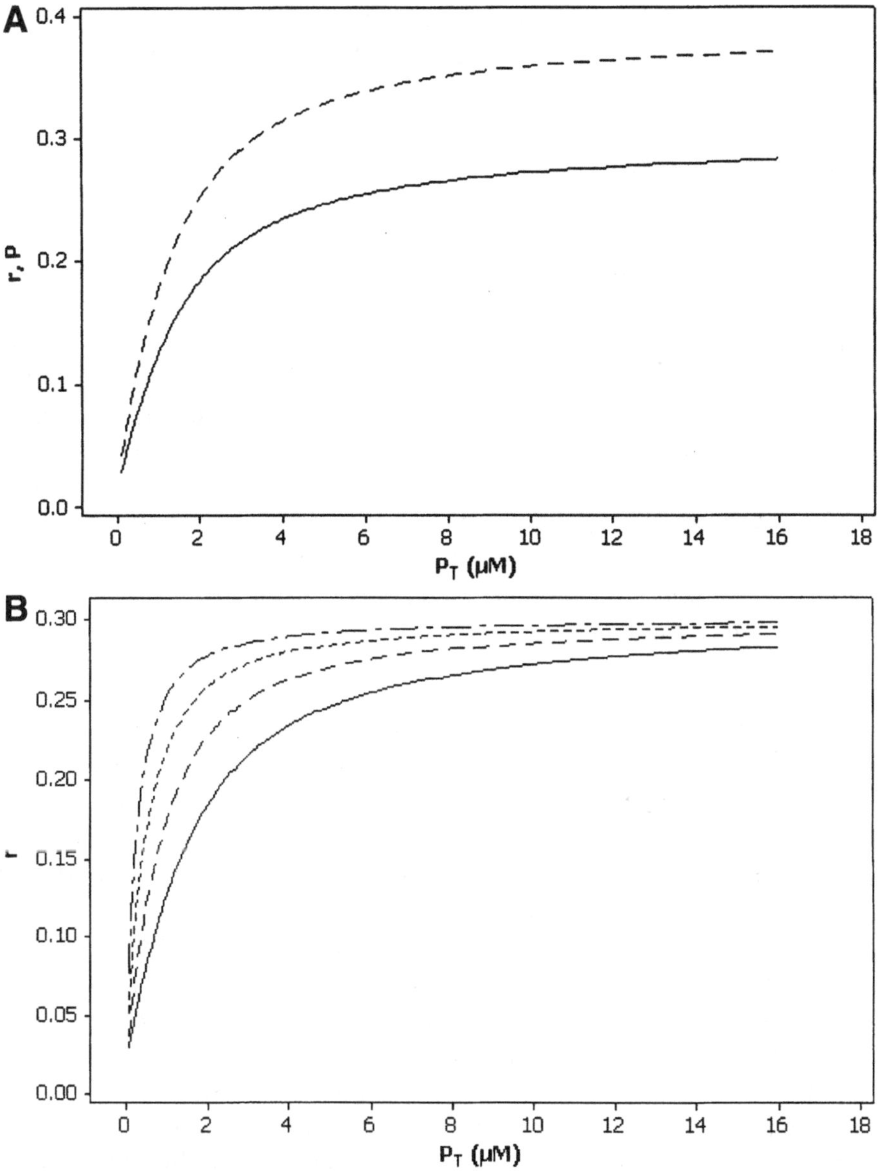

Fig. 2. Simulated titrations of a fluorescent ligand with a protein: anisotropy (r) and polarization (P) vs total protein concentration (P_T). The parameters used in all simulations were: $K_d = 1\ \mu M$, $L_T = 1\ \mu M$, $P_T = 0.0625\ \mu M$ to $16\ \mu M$. $r_f = 0.02$, $r_b = 0.3$. **(A)** solid line: anisotropy, $g = 1$; long dashes: polarization, $g = 1$; **(B)** solid line: anisotropy, $g = 1$; long dashes: anisotropy, $g = 2$; dashes: anisotropy, $g = 4$; dash-dot: anisotropy, $g = 8$. The computer program for all simulations was written in Prospero Extended Pascal in this laboratory. Graphs were generated using the MINITAB Release 14 Statistical Software.

ization measurements. He pointed out that according to the definition of polarization, in order to measure a polarization of 0.05 with a precision of 1%, I_{II} and $I_{\perp}$ must be known to within 1 part in 1000. Weber classified the systematic errors into three categories, namely:

1. Errors resulting from faulty settings of the parts (by which he specifically meant orientation of the polarizers).
2. Errors resulting from nonnegligible size of the source (by which he meant the aperture effect; *see* **ref. *8*** for a recent discussion of this effect).
3. Errors resulting from stray light (*see* **ref. *8*** for a recent discussion).

Weber applied rigorous propagation of error treatment to all of these potential errors, and the serious student of polarization determinations should read this classic paper. The effect of different types of polarizers and photon counting statistics on polarization measurements was discussed by Jameson et al. ***(22)*** in their description of the first photon-counting polarization instrument. In this chapter, we shall extend these discussions of measurement errors by explicitly considering the effect of specific errors or uncertainties on the final estimate of the fraction of bound ligand, and hence on the estimated dissociation constant of the ligand–protein complex.

3.2. Modeling of Uncertainties in Determination of Dissociation Constants

One possible way to assess the accuracy and reliability of fluorescence anisotropy and polarization data is the development of an error propagation model. Such a model can provide information about the sensitivity of individual input parameters, as well as a quantitative measure of the quality of the output, i.e., the binding parameters. We note that Tetin and Hazlett ***(23)*** have presented an excellent discussion of ligand binding in the context of antibody/hapten interactions, and have pointed out the effect of errors in the concentration of bound ligand upon the dissociation constant. Our present treatment shall extend their analysis and, in particular, shall consider the effect of errors in the experimental parameters on the resolved binding parameters.

First, we will formalize the notion of error propagation in general, and then we will employ the generalized model to describe the uncertainties for calculation of binding constants in particular. In this chapter, we use the synonymous terms error, deviation, and uncertainty to represent the variation in measured data. Thus, the term uncertainty indicates absolute error, whereas fractional uncertainty denotes relative error. Percentage uncertainty is the fractional uncertainty multiplied by 100%.

Based on Stoer and Bulirsch ***(24)***, we consider a multivariate vector function ϕ, where ϕ is given by m real functions ϕ_i whose values are $y_i = \phi_i(x_1, \ldots, x_n)$, $i = 1, \ldots, m$. We must investigate how the input uncertainties ϕ of Δx affect the final result y = $\phi(x)$. Suppose component functions ϕ_i have continuous first derivatives. Let $x^{\sim}$ be an approximate value for x. Then we denote the uncertainty of $x_i^{\sim}$ and $x^{\sim}$, respectively by:

$$\Delta x_i = x_i^{\sim} - x \tag{17}$$

and

$$\Delta x = x^{\sim} - x \tag{18}$$

The fractional uncertainty of $x_i^{\sim}$ is defined as the quantity:

$$\varepsilon_{xi} = \frac{\Delta x_i}{x_i} \tag{19}$$

Replacing the input data x by $x^{\sim}$ leads to the result $y^{\sim} = \phi(x^{\sim})$ instead of $y = \phi(x)$. Expanding in a Taylor series and disregarding higher order terms gives

$$\Delta y_i = \sum_j \frac{\Delta x_j \, \partial \varphi_i(x)}{\partial \chi_i} \tag{20}$$

The quantity $\partial\phi_i(x)/\partial x_j$ represents the sensitivity with which y_i reacts to the uncertainty Δx_j of x_j. If $y_i \neq 0$ for $i = 1,\ldots,m$ and $x_j \neq 0$ for $j = 1,\ldots,n$ then a similar error propagation formula holds for fractional uncertainties:

$$\varepsilon_{yi} = \sum_j \varepsilon_{xj} \frac{x_j}{\varphi_i(x)} \frac{\partial \varphi_i(x)}{\partial \chi_j} \tag{21}$$

The $(x_j/\phi_i)\partial\phi_i/\partial x_j$ indicate how strongly a fractional uncertainty in x_j affects the fractional uncertainty in y_j. The amplification factors have the advantage of not depending on the scales of y_i and x_j.

The amplification factors for relative uncertainties are customarily called condition numbers. If any condition numbers are present which has large absolute values, then the problem is ill-conditioned, otherwise it is well-conditioned. For ill-conditioned problems, small relative errors in the input data x can cause large relative errors in the results $y = \phi(x)$. The condition number is meaningful only for nonzero y_i, x_i. It cannot be easily realized for many purposes, because the condition of ϕ is described by m x n numbers. For these reasons, we will present the conditions of particular problems in a more convenient fashion using contour plots to describe the uncertainties in the determination of dissociation constants.

Because fluorescence anisotropy is an additive property, we also have to consider the propagation of uncertainties for additive operations, $\phi(u,v) = u \pm v$. The fractional uncertainty specializes to

$$\varepsilon_{u\pm v} = \frac{u\varepsilon_u}{u \pm v} \pm \frac{v\varepsilon_v}{u \pm v} \tag{22}$$

The fractional uncertainties of the components do not propagate strongly into the results, provided the components u and v have the same sign. The condition numbers $u/(u+v)$, $v/(u+v)$ then lie between 0 and 1and they add up to 1, whence

$$\left|\varepsilon_{u \pm v}\right| <= \max\left\{\left|\varepsilon_u\right|,\left|\varepsilon_v\right|\right\} \tag{23}$$

If one component is small compared to the other, but carries a large fractional uncertainty, the result $u + v$ will still have a small fractional uncertainty so long as the other component has only a small fractional uncertainty, i.e., error damping occurs. However, if two components of different sign are added (a rare case in anisotropy and polarization measurements) then at least one of the factors [$u/(u+v)$ or $v/(u+v)$] is bigger than 1, and at least one of the fractional uncertainties will be amplified. This amplification will be extreme if $u = -v$ holds, and therefore cancellation occurs.

3.3. Measurement of Binding Constants

As discussed earlier, if a fluorescent ligand binds to a protein its fluorescence polarization and anisotropy will typically increase to reflect the slower rotation of the ligand–protein complex relative to free ligand. Using either polarization or anisotropy data one can calculate the fraction of ligand bound (x) at any protein concentration. The dissociation constant, K_d, for n identical sites without interaction can then be calculated from x by the expression:

$$K_d = \frac{(1-x)\left(nP_T - xL_T\right)}{x} \tag{24}$$

where P_T and L_T are the total concentration of protein and ligand, respectively. When $n = 1$, the expression reduces to the special (although most common) case of a single binding site. P_T and L_T are value constants in a binding experiment and the primary variable is x on which K_d depends. We must now investigate how uncertainties Δx of x affect the final result, K_d. We start this investigation by considering systematic deviations and random uncertainties alone.

3.4. Systematic Deviations

In recent literature, in particular the literature associated with clinical chemistry applications of the polarization/anisotropy method, we have noted that some investigators analyze their polarization data using the formulation associated rigorously only with the anisotropy function (*see* **ref. 25**). This practice leads to systematic deviations in determination of x and subsequently of K_d. Specifically, direct substitution of polarization values into **Eq. 15**, derived for anisotropy measurements, instead of converting to anisotropies using **Eqs. 7** or

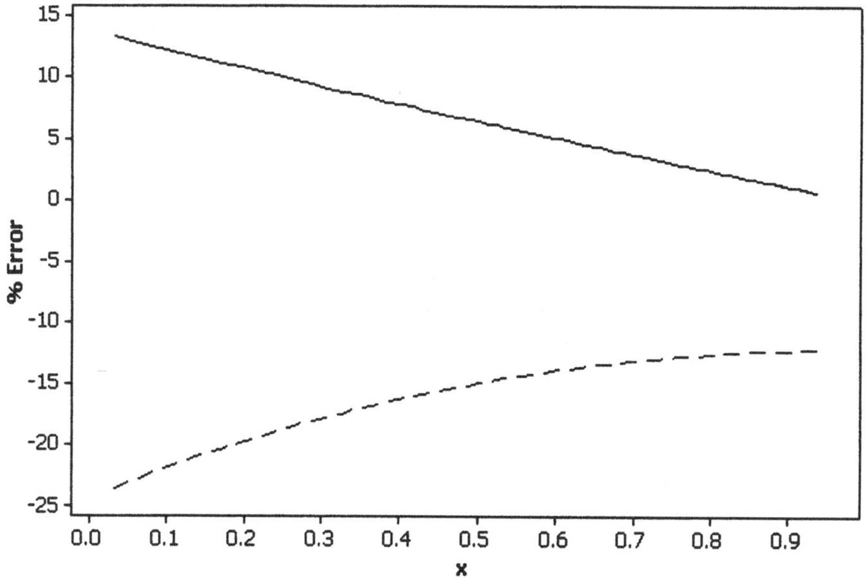

Fig. 3. Percentage error in K_d with respect to direct substitution of polarization in place of anisotropy in the anisotropy additivity equation (**Eq. 15**). Perfect measurement (no input error) is assumed except for the improper substitution. X, fraction of ligand bound; $g = 1$. Solid line, % error in X (direct substitution results in 1–14% overestimation of X); long dashes, % error in K_d (direct substitution results in 12–24% underestimation of K_d). Model parameters are as in **Fig. 2**.

8, or instead of using **Eq. 16** derived for polarization measurements, results in *systematic overestimation of* x *and underestimation of* K_d. This practice is sometimes justified by the rationalization that the attendant errors are not large, yet a systematic examination, using error propagation analysis (analogous to the original treatment of Weber on systematic instrument errors), has not been heretofore presented. In such error propagation treatments, it is customary to present results with a suitable norm, i.e., without their sign. In this chapter, however, we examine the sign of a particular error term as it may give information whether the uncertainty in the result is a case of overestimation or underestimation.

Figure 3 shows ε_K, the percentage deviation in K_d as a function of x and ε_x, the relative deviation in x. ε_x and ε_K were calculated as $(x_p - x_r)/x_r$, and $(K_p - K_r)/K_r$, respectively, and expressed as a percentage. Clearly, it is not enough to simply substitute P for r without taking into account how the particular errors enter into the respective results. Systematic deviations are cumulative in nature. The percentage deviation in x resulting from direct substitution is 1% to 14%

over the whole x range, and the percentage deviation in K_d conditioned on the deviation in x is –12% to –24%. Such deviations may be easily corrected for or eliminated by using the appropriate function to calculate x.

3.5. Precision of Anisotropy and Polarization Determinations

We now will consider uncertainties in the input parameters. As usually practiced, the precision of anisotropy and polarization data is approximately ± 0.002 and ± 0.003, respectively (we note that these are conservative estimates which will vary depending upon the instrumentation utilized, signal strength, and background). Applying the generalized error propagation model described above, ε_x the fractional uncertainty in x with respect to ε_r the fractional uncertainty in r can be calculated by the function:

$$\varepsilon_x = \frac{\varepsilon_r r}{r - r_f} + \frac{\varepsilon_r r(g-1)}{r_b - r_f + (g-1)(r_b - r)} \tag{25}$$

ε_x can be used then to calculate the fractional uncertainty in K_d as follows:

$$\varepsilon_{Kd} = \frac{\varepsilon_x \left(x^2 L_T - nP_T\right)}{(1-x)\left(nP_T - xL_T\right)} \tag{26}$$

We must give a range of possible true values for K_d based on the uncertainties in x. **Figure 4** shows ε_{Kd} as a function of x and ε_x expressed as percentage. As it can be seen, propagation of seemingly small uncertainties in the input parameters can cause larger variations in the results. At low and high values of x, the percentage uncertainty in K_d can be as high as 24% conditioned on a fixed precision of ± 0.002 anisotropy units. Unfortunately, many fluorescence practitioners are not aware of these constraints and mistakenly believe that the precision achievable with modern spectrofluorimeters equates to accuracy in the results. A measurement may be accurate without being precise and vice versa.

For polarization measurements, the fractional uncertainty in x can be derived similarly:

$$\varepsilon_x = \frac{\varepsilon_p P}{P - P_f} + \frac{\varepsilon_p \left[P_b - P_f + (g-1)(3-P_f)\right]}{(3-P)\left[P_b - P_f + (g-1)(3-P_f)(P_b - P)\right]} \tag{27}$$

This function is more complex than the one for anisotropy measurements and the uncertainty propagates through polarization is slightly larger. The percentage deviation between ε_x calculated from anisotropy and polarization readings is less than 0.5% over the whole x interval though. Therefore, only anisotropy will be considered in the rest of the chapter because of its less convoluted error propagation properties.

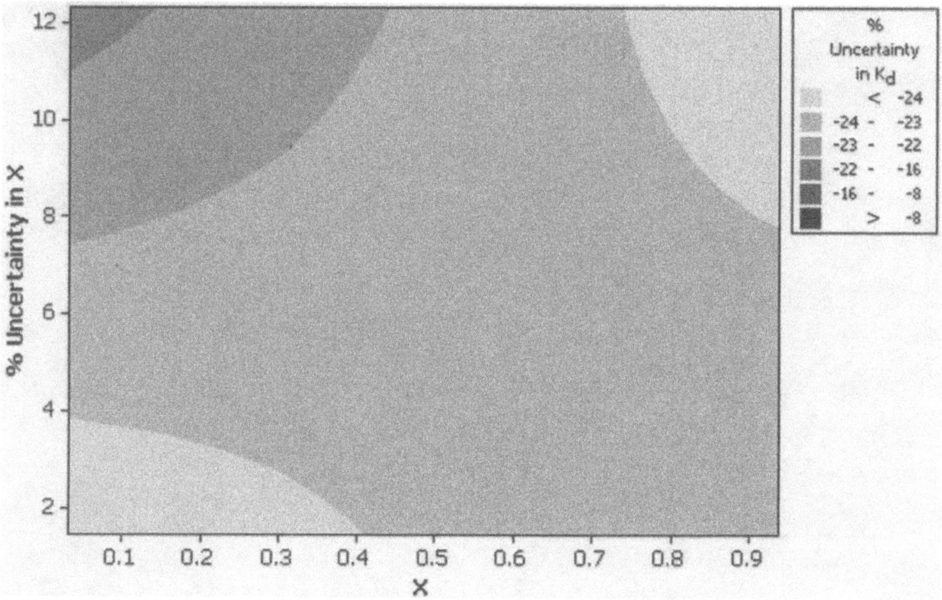

Fig. 4. Percentage uncertainty in K_d with respect to a fixed uncertainty in anisotropy readings. In an otherwise perfect measurement, a fixed uncertainty of +0.002 anisotropy unit is considered, which results in a 8–24% underestimation of K_d. An uncertainty of –0.002 anisotropy unit leads to 8–24% overestimation of K_d. Model parameters are as in **Fig. 2**. X, fraction of ligand bound; $g = 2$.

3.6. Uncertainty in the Anisotropy of the Ligand Free in Solution

Typical anisotropy values for fluorescent ligands free in solution are in the range of 0.02 (*see* **Note 6**), whereas the precision of an anisotropy reading is normally around ± 0.002. Realistically then, approx 10% uncertainties in the value of r_f can be assumed. Again, propagating the errors in r_f, the relative uncertainty in x with respect to ε_f can be calculated by the function:

$$\varepsilon_x = \frac{-\varepsilon_{rf}\, r_f}{r - r_f} + \frac{\varepsilon_f\, r_f}{r_b - r_f + (g-1)(r_b - r)} \tag{28}$$

and subsequently ε_K can be computed by **Eq. 26**. **Figure 5** shows the effect of uncertainty in the anisotropy of the free ligand on K_d. Since r_f is typically low, K_d is mostly affected at low x values where the observed anisotropy is close to r_f and the amplification factor is higher. The percentage uncertainty in K_d is 5–20% over the entire x range.

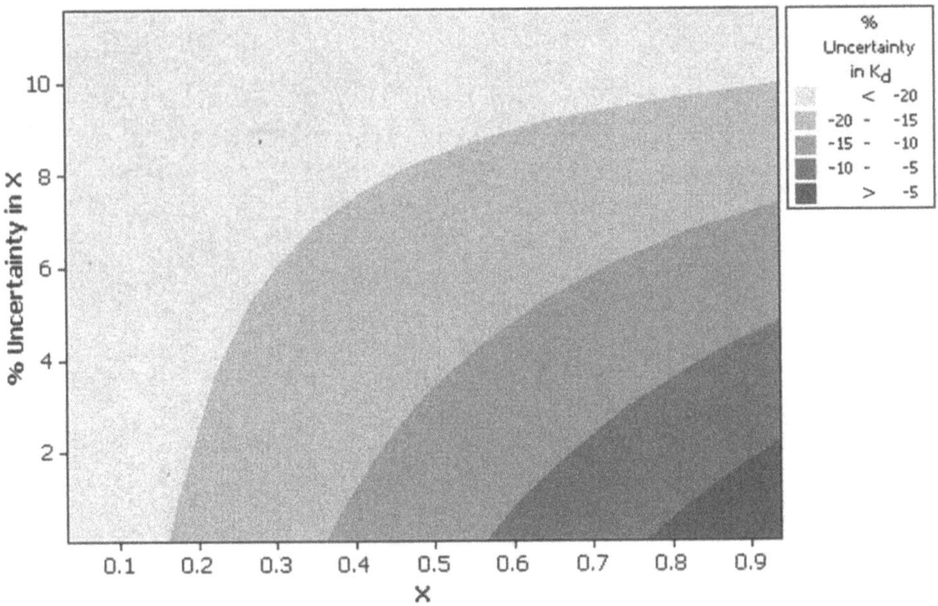

Fig. 5. Percentage uncertainty in K_d with respect to input uncertainties in the anisotropy of the free ligand. A fixed uncertainty of –10% in r_f is considered which results in approx 5–20% underestimation of K_d. An uncertainty of +10% in r_f leads to 5–20% overestimation of K_d. Model parameters are as in **Fig. 2**. X, fraction of ligand bound; $g = 2$.

3.7. Uncertainty in the Anisotropy of the Bound Ligand

Often other uncertainties are larger than the cases considered previously. As the limiting anisotropy is an asymptotic property, relatively small uncertainties in ε_b can propagate into larger deviations in K_d as the observed anisotropy approaches the anisotropy of the bound ligand (*see* **Note 5**). The uncertainty in ε_x with respect to ε_b is given by the function:

$$\varepsilon_x = \frac{-\varepsilon_b \, g r_b}{r_b - r_f + (g-1)(r_b - r)} \tag{29}$$

where the upper bound of the amplification factor equals to g if $r \rightarrow r_b$. This fact may lead to considerable uncertainties in ε_x even at moderately low values of g. **Figure 6** shows the dependence of uncertainties in K_d with respect to –5% underestimation of r_b. At x values larger than 0.7, K_d appears to be underestimated by 120–150 %. A + 5% overestimation of r_b would result in 120–150%

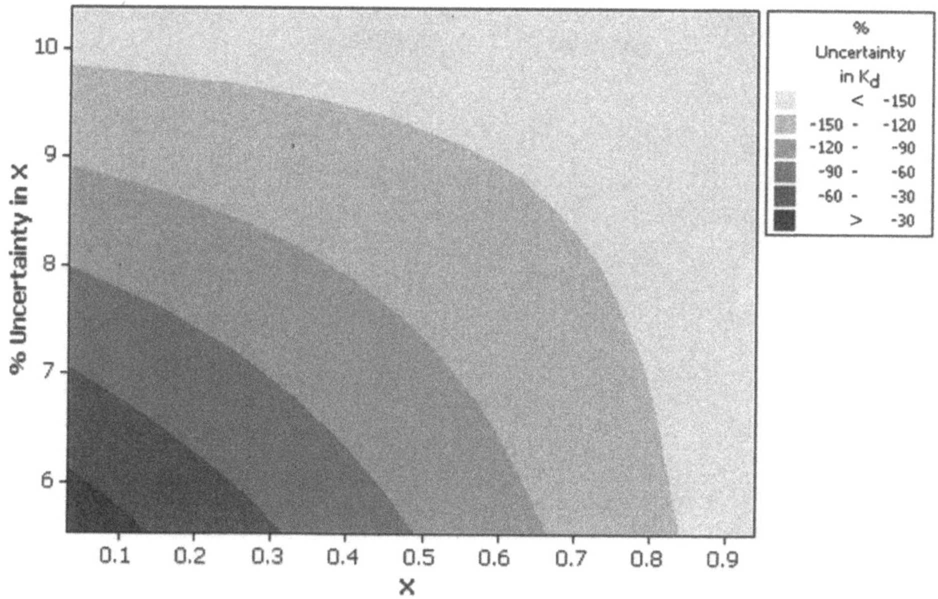

Fig. 6. Percentage uncertainty in K_d with respect to input uncertainties in the anisotropy of the bound ligand. A fixed uncertainty of –5% in r_b is considered which results in 30–150% underestimation of K_d . Similarly, an uncertainty of +5% in r_b leads to overestimation of K_d to the same extent. Model parameters are as in **Fig. 2**. X, fraction of ligand bound; $g = 2$.

overestimation of K_d. Clearly, the fixed value of r_b should be as accurate as possible to minimize uncertainties in K_d.

3.8. Uncertainty in the Fluorescence Enhancement Factor

A further understanding of the accuracy of the calculated binding constants can be obtained by propagating the uncertainties in the enhancement factor. ε_x as a function of ε_g can be calculated by the function:

$$\varepsilon_x = \frac{-\varepsilon_g \, g(r_b - r)}{r_b - r_f + (g-1)(r_b - r)} \tag{30}$$

The amplification factor is small at higher x values and is only moderately high at x values less than 0.3. A 10% underestimation of g results in 12 to 18% underestimation of K_d, whereas a 10% overestimation of g results in a similar 12 to 18% overestimation of K_d. The accuracy of g does not have a major effect

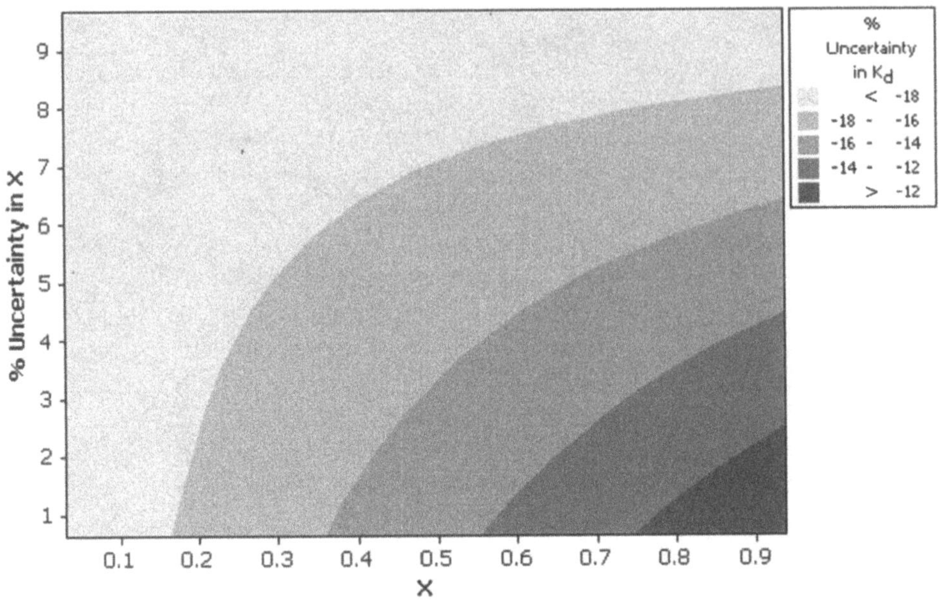

Fig. 7. Percentage uncertainty in K_d with respect to input uncertainties in the fluorescence enhancement factor. A fixed uncertainty of –10% in g is considered which causes 12–18% underestimation of K_d. An uncertainty of +10% in g results in 12–18% overestimation of K_d. Model parameters are as in **Fig. 2**. X, fraction of ligand bound; $g = 2$.

on the final results for K_d when g is the only source of uncertainty. **Figure 7** shows the dependence of uncertainties in K_d with respect to –10% uncertainty in g.

3.9. Modeling of Multiple Uncertainties Together

We now employ the generalized error propagation model to each input parameter jointly. Because the individual input terms has opposing signs and their uncertainties are in a ± value range, the uncertainties in the output cannot be described by a predetermined mathematical model. Rather, it is possible to calculate an upper bound for the joint maximum uncertainty assuming that all input certainties propagate into the same direction, i.e., overestimation or underestimation, respectively. Technically, using absolute norms will have the same effect. The uncertainty of results from any real experiment will be then smaller than the simulated upper bound. Random uncertainties obey the law of probability and tend to cancel or damp each other as discussed earlier. By substituting all individual error terms into **Eq. 26**, ε_{Kd} for the joint uncertainty can

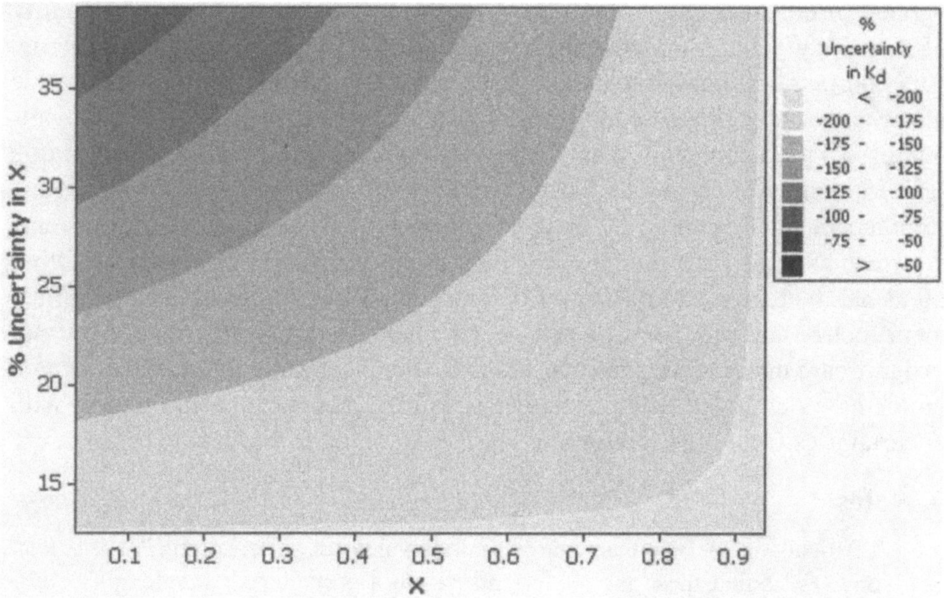

Fig. 8. Upper bound percentage uncertainty in K_d with respect to input uncertainties jointly present in all input parameters. A fixed uncertainty of +0.002 in r, –10% in r_f, –5% in r_b and –10% in g is considered which all propagates to underestimate K_d. At worst case, K_d appears 50–200% lower than its true value. Reversing all signs in the individual error terms leads to overestimation of K_d to the same extent. Model parameters are as in **Fig. 2**. X, fraction of ligand bound; $g = 2$.

be calculated. This effect is shown in **Fig. 8** for K_d using the same input parameters and their respective accuracies as before. Interestingly, the joint upper bound attains a maximum value when x is higher than 0.6 and ε_x is lower then 20% caused by the influence of g on the joint output. In combination with other errors, the uncertainty in g has a larger and biased effect on the common amplification factor, at least with the particular input parameters used for the present modeling.

3.10. Conclusion

An error propagation model has been developed for modeling of uncertainties in determination of dissociation constants. The main parameters are the multivariate functions of typical input constraints for anisotropy measurements. The values of all input parameters as well as their accuracies are needed for the simulation. Each single parameter is investigated separately. It is also possible to perform the error propagation with multiple parameters together. The development of the error propagation provides the opportunity to assess the quality of output data conditioned on the uncertainties in the input parameters alone. An important finding is that uncertainties in the estimated or experimental value of the anisotropy of the bound probe cause the largest inaccuracies in the output dissociation constants. This consideration is especially important for large scale screening experiments where a number of unknown proteins are tested simultaneously. Assuming the same anisotropy for the bound probe in all systems may lead to large inaccuracies in the determination of binding constants. Other input parameters such anisotropy of the free ligands, fluorescence enhancement factor, and precision of anisotropy readings propagate smaller but not insignificant uncertainties into the results. The uncertainty in the enhancement factor has a complex effect when it is jointly present in combination with uncertainties from other factors.

4. Notes

1. A critical consideration as regards limiting values of polarization (P_0) or anisotropy (r_0) is that these parameters may have a pronounced wavelength dependence. *See* **refs. *5–8*** for example.
2. For a spherical protein of molecular weight 44 kD, with a partial specific volume of 0.74 and 0.3 mL/mg hydration, at room temperature, one then calculates $\rho = (3)(0.01)(44000)(0.74 + 0.3)/(8.31 \times 10^7)(293) = {\sim}56$ ns. So, to a rough approximation the Debye rotational relaxation time (in ns) for a spherical protein is close to its molecular weight in units of kD. In fact, to a rough approximation one can apply this *rule-of-thumb* to systems other than proteins, e.g., if a fluorophore has a molecular weight of 500 daltons, its rotational relaxation time will be (very) approx 0.5 ns. We should also comment here on the term *rotational correlation time*, often denoted as τ_c, is often used in conjunction with anisotropy determina-tions. In fact, $\rho = 3\tau_c$, a fact that stems form the original definitions of these terms ***(20)***.
3. The term *fractional contribution* actually refers to the fractional contribution to the photocurrent (or number of photons), which each particular molecular species provides. This contribution does not then strictly refer to the fraction of molecular species present, but will depend upon the absorption and fluorescence spectra of each component, the particular region of the spectrum being monitored, and the response characteristics of the instrument at the relevant wavelength regions. For a discussion of such instrument response characteristics *see* **ref. *3***.
4. Consider a typical small fluorescent ligand with a rotational relaxation time of 1 ns. If the fluorescent lifetime is 0.1 ns, then the polarization observed for the free ligand will be quite high—near the limiting polarization—because the probe does not have time to rotate appreciably before emission. On the other hand, if the probe lifetime is very long (e.g., some pyrene probes can be more than a hundred ns and ruthenium can be hundreds of ns) then both free and bound probes could give very low polarization values.

5. Companies such as Molecular Probes (now part of Invitrogen), for example, specialize in creating literally thousands of fluorescence probes that can be reacted with amines, thiols, hydroxyls, etc. The highest polarization or anisotropy values possible for a fluorescence probe in solution is 0.5 (polarization) and 0.4 (anisotropy), for the case of one photon excitation *(7)*. However, even for the most favorable ratio of probe lifetime to rotational relaxation time, this limit is rarely achieved as a result of the often extensive local mobility of the probe about its point of attachment to the ligand *(9)*. This effect will often lower the observed anisotropy of bound probe to values in the range of 0.3 or lower. Determination of the anisotropy of the bound probe can best be accomplished by measuring the anisotropy when the protein concentration greatly exceeds the ligand concentration and is, furthermore, well above the dissociation constant of the ligand–protein complex *(1)*.

Acknowledgments

DMJ wishes to acknowledge support from the American Heart Association (GIA0151578Z). GM is supported in part by Grant G12-RR03061 (Cadman, Core Support) from the National Institutes of Health.

References

1. Weber, G., 1992, *Protein Interactions*, Chapman and Hall, New York.
2. Winzor, D. J. and Sawyer, W. H. (1995) *Quantitative Characterization of Ligand Binding*, Wiley-Liss, New York.
3. Jameson, D. M., Croney, J. C., and Moens, P. D. (2003) Fluorescence: basic concepts, practical aspects and some anecdotes. *Methods Enzymol.* **360,** 1–43.
4. Oiwa, K., Jameson, D. M., Croney, J. C., Eccleston, J. F., and Anson, M. (2003) The 2'-O- and 3'-O-Cy3-EDA-ATP(ADP) vanadate complexes with myosin S1 are spectroscopically distinct. *Biophys. J.* **84,** 634–642.
5. Valeur, B. (2002) *Molecular Fluorescence*, Wiley-VCH Publishers, Weinheim, Germany.
6. Lakowicz, J. R. (1999) *Principles of Fluorescence Spectroscopy*, Kluwer Academic/Plenum Publishers, New York.
7. Weber, G. (1966) Polarization of the fluorescence of solutions. In *Fluorescence and Phosphorescence* (D. Hercules, ed.), Wiley, New York, pp. 217–240.
8. Jameson, D. M. and Croney, J. C. (2003) Fluorescence polarization: past, present and future. *Comb. High Throughput Chem.* **6,** 167–173.
9. Jameson, D. M. and Seifried, S. E. (1999) Quantification of protein-protein interactions using fluorescence polarization. *Methods* **19,** 222–233.
10. Weigert, F. (1920) Uber polarisiertes fluoreszenzlicht. *Verh. d.D. Phys. Ges.* **1,** 100–102.
11. Perrin, F. (1926) Polarisation de la lumière de fluorescence. Vie moyenne des molécules dans l'etat excité. *Jour de Phys, VIe Série* **7,** 390–401.
12. Laurence, D. J. R. (1952) A study of the adsorption of dyes on bovine serum albumin by the method of the polarization of fluorescence. *Biochem. J.* **51,** 168–177.

13. Dandliker, W. B. and Feijen, G. A. (1961) Quantification of the antigen-antibody reaction by the method of polarization of fluorescence. *Biochem. Biophys. Res. Comm.* **5,** 299–304.
14. Dandliker, W. B. and De Saussure, V. A. (1970) Fluorescence polarization in immunochemistry. *Immunochemistry.* **7,** 799–828.
15. Jolley, M. E., Stroupe, S. D., Schwenzer, K. S., Wang, C. J., Lu-Steffes, M., Hill, H. D., Popelka, S. R., Holen, J. T., and Kelso, D. M. (1981) Fluorescence polarization immunoassay. iii. An automated system for therapeutic drug determination. *Clin. Chem.* **27,** 1575–1579.
16. Weber, G. (1952) Polarization of the fluorescence of macromolecules. 1. Theory and experimental method. *Biochem. J.* **51,** 145–155.
17. Jablonski, A. (1960) On the notion of emission anisotropy. *Bull. Acad. Polon. Sci. Serie des sci. math. astr. et phys.* **6,** 259–264.
18. Mocz, G., Helms, M.K., Jameson, D.M., and Gibbons, I.R. (1998) Probing the nucleotide binding site of axonemal dynein with the fluorescent nucleotide analog 2'(3')-O-(N-methylanthrananiloyl)-adenosine 5'-triphosphate. *Biochemistry* **37,** 9862–9869.
19. Jameson, D. M. and Sawyer, W. H. (1995) Fluorescence anisotropy applied to biomolecular interactions. *Methods Enzymol.* **246,** 283–300.
20. Jameson, D. M. and Eccleston, E. F. (1997) Fluorescent nucleotide analogs: synthesis and applications. *Methods Enzymol.* **278,** 363–390.
21. Weber, G. (1956) Photoelectric method for the measurement of the polarization of the fluorescence of solutions. *J. Opt. Soc. Amer.* **46,** 962–970.
22. Jameson, D.M., Weber, G., Spencer, R.D., and Mitchell, G. (1978) Fluorescence polarization: measurements with a photon-counting photometer. *Rev. Sci. Instru.* **49,** 510–514.
23. Tetin, S.Y. and Hazlett, T.L. (2000) Optical spectroscopy in studies of antibody-hapten interactions. *Methods* **20,** 341–361.
24. Stoer J. and Bulirsch R. (1980) Error propagation. In *Introduction to Numerical Analysis*, Springer-Verlag, New York, pp. 9–20.
25. Prystay, L., Gosselin, M., and Bandks, P. (2001) Determination of equilibrium dissociation constants in fluorescence polarization. *J. Biomol. Screen.* **6,** 141–150.

16

Ligand Binding With Stopped-Flow Rapid Mixing

Mark S. Hargrove

Summary

Stopped-flow rapid mixing is a common, direct technique for the study of ligand-binding reactions. In this method, protein and ligand are mixed together at relatively high velocities directly into an observation chamber, so that time courses for reactions occurring on time scales as short as a few milliseconds can be measured. This chapter presents an introduction to this technique, including a discussion of experimental and technical issues that must be addressed when designing stopped-flow experiments. Simple experiments for measuring reaction dead time and flushing volume are also described along with the details of several common reaction schemes and methods for data analysis.

Key Words: Stopped-flow mixing; kinetics; ligand binding.

1. Introduction

Ligand-binding reactions have been the focus of biochemistry since its inception. The quantification of these reactions reveals molecular details of enzyme function, protein–protein interactions, molecular transport, and many other fundamental aspects of biochemistry. All ligand-binding reactions can be separated into two parameters: equilibrium and kinetic. Equilibrium parameters describe both the propensity for a reaction to occur and the concentrations of products and reactants at equilibrium, but do not predict how fast a reaction will occur or the mechanism of the reaction. Kinetic parameters provide reaction rates, which can be used to investigate mechanisms and calculate equilibrium parameters. Many methods for kinetic analysis have been developed that have strengths and weaknesses depending on the nature of the reaction in question (*see* **Note 1**).

Kinetic experiments can be placed into two general categories: 1) those that are slow enough that the times required for mixing and initiation are insignificant compared to their time courses, and 2) those that are fast enough to require

From: *Methods in Molecular Biology, vol. 305: Protein–Ligand Interactions: Methods and Applications*
Edited by: G. U. Nienhaus © Humana Press Inc., Totowa, NJ

special attention to ensure that kinetic events are not lost prior to data collection. In many cases, the concentrations of reactants can be adjusted to yield reaction time-courses that are measurable without special consideration for mixing and initiation. This is true for many enzyme reactions because it is not necessary to actually observe the enzyme–substrate complex; time-courses can measure the buildup of product (or removal of substrate), and enzyme concentrations can be lowered until reaction velocities are manageable (*see* **Note 2**). However, for nonenzymatic reactions, one must observe the complex between reactants directly, which constrains the concentrations to those that achieve measurable signals. For this reason, ligand-binding reactions frequently have rapid time-courses that fall into the latter category, requiring faster methods for initiating reactions and data collection.

For fast reactions, which occur in a minute or less, many experimental techniques have been developed for rapid initiation so that the time-courses are not lost in the *dead time* of the reaction (the time that elapses between initiation and data collection). The first and most direct technique designed to circumvent this problem is rapid mixing of reactants directly from two syringes into an observation chamber or cuvet. If flow is continuous, the reaction matures as a function of flow rate and distance from the point of mixing. In this *continuous flow* method, one can measure the signal associated with binding as a function of distance from the mixer to obtain a time-course for the reaction ***(1)***. The strengths of the continuous flow method are that measurement of the earliest time points is limited only by the physical process of mixing, and a rapid response time is not required for detection. The drawbacks include the requirements for a continuous supply of reactants and a movable detection system.

Alternatively, reactants can be mixed and the flow rapidly stopped ***(2)***. One can then measure the change in signal over time in this *stopped-flow* reaction to obtain a time course directly. Stopped-flow methods are convenient because only very small volumes of reactants are required, and the detection system needs only to focus on an isolated area. The drawbacks are a finite dead time associated with both mixing and transport to the cuvet and a requirement for rapid detection because time-courses are recorded in real time. Modern instruments have shortened the dead time in stopped-flow reactions to the point that they are no longer inferior to continuous flow instruments in this regard, and detection systems are available for many signals that are not limiting in time resolution.

Stopped-flow mixing has been combined with the measurement of a variety of signals for the observation of reaction time-courses. A few examples are optical absorbance, fluorescence, circular dichroism (CD), infrared (IR), electron paramagnetic resonance (EPR), and light scattering. Any signal that can be measured through a cuvet is in principle applicable to stopped-flow mixing.

Commercially available stopped-flow equipment can be purchased *off-the-shelf* for detection of absorbance, fluorescence, CD, and flow-flash experiments (*see* **Note 3**). If detection requires a slower form of analysis, *quenched-flow* methods can be used in which the reactants are mixed, allowed to incubate for a predetermined period of time, and then mixed with a quenching reagent or frozen to stop the reaction. The resulting samples represent stable time points in the progression of a reaction that can be analyzed later.

2. Experimental and Technical Issues

Stopped-flow mixing is achieved by rapidly forcing reactants together into a cuvet (**Fig. 1**). Mixing starts when the reactants come together at a mixing point and continues as the sample flows into the cuvet. Technological advances over the past few decades have improved every aspect of the stopped-flow mixer, and these details along with a thorough perspective on technical issues are available elsewhere *(3)*. Once mixing is achieved, detection is initiated at the cuvet using specifications consistent with the signal change observed for the reaction. If the reaction is monitored optically, the detection system is a spectrophotometer capable of collecting absorbance values in a rapid manner, and details associated with detection in an absorbance stopped flow system are those applicable to absorption spectroscopy (*see* **Note 4**).

Modern instruments provide independent control over the volume and flow rate of each reactant dispensed, enable easy replacement of a variety of cuvets for use in different experiments, and use computer control over every aspect of the experiment including filling the reaction syringes, triggering the detection system, and collecting the reaction time-course. However, regardless of the apparatus used for mixing and data collection, there are a number of experimental issues one must address that provide a helpful introduction to stopped-flow techniques. In **Subheading 2.1.**, some kinetic schemes are described that are frequently encountered in stopped-flow experiments that also provide the basic equations used to characterize the performance of the instrument. Specific instrumental parameters such as *dead time* and *flushing volume* are addressed along with experiments for their determination. Multiwavelength experiments are also described along with a simple method for collecting kinetic difference spectra. Finally, the basic elements of data analysis are described in the context of stopped-flow experiments.

2.1. Kinetic Considerations

Time-courses resulting from stopped-flow mixing consist of the change in time of a signal associated with the progression of a reaction. Reaction velocities or rate constants are extracted from these data and used in defining reaction schemes. The most useful time-course is one in which the reaction goes to

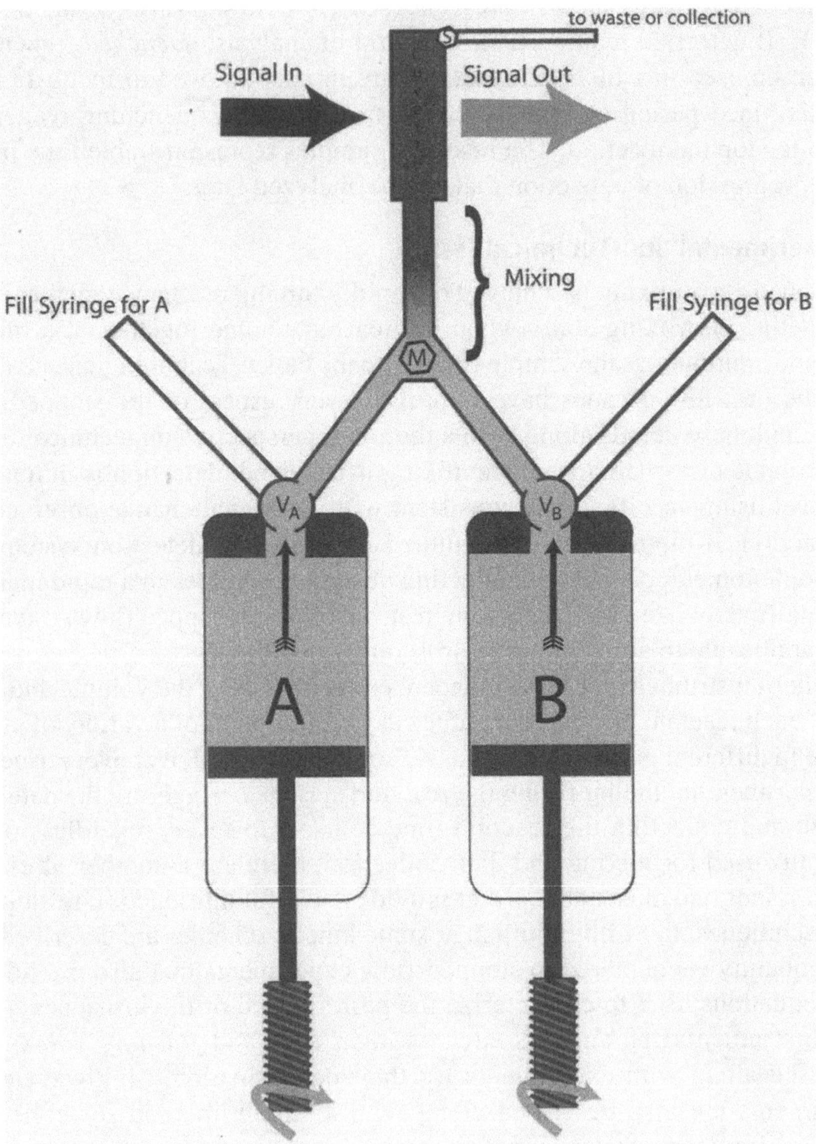

Fig. 1. A diagram of a stopped-flow mixer. Reactants A and B are drawn into the reaction syringes from the fill syringes through valves V_A and V_B. The valves are then rotated to direct the reactants toward a common mixing point (M) when the driving mechanism exerts pressure. The previous contents of the tubes and cuvet are flushed out of the stopping valve S that is open during flow. When flow is stopped, the stopping value is closed and the contents of the cuvet can be probed during maturation of the reaction mixture.

completion, because the rate constant can be extracted from the decay in signal independently of other variables, and even if some signal is lost in the dead time (*see* **Note 5**). There are many texts dedicated to reaction kinetics that should be consulted for an in-depth description of kinetics and time-course analysis *(7–9)*. However, there are a few basic kinetic schemes useful for designing experiments and interpreting time courses that must be considered in an introduction to the stopped flow method.

2.1.2. First-Order Reactions

The simplest reaction frequently encountered in kinetic experiments is the first-order reaction.

$$P_U \xrightarrow{k_1} P_F \tag{1}$$

Equation 1 describes many reactions, such as some protein-folding reactions, that do not involve binding of exogenous ligands. Reaction conditions must be established in which the transition from the unfolded (U) to the folded (F) state of the protein is initiated by mixing, and the final equilibrium lies in the F conformation (or the U conformation if the unfolding reaction is being analyzed). The rate law for this reaction is:

$$\frac{d[P_U]}{dt} = k_i[P_U] \tag{2}$$

Integration of this differential equation gives the familiar expression describing the time-course for a first-order reaction

$$A_t = A_0 e^{-k_1 t} \tag{3}$$

where A_t is the change in signal at time t, A_0 is the total amplitude expected for the reaction, and k_1 is the observed rate constant in units of s^{-1}. By setting A_t = 0.5, and A_0 = 1, one can calculate the half-life of a first-order reaction as:

$$t_{1/2} = \frac{0.693}{k_1} \tag{4}$$

These equations lead to single, exponentially decaying time-courses where A_t decays from a maximum value (A_0). **Figure 2** shows an example of time-courses where $A_0 = 1$, and $k_1 = 0.0693$, 0.693, and 6.93 s^{-1} . Time-courses are well defined when the signal change is large compared to the noise and a clear endpoint is reached. In **Fig. 2**, the time-course with $k_1 = 0.693$ s^{-1} is ideal; the others either do not define the endpoint of the reaction (with $k_1 = 0.0693$) or waste too many data points defining the endpoint (with $k_1 = 6.93$). A time-course collected for ten half-lives is generally well defined, as depicted by the middle trace in **Fig. 2**.

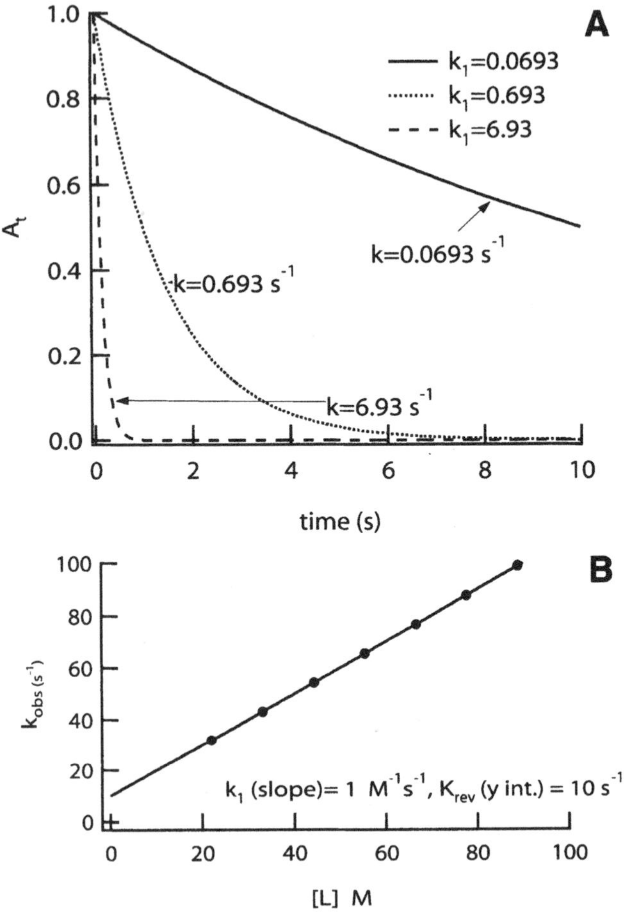

Fig. 2. **(A)** Exponential decay time-courses associated with first-order reactions. For each time-course, $A_0 = 1$, and the rate constant k_1 is varied. The dotted line, with $k_1 = 0.693\ s^{-1}$, is best defined by the time scale in this example in which the total time over which collection occurs is ten times the half life of the reaction. Proper definition of the other two time-courses would require longer (for $k_1 = 0.0693\ s^{-1}$) and shorter (for $k_1 = 6.93\ s^{-1}$) time scales, respectively. **(B)** The observed reaction rate is plotted vs $[L]$ for a bimolecular reaction carried out under pseudo first-order conditions with L in excess. The slope is equal to the bimolecular rate constant and the y-intercept provides the rate constant for the reverse reaction (**Eq. 8**).

2.1.3. Second- and Pseudo First-Order Reactions

Simple ligand-binding reactions are described by a second-order process of the following form:

$$P + L \xrightarrow{k_1} P_L \tag{5}$$

This reaction scheme has an integrated rate expression that is complex and involves the starting concentrations of both reactants (P_0 and L_0), and each reactant at time t (P_t and L_t).

$$\frac{1}{[P_0]-[L_0]}\ln\left|\frac{[L_0][P_T]}{[P_0][L_T]}\right| = k_1 t \qquad (6)$$

Thus, many factors must be accounted for in analyzing second-order reactions. A common way around this complexity is to carry out reactions with one of the reactants (usually the ligand) in vast excess. If $L_0 >> P_0$, then L_t will be approximately L_0 at all times. Under these conditions, **Eq. 6** reduces to

$$P_T = P_0 e^{-k_1[L]t} \qquad (7)$$

where the observed rate constant at each concentration of ligand is $k_{obs} = k_1[L]$. These conditions are called *pseudo first-order* because the reaction depends on the concentration of only one reactant in spite of the fact that it is bimolecular. As many binding reactions are carried out with excess ligand, these conditions are very useful; a plot of k_{obs} vs $[L]$ will yield a straight line with a slope equal to the bimolecular rate constant for the reaction (k_1) (**Fig. 2B**). The half-life of the reaction can be calculated by **Eq. 4** for each concentration of ligand, as is the case for first-order reactions.

2.1.4. Consideration of the Reverse Reaction

For first-order and pseudo first-order reactions, the reverse reaction (k_{rev}) can only be ignored if it is much smaller than k_1. However, if it is appreciable, the observed rate constant will be a sum of $k_1[L]$ and k_{rev}.

$$k_{obs} = k_1[L] + k_{rev} \qquad (8)$$

Therefore, a plot of k_{obs} vs $[L]$ will yield a line with a slope equal to k_1 (as above) and the y intercept will equal k_{rev} with units of s^{-1}. An example of data for a reaction like this one is shown in **Fig. 2B**, which represents k_{obs} at different concentrations for binding of a ligand (L) to a protein when k_1 = 1 $\mu M^{-1}s^{-1}$ and k_{rev} = 10 s^{-1}. In this case, the value of k_{rev} is large enough compared to $k_1[L]$ that it is observable as the y-intercept of this plot (*see* **Note 6**).

2.1.5. Consecutive Reactions

In many biochemical pathways, one reaction leads to another and the observed time-courses have contributions from each constituent process. In these cases, observed reaction rates can be complex combinations of each of the rate constants characterizing the reactions. Some reaction schemes have integrated solutions for their time-courses (like **Eqs. 3** and **6**), but many others must be analyzed numerically to determine the time dependence of the concentration of each reactant (this is discussed further in **Subheading 2.6**). However, in many cases the expected *observed* rate constant for a complex reaction can be estimated using *steady-state* and other approximations ***(7,8)***. Three reaction schemes that occur frequently in biological ligand binding and stopped-flow reactions are discussed below.

2.1.6. A First-Order Followed by a Second-Order Reaction

One common scheme is a first-order reaction followed by ligand binding. This occurs when only one conformation of a protein is capable of ligand binding (P_F) and the other (P_U) must transition to this state for the reaction to proceed.

$$P_U \underset{k_{-1}}{\overset{k_1}{\rightleftharpoons}} P_F + L \xrightarrow{K_L} P_L \tag{9}$$

An integrated solution to this scheme is available ***(9)***, but only when the reaction starts with $P_{total} = P_F$. However, in stopped-flow experiments, the reaction would be initiated from an equilibrium mixture of P_U and P_F reacting with L. A reaction of this form is best carried out under pseudo first-order conditions with L. If so, k_{obs} can be calculated by assuming an *improved* steady-state concentration of P_F ***(7,10)***.

$$k_{obs} = \frac{k_1 k_L [L]}{k_1 + k_{-1} + k_L [L]} \tag{10}$$

Equation 10 predicts time-courses that vary with $[L]$ in an intuitive manner. As expected from **Eq. 9**, if bimolecular binding of L is very rapid compared to the conformational transition (i.e., $k_L[L] >> k_1$), the first-order event will be the rate limiting step and k_1 will be observed at each ligand concentration. Alternatively, if $k_1 >> k_L[L]$, k_{obs} will be equal to $k_L[L]$, and thus linearly dependent on ligand concentration. When neither relationship of rate constants exists for the reaction, k_{obs} is nonlinear and gradually approaches an asymptote equal to k_1 as $[L]$ increases (**Fig. 3**).

2.1.7. Second-Order Followed by a First-Order Reaction

Another common reaction scheme is one in which bimolecular binding generates a complex that is then converted into a final product through a first-order reaction. This reaction is common to many enzymes that undergo conformational changes after binding substrate, and to redox enzymes that first form a complex and then transfer electrons.

$$P + L \underset{k_{-L}}{\overset{k_L}{\rightleftharpoons}} P:L \xrightarrow{k_R} P_L \tag{11}$$

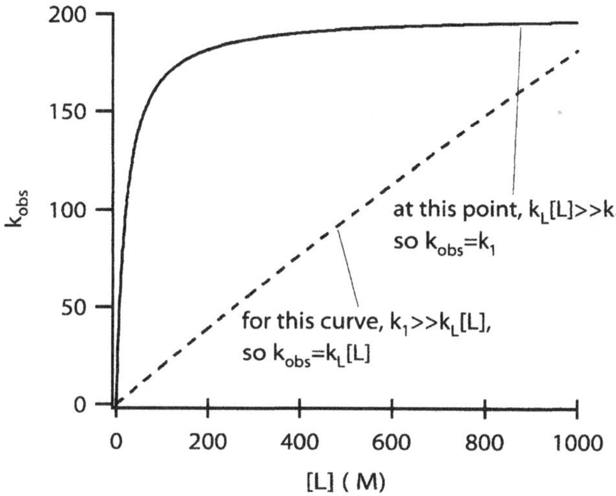

Fig. 3. Examples of k_{obs} as a function of $[L]$ for reactions obeying the form of **Eqs. 9** and **11**. For the solid curve, $k_L = 10\ \mu M^{-1}s^{-1}$, $k_{-L} = 1\ s^{-1}$, and $k_1 = 200\ s^{-1}$. Under these conditions, an asymptote with $k_{obs} = k_1$ is observed. For the dashed line, with $k_L = 0.2\ \mu M^{-1}s^{-1}$, $k_{-L} = 1\ s^{-1}$, and $k_1 = 2{,}000\ s^{-1}$, $k_{obs} = k_L[L]$ is not limited by the first order reaction.

When the reaction is carried out under pseudo first-order conditions with $[L]$ (as in the case previously), the observed rate constant depends on ligand concentration as described by the following equation:

$$k_{obs} = \frac{k_R k_L [L]}{k_R + k_{-L} + k_L [L]} \qquad (12)$$

In spite of the fact that the sequence of first- and second-order reactions in **Eqs. 9** and **11** are reversed, **Eqs. 10** and **12** are very similar in form. As described for **Eq. 10, Eq. 12** predicts $k_{obs} = k_R$ if $k_L[L] >> k_R$ and $k_{obs} = k_L[L]$ if $k_R >> k_L[L]$ (analogous to **Fig. 3**). Thus, the sequences of these reactions cannot be determined by observation of k_{obs} alone. However, one can establish some relationships between the values of each constituent rate constant, which provides insight into the mechanism of the reaction by identifying the rate-limiting step.

2.1.8. Displacement Reactions

Many experiments are designed in which two ligands compete for a protein. The reaction is started with the protein bound to one ligand (P_L) and the second acts as an irreversible scavenger (S) so that the original ligand is displaced during the reaction.

$$PL \underset{k_L}{\overset{k_{-L}}{\rightleftarrows}} L + P + S \xrightarrow{k_S} P_S \tag{13}$$

This reaction scheme is commonly used to measure the dissociation rate constant of the P_L complex (k_{-L}). Under these conditions, k_{-L} can be calculated from k_{obs} using pseudo first-order conditions for L and S, and a steady-state approximation for $[P]$:

$$k_{-L} = k_{obs}\left(1 + \frac{k_L\,[L]}{k_S\,[S]}\right) \tag{14}$$

Thus, stopped-flow reactions can be used to calculate association and dissociation rate constants by initiating reactions appropriately and analyzing the resulting time-courses in the context of reaction schemes and the observed rate constants expected for them.

2.2. Dead Time

The dead time of a stopped-flow reaction is important if the time-course to be measured is rapid compared to the time it takes for the reactants to mix and enter the cuvet. Many factors can affect dead time, including those that affect mixing (described later), flow rate, and cuvet volume. A rough estimate comes from the following equation:

$$\text{Deadtime} = \frac{V_D}{\text{Flowrate}} \tag{15}$$

where V_D is the total post-mixing volume that is displaced during flow, including that of the cuvet (*see* **Note 7**). Modern conventional stopped-flow mixers can achieve dead times around 1 ms with small volume cuvets. For absorbance measurements, where a 1 cm path is usually required to obtain good signal, dead times are typically around 3 ms. However, because so many factors determine dead time, it is best to measure it empirically if its value is required.

To measure dead time, one needs a test reaction for which the total change in absorbance is known. In this example, the reaction is ferric horse heart myoglobin (Mb) binding to azide *(11)*. The Mb concentration is 4 μ*M*, predicting a change in absorbance of 0.3 at 409 nm (**Fig. 4A**). A time-course for the reaction is measured at 0.1 *M* azide (after mixing), and the observed amplitude is compared with the expected amplitude to provide the dead time using the following relationship:

$$t_D = \frac{\left(\ln \frac{A_{\text{expected}}}{A_{\text{observed}}}\right)}{k} \tag{16}$$

As shown in **Fig. 4**, the fitted value for k at 0.03 *M* azide is 101 s^{-1} and A_{observed} is 0.21 (the bimolecular rate constant for this reaction is 3100 $M^{-1}s^{-1}$).

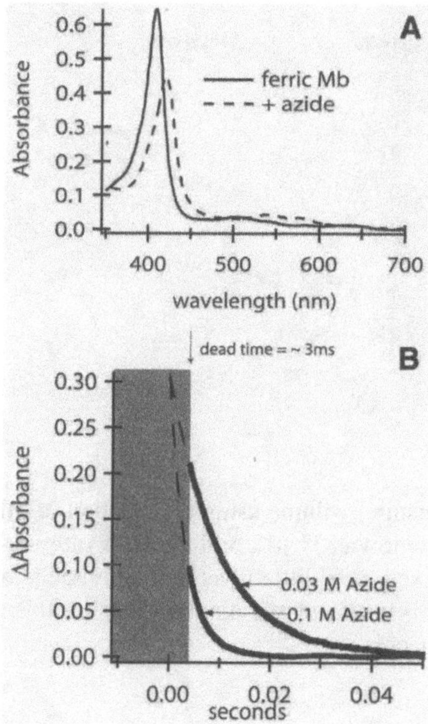

Fig. 4. Determination of dead time. 4 μ*M* horse heart myoglobin (Mb) is reacted with sodium azide. **(A)** Spectra of 4 μ*M* ferric Mb in the presence and absence of azide shows the expected absorbance change at each wavelength. **(B)** Time-courses for binding collected at 409 nm, observing the disappearance of unliganded Mb, where the expected absorbance change is 0.3. The time-course for 0.03 *M* azide is slower, and shows a larger (but still incomplete) amplitude than that for the reaction with 0.1 *M* azide. In this case, over half of the expected amplitude is lost in the dead time. The dashed lines show what these time courses would look like if they were complete, and the shaded section indicates the portion of the time-course not observed in the reaction.

From **Eq. 16**, the dead time for this reaction is calculated to be 3.5 ms. The reaction was carried out at a second concentration of azide (0.1 *M*) providing a similar value for dead time ($k = 296\ s^{-1}$, and $A_{observed} = 0.1$). The dashed lines in **Fig. 4B** demonstrate the complete time-courses as they would appear with no amplitude loss as a result of dead time. From these examples, it is evident that significant portions of the time-course are lost as the reaction half-life approaches the dead time of the instrument (at 0.1 *M* azide in this example, $t_D \sim t_{1/2}$). In general, if the half-life of the reaction is much shorter than the dead time, analysis of the reaction becomes impossible.

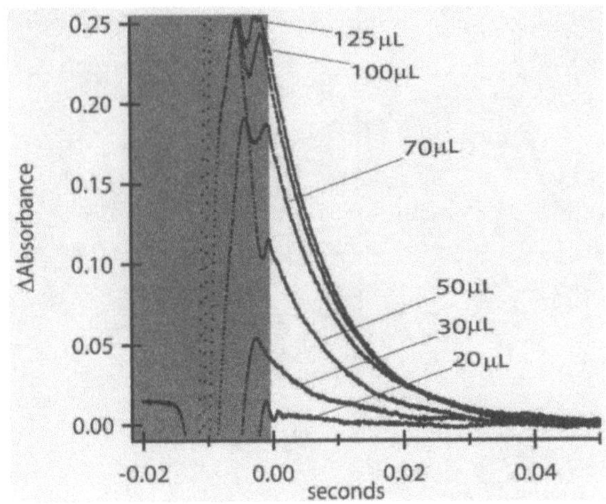

Fig. 5. Determination of flushing volume using the reaction of Mb with azide. The cuvet volume in this experiment was 31 μL, and the flow rate was 15 mL/s. When 20 μL of each reactant are mixed, very little reaction amplitude is observed. It is not until 100 μL of each reactant is used that the maximum amplitude is observed. Volumes greater than this are unnecessary.

2.3. Flushing Volume

Another experimental concern is the volumes of reactants used in each mixing experiment. Modern mixers have reduced this amount to tens of microliters in many cases. In all cases, the minimum volume required for each reactant is such that the combined volume displaces the contents of the mixing line and the cuvet. These volumes will be different for each apparatus. Many other factors influence mixing, including flow rates, tube and cuvet materials and dimensions, and anything affecting sample viscosity such as temperature. In general, there is an inverse relationship between flow rate and flushing volume. At higher flow rates, flow becomes more turbulent and flushing is less efficient because sample displacement is no longer linear. Therefore, flushing is improved by slowing the sample flow rate or increasing the sample volume used in each shot.

If the goal is to minimize reactant volumes, one should measure the required flushing volumes (at a particular flow rate) in the system. If dead time is not an issue, a slower flow rate should be used to increase flushing efficiency. The same reaction used for determination of dead time can be used to determine flushing volume. In the example shown in **Fig. 5**, the flow rate for each reactant was set to 15 mL/s. Reactions were then measured at a variety of reactant volumes ranging from 20 to 125 μL per reactant, per shot. At 20 μL the amplitude of the reaction is barely detectable, indicating that the previous contents of the cuvet had not been adequately displaced with fresh reactants (in spite of the fact that the total volume delivered is greater than the dead volume). As reactant volume is increased, the amplitude increases. This increase in amplitude stops at 100 μL, indicating that this is the minimum volume required at this particular flow rate. This volume would be slightly lower at slower flow rates, and higher at higher flow rates, and should be measured for each system if sample is limiting.

2.4. Multimixing

Many newer stopped-flow reactors have more than two sample syringes, providing the ability to mix more than two samples in one reaction. The benefit of this arrangement includes varying ligand concentration without changing syringes and pre-incubation experiments in which two reactants are mixed, allowed to incubate, then mixed with a third reactant. **Figure 6** shows a set of data using multimixing to measure a bimolecular rate constant using only three syringes. This experiment shows ferrous horse heart Mb binding CO. Syringe 1 contains 1000 μ*M* CO, syringe 2 contains buffer alone (0.1 *M* KPO_4, pH 7.0), and syringe 3 contains 8 μ*M* Mb. All syringes contain 0.05 μ*M* sodium dithionite to reduce the heme iron and scavenge oxygen. The total reaction volume is 200 μL, and the concentrations of CO are provided by mixing the volumes in **Fig. 6B** for each shot. A few time-courses are shown in **Fig. 6**, and k_{obs} values for each CO concentration are plotted in the inset. Fitting a line to k_{obs} gives a slope of 0.6 $\mu M^{-1}s^{-1}$ and a y-intercept near zero. These values reflect the bimolecular rate constant for CO binding and a very slow CO dissociation rate constant (the actual value is 0.02 s^{-1}).

The ability of a particular stopped-flow system to carry out multimixing experiments requires multiple reaction syringes and the ability to deliver different volumes from each syringe. This can be achieved by either having independent control over the driving system for each syringe or, if the system has only one drive device, by varying the diameter of the reactant syringe. Independent drive systems for each syringe provide the most efficient and flexible mechanism for multimixing experiments.

2.5. Multiple Wavelength Data Collection

Many stopped-flow systems operating in absorbance mode allow the user to collect time-courses at multiple wavelengths. The benefit of having multiwavelength data is that the spectral signature of each reactant and any intermediates can be tracked more easily. Ideally, one could collect an absorbance spectrum at each time point. Several detection devices have been developed for this purpose and are available on commercial instruments including a *rapid scanning*

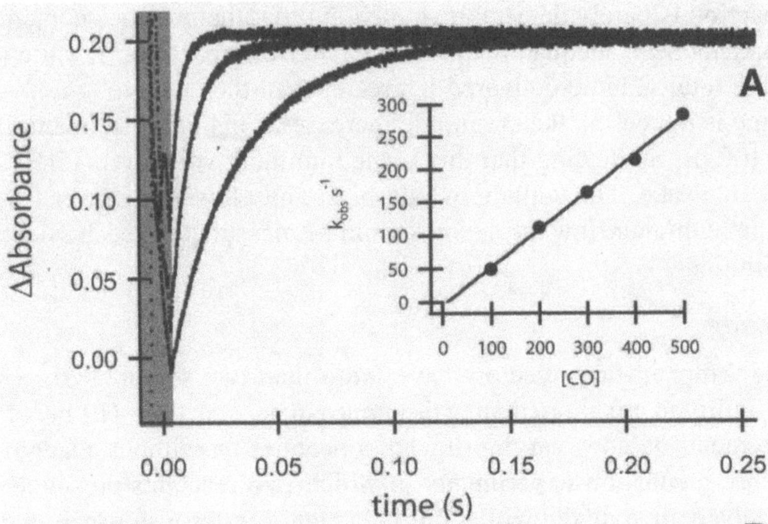

B

syringe 1 (CO) μL	syringe 2 (buffer) μL	syringe 3 (Mb) μL	resulting [CO] μM
0	100	100	0
20	80	100	100
40	60	100	200
60	40	100	300
80	20	100	400
100	0	100	500

Fig. 6. (**A**) An example of a three-syringe multimixing experiment. The reaction is CO binding to ferrous Mb. This bimolecular reaction is carried out under pseudo first-order conditions in CO, and time-courses are observed at each concentration of CO (only those for 100, 200, and 400 μ*M* CO [after mixing] are shown for clarity). The observed rate constants at each (CO) are shown in the inset. The slope of these data indicate the bimolecular rate constant for the reaction (as described for **Fig. 2B**), but k_{rev} is in this case too small to be observed as a y-intercept. (**B**) This table shows the reaction volumes delivered from each syringe to achieve the (CO) in each reaction.

monochromator, charge coupled device (CCD), and diode array detectors. Alternatively (and at significantly lower cost), independently functioning CCD detectors can be interfaced with many stopped flow systems (*see* **Note 8**).

In the absence of detectors specifically designed for this purpose, one can manually collect time-courses while incrementing the wavelength used for detection. This will generate a *kinetic difference spectrum* as demonstrated in **Fig. 7**. The reaction of ferric Mb with azide is again used in this experiment. **Figure 7A** shows the absorbance spectra for Mb before and after azide bind-

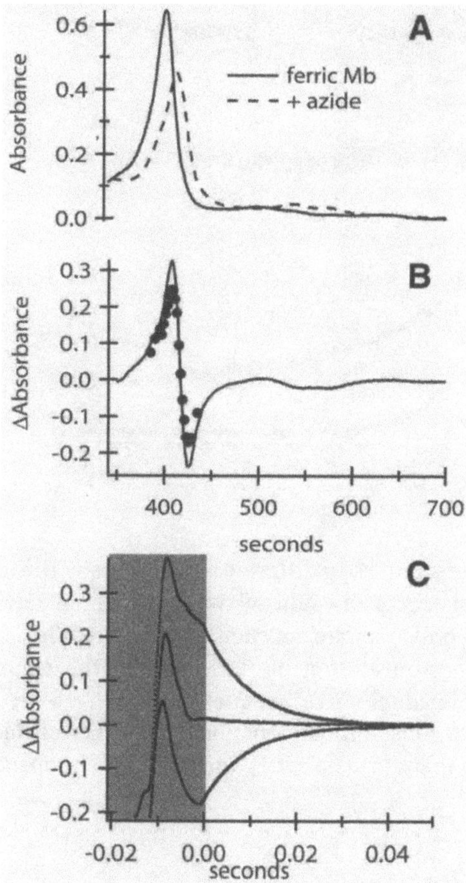

Fig. 7. A simple method for determining a kinetic difference spectrum. **(A)** The reaction of ferric Mb is represented by the absorbance spectra of each species (as in **Fig. 4A**). **(B)** The difference spectrum can be calculated from **(A)** by subtracting the spectra. **(C)** Time-courses can be collected at different wavelengths. Shown here are those at 409 nm (upper), 417 nm (middle), and 422 nm (lower). The amplitudes at several wavelengths are overlaid with the expected difference spectrum in **(B)**, showing that the kinetic difference spectra represents that expected for the reaction.

ing. **Figure 7B** shows the difference spectrum associated with this reaction. To collect the kinetic difference spectrum, the amplitude of the reaction was measured as a function of wavelength by manually varying the monochromator setting and measuring the reaction at each wavelength (**Fig. 7C**). These amplitude values are overlaid with the observed difference spectra in **Fig. 7B**, showing a close relationship between the expected and observed values.

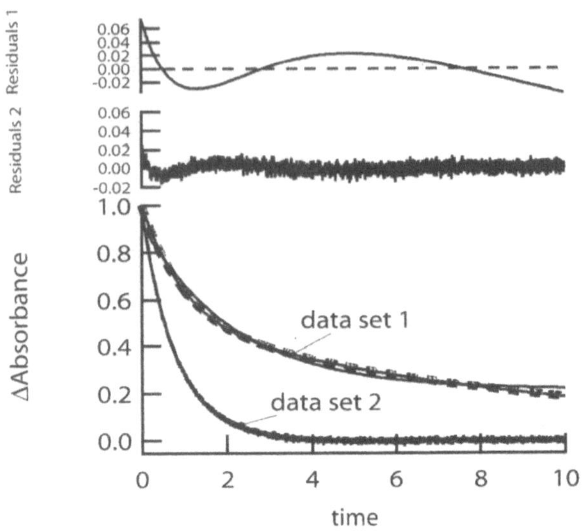

Fig. 8. The fitting of time-courses to different sets of data is illustrated. In data set 1, the time-course is calculated as the sum of two exponential rate constants (solid line). If these data are fit to only one rate constant, the solid residuals line is generated (top graph), which shows nonrandom deviation from the time-course. If two exponentials are used, the residual curve is equal to zero. Data set 2 is real binding data that represents one rate constant, but contains noise. The residuals plot for these data is much closer to zero than for data set 1, where two rate constants was required but only one was used.

2.6. Data Analysis

There are two aspects of data analysis that one must address in ligand-binding experiments. First, rate constants must be extracted from time-courses. Second, rate constants and their dependence on reactant concentrations are interpreted based on a reaction mechanism. Alternatively, time-courses can be simulated from reaction mechanisms using integrated or numeric methods to calculate the time dependence of each reactant.

Extracting rate constants from kinetic experiments uses nonlinear least squares fitting of observed time-courses to exponential functions such as **Eq. 3**. This is usually accomplished using a graphics software package like Igor Pro (www.wavemetrics.com/) or one of many that are commercially available. In some cases, multiple exponential terms or distributions of exponents are required to fit more complex data. It is best to start with the simplest possible method for analysis, because a multiple exponential expression will always fit just as well or better to the time-course than a single exponent. Residuals calculated as the difference between the fitted curve and the time-course can be used to evaluate goodness of fit. If a fitted curve describes the time-course well, the residuals should be small and evenly distributed (**Fig. 8**). A poorer fit gives larger residual values that usually vary with time. To decide if multiple terms are needed in a fit, start with the simplest expression and plot the residual curve. Then try a more complex model. If the residual values decrease and become more random in sign, the more complex fitting model is probably required.

Relating your observed rate constants to a model requires a reaction scheme. As described in **Subheading 2.1.**, some reaction schemes can be analyzed by integration to calculate time-courses for each reactant, or by approximations to yield useful equations predicting rate constants for the reaction. However, many reaction differential equations do not have convenient integrated solutions or approximations. Under these conditions, analysis of the reaction by numeric integration is required. Several excellent computer programs for this purpose are available including KINSIM and FITSIM ***(12,13)***. The strength of numeric integration is that it is easy to model any reaction scheme and compare the expected time-courses or reaction rates to observed data. This method can also be used for data fitting if enough parameters are known. One weakness of numerical analysis is that it is sometimes abused because so many parameters are available for manipulation ***(14)***. Whether or not integrated expressions are available, the best usage of kinetic data for learning about reaction mechanisms comes from a combination of careful extraction of rate information from time-courses, fitting of rate constants to those predicted from models, and comparison of simulated and expected time-courses.

3. Notes

1. **Reference** ***3*** provides a detailed review of many methods along with a historical perspective on their development. The range of methods employed to measure ligand binding can be divided into two general groups: *mixing methods* and *relaxation methods*. Mixing methods are the most obvious, as the two reactants are simply combined to initiate the reaction. Relaxation methods require a rapid method for perturbing the equilibrium in a system using some environmental parameter that could include, pressure, temperature, or a flash of light or other pulse of electromagnetic radiation. The general idea of relaxation methods have led to many standard techniques including nuclear magnetic resonance (NMR), flash photolysis, and temperature-jump.
2. In Michaelis–Menten enzyme reactions the observed reaction velocity (V_{max}) depends on total enzyme concentration (E_T) as described by the following equation: $V_{max} = k_{cat}[E_T]$, where k_{cat} is the rate constant for catalysis once the substrate is bound (assuming the substrate dissociation reaction is slow). Therefore, when measuring the kinetics of enzyme catalysis, one can in most cases experimentally lower $[E_T]$ until observed velocities occur at manageable rates. The ability to

vary $[E_T]$ results from the fact that in most cases the breakdown of substrate or formation of products is observed to monitor the progression of reaction.

3. Major manufacturers are Applied Photophysics (www.photophysics.com/), Bio-Logic (www.bio-logic.fr/), Kintek (www.kintek-corp.com/), and Olis (www.olis-web.com). Many other manufacturers make stopped-flow modules designed for rapid mixing in their analysis equipment.
4. Detailed reviews of stopped-flow methods and reactions specific to absorbance ***(4)***, fluorescence ***(5)***, and CD signals ***(6)*** are readily available.
5. The velocity of a reaction is the change in signal as a function of time. To convert to units of concentration, one must know the expected value for the observed signal for both products and reactants. For absorbance measurements, the extinction coefficients for each must be known at the wavelength of observation. In many cases, for signals like fluorescence, this can be nonlinear and difficult to determine. Under most circumstances, complete time-courses will obey **Eq. 3**, from which the rate constant is extracted from the curvature of the data and not the absolute value of the amplitude of the reaction. Therefore, it is not necessary to know the signal associated for the starting and endpoint of the reaction in order to obtain a rate constant from a complete time-course.
6. It should also be noted that if the reverse rate constant is large enough to observe as an appreciable y-intercept in a plot of k_{obs} vs $[L]$, the reaction will not go completely to the ligand bound form at lower $[L]$ because the on- and off-rates are of similar values. Therefore, as $[L]$ is increased (and k_{obs} becomes $\gg k_{rev}$), the reaction amplitude will increase as the equilibrium fractional saturation approaches 1.
7. For example, if the flow rate is 15 mL/s, the cuvet volume is 0.1 mL, and there is 0.02 mL between the mixing point and the cuvet, the dead time will be approx 6.8 ms.
8. For example, the USB2000 spectrometer from Ocean Optics (www.ocean-optics.com/) can collect spectra every 3 ms and will interface with a personal computer through a USB port.

References

1. Gutfreund, H. (1969) Rapid mixing: continuous flow. *Meth. Enzymol.* **16,** 229–249.
2. Gibson, Q. (1969) Rapid mixing: stopped flow. *Meth. Enzymol.* **16,** 187–228.
3. Hammes, G. (1974) in *Techniques of Chemistry, Vol. VI, Part II*, Wiley-Interscience, New York, pp. 1–61.
4. Wilson, M. and Torres, J. (2000) In *Spectrophotometry & Spectrofluorimetry* (Gore, G., ed.), Oxford University Press, Washington DC, pp. 209–239.
5. Gore, G. and Bottomley, S. (2000) in *Spectrophotometry & Spectrofluorimetry* (Gore, G., ed.), Oxford University Press, Washington DC, pp. 241–264.
6. Rodger, A. and Carey, M. (2000) in *Spectrophotometry & Spectrofluorimetry* (Gore, G., ed.), Oxford University Press, Washington DC, pp. 265–281.
7. Espenson, J. H. (1995) *Chemical Kinetics and Reaction Mechanisms*. McGraw-Hill Series in Advanced Chemistry, McGraw-Hill.

8. Pilling, M. and Seakins, P. (1997) *Reaction Kinetics*, Oxford University Press, Washington DC.
9. Capellos, C. and Bielski, B. (1972) *Kinetic Systems*, John Wiley & Sons, Inc., New York.
10. Trent, J. T., III., Hvitved, A. N., and Hargrove, M. S. (2001) A model for ligand binding to hexacoordinate hemoglobins, *Biochemistry* **40,** 6155–6163.
11. Brancaccio, A., Cutruzzolá, F., Allocatelli, C. T., Brunori, M., Smerdon, S. J., Wilkinson, A. J., Dou, Y., Keenan, D., Ikeda-Saito, M., Brantley, R. E., Jr., and Olson, J. S. (1994) Structural factors governing azide and cyanide binding to mammalian metmyoglobins, *J. Biol. Chem.* **269,** 13,843–13,853.
12. Zimmerle, C. and Frieden, C. (1989) Analysis of progress curves by simulations generated by numerical integration, *Biochem J.* **258,** 381–387.
13. Barshop, B., Wrenn, R., and Frieden, C. (1983) Analysis of numerical methods for computer simulation of kinetic processes: development of KINSIM-a flexible, portable system. *Anal. Biochem.* **130,** 134–145.
14. Johnson, K. A. (1998) Advances in transient-state kinetics, *Curr. Opin. Biotech.* **9,** 87–89.

17

Circular Dichroism Spectroscopy for the Study of Protein–Ligand Interactions

Alison Rodger, Rachel Marrington, David Roper, and Stuart Windsor

Summary

Circular dichroism (CD) is the difference in absorption of left and right circularly polarized light, usually by a solution containing the molecules of interest. A signal is only measured for chiral molecules such as proteins. A CD spectrum provides information about the bonds and structures responsible for this chirality. When a small molecule (or ligand) binds to a protein, it acquires an induced CD (ICD) spectrum through chiral perturbation to its structure or electron rearrangements. The wavelengths of this ICD are determined by the ligand's own absorption spectrum, and the intensity of the ICD spectrum is determined by the strength and geometry of its interaction with the protein. Thus, ICD can be used to probe the binding of ligands to proteins. This chapter outlines protein CD and ICD, together with some of the issues relating to experimental design and implementation.

Key Words: Circular dichroism; proteins; chirality; ligand binding; induced circular dichroism.

1. Introduction

A key feature of any biological system is its chirality or asymmetry or handedness: a chiral molecule has a mirror image that is not superposable on itself. This means the two mirror images cannot be rotated so that they look exactly the same. Macroscopic as well as smaller scale chirality is ultimately dependent on the molecular level. Because many molecules in biological systems are chiral and are present in only one enantiomeric form, the macroscopic structures they build are also chiral. Molecular chirality is perhaps most obvious with a helical molecule such as the double helical structure of B–DNA, but it is true of all proteins and nucleic acids.

From: *Methods in Molecular Biology, vol. 305: Protein–Ligand Interactions: Methods and Applications*
Edited by: G. U. Nienhaus © Humana Press Inc., Totowa, NJ

Circular dichroism (CD), which is the difference in absorption of left and right circularly polarized light, is probably the simplest technique for non-destructively providing solution phase structural information about chiral molecules. Many ligands (usually small molecules that bind to a macromolecule) in biological systems are also chiral, in which case they have their own CD spectrum, which will probably be perturbed when the ligands bind to a protein. If a ligand is achiral, then it will have no intrinsic CD but will gain an induced CD (ICD) signal in its transitions when it binds to a protein. It is this ICD signal that contains the information about the asymmetry of the protein–ligand interaction. In this chapter, we will focus on how to measure CD spectra of proteins and protein–ligand complexes and how to analyze the data.

1.1. Protein Absorbance Spectroscopy

In order to understand CD spectroscopy and use the data intelligently, it is essential that one measures the absorbance spectrum of one's sample since this shows where to expect CD signals. The Beer–Lambert law for the absorption of light by a sample of concentration C is

$$A = \varepsilon C \ell \tag{1}$$

where ℓ is the length of the sample through which the light passes, and ε is known as the molar extinction coefficient and depends on the wavelength at which the absorbance is being measured. If ℓ is measured in cm and C in M = mol dm^{-3}, then ε has units of mol^{-1} dm^3 cm^{-1}. The Beer–Lambert law is valid as long as the spectrometer can measure the intensity of photons passing through the sample (i.e., the concentration is not so large that essentially all photons are absorbed), *and* there are no concentration-dependent intermolecular interactions.

In the case of peptides and proteins, the spectroscopy of the amide bonds, the side chains, and any prosthetic groups (such as haems) determines the observed UV/visible absorption spectra with their intensities and wavelengths often being affected by the local environment of the groups. UV spectra of proteins are usually divided into the *near* and *far* UV regions. The near UV in this context means 250–300 nm and is often described as the aromatic region because of the absorption of the aromatic amino acids, though transitions of disulfide bonds (cysteine–cysteine bonds) also contribute to the total absorption intensity in this region. The far UV (< 250 nm) is dominated by absorption as a result of the peptide backbone of the protein (*see* **Fig. 1**), but transitions from some side chains also contribute. The far UV absorbance of a protein is typically of the order of a factor of 100 stronger than the near UV absorbance.

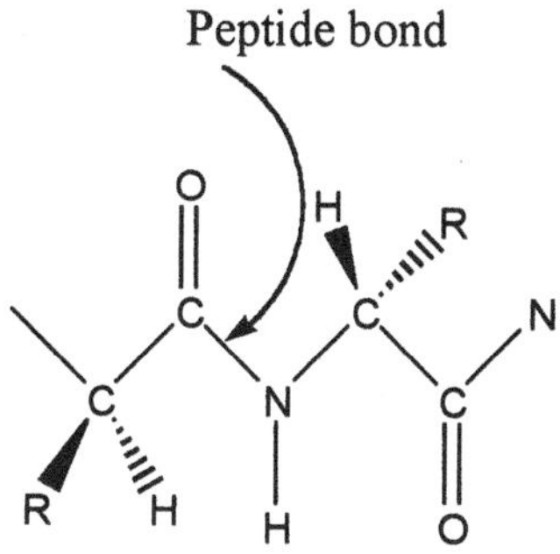

Fig. 1. Peptide units of amino acids linked to form a peptide backbone. R denotes side chains.

1.2. Protein Circular Dichroism

CD may be defined as the difference in absorption of left and right circularly polarized light:

$$CD = \Delta A = A_l - A_r \tag{2}$$

where the subscripts indicate the polarization of light incident on the sample. The CD of macromolecules differs from that of small molecules in that the spectroscopy of the component units (amino acids, in the case of proteins) is at least partly understood: the form of the CD spectra result largely from the arrangement of the units in space, and the coupling between the transitions of those units. The CD of proteins is most commonly used in one of two ways: (i) to probe the secondary structure of the protein itself, and (ii) to probe its binding interaction with small molecules (usually referred to as ligands). Even though both of these applications are in practice largely empirical, it is useful to have some understanding of the origin of the signal being measured.

Proteins are long chains of amino acids. If the side chain on an amino acid is not H (i.e., not glycine), then the tetrahedral carbon of that amino acid is a chiral center and we should expect a CD signal in transitions of the neighboring amide groups and side chains. If, however, there is free rotation about the bonds of the main chain, the observed amide $n \rightarrow \pi^*$ (a weak transition with $\varepsilon \sim 100$ mol^{-1} dm^3 cm^{-1} occurring at 210–230 nm) and $\pi \rightarrow \pi^*$ (a stronger transition with $\varepsilon \sim 7000$ mol^{-1} dm^3 cm^{-1} centered at 190–200 nm) ***(1,2)*** CD will be relatively small. Thus, a truly random or denatured protein will have only a small CD signal at the wavelengths accessible to most CD machines. The most stable conformation of a protein under physiological conditions, however, is not random but composed of well-defined structures that give macro-chiral units (including the well established α-helix and β-sheet) with significant CD intensities that add together to give the observed spectrum. A wide range of

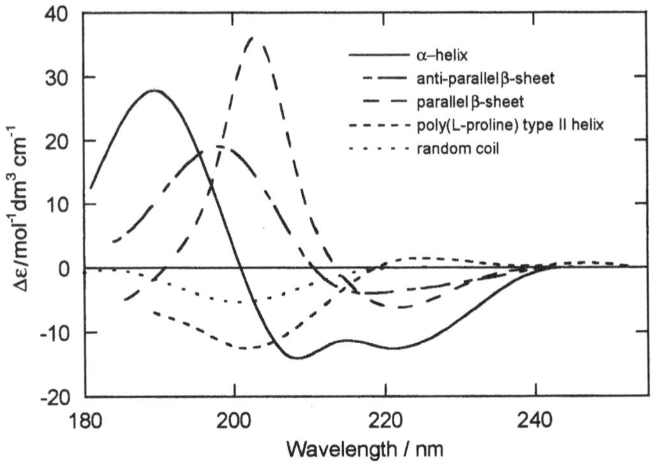

Fig. 2. Backbone (i.e., 180–250 nm) CD spectra for different protein secondary structural motifs ***(14,15)***. α-helix (solid line, average of poly(γ-methyl-L-glutamate) in F_6iPrOH and poly(L-alanine) as a film), antiparallel β-sheet (dashed line, BOC (L-alanine)7OMe as a film), parallel β-sheet (dash dot dot line, BOC (L-valine)7OMe as a film), poly(L-proline) II type helix (dotted line, systemin in 10 m*M* phosphate buffer at 5°C, pH 7.0), and random coil (dash dot line, collagen at 45°C in 0.01 *M* phosphate buffer, pH 3.5).

fitting programs are available to deconvolute the experimental data into percentages of the structural motifs *(3)*. Some work in terms of the spectra of a set of identified structural motifs, others use a basis set of real protein spectra where the percentage of different structural units is known. The approach of Johnson et al. *(4)* uses a large basis set of such proteins, chooses a subset of these randomly until a good fit is obtained. Almost all approaches will give a good estimate of α-helix content *if* the protein concentration is known (*see* **Note 1**). This is because the α-helix profile is the most distinct and also largest in magnitude as illustrated in **Fig. 2**. The distinctive shape arises from the coupling of the $\pi \rightarrow \pi^*$ transition moments in each amide chromophore and results in a component at about 208 nm which contributes to the characteristic α-helix CD spectrum. Good estimates will also be obtained for proteins without α-helices. Reasonable estimates will be obtained for mixed structure proteins, though here accurate values of the amino acid residue concentration are essential.

1.3. Units of Circular Dichroism Spectroscopy

If equal intensities of left and right handed circularly polarized light (i.e., linearly polarized light) were passed through a CD active sample at the same time, the result of differential absorption would be slightly elliptically polarized light. For historical reasons many CD spectropolarimeters therefore produce a CD spectrum in units describing this degree of *ellipticity*, θ, in millidegrees even though they measure differential absorbance (ΔA). The conversion between these two is ***(3)***:

$$CD = \Delta A = \frac{4\pi\theta\,(\text{deg})}{180 \ln 10} = \frac{\theta\,(\text{mdeg})}{32{,}982} \tag{3}$$

The CD Analog of the Beer–Lambert Law,

$$CD = \Delta A = \Delta\varepsilon C\ell \tag{4}$$

where $\Delta\varepsilon$ is the differential molar extinction coefficient (in units of $\text{mol}^{-1}\,\text{dm}^3\,\text{cm}^{-1}$), C is concentration in mol dm^{-3}, and ℓ is the path length of sample through which the light passes, in cm. This allows conversion to molar absorbance (or if millidegrees have been retained, molar ellipticity). In this context, it is important to have clear in one's mind what concentration is being used. For example, protein concentrations are sometimes in mg/mL, sometimes in moles of protein molecules per dm^3, and sometimes (as for protein structure fitting programmes) in moles of amino acid residues per dm^3.

2. Materials

2.1. Samples and Solvents

The materials for protein–ligand CD spectroscopy are most obviously the protein and ligand samples of interest. In addition, there is the solvent in which the sample is prepared. For proteins this usually means a buffer, as many proteins need buffers to retain their structure. However, if the buffer of choice has an absorbance above 1 at the concentrations to be used anywhere in the wavelength range of interest for CD, then the buffer will be absorbing most of the light and the quality of your CD spectrum will be poor or even rubbish. Phosphate is a good spectroscopy buffer; Tris is satisfactory; cacodylate is spectroscopically good but is an arsenic salt; acetate is far from ideal and may limit your accessible range to above approx 215 nm; and any buffers with aromatic groups are impossible. It must be remembered that the solvent and buffer can sometimes affect protein structure, and therefore the CD spectra obtained. Even if there are no structural changes, changing the chemical environment can affect electronic transitions, and therefore also affect the CD spectra.

2.2. Instrument Calibration

It is important to calibrate the CD machine for both intensity and wavelength. It is straightforward to carry out a check on the state of calibration of a CD instrument. If the intensity calibration is outside specification, it is usually easy to reset the intensity yourself, but it is more difficult to reset the wavelength, and a service engineer should be called.

Wavelength calibration (*see* **Note 2**) requires a wavelength standard. One option is a solid neodymium filter, which usually comes with the CD spectrometer. The absorbance maximum of this should be 586 ± 0.8 nm. Another is a Holmium Oxide filter, or solution, which has a characteristic spectrum across the near UV and visible regions. The absorbance maximum in a CD machine can usually be determined by determining where the gain voltage on the photomultiplier tube (often called the high tension voltage) is maximum.

Intensity calibration (*see* **Note 3**) is conventionally carried out using aqueous ammonium d-10-camphor sulfonate (ACS), which has been related to the established hydroscopic primary standard camphor sulfonic acid (CSA) *(5,6)*. ACS is available from a number of suppliers (including Katayama Chemical Co. 05-1251), although care should be taken to ensure its enantiomeric as well as chemical purity. The concentration of ACS usually used is 0.06% w/v, i.e., 60.00 mg in 100 mL or 6.000 mg in 10 mL, which should give a CD signal of 190.4 ± 1 mdeg at 290.5 nm (the CD maximum). (Note: CSA has a different molecular weight from ACS so a 0.06% solution has different molar concentrations, and hence different CD signals for the two standards.) It is important to weigh the solid material accurately (to 4 significant figures). For 6 mg this requires a 6 figure balance. The standard solution can be stored at 4°C. However, the storage time seems to be dependent on the container in which it is stored and storage of longer than 2 wk may be problematic.

2.3. Path Length Calibration

In order to use and compare CD data, it is important that the path length of the cell is measured—often the nominal path length of a cell is far from the actual path length. This is particularly important for short path length cells often used in the far UV (secondary structure) region. One approach is to use an aqueous potassium chromate solution of known concentration (*see* **Note 4**). The Beer–Lambert Law can then be used to calculate the path length from a measurement of absorbance. For 0.01 mm path length, a 0.2 *M* potassium chromate solution is required. For longer path lengths, a more dilute solution is required. To prepare a 0.2 *M* potassium chromate solution, accurately weigh 0.971 g potassium chromate (Aldrich 21,661-5, 99% ACS reagent) and transfer it to a 25 mL volumetric flask. Add approx 20 mL water and one pellet of potassium hydroxide (e.g., AnalaR BDH 102104V). Make the solution up to volume and mix well. This solution may be stored indefinitely at 4°C. If lower concentration samples are prepared by diluting the 0.2 *M* solution the potassium hydroxide is also diluted, and so the solution stability is reduced.

3. Methods

3.1. Collecting Circular Dichroism Data

Measuring a CD spectrum is a routine procedure assuming one has access to a CD spectropolarimeter. If your sample gives a good UV-visible absorbance spectrum then it is highly likely that (if it is chiral) you will get a good CD spectrum. The essential features of a CD spectrometer are a source of (more or less) monochromatic left and right circularly polarized light and a means of detecting the difference in absorbance of the two polarizations of light. CD machines are much more expensive than UV machines, as one is typically expecting absorbance differences of the order of 10^{-3} to 10^{-4} to be accurately measured against a background absorbance of just less than 1. For protein CD the spectropolarimeters also need to be nitrogen purged, not only to prevent the high energy lamps producing ozone and therefore damaging the delicate optics within the instrument, but also to avoid having O_2 in the sample compartment absorbing the incident radiation and limiting the lowest wavelengths that can be measured. In practice this means a moderate nitrogen flow rate (3–5 cm^3/min) at all times, with an increase to 20 or more cm^3/min when collecting data below 190 nm. **Subheading 3.1.1.** describes the important steps you need to consider while making CD measurements

3.1.1. Before You Start

In order to collect good quality and reliable CD data you should ensure that your instrument is properly calibrated both in terms of wavelength (*see* **Note 2**) and intensity (*see* **Note 3**). You should also carefully select a sample cell that is suitable for CD measurements (*see* **Note 5**), and is an appropriate path length for the concentration of the sample that you are measuring (*see* **Note 6**). Because the actual path length of cells can differ significantly from the nominal values, you should always ensure that you calibrate the path length of the cell that you are using (*see* **Note 4**). For fixed path length sealed cells, it is useful to calibrate all of your cells, and keep a calibration table handy for easy reference.

3.1.2. Sample Absorbance

Because the wavelength and absorbance ranges are the same for both normal absorption and CD it is advisable to run a normal absorption spectrum of the sample for which you wish to measure the CD spectrum first. *Always* leave the reference beam of the absorbance spectrometer empty for this experiment, because we are concerned with the total absorbance of the sample including any achiral molecules in the buffer. It is preferable to have the maximum absorbance less than 1.5 (and certainly less than 2).

3.1.3. Wavelength Range

CD spectrometers usually scan from longer wavelengths to shorter ones. Ensure you collect data for at least 20 nm to the longer wavelength side of any absorbance band. Select a wavelength range starting so that there is at least 20 nm of zero absorbance beyond the normal absorption envelope(s) of interest. When the baseline spectrum (*see* **Subheading 3.1.4.**) is subtracted from the sample spectrum, the region outside the absorption envelope should be flat. If it is not then this probably means either there is a very weak absorbance band that has a large dissymmetry factor, and hence large CD signal compared with its normal absorbance intensity, or, more probably, there is some light scattering by the sample. Sources of light scattering include dirty cuvets (inside or outside), undissolved sample, condensation of samples, and particulate samples (*see* **Note 7**). Given that the wavelengths of light being scattered are less than 1 μm, one does not necessarily expect to be able to see by eye the presence of such particles.

3.1.4. Baseline and Zeroing

A machine baseline (i.e., CD spectrum of air) measured on a standard CD spectrometer will not be flat as the optical components in the instrument are birefringent. This can be ignored by storing a machine baseline in the instrument. However, the cuvet used for an experiment will also have its own CD spectrum—CD matched cuvets usually have slightly different intrinsic spectra. So always collect a baseline spectrum of your solvent/buffer under the same conditions as the sample spectrum using the same cuvet in the same orientation with respect to the light beam. Subtract the baseline spectrum from the sample spectrum to produce the final CD plot. Although one can often have the cuvet baseline automatically subtracted by the software, this may not be a good idea for small signals or sticky samples that could be retained on the cuvet (*see* **Note 7**)—it is better to see any problems in the baseline spectrum. It is often the case that even when the baseline is subtracted, the CD is not exactly zero outside the absorption envelope. However, if it is flat outside the absorption envelope, the spectrum may be zeroed by adding or subtracting a constant (either within the spectrometer software or using your chosen data plotting software).

3.1.5. Signal-to-Noise Ratio

The signal-to-noise ratio in a CD spectrum increases with: $\div n$, where n is the number of times the spectrum is accumulated (and the data averaged); $\div t$, where t is the time over which the machine averages each data point; and $\div I$ where I is the intensity of the light beam. The actual definition of t depends on whether your instrument moves to a wavelength and sits there to collect data or whether it continuously scans. The best way to deal with this issue is to try some scans with different time constants or response times to see if the spectrum is distorted. I is usually influenced by bandwidth and also wavelength, because the lamp does not have uniform intensity at all wavelengths. In fact, when spectra become noisy one of the reasons is often that the lamp is old and its light intensity decreased. Most CD spectrometers have both short timescale (millisecond to minutes) and long timescale (minutes to hours) baseline variations. To average over short timescale fluctuations collect a number of spectra or have a longer response time. Longer timescale fluctuations are usually dealt with by alternating collection of sample and baseline spectra or by assuming the fluctuation is a wavelength independent shift of the zero point. While appearing to be wishful thinking, the assumption of linear baseline drift seems to be valid especially with more recent instruments.

3.1.6. Parameter Sets

CD spectrometers give the operator considerable control over response time (τ), scan speed (s), bandwidth (b) [the wavelength range (error) of the incident light], and data interval (d). To optimize signal-to-noise effects (*see* **Subheading 3.1.5.**) t should be selected to be as large as possible subject approximately to

$$\tau \times s \leq \frac{b}{2}$$

If τ is too long for the chosen s and b, then maxima of peaks (both positive and negative) will be cut off and their wavelengths shifted. A control scan using

$$\tau' = \frac{\tau}{2} \text{ (or } s' = 2s\text{)}$$

should be used to check that spectra are not being distorted by the chosen parameters. The data interval determines the wavelength interval between data points. This may or may not have implications for the time a scan takes depending on whether the instrument operates in a stepped scan mode or continuous scan mode.

Scan speeds of 50 nm/min, $\tau = 1$ s, $b = 1$ nm, and a data step of 0.5 nm seem to be a good starting point as a parameter set for most protein–ligand experiments where the samples have the broad band shapes usually found for protein samples and most ligands. It is often advisable to perform a fast preliminary scan to determine whether there is any point in collecting an accurate spectrum. The most significant consideration for reliable data is whether the sample has too high an absorbance—this is summarized by the high-tension voltage (or equivalent label) of the photomultiplier tube. It may also be necessary to select an appropriate sensitivity scale—the need for this will be indicated by a flat straight line across peaks in your spectrum.

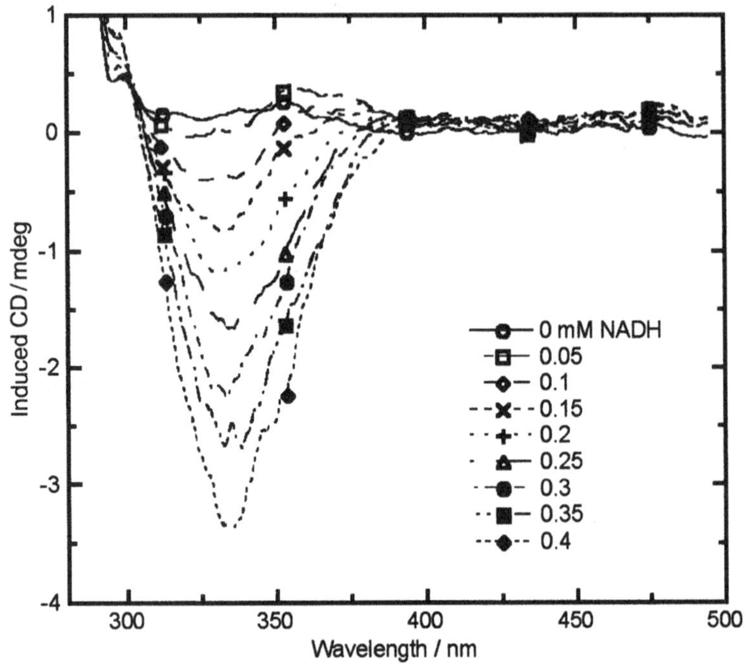

Fig. 3. CD of lactate dehydrogenase (LDH, 22.5 μ*M*) with increasing concentrations of NADH. LDH concentration was kept constant by adding equal volumes of NADH (from a 10 m*M* stock solution) and LDH from a 45 μ*M* stock solution. The measured spectrum is the sum of the intrinsic spectra of the two components, so there is no evidence of interaction by CD.

3.1.7. Linearity

It is good practice to check that the CD signal is proportional to sample concentration by running a spectrum of a sample and a diluted sample and checking that the signal scales with concentration. If it does not then you have evidence of solute–solute interactions in your sample or too high an absorbance for the instrument to be able to cope with.

3.2. Protein–Ligand Interactions Probed by Circular Dichroism: Titration Experiments

When a ligand binds to a protein molecule, its transitions will acquire an ICD spectrum. If it is itself achiral then this will change its CD spectrum from zero to something; if it is chiral then its ICD is the difference between its own CD spectrum and the spectrum in the presence of the protein. **Figure 3** shows a titration series where it appears that a large ICD is being observed when NADH binds to lactate dehydrogenase. However, the ligand is chiral and we are only measuring the ligand's intrinsic CD. The protein will also acquire an ICD (though with a large protein it may only be a very small change). The ICD in the protein regions of the spectrum are usually a combination of protein and ligand ICD as most ligands have absorption intensity below 300 nm. A number of examples are given in **Subheading 3.3.**

If there is just one binding mode (or equal occupancy of more than one) between protein and ligand then the magnitude of the ICD will be proportional to the concentration of protein–ligand complex (*see* **Subheading 3.3.**). The challenge is to determine the proportionality constant, especially if high or low loadings cannot be achieved experimentally. If there is a succession of binding modes being occupied, the ICD spectral shape usually changes.

Apart from the challenges of analyzing the data once collected, there are a number of issues that must be considered for ligand binding experiments. One typically proceeds by titrating (or adding) one solution to another. There are an infinite number of ways of doing this. Conceptually, the simplest is to make up a series of independent solutions. However, this is usually the least accurate method as pipeting errors are most significant. Alternatively, one can add very small volumes of a highly concentrated stock solution of ligand or protein, and either ignore the dilution effect or account for it assuming it has no structural effect on the protein. A simple way to avoid dilution effects is as follows. Consider a starting sample that has concentration x M of species X. Each time y μL of Y is added, also add y μL of a $2x$ M solution of X. The concentration of X remains constant at x M. An infinite number of variations on this theme are possible, one of which is to add a solution already containing x M of X. One must always consider whether the order of addition of components or the stock solution concentrations (particularly local high concentrations during addition of the titrant) in such an experiment has any effect on the nature of the interaction.

Whether the protein or the ligand concentration is kept constant during a titration series depends on a number of issues. If the protein CD spectrum changes as a function of its own concentration then it is usually wise to keep it constant. However, if it does not, by keeping the ligand concentration constant one can more easily monitor any changes in ligand ICD as a function of protein:ligand ratio. Any ICD in the protein region of the spectrum is likely to be composed of both protein and ligand ICD spectra as it is hard to find ligands without any spectroscopy below 300 nm.

3.3. Examples of Protein-Ligand Circular Dichroism Spectra

3.3.1. Protein–Solvent Interactions

An illustration is given in **Fig. 4** of the effect on a protein CD spectrum of a ligand with no spectroscopy. In this case solvent induced conformational

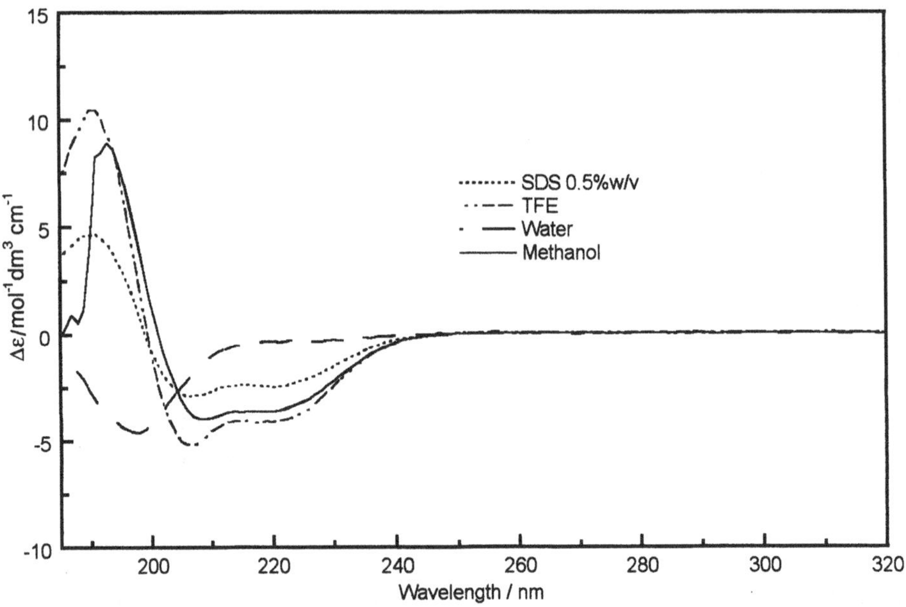

Fig. 4. CD of a peptide (MSLSRRQFIQASGIALCAGAVPLKASA) in different solvents *(16)*. Water shows a random coil structure. The TFE spectrum is approx 50% α-helical.

changes in a short peptide (MSLSRRQFIQASGIALCAGAVPLKASA in single letter code) are illustrated. A change from approx 50% α-helical (in TFE) to random coil (in water) is observed—this is an extreme change and not to be expected for most ligand binding.

3.3.2. Protein–Protein Interactions

In many instances, the ligand that binds to a protein is in fact another protein. **Figure 5** shows the CD spectra measured for F-actin (the polymerized form of actin), a myosin molecule (S1) and the mixture of the two. The theoretical spectrum calculated from adding the actin and myosin spectra is also shown. The difference between this theoretical spectrum and the experimental mixture is small but clear and reproducible, showing that either the actin or the myosin secondary structure is perturbed by their interaction or there is some electronic perturbation of their transitions. It is not possible to distinguish these possibilities.

3.3.3. ICD of a Protein–Ligand System

Bovine serum albumin binds almost anything that can be put into solution with it, however, one of its main roles is as a steroid transporter protein in

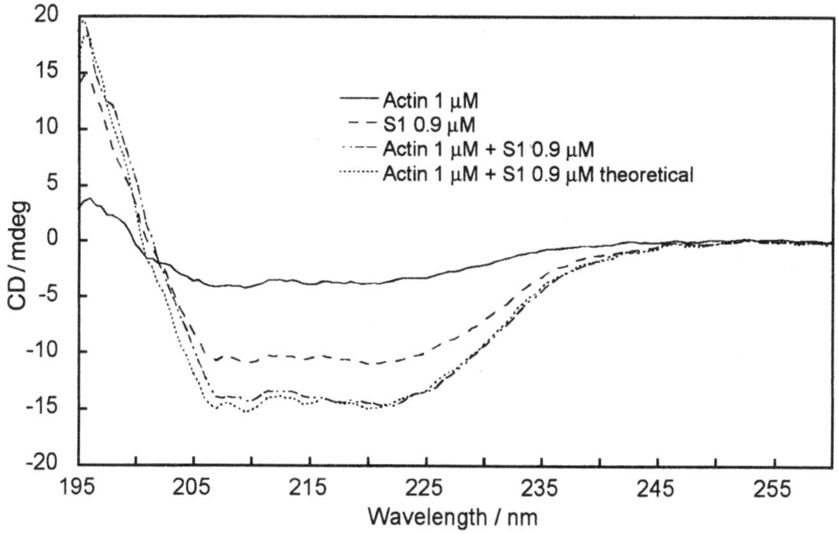

Fig. 5. CD spectra of F-actin (1 μ*M* in actin monomer), S1 myosin (0.9 μ*M*), the mixture of the two proteins and the sum of the two independent spectra. All samples were made up in 5 m*M* Tris buffer, pH 8.0, 20 μ*M* $CaCl_2$, 1 m*M* ATP, 1 m*M* $MgCl_2$, and 50 m*M* KCl. Pathelength used was 1 mm.

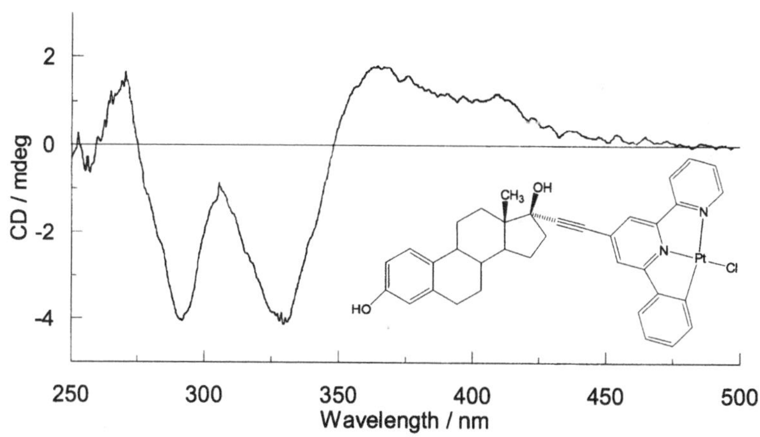

Fig. 6. ICD for PtEEtpy (30 μ*M*, structure is shown in the figure) bound to bovine serum albumin (30 μ*M* in 5 m*M* sodium cacodylate buffer).

blood. An example of the CD it induces in a steroid coupled to a platinum metal complex is shown in **Fig. 6**. The signal above 300 nm is caused by the perturbations to the ligand spectroscopy. That below 300 nm is a combination of protein and ligand ICD signals.

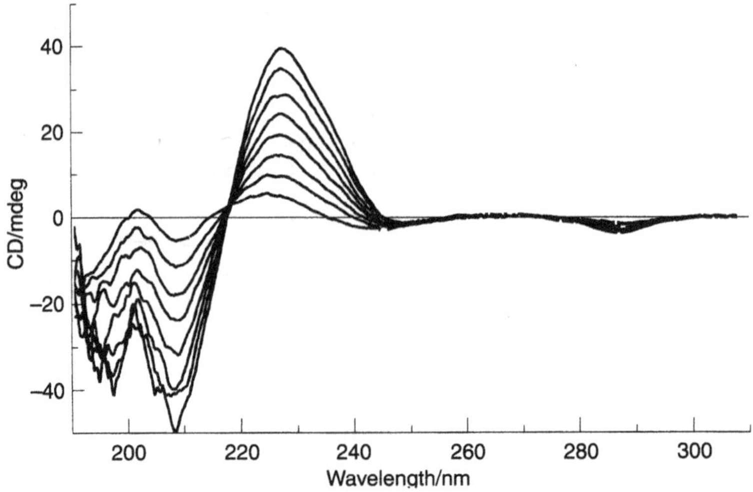

Fig. 7. ICD measured when increasing volumes of a concentrated solution of vancomycin containing 25 μ*M* ristocetin are added to a 25 μ*M* ristocetin solution. The intrinsic CD spectra of the peptides have been subtracted to give only the ICD. The 0 μ*M* vancomycin spectrum is therefore a flat line and has been omitted. Vancomycin concentrations (increasing ICD magnitude) are: 25 μ*M*, 50 μ*M*, 75 μ*M*, 100 μ*M*, 125 μ*M*, 150 μ*M*, 175 μ*M*, 200 μ*M*. (Data from **ref.** ***17***).

3.3.4. Peptide–Peptide Titration Series

Vancomycin and ristocetin are glycopeptide antibiotics that prevent cross-linking and transglycosylation during bacterial cell wall formation. Noncovalent dimerization plays a key role in their activity, and CD can be used to give binding constants for such an interaction. The data in **Fig. 7** are for the equilibrium

$$V + R \rightleftharpoons V - R$$

where V denotes vancomycin, R ristocetin and V–R their noncovalent complex. The ICD is assumed to be proportional to the concentration of dimer. **Figure 7** shows the data collected as increasing concentrations of vancomycin are added to a solution where the concentration of the ristocetin is maintained at 25 μ*M*. The equilibrium constant for the dimerization can then be calculated to be $K_{\text{dimerisation}} = 20 \pm 5\ (\text{m}M)^{-1}$ (*see* **Notes 8–10** for methods for determining equilibrium binding constants).

3.4. Other Techniques

CD measurements of protein–ligand binding are in fact not always sensitive enough to detect ligand binding, if there is little change in the protein secondary structure. The near UV CD is only perturbed if the ligand binds near an aromatic residue; and the ligand ICD signal may be very small. Thus, ligand binding may be more effectively probed by other techniques such as fluorescence or even absorbance spectroscopy. Measuring fluorescence polarization anisotropy (FPA) may in fact achieve some of the advantages of CD spectroscopy in that it is the difference between two polarizations of radiation so much of the background signal gets subtracted off.

CD is particularly suited to probing chiral molecules because achiral effects are canceled out. Vibrational circular dichroism (VCD), Raman optical acitivity (ROA), and optical rotatory dispersion (ORD) do the same thing. VCD and ROA are attractive as they probe the vibrations of the molecule, which can usually be readily assigned and there may well be many more transitions than in UV–CD providing complementary information. Both VCD and ROA currently require much higher sample concentrations and generally much longer data accumulations than UV–CD, so they are not nearly as widely available. ORD is related to CD by the Kramers–Kronig transformation, so in principle contains the same information. However, the information is much harder to extract and data over the full wavelength range is at least in principle required.

For fibrous proteins such as actin (*see* **Subheading 3.3.2.**) the binding of a ligand is usually readily probed by flow oriented linear dichroism *(7)* (the difference in absorption of light linearly polarized along the orientation axis and perpendicular to that axis) because the ligand is invisible until it is oriented by the fiber to which it binds.

4. Notes

1. Any quantitative application of CD spectroscopy to protein structure fitting or ligand binding needs a fairly accurate estimate of protein concentration. For a well-known protein the extinction coefficient at 280 nm may well be available on one of the many protein data base websites or in the literature. It is then a fairly simple matter to take a sample (probably of approx 1 mg/mL), measure its absorbance spectrum in a 1 cm cuvet, and use the Beer–Lambert law. However, you should note that the extinction coefficient from a database is almost certainly a theoretical one determined from adding the contributions of aromatic amino acids and disulfide bonds. Thus, it ignores environmental effects. At the very least one should completely denature one's protein to use the theoretical value, then determine a native ε value for later use. Denaturing the protein is usually carried out with high concentrations of guanidinium chloride, whose purity may be sufficiently suspect that it should be determined using refractive index measurements *(8)*. Alternatively, for a totally new protein, amino acid analysis may be recommended *(9)*.
2. After the CD machine has been on for more than 30 min run a spectrum of a neodymium filter from 610 nm to 560 nm with instrument parameters set for a

fairly slow scan with small data pitch, for example: 0.1 nm data pitch, 20 nm/min scan speed, 1 accumulation, 0.25 s response time, 1.0 nm band width. Note, the photomultiplier tube voltage—the absorbance maximum is also the HT voltage maximum and it should occur at 586 ± 0.8 nm. It is a good idea to note the wavelength accuracy in an equipment maintenance log. If the wavelength accuracy is not within specification, but the shift is constant across the wavelength range (check for the same variation with ACS, *see* below), then you can recalibrate the spectrum accordingly. However, it is advisable to call in an engineer.

3. The intensity calibration of a CD machine is usually carried out by collecting a spectrum from 350 nm to 250 nm of 0.06% aqueous ammonium d-10-camphor sulfonate (ACS) in a 1 cm path length cuvet. A typical set of instrument parameters is: 0.1 nm data pitch, 50 nm/min scan speed, 1 accumulation, 1 s response time, 1.0 nm band width. Subtract a water baseline run with the same cuvet and parameters. The wavelength and intensity of the peak should be 190.4 ± 1 mdeg at 290.5 nm. It is a good idea to record the values in the instrument log. If the intensity is not within stated limits, use an independently made fresh ACS standard and repeat the calibration test. If the value is reproducible all subsequent data may be scaled to bring the intensity to the correct value. Alternatively, the instrument may be recalibrated.

   This approach to instrument calibration is based on the assumption that a single point calibration is sufficient for the whole spectrum. This is transparently not the case and work is in progress to try to establish a single solution calibrant for the full wavelength range. Of particular concern is the instrument performance below 200 nm where light scattering effects become significant as the lamp intensity is reduced. The 0.06% ACS solution in a 1 mm cuvet can be used to monitor this as the ACS has a negative CD maximum at 191 nm. The magnitude of the ratio of the negative to positive peaks is probably approx 2.1. Certainly, a value below 2.0 indicates poor instrument performance ***(10)***.

4. It is important to determine the cuvet path length if a short path length (less than 1 mm) is being used. One millimeter and longer path length cuvets can usually be assumed to be that specified by the manufacturer (though if pathlength is important this should be confirmed). 0.01 mm path lengths are almost never close to that specified, and indeed the path length of a filled demountable cuvet varies from fill to fill and user to user.

   Short path lengths (0.1 mm and shorter) can be determined using interference fringes on an empty cell. However, for the shorter path length demountable cuvets how a given operator fills the cuvet will affect the path length. The best method we have found is for each user to fill the demountable cell with the potassium chromate solution of appropriate concentration and measure the UV/visible absorption spectrum from 600 nm to 350 nm. The path length is then calculated using the Beer–Lambert law ($\varepsilon = 4830\ mol^{-1}\ dm^3\ cm^{-1}$ at 372 nm). It is important to always assemble the cell the same way (mark the cuvet at one end with a pencil and note which edge is the beveled edge). Path lengths of demountable cuvets do vary over time (as the edges of the cell get worn). It is important for at least three measurements to be performed. A new user of 0.01 mm cuvets, in our experience, takes hours of reloading and remeasuring to obtain a reproducible path length.

5. Either cylindrical or rectangular cuvets may be used for CD. Cylindrical cells are usually deemed to have lower birefringence (baseline CD) than rectangular cuvets, however, if UV and CD *matching* is requested when the cuvets are purchased rectangular cuvets seem to be equally good. Water-jacketed cylindrical cells enable the sample to be thermostatted most simply, and also take the least sample volume for a given path length. With these cuvets, you must check that the configuration of your light beam and cuvet holder to ensure that the light beam passes through the sample and not the quartz walls and cooling water parts of the cuvet. Rectangular cells have a number of advantages over cylindrical cuvets for the 1 mm and longer path length experiments: they are cheaper, may be used in standard absorption spectrophotometers (so CD and normal absorption data may be collected on exactly the same sample), and may be used for a protein-ligand serial titration experiment as approx 60% of a rectangular cell can be empty for the first spectrum and gradually filled.

   If path lengths of 0.1 mm or less are required it is probably best to use demountable cuvets where the sample is dropped onto a quartz disk or plate that is etched to a predefined depth, and then another quartz disk/plate is carefully placed on top. Titrations are not possible in demountable cuvets unless independent samples are made.

   All of the light beam incident upon the cuvet must pass through the sample and not be clipped or reflected by the walls or base of the cell or the meniscus of the solution, otherwise the measured spectrum is affected by scattered light. Thus, the narrow cells often used to minimize sample volume in a normal absorption spectrophotometers *cannot* be used for CD unless the light beam is chopped or focused or is intrinsically small. While focusing of the light beam is possible, one must ensure that (i) the lenses used for the focusing are not themselves significantly birefringent (CD active), (ii) the light beam does not diverge and hit the sides of the cuvet while passing through the sample, and (iii) the whole light beam incident on the sample is collected by the photomultiplier tube (PMT). The light beam must not be focused too tightly on the PMT itself, otherwise the PMT may be damaged.

   For UV/visible (*see* above note) CD, high-quality quartz cuvets that transmit the full wavelength range of UV/visible (*see* above note) light are required. In the visible region glass may be used, but it is generally advisable to use quartz even here. Plastic cuvets typically have high intrinsic birefringence so should be avoided. In any case, the need to run a baseline of each cuvet used (*see* below) removes the usual attraction of disposable plastic cuvets.

6. The required path length for the protein backbone region spectrum (from 260 nm to 190 nm) may be estimated on the basis that a 1 mm cuvet probably requires a approx 0.1–0.2 mg/mL protein solution. Sometimes it is desirable to adjust concentrations to use an available cuvet, sometimes it is desirable to choose a path

length to avoid dilution of a sample (e.g., monoclonal antibodies usually have a slightly concentration dependent CD spectrum). The path length required for the aromatic region (from 300 nm to 250 nm) depends on the concentration of aromatic chromophores in the protein. For a protein with no aromatic groups and no disulfide bonds, there will be no aromatic region CD signal whatever the concentration or path length used. Typically, 100 times more protein needs to be in the light beam for near UV measurements than for far UV measurements. The path length required to measure the CD induced into ligand transitions upon binding to a protein is chosen to give an absorbance of approx 1 at the wavelength of interest (usually around the absorption peaks of the ligand).

7. It is essential that the cuvet is cleaned well. Any deposit of chiral material on the quartz will have a CD spectrum. Sometimes one just hopes this subtracts off with a baseline, however, this is not good practice. To clean a cuvet one may proceed as follows. Rinse it well at least three times with high purity water (18.2 MΩ) followed by ethanol or acetone. Dry the inside of a nondemountable cuvet with nitrogen or compressed air (but beware of oil deposits from the compressor) or a hairdryer. Dry the outside of a cuvet with a tissue, wipe with a lens cloth, and remove any fibers with a nitrogen line. If the cuvet shows traces of protein residue (as most easily shown by a protein CD spectrum being observed for the baseline), wash well with detergent (e.g., Hellmanex), and rinse with water. If the residue still remains, place the cuvet in a solution of 6 *M* nitric acid (beware of local safety issues here, e.g., acetone and nitric acid are explosive) or Hellmanex (make sure the cleaning agent gets inside the cuvet) and allow it to stand for 10 min or longer before removing and rinsing thoroughly with water.
8. When measuring protein–ligand interactions, one is frequently interested in the strength of the interaction. The simplest measure of the binding strength between a protein and a ligand is the equilibrium binding constant,

$$K = \frac{L_b}{L_f S_f} \tag{5}$$

for the equilibrium

$$\text{free ligand} + \text{empty protein binding site} \rightleftharpoons \text{protein} - \text{ligand complex} \tag{6}$$

where L_b is the concentration of bound ligand, L_f is the concentration of free ligands, and S_f is the free site concentration. The total binding site concentration is

$$S_{tot} = nC_M \tag{7}$$

where C_M is the macromolecule concentration and n is the number of sites per protein. For proteins, it is usually safe to assume $n = 1$ for the first binding site. CD can be used to determine binding constants if the strength is of the order of $(\mu M)^{-1}$ or a little weaker or stronger simply because of the concentration of the samples required to get a reasonable signal. If no ICD is observed then it is likely that the interaction is weaker. If the ICD is proportional to concentration of ligand (or protein) added then it is too high to be measured by CD.

Assuming that CD is a useful method for determining K, methods are outlined below which are particularly appropriate for use with CD data where the starting point for equilibrium constant determination is usually:

$$L_b = \alpha\rho \tag{8}$$

where ρ is the CD signal at a chosen wavelength and α (which is a function of wavelength) is a constant over the range of binding ratios being considered. The simplest means of determining a is usually from the low binding ratio limit where all the ligand is assumed to be bound so L_f is assumed to be zero. If this is indeed the case, then a plot of ρ vs L (total ligand concentration) at constant protein concentration should be a straight line with slope α. Alternatively, the maximum ICD signal may be used to determine α if n is known (in this limit $L_b = C_M$). One can then calculate K from a point-by-point analysis of the data. Alternatively, the Scatchard plot is probably still the most widely-used method of combining data from a titration series.

The method most widely used in one form or another for determining K is the one developed by Scatchard *(11)*. **Equation 5** is rearranged as follows:

$$\frac{r}{L_f} = \frac{KS_f}{C_M} = Kn - rK \tag{9}$$

where

$$r = \frac{L_b}{C_M} \tag{10}$$

because $S_f = S_{tot} - L_b$. So, a plot of r/L_f vs r has slope $-K$ and y-intercept K/n. The x-intercept occurs where $r = n$. L_b, and hence L_f, may be determined directly from the CD if a has been determined as discussed earlier.

9. It is not always possible to get data for either high- or low-binding ratio limits to determine a. In such cases the intrinsic method is used:

Equation 5 may be written *(12)*

$$K = \frac{\alpha\rho}{(S_{tot} - \alpha\rho)(L_{tot} - \alpha\rho)} \tag{11}$$

which can be rearranged to give

$$L_{tot} = \frac{L_{tot}S_{tot}}{\alpha\rho} - S_{tot} + \alpha\rho - \frac{1}{K} \tag{12}$$

for two different total ligand concentrations, and the same protein site concentration, i.e.,

$$\frac{L_{tot}^k - L_{tot}^j}{\rho^k - \rho^j} = \frac{S_{tot}}{\alpha}\left(\frac{\frac{L_{tot}^k}{\rho^k} - \frac{L_{tot}^j}{\rho^j}}{\rho^k - \rho^j}\right) + \alpha \tag{13}$$

Thus, a plot of

$$y = \frac{L_{tot}^k - L_{tot}^j}{\rho^k - \rho^j} \quad \text{vs} \quad x = \left(\frac{\dfrac{L_{tot}^k}{\rho^k} - \dfrac{L_{tot}^j}{\rho^j}}{\rho^k - \rho^j} \right)$$

should be a straight line with slope $C_M(\alpha/n)^{-1}$ and intercept α. The concentration of bound molecules in any sample may then be determined as may n. The equilibrium binding constant, K, may then be calculated. Alternatively, using these accurate values of n and a Scatchard plot (*see* previously) may be used to determine the best value of K using all the data points.

It is sometimes convenient to perform experiments with constant ligand and varying macromolecule concentration. In this case C_M, and hence S_{tot}, are the variables and L_{tot} is fixed. Rather than **Eq. 13** we then use:

$$\frac{S_{tot}^k - S_{tot}^j}{\rho^k - \rho^j} = \frac{L_{tot}}{\alpha} \left(\frac{\dfrac{S_{tot}^k}{\rho^k} - \dfrac{S_{tot}^j}{\rho^j}}{\rho^k - \rho^j} \right) + \alpha \tag{14}$$

or equivalently

$$\frac{C_M^k - C_M^j}{\rho^k - \rho^j} = \frac{L_{tot}}{\alpha} \left(\frac{\dfrac{C_M^k}{\rho^k} - \dfrac{C_M^j}{\rho^j}}{\rho^k - \rho^j} \right) + \alpha / n \tag{15}$$

10. A glance at almost any reference dealing with titrations will make one realize that the simple equilibrium model we have assumed is quite probably invalid and also the options for data analysis are almost endless ***(13)***. In particular, whereas the traditional approaches mentioned previously involve linearizing the data in some way as this enables one to see by eye whether such an approach is valid, with available computers and packages it is fairly simple to convert almost any model into a plot and determine constants for it without weighting data in any way.

References

1. Nakanishi, K., Berova, N., and Woody, R. W. (ed.) (1994) *Circular Dichroism: Principles and Applications*. VCH, New York.
2. Rodger, A. and Nordén, B (1997) *Circular and Linear Dichroism*. Oxford University Press, Oxford.
3. Johnson, W. C. Fitting programs are available to deconvolute the experimental data into percentages of the structural motifs. Website: e.g., http://www.cryst.bbk.ac.uk/cdweb/html/home.html; Dicroweb: a facility of the BBSRC Centre for Protein and Membrane Structure and Dynamics. http://oregonstate.edu/dept/biochem/faculty/johnson.html. Date accessed: November 20, 2004.
4. Johnson, W. C. (1999) Analyzing protein circular dichroism spectra for accurate secondary structures. *Proteins: Structure, Function and Genetics* **7,** 307–312.

5. Chen, G. C. and Yang, J. T. (1977) Two-point calibration of circular dichrometer with D-10-camphorsulfonic acid. *Anal. Lett.* **10,** 1195–1207.
6. Takakuwa, T., Konno, T., and Meguro, H. (1985) A new standard substance for calibration of circular dichroism: ammonium D-10-camphorsulfonate. *Anal. Sci.* **1,** 215–218.
7. Dafforn, T. R., Halsall, D. J., Serpell, L. C., Rajendra, J., and Rodger, A. (2004) The use of linear dichroism to determine the orientation of secondary structural elements within protein fibres. *Biophys. J.* **86,** 404–410.
8. Pace, C. N. (1986) Determination and analysis of urea and guanidine hydrochloride denaturation curves *Methods Enzymol.* **131,** 266–280.
9. Gill, S. C. and von Hippel, P. H. (1989) Calculation of protein extinction coefficients from amino acid sequence data. *Anal. Biochem.* **182,** 319–326.
10. Miles, A. J., Wien, F., Lees, J. G., Rodger, A., Janes, R. W., and Wallace, B. A. (2003) Calibration and standardisation of synchrotron radiation circular dichroism and conventional circular dichroism spectrophotometers. *Spectroscopy* **17,** 653–661.
11. Scatchard, G. (1949) The attraction of proteins for small molecules and ions. *Ann. N.Y. Acad. Sci.* **51,** 660–672.
12. Rodger, A. (1993) Linear dichroism. *Methods Enzymol.* **226,** 232–258.
13. Polster, J. and Lachman, H. (1989) *Spectrometric Titrations: Analysis of Chemical Equilibria.* VCH Verlagsgesellschaft, Weinheim, Germany.
14. Johnson, W. C., Jr. (1988) Secondary structure of proteins through circular dichroism spectroscopy. *Ann. Rev. Biophys. Biophys. Chem.* **17,** 145–166.
15. Johnson, W. C., Jr. (1985) Circular dichroism and its empirical application to biopolymers. *Methods Biochem. Anal.* **31,** 61–163.
16. Miguel, M. S., Marrington, R., Rodger, P. M., Rodger, A., and Robinson, C. (2003) An *Escherichia coli* twin-arginine signal peptide switches between helical and unstructured conformations depending on hydrophobicity of the environment. *Euro. J. Biochem.* **270,** 1–8.
17. Green, P. (1999) PhD Thesis, University of Warwick, UK.

18

High-Throughput Screening of Interactions Between G Protein-Coupled Receptors and Ligands Using Confocal Optics Microscopy

Lenka Zemanová, Andreas Schenk, Martin J. Valler, G. Ulrich Nienhaus, and Ralf Heilker

Summary

Interactions of extracellular ligands with proteins in the cellular plasma membrane are the starting point for various intracellular signaling cascades. In the pharmaceutical industry, particular attention has been paid to G protein-coupled receptors (GPCRs), which are involved in various disease processes. In so-called high-throughput screening (HTS) campaigns, large medicinal chemistry compound libraries were searched for bioactive molecules that would either induce or inhibit the activity of a specific disease-relevant GPCR. In the respective drug discovery assays, the test compound typically competes with the physiological ligand for a binding site on the receptor. The transmembrane receptor is prepared in the form of membrane fragments or, as described here, in so-called virus-like particles (VLiPs). As hundreds of thousands of test compounds must be analyzed, there is a strict need for low volume binding assays to save the expensive bioreagents, and to reduce the consumption of the test compounds. In this chapter, we describe the application of confocal optics microscopy to measure GPCR ligand interactions in low microliter assay volumes.

Key Words: Ligand; receptor; high-throughput screening; fluorescence spectroscopy; confocal optics; miniaturization; binding assay.

1. Introduction

1.1. Receptor–Ligand Interaction

Receptors are specialized protein molecules on the surfaces of cells with a specific binding site for a small molecule, a ligand, which may be an intrinsic molecule (such as a hormone or neurotransmitter) or an extrinsic drug mol-

From: *Methods in Molecular Biology, vol. 305: Protein–Ligand Interactions: Methods and Applications*
Edited by: G. U. Nienhaus © Humana Press Inc., Totowa, NJ

ecule. This interaction is often likened to the function of a lock (receptor) and a key (ligand). Binding of the ligand leads to a specific biochemical or physiological response. Ligands may bind to receptors, but may also interact with other molecules without producing a specific effect. The receptor–ligand interaction is a crucial event in cell-to-cell communication, which is necessary for the organization of multicellular organisms. Receptors are important targets for drugs in the pharmaceutical industry.

The development of new and effective drugs is a complex multistep procedure. It starts with a process called lead discovery, which provides unique molecule leads against a biochemical or cellular target. These compounds may not necessarily be drug-like lead compounds, but rather serve as information-rich probes. They are subsequently analyzed to provide detailed information about the target under study, and will ultimately facilitate the design of a variety of new compounds satisfying the desired drug-like properties. High-throughput screening (HTS), described in **Subheading 1.2.**, is an essential part of the lead discovery process.

1.2. High-Throughput Screening: Historical Development and Industry Trends

Early lead discovery activities in the pharmaceutical industry were commonly based around extended, sequential, and chemical optimization of an initial structure of interest. Often, the initial structures were publicly known or in some way associated with the target itself (i.e., the ligand of a G protein-coupled receptors [GPCR]). Whereas successful in the context of the industry at this time, with increasing pressure for novelty and productivity came the quest for improving the efficiency of these early stages.

In the 1990s, the concept of *random* testing developed gradually as an attempt to capitalize on the pre-existing compounds made and retained within most longstanding pharmaceutical companies. The premise was, and remains, that testing of such sample collections will reveal previously unrecognized biological activities of these compounds, which can enter the synthesis-driven optimization process as valuable new lead structures. In addition, multiple, alternative lead structures for one target may become available through this approach, offering a choice of chemical starting points for the project. Whereas the concepts of diversity-based screening have remained, the procedures have developed dramatically.

It was recognized early in the history of HTS that such a process could be technically difficult, time-consuming, and expensive. In the intervening time, typical sample numbers in compound collections have increased enormously, among other reasons due to the advent of combinatorial chemistry. However, through the development of new screening technologies, advanced engineering and automation solutions, and through assay miniaturization, these early fears have been substantially overcome.

The scope of assay biology available to HTS has also greatly expanded, and the discipline continues to pioneer efficient bioassay methodologies. Most recently, a rapid and substantial expansion of our knowledge about novel genes has come through the Human Genome sequencing project. Functional validation of the myriad of new gene products has become a new bottleneck in applying this knowledge to drug discovery. This has led to the rise in popularity of chemical genomics; that is, the use of screening to discover small molecule effectors which can be used as tools for target validation: to investigate the functional relevance of target sequences and the associated expression products.

The improvements in efficiency that can be achieved in primary screening are also applicable to secondary assay stages. The range of assay technologies that can be addressed by automated, high capacity approaches is constantly expanding. This gives the opportunity for removing most in vitro project bioassay limitations, at least up to the stage of whole animal studies.

1.3. Automation and Miniaturization

Ten to 15 yr ago, the screening for bioactive compounds was limited to the use of bench-top devices for reagent dispensing and compound pipetting. Thus, compound testing was very much a manual, or at best a semimanual process. The introduction of robotics and laboratory automation in the early 1990s has led to dramatic changes of the compound screening process. State-of-the-art screening systems are based on the use of 384-well or 1536-well microtiter plates (MTPs). The MTPs are moved from one device to the next in an automated and time-scheduled way. Bioactivity detection of compounds is based on novel fast measuring techniques such as scintillation imaging, fluorescence, or luminescence. Several *homogeneous* bioassay formats have been established (in this context *homogeneous* means that the assays do not depend on a washing or filtration step).

Driven by bioreagent cost and by the need to reduce test compound depletion, HTS labs worldwide have attempted to reduce the assay volume. Thus, the 96-well MTP standard of ten years ago has widely been replaced by the 384-well or 1536-well MTP. In parallel, bioassay volumes have typically dropped from about 200 µL to the low-µL range (5–15 µL). This development has been enabled by the design of precise low-microliter dispensing and pipetting devices.

1.4. Nonconfocal High-Throughput Screening Technologies to Study Ligand–G Protein-Coupled Receptors Interactions

GPCRs are the most frequently addressed drug targets in the pharmaceutical industry. Nearly 2000 GPCRs have been reported since bovine opsin was

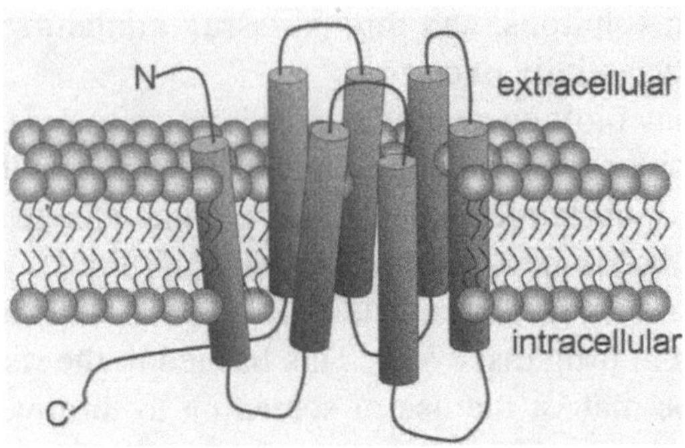

Fig. 1. Schematic representation of the general structure of GPCRs. GPCRs contain an extracellular N-terminal segment, seven transmembrane domains (TM), three exoloops, three cytoloops, and an intracellular C-terminal segment.

cloned in 1983 *(1)* and the β-adrenergic receptor in 1986 *(2)*. They are classified into more than 100 subfamilies according to sequence homology, ligand structure, and receptor function. All GPCRs (**Fig. 1**) have in common an extracellular N-terminal segment, seven transmembrane domains (TM), three exoloops, three cytoloops, and a C-terminal segment *(3)*.

The individual segments vary markedly in size among different GPCRs. A considerable diversity exists in the ligands that bind to GPCRs. They include proteins, small peptides, amino acid and fatty acid derivatives, monoamines, and purines. The majority of hormones and neurotransmitters exert their physiological responses by binding to cell-surface receptors belonging to the family of GPCRs. Likewise, GPCRs act as important sensors of exogenous stimuli, such as light and odors. The increased throughput in the HTS area has raised the need for fast and possibly homogeneous detection technologies. To study ligand–GPCR interactions, mainly two detection techniques have been employed in recent years: scintillation proximity assays (SPA) and, to a lesser extent, fluorescence polarization (FP).

1.4.1. Scintillation Proximity Assay

Despite all efforts of the pharmaceutical drug discovery units to reduce the use of radioactive detection technologies, the scintillation proximity assay (SPA) is still an important HTS format *(4)*. SPA employs so-called scintillation beads, microspheres that are loaded with a β-particle sensitive scintillant. The surface of these beads is chemically treated to enable the coupling of reagents such as wheat germ agglutinin (WGA). WGA binds glycolipids and glycoproteins in typical cellular biomembranes. Thus, membrane fragments containing a target transmembrane receptor of interest may be immobilized to the scintillant surface. If a radioisotopically-labeled ligand binds to the target receptor, the emitted radiation will activate the scintillant and produce light. The light intensity can be measured by a scintillation counter and is proportional to the amount of scintillant-proximal radioligand. If the ligand is labeled with a low energy β-emitting radioisotope (such as ^3H) and is not bound to the target molecule on the scintillant surface, its radiation energy is absorbed by the surrounding aqueous environment and no light is produced. Thus, bound radioligands give a scintillation signal and free radioligands do not, which renders the SPA format homogeneous.

In the past, photomultiplier tube (PMT)-based scintillation counters were used to measure the scintillation from a single well by a single PMT. A few years ago, several companies have introduced low temperature CCD camera-based imaging devices for scintillation measurements ***(5,6)***. These imaging readers achieve read times of about 5 min for a 384-well plate with comparably good scintillation count statistics as the same assay measured in a PMT-based device with read times of about 40 min for a 384-well plate ***(7)***. This time saving is particularly important in an HTS environment.

As CCD-based cameras are more sensitive for red wavelengths, novel SPA Imaging Beads have been designed that contain europium chelates as scintillants ***(8)***. The europium-containing Imaging Beads emit light in the red region at about 615 nm. As a further advantage, the interference of the red-shifted scintillation with colored compounds is much less of a problem, because most organic compounds produced in medicinal chemistry are in the yellow to brown range and absorb light in the blue region of the spectrum.

1.4.2. Fluorescence Polarization

In an attempt to reduce the number of radioactivity-based GPCR screening campaigns, the FP assay format has been incorporated into the HTS platforms of some pharmaceutical companies ***(9)***. FP measurements are based on the fact that fluorophores absorb light along a particular direction with respect to the molecular axes. The extent to which the fluorophore rotates during the excited-state lifetime determines its polarization. This physical principle can be employed to study the binding of a small fluorophore-labeled ligand to a large receptor. A large entity, such as a transmembrane receptor in a membrane fragment, rotates more slowly than the small fluorophore-labeled ligand. If the ligand binds to the receptor, the rotational rate of the fluorophore is reduced, and the polarization of the fluorescence emission is high. If the binding of the ligand is inhibited by a test compound the fluorophore maintains the high rotational rate of the free ligand-fluorophore conjugate: the fluorescence emission becomes depolarized. Advances in dye chemistry and the introduction of sensitive FP readers have led to the more frequent use of FP assays in HTS to study receptor ligand interactions *(10)*.

1.5. Confocal Optics Microscopy for High-Throughput Screening of G Protein-Coupled Receptors

In recent years, both academia and pharmaceutical industry have produced significant advances in confocal detection and spectroscopy by laser-induced fluorescence. Confocal fluorescence studies provide information on identity, size, diffusion coefficient, and concentration of the fluorescently labeled entity. This enables the establishment of sophisticated biochemical drug screening assays using the multitude of fluorescence parameters that can be observed (e.g., molecular brightness, fluorescence lifetime, anisotropy, resonance energy transfer). Confocal HTS systems focusing on femtoliter-sized observation volumes allow for assay volumes far beyond current limits.

For the reasons stated above, HTS groups worldwide have made attempts to reduce the sample volume in drug screening assays dramatically. This miniaturization faces two major challenges, handling of minute volumes and the sensitivity of the typically optical detection. Optimization of liquid handling tools has preceded the development on the detection side: several submicroliter pipeting and dispensing devices have been established in HTS routine use. On the other hand, all macroscopic fluorescence methods face the problem of increasing background with decreasing assay volumes. In contrast, the femtoliter-sized confocal observation volume enables miniaturization without loss of data quality *(11,12)*. In this chapter, we demonstrate the applicability of confocal fluorescence spectroscopy to study GPCR–ligand interactions. In particular, one method referred to as photon counting histogram (PCH) analysis or fluorescence intensity distribution analysis (FIDA) has been investigated.

2. Materials

All chemical reagents were purchased from Sigma-Aldrich (Taufkirchen, Germany) and Roth (Karlsruhe, Germany), if not otherwise specified. All used buffers should be tested for their fluorescence because the background fluorescence in the assay should be kept minimal (*see also* **Note 1**).

2.1. G Protein-Coupled Receptors in Membrane Fragments

1. Cloned human GPCRs produced, e.g., in transfected Chinese hamster ovary (CHO) cells are provided as membrane fragments by vendors such as BioSignal Packard (Montréal, Canada) or Euroscreen (Brussels, Belgium).
2. The membrane fragments were stored at –80°C in 50 m*M* Tris-HCl pH 7.4, 10% glycerol, 1% BSA. Some GPCR membrane fragments were delivered in buffer with $MgCl_2$ or sucrose instead of BSA.

3. For the binding assay, the membranes were thawed on ice and diluted in receptor diluting buffer (50 m*M* Tris-HCl pH 7.4, 1 m*M* $CaCl_2$, 0.2% BSA). Some manufacturers recommend a different buffer.

2.2. G Protein-Coupled Receptors in Virus-Like Particles

Evotec OAI has established a method that selectively incorporates transmembrane proteins (e.g., GPCRs) into virus-like particles (VLiPs). For this purpose, the GPCR of interest (the *target molecule*) is co-expressed in insect cells (Spodoptera frugiperda 9 [Sf9] cells) together with the Gag protein from Moloney murine leukemia virus, which leads to viral particle formation and budding at the cell surface. The target molecule at its cytoplasmic portion and the Gag protein at its C-terminus contain amino acid sequences that enable their interaction (Hunt et al., personal communication). By virtue of this interaction, the Gag protein recruits the target molecule into the budding VLiP. The resulting chimeric VLiPs are released into the extracellular medium (**Fig. 2A**). The chimeric VLiPs (**Fig. 2B**) constitute a concentrated source of the GPCR of interest and thereby show great potential for fluorescence fluctuation spectroscopy techniques, which require very high purity of the samples.

1. The supernatant of the GPCR-/gag-cotransfected Sf9 cell culture, which contains the chimeric VLiPs (**Fig. 2B**), was concentrated and stored in 50 m*M* Tris-HCl pH 7.4, 1 m*M* $CaCl_2$, 0.2% BSA, and 0.1% NaN_3 at –80°C.
2. Before use, it was thawed on ice and homogenized by ultrasonication in sonication bath for 1 min, as recommended by the manufacturer, and diluted with receptor diluting buffer (50 m*M* Tris-HCl pH 7.4, 1 m*M* $CaCl_2$, 0.2% BSA).
3. After centrifugation at 3000*g* for 1 min, the supernatant was used for the binding assay.

2.3. G Protein-Coupled Receptors Ligands and Competitor Compounds

1. The peptide ligand was N-terminally labeled with 5-carboxy TMR. TMR-ligand was provided as a lyophilized solid by Evotec OAI (Hamburg, Germany).
2. Unlabeled ligand was purchased from Bachem Feinchemikalien (Bubendorf, Switzerland) and competitors were from Tocris Cookson (Avonmouth Bristol, UK). All ligands and competitors were received in lyophilized form, then dissolved in dimethyl sulfoxide (DMSO), aliquoted, and stored at –80°C.
3. For the binding or competition assay, they were diluted in ligand diluting buffer (50 m*M* Tris-HCl pH 7.4, 1 m*M* $CaCl_2$, 0.2% Tween-20). The residual concentration of DMSO should be kept lower than 1% in the assay (*see* **Note 2**).

2.4. Confocal Microscope

In confocal microscopy, the illumination light is focused onto the sample by a microscope objective, and only a tiny volume of the sample is illuminated in the focal plane. Fluorescence light from this volume is imaged onto the confocal pinhole and reaches the detector. Light from outside of the focal plane is

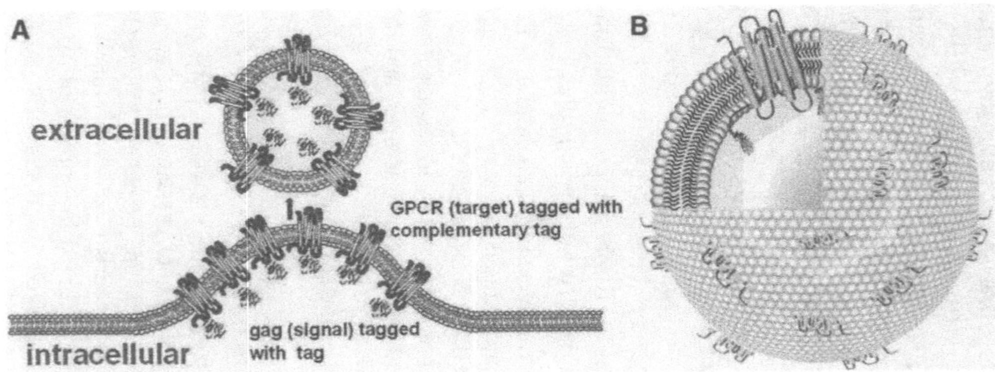

Fig. 2. (**A**) Schematic representation of a VLiP budding from a host cell incorporating a GPCR of interest. A Gag-tag fusion protein is co-expressed in a cellular system with the GPCR, which carries the complementary peptide tag at the C-terminus. Expression of a retroviral Gag protein in the host cells results in the accumulation of the Gag protein at the plasma membrane due to signals present within the N-terminal portion of the Gag protein. High concentrations of this protein at the plasma membrane induce a budding process, by which the VLiPs are released into the extracellular medium where they can be harvested. The incorporation of the respective target GPCR within the envelope of the VLiPs is the result of a strong specific interaction of the peptide tag covalently attached to the C-terminus of the Gag protein with the complementary, specific peptide tag attached to the C-terminus of the GPCR. (**B**) Three-dimensional view of a GPCR-chimeric VLiP. A lipid bilayer, which is derived from the plasma membrane of the host cell, surrounds a central viral capsid, which is a specific aggregate of Gag proteins. The GPCR of interest is physiologically integrated into the viral lipid bilayer, the extracellular portion of the GPCR facing outside.

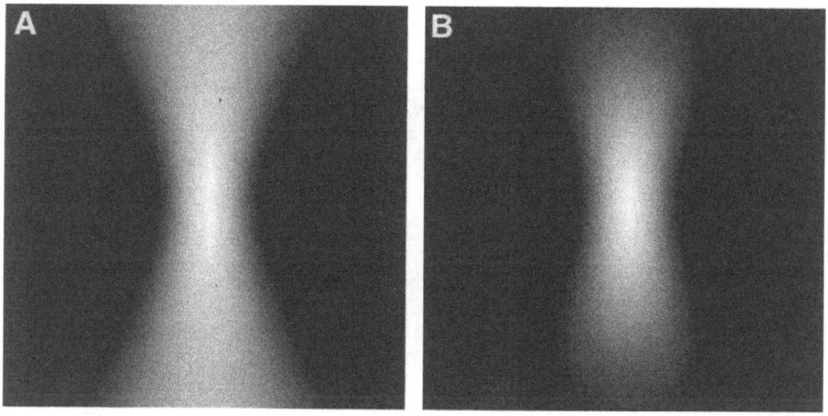

Fig. 3. Illumination and observation volume in confocal microscopy. **(A)** Simulated cross section of the Gaussian profile of a laser beam, focused to the centre by the objective lens. **(B)** The effect of a confocal pinhole is simulated by an additional Gaussian with 5-times larger extension along the light propagation axis. In both panels, the light intensity is shown using a logarithmic gray scale.

blocked by the pinhole, and thus cannot be detected. The difference between the illuminated volume and the observation volume in confocal microscopy is visualized in **Fig. 3**.

1. In our confocal microscopy setup, depicted schematically in **Fig. 4**, light from an Argon/Krypton ion laser (model 164, Spectra-Physics) is focused on the sample by using a dichroic mirror (Q525LP, AHF, Tübingen, Germany) and a water immersion objective (UPAPLO 60_/1.2W, Olympus, Hamburg, Germany) in an inverted microscope (Axiovert 35, Carl Zeiss, Göttingen, Germany). The emitted photons are collected through the same objective, focused on a confocal pinhole (50 μm diameter), and detected by an avalanche photodiode (SPCM-AQR-14, Perkin-Elmer, Vaudreuil, Canada).
2. For the experiments described here, we used 514 nm, and the observed spectral band was limited to wavelengths between 557 and 607 nm using a band pass filter (HQ 582/50, AHF) and a long pass filter (HQ 700/300, AHF).
3. Movement of the confocal volume through the sample was achieved by scanning the sample in a circle of 30 μm radius with a velocity of 0.8 μm/ms using a piezoelectric scanning stage (Tritor 102 Cap, Piezosysteme Jena, Jena, Germany).
4. The photodiode output signals were autocorrelated in a digital correlator (ALV-5000/E, ALV, Langen, Germany), acquired by using a computer card for single-photon counting (TimeHarp100, Pico-Quant, Berlin, Germany) or simply collected as a function of time using a home-made computer card. The latter card enables data analysis with our own data acquisition software.

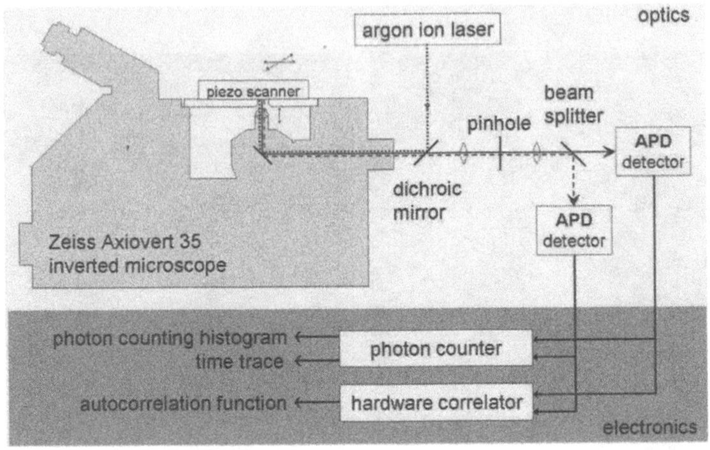

Fig. 4. Schematic of the confocal microscope with its optical and electronical components. The system consists of a confocal laser scanning microscope with two detection channels for different spectral ranges. To achieve single-molecule sensitivity, a high numerical aperture (N.A.) water immersion objective and avalanche photodiode (APD) detectors are used, as well as confocal detection with a confocal volume of approx 1 fl. The electronic part consists of a correlator and a photon counter calculating autocorrelation function and photon counting histogram (PCH), respectively.

2.5. Instrumentation for Fluctuation Fluorescence Spectroscopy in a High-Throughput Screening Environment

Whereas several companies supply devices that enable fluctuation fluorescence spectroscopy measurements, Evotec OAI is most advanced in applying this technology to HTS. The Insight reader (Evotec Technologies GmbH, Germany) detects fluorescent molecules with single molecule resolution. Submicroliter miniaturization without loss of signal quality becomes possible by the use of confocal optics. Parallel analysis of multiple fluorescent dyes is enabled by multiple laser sources and detectors. Detection of single molecules at different wavelengths or polarization states with nanosecond time resolution is made possible using two highly sensitive single photon detectors. High-speed signal processing boards support a real-time calculation of all incorporated methods. Measurement times for large particles such as membrane fragments are reduced by two-dimensional beam scanning. Evotec Technologies GmbH reports a typical readout time per well of about 1 s, which fulfils the requirements of ultra-HTS.

The basic version of the Insight reader is equipped with three lasers: an Ar^+ ion laser for excitation at 488 and 514 nm, a green HeNe laser (543 nm), and a

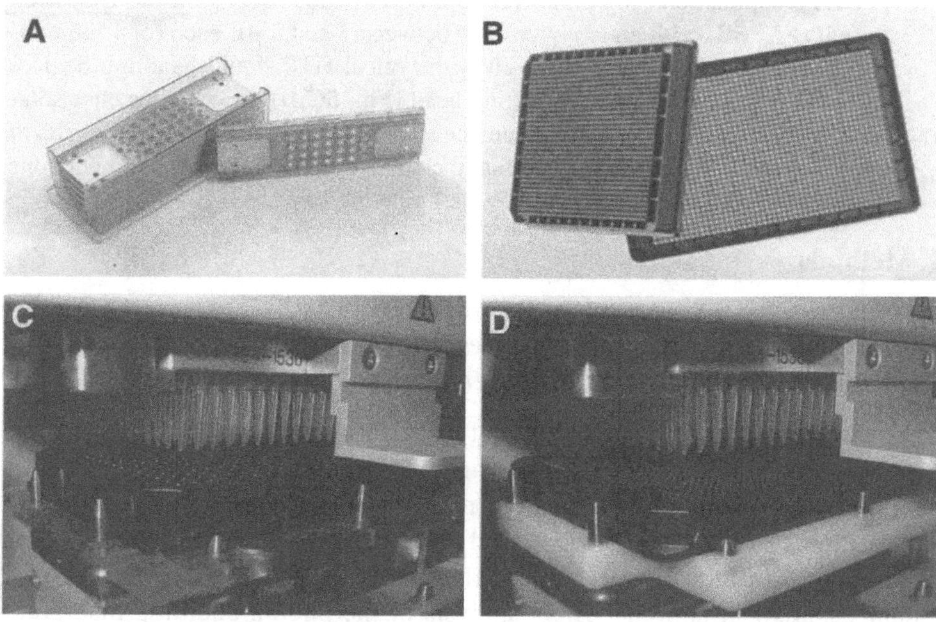

Fig. 5. Sample cells and pipeting devices. **(A)** AssayChip with 24 wells (Evotec Technologies); **(B)** 1536-well plate with glass bottom (GreinerBioOne); **(C)** pipetor head with 384-tips (CyBiWell) with 384-well plate; and **(D)** pipetor head with 384-tips (CyBiWell) with 1536-well plate.

red HeNe laser (633 nm). It supports the FCS, FIDA, FCCS, and 2D-FIDA modes described **in Subheading 3.1.**

For fluorescence lifetime measurements, another version of the Insight reader is available, which additionally contains a mode-locked, frequency-doubled Nd:YAG green laser (532 nm) and a pulsed red laser diode (635 nm). The latter reader also contains a software extension that supports the below described fluorescence lifetime analysis and the FILDA technique.

2.6. Sample Cells

1. For microscopic measurements, a sandwich was assembled from two microscope cover slips and two pieces of double sided, self-adhesive tape. The 200 μm thick tape was attached to the glass such that a 2 mm wide channel was formed in the middle into which the sample solution was infused by capillary action.
2. For HTS assay development, prefabricated polypropylene glass bottom sample carriers such as the AssayChip 24/25 (**Fig. 5A**) with 24-wells from Evotec Technologies GmbH (Germany) were used. The sample volume was between 6 and 8 μL.

3. For HTS applications, we employed microtiter plates (MTPs) in the 1536-well format (**Fig. 5B**) with an assay volume between 3 and 5 µL each for 1536 individual samples. This MTP fits well with typical HTS liquid handling devices such as the standard 384-tip pipetting head (**Fig. 5C,D**) that can dispense solutions in the low µL range (for the device shown down to 0.5 µL) to 384 wells of the 1536-well MTP in parallel, so that the 1536-well MTP can be filled with one bioreagent solution in only four 384-well pipeting steps.

3. Methods

3.1. Fluorescence Correlation Spectroscopy, Photon Counting Histogram/ Fluorescence Intensity Distribution Analysis, and Other Techniques

The primary data generated by a confocal fluorescence spectroscopy experiment is a record of fluorescence intensity as a function of time (**Fig. 6A**). This data contains information about the number of molecules in the confocal observation volume at a given time and about their fluorescent properties. The information can in principle be extracted with two different, partly complementary statistical approaches, 1) analysis of the photon counting histogram (**Fig. 6B**), and 2) analysis of the autocorrelation function (**Fig. 6C**). The latter method is called fluorescence correlation spectroscopy (FCS). It was introduced by Magde, Elson, and Webb in 1972 *(13–17)* and applied for the first time by the same group in 1974. FCS allows us to distinguish between different molecules in the sample with regard to their size and to measure their concentrations.

Instead of analyzing the temporal fluctuations of the fluorescence signal (as done in FCS), one can also analyze the statistics of the intensities of the generated photon bursts. Photons emitted during passage of the fluorophore through the confocal observation volume are counted for a short time interval (time bin) and plotted as a histogram. The distribution of the numbers of photons per bin can be analyzed using two slightly different statistical methods, the PCH analysis developed by Chen and coworkers *(18)*, and FIDA developed by Kask and coworkers from Evotec OAI *(19)*.

FIDA/PCH analyses allow us to distinguish between different molecules in the sample by their brightness and to measure their concentrations. Thus, the application of FIDA/PCH to an HTS ligand-receptor binding assay necessitates changes in the molecular brightness upon binding, which may arise from changes in the fluorescent properties of the fluorophore-labeled species, binding to a multivalent receptor, or binding to vesicle/membrane fragment/particle with multiple receptors (*see* **Note 1**).

Further techniques have been developed that may include additional observables of fluorescence fluctuation spectroscopy, which help to identify

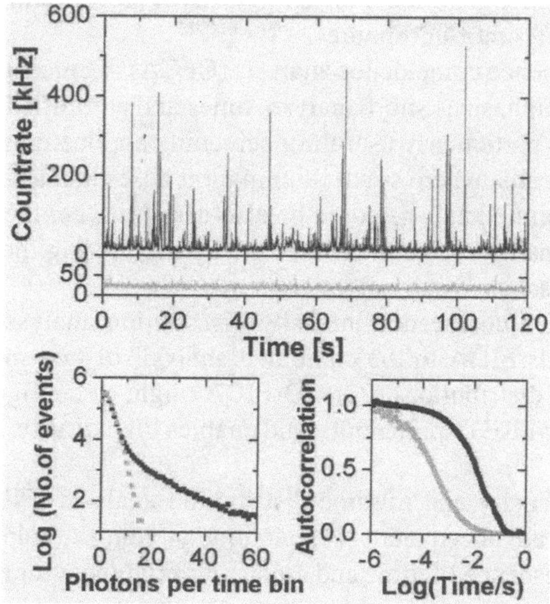

Fig. 6. Primary data and their processing for a binding measurement. A solution of 15 n*M* ET_A-Receptors in VLiPs was incubated with 2 n*M* TMR-ET-1 at room temperature for = 1 h (black). To confirm specific binding, an excess of unlabeled ligand ET-1 was added to the control samples (gray). **(A)** Time traces: The time-dependent fluorescence intensity signal of VLiPs incubated with labeled ligand shows strong fluctuations based on high brightness and slow diffusion of the VLiP with bound TMR-ET-1 (black). These fluctuations are absent from samples with an excess of unlabeled ligand (gray). **(B)** Photon counting histograms: The distribution of number of photons detected per time bin (50 μs) is calculated from the time trace. The histogram for VLiPs with bound TMR-ET-1 (black) shows events with high number of photons, whereas in the case of an excess of unlabeled ET-1 (gray) only events with lower number of photons could be seen. **(C)** Autocorrelation functions: The autocorrelation function could also be calculated from the time trace. The autocorrelation function of the fluorescent intensity of VLiPs with TMR-ET-1 shows the slow diffusion time of the vesicles (black). In the presence of an excess of unlabeled ligand, the diffusion time of the freely diffusing TMR-ET-1 could be measured (gray).

different species in complex biological samples and provide additional information about the properties of the species or their binding.

Fluorescence cross correlation spectroscopy (FCCS) extends FCS by using two-color cross correlation analysis ***(16,20)***. This approach detects the simultaneous appearance of fluorescence at two different wavelengths, which is expected to occur for ligand-receptor complexes if ligand and receptor are labeled with two different fluorophores.

Confocal fluorescence coincidence analysis (CFCA) is a recently developed technique, which emphasizes short analysis times and simplified data evaluation *(21)*. Thus, it is particularly useful for screening applications and/or measurement on living cells, where small illumination doses are desirable.

Fluorescence lifetime analysis could be also used with confocal optics and gives insight into changes of the excited state by monitoring the fluorophore lifetime in the nanosecond time range.

Two-dimensional fluorescence intensity distribution analysis (2D-FIDA; Evotec OAI) extends FIDA to the combined analysis of two simultaneously recorded brightness distributions *(22)*. 2D-FIDA might be configured for two-color, anisotropy, or FRET applications and enables high quality and high content data.

Fluorescence intensity and lifetime distribution analysis (FILDA; Evotec OAI) represents an advanced analysis technique yielding simultaneous information on the fluorescence lifetime and molecular brightness for multiple fluorescent species *(23)*.

3.2. Binding Assay

1. A solution of GPCRs in VLiPs or in membrane fragments of CHO cells was mixed in equal ratio with fluorophore-labeled ligand and incubated at room temperature for 1–2 h. For stability measurements of VLiPs and membrane fragments, extended incubation times up to 24 h were applied.
2. To measure nonspecific binding, a 1000-fold excess of unlabeled ligand was added to the control samples.
3. For the competition assay, the fluorophore-labeled ligand was premixed with an increasing concentration of competitor and then incubated with a solution of the respective GPCR in VLiPs or in membrane fragments at room temperature for 1–2 h.

3.3. Application of Fluorescence Intensity Distribution Analysis/Photon Counting Histogram to a Drug Discovery Receptor–Ligand Binding Assay

As described previously, FIDA/PCH analyses allow us to distinguish between different molecules in the sample by their brightness and to measure their concentrations. A dimmer fluorophore emits less photons per time bin during its passage through the confocal volume than a brighter fluorophore (*see* inset in **Fig. 7**). The necessary change in the molecular brightness upon binding arises in our example from binding of the fluorophore-labeled ligand to vesicles/membrane fragments with multiple receptors. In **Fig. 7**, a PCH dia-

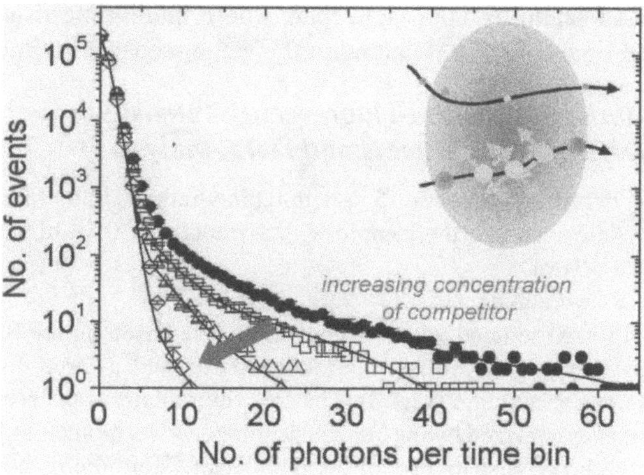

Fig. 7. PCH set for competitor X. A solution of 1.5 n*M* GPCR in VLiPs was incubated with 2 n*M* labeled ligand and with indicated concentrations of competitor. Photon counting histograms (symbols) and data analysis (lines) are shown for representative set of a competition experiment with competitor X. With increasing competitor concentration (arrow) the bound ligand (corresponding to bright species) is replaced. The PCH analysis yields the concentrations of free and bound ligand.

gram is shown as an example of a peptide binding to its receptor in a VLiP (closed data points). As multiple receptors are incorporated into the VLiP or the membrane fragments, the lipid vesicle carries more fluorophore-labeled ligands upon binding of the ligand to its receptor. The vesicle with multiple bound ligands emits many more photons during its passage through the observation volume than the free ligand (*see* **Note 3**). If a reference inhibitor is added at increasing concentrations, as shown in **Fig. 7**, the fluorophore-labeled ligand is displaced from the receptor and appears as an unbound, less bright entity. The concentrations of bound and unbound ligand can be derived from PCH analysis (*see* **Note 4**).

Applications of FIDA to membrane receptor assays were published previously ***(24–26)***. Most known sources of membrane receptors bear multiple receptors, and therefore FIDA appears as an ideal method for such assays. Membrane receptors, like GPCRs, can be obtained in cell membrane fragments from cell culture or partly purified and reconstituted in lipid vesicles or enriched in the previously described VLiPs.

The measuring time may be reduced by an active search for membrane fragments in the sample by moving the focused laser beam or the entire sample.

Using this so-called scanning approach, many more membrane fragments or VLiPs with bound ligand can be detected in the same measuring time.

3.4. Photon Counting Histogram/Fluorescence Intensity Distribution Analysis Measurements and Data Analysis

1. PCHs were recorded 10 times for 15 s, with a bin width of 50 μs (unless otherwise stated), while scanning the sample with a velocity of 0.8 μm/s on a circle with a radius of 30 μm.
2. The ten data sets were averaged.
3. Averaged data were modeled with a theoretical curve based on the FIDA algorithm *(19)*, using the generating function approach and a three-dimensional Gaussian observation volume (**Fig. 3**). The free fluorophore was described by a single species (characterized by the concentration c_1 and its molecular brightness η_1), whereas the broad distribution of the number of fluorophores bound to the VLiPs was modeled by three species ($i = 2 - 4$), with their molecular brightness parameter being a multiple of the brightness of an individual fluorophore, $\eta_i = n_i \cdot \eta_1$. This four-species model was complemented by two contributions, one accounting for background depending on excitation light (e.g., scattered excitation light), and the other one independent of the excitation light (e.g., electronic noise).
4. The concentration of free ligand, c_L, was identified with the concentration of the first species, c_1, while the concentration of bound ligand, c_B, is given by the sum of the concentrations of species 2 – 4, weighted by their relative molecular brightness values, $n_i = \eta_i/\eta_l$:

$$C_B = \sum_{i=2}^{4} c_i \cdot \frac{\eta_i}{\eta_1} \tag{1}$$

5. The concentration data were averaged over three independent measurements.

4. Notes

1. The fundamental problem of fluorescence-based methods is to keep the fluorescence background low. In HTS, highly fluorescent compounds in the screened-compounds library may disturb the evaluation. For potential false-positive results, a complementary method is necessary to confirm the results. This problem can be alleviated or entirely avoided by using fluorophores emitting in the red spectral region.
2. Fluorophores and peptides in aqueous solutions often tend to adsorb to the surface of the measuring cell. To avoid the adsorption of the labeled peptide onto the surface, the addition of Tween or other detergent to the buffer is necessary.
3. Binding assays can be arranged in two ways, either as a ligand titration at constant receptor concentration or as a receptor titration at constant ligand concentration. The first approach saves the expensive receptor. However, at a high concentration of ligand, there are many free (fluorescent!) ligand molecules in

the confocal volume. The PCH/FIDA analysis becomes difficult because one cannot distinguish between multiple bound ligands and many free ligand molecules. The second approach, titration with receptor, can be unfeasible because the concentration of receptor stock solution is too low and one cannot achieve saturation. Additionally, pipetting of the viscous receptor solution may not yield the desired accuracy in the dilution steps.

4. In the HTS environment, the typical binding assays are done as competition experiments searching for new competitors of the ligand. It means that compounds are added while both ligand and receptor concentrations are kept constant. The concentration of compounds is constant in the primary screen, where the answer yes/no is expected. In a later characterization of a competitor (i.e., determination of IC50, which is the concentration required for 50% binding inhibition), the compound concentration varies but the receptor and ligand concentration still remains constant.

Acknowledgments

We thank Elza Amirgoulova, Don C. Lamb, Carlheinz Röcker, Vladimir Ens for assistance throughout the project. We are greatly indebted to Nicholas Hunt, Birgit Hecks, Rudolf Zirwes, Timm Jessen, Leif Brand, and Martin Pitschke for technical advice and helpful discussions.

References

1. Nathans, J. and Hogness, D. S. (1983) Isolation, sequence analysis, and intron-exon arrangement of the gene encoding bovine rhodopsin. *Cell* **34,** 807–814.
2. Dixon, R. A., Kobilka, B. K., Strader, D. J., Benovic, J. L., Dohlman, H. G., Frielle, T., Bolanowski, M. A., Bennett, C. D., Rands, E., and Diehl, R. E. (1986) Cloning of the gene and cDNA for mammalian beta-adrenergic receptor and homology with rhodopsin. *Nature* **321,** 75–79.
3. Ji, T. H., Grossmann, M., and Ji, I. (1998) G protein-coupled receptors. I. Diversity of receptor-ligand interactions. *J. Biol. Chem.* **273,** 17,299–17,302.
4. Alouani, S. (2000) Scintillation proximity binding assay. *Methods Mol. Biol.* **138,** 135–141.
5. Ramm, P. (1999) Imaging systems in assay screening. *Drug Discov. Today* **4,** 401–410.
6. Sorg, G., Schubert, H. D., Buttner, F. H., and Heilker, R. (2002) Automated high throughput screening for serine kinase inhibitors using a LEADseeker scintillation proximity assay in the 1536-well format. *J. Biomol. Screen.* **7,** 11–19.
7. Sorg, G., Schubert, H. D., Buttner, F. H., Valler, M. J., and Heilker, R. (2002) Comparison of photomultiplier tube and charge coupled device-based scintillation counting. *Life Science News* **11,** 1–3.
8. Jessop, R. A. (1998) Imaging proximity assays. *Proc. SPIE* **3259,** 228–233.
9. Banks, P. and Harvey, M. (2002) Considerations for using fluorescence polarization in the screening of g protein-coupled receptors. *J. Biomol. Screen.* **7,** 111–117.

10. Harris, A., Cox, S., Burns, D., and Norey, C. (2003) Miniaturization of fluorescence polarization receptor-binding assays using CyDye-labeled ligands. *J. Biomol. Screen.* **8,** 410–420.
11. Auer, M., Moore, K. J., Meyer-Almes, F. J., Guenther, R., Pope, A. J., and Stoeckli, K. (1998) Fluorescence correlation spectroscopy: lead discovery by miniaturized HTS. *Drug Discov. Today* **3,** 457–465.
12. Zemanova, L., Schenk, A., Valler, M. J., Nienhaus, G. U., and Heilker, R. (2003) Confocal optics microscopy for biochemical and cellular high-throughput screening. *Drug Discov. Today* **8,** 1085–1093.
13. Ehrenberg, M. and Rigler, R. (1974) Rotational brownian motion and fluorescence intensity fluctuations. *Chem. Phys.* **4,** 390–401.
14. Magde, D., Elson, E. L., and Webb, W. W. (1972) Thermodynamic fluctuations in a reacting system—measurement by fluorescence correlation spectroscopy. *Phys. Rev. Lett.* **29,** 705–708.
15. Magde, D., Elson, E. L., and Webb, W. W. (1974) Fluorescence correlation spectroscopy. II. An experimental realization. *Biopolymers* **13,** 29–61.
16. Rigler, R. (1995) Fluorescence correlations, single molecule detection and large number screening. Applications in biotechnology. *J. Biotechnol.* **41,** 177–186.
17. Thompson, N. L. (1991) Fluorescence correlation spectroscopy, in *Topics in Fluorescence Spectroscopy, Vol. 1* (Lakowicz, J. R., ed.), Plenum Press, New York, pp. 337–378.
18. Chen, Y., Muller, J. D., So, P. T., and Gratton, E. (1999) The photon counting histogram in fluorescence fluctuation spectroscopy. *Biophys. J.* **77,** 553–567.
19. Kask, P., Palo, K., Ullmann, D., and Gall, K. (1999) Fluorescence-intensity distribution analysis and its application in biomolecular detection technology. *Proc. Natl. Acad. Sci. USA* **96,** 13,756–13,761.
20. Schwille, P., Meyer-Almes, F. J., and Rigler, R. (1997) Dual-color fluorescence cross-correlation spectroscopy for multicomponent diffusional analysis in solution. *Biophys. J.* **72,** 1878–1886.
21. Winkler, T., Kettling, U., Koltermann, A., and Eigen, M. (1999) Confocal fluorescence coincidence analysis: an approach to ultra high-throughput screening. *Proc. Natl. Acad. Sci. USA* **96,** 1375–1378.
22. Kask, P., Palo, K., Fay, N., Brand, L., Mets, U., Ullmann, D., Jungmann, J., Pschorr, J., and Gall, K. (2000) Two-dimensional fluorescence intensity distribution analysis: theory and applications. *Biophys. J.* **78,** 1703–1713.
23. Palo, K., Brand, L., Eggeling, C., Jager, S., Kask, P., and Gall, K. (2002) Fluorescence intensity and lifetime distribution analysis: toward higher accuracy in fluorescence fluctuation spectroscopy. *Biophys. J.* **83,** 605–618.
24. Klumpp, M., Scheel, A., Lopez-Calle, E., Busch, M., Murray, K. J., and Pope, A. J. (2001) Ligand binding to transmembrane receptors on intact cells or membrane vesicles measured in a homogeneous 1-microliter assay format. *J. Biomol. Screen.* **6,** 159–170.
25. Rudiger, M., Haupts, U., Moore, K. J., and Pope, A. J. (2001) Single-molecule detection technologies in miniaturized high throughput screening: binding assays

for G protein-coupled receptors using fluorescence intensity distribution analysis and fluorescence anisotropy. *J. Biomol. Screen.* **6,** 29–37.

26. Scheel, A. A., Funsch, B., Busch, M., Gradl, G., Pschorr, J., and Lohse, M. J. (2001) Receptor-ligand interactions studied with homogeneous fluorescence-based assays suitable for miniaturized screening. *J. Biomol. Screen.* **6,** 11–18.

19

Single-Molecule Study of Protein–Protein and Protein–DNA Interaction Dynamics

H. Peter Lu

Summary

Protein–protein and protein–DNA interactions play critical roles in biological systems, and these interactions often involve complex mechanisms and inhomogeneous dynamics. Single-molecule spectroscopy is a powerful and complimentary approach to decipher such spatially and temporally inhomogeneous protein interaction systems, providing new information that are not obtainable from static structure analyses, thermodynamics characterization, and ensemble-averaged measurements. To illustrate the single-molecule spectroscopy and imaging technology and their applications on studying protein–ligand interactions, this chapter focuses on discussing two recent single-molecule spectroscopy studies on protein–protein interaction in cell signaling process and on protein–DNA interactions in DNA damage recognition process.

Key Words: Single-molecule spectroscopy and imaging; single-molecule protein conformational dynamics; protein–protein interaction dynamics in cell signaling; protein–DNA recognition dynamics in DNA damage recognition and repair.

1. Introduction

Protein–protein and protein–DNA interactions play critical roles in biological functions of living cells, such as cell signaling, receptor-ligand activation, cellular metabolism, DNA damage recognition and repair, gene expression, replication, recombination, etc. These protein interactions often involve complex mechanisms and inhomogeneous dynamics with significant conformational changes. Protein–protein, protein–ligand, and protein–DNA interactions are often intrinsically single-molecule processes at an induction stage associ-

From: *Methods in Molecular Biology, vol. 305: Protein–Ligand Interactions: Methods and Applications*
Edited by: G. U. Nienhaus © Humana Press Inc., Totowa, NJ

ated with the initiation of crucial early events in living cells. For example, cell-signaling processes are often initiated through a few copies of protein-interaction complexes, being amplified along the signaling pathway. Protein interactions in living cells are typically inhomogeneous, both spatially and temporally, and protein interaction dynamics can be different from molecule to molecule and from site to site. Furthermore, the dynamics can be different from time to time for the same individual molecules ***(1–12)***, which is beyond the scope of the conventional kinetics. Such differences are mainly in changes of mechanisms and fluctuations of rate processes. Inhomogeneous and complex protein interactions are extremely difficult to identify and analyze in an ensemble-averaged measurement because they depend upon averaged properties. When many molecules are measured simultaneously, nonsynchronizability, stochastics, and multiple-state conformations ***(5–12)*** involved in protein–protein interactions present a significant difficulty for a molecular-level mechanistic characterization. To characterize dynamic and inhomogeneous chemical and biological events, single-molecule approaches are powerful because they remove the inhomogeneity and characterize the complex dynamics by studying one molecule at a time ***(9,11,13–23)***. Furthermore, the high temporal and spatial resolution obtainable in single-molecule fluorescence spectroscopy and imaging ***(5–12)*** makes them ideal for studying protein interaction dynamics under physiological conditions.

This chapter focuses on one important area in this rapidly developing field: protein–DNA and protein–protein interactions. Specifically, we focus our discussion on applying single-molecule fluorescence spectroscopy to protein–DNA interactions in DNA damage recognition ***(9)*** and protein–protein interactions in a cell signaling system ***(10)***.

Among the major new findings is that protein–protein and protein–DNA interactions and recognitions often involves highly flexible protein tertiary structures under slow conformational fluctuations at a time-scale of submilliseconds to seconds. Large-amplitude conformational fluctuations have been observed that underlie significant inhomogeneities in the interactions of protein complexes. To a certain extent, recent nuclear magnetic resonance (NMR) static and ensemble-averaged structural studies of protein complexes also indicate that binding domains undergo dramatic conformational changes from disordered to ordered states upon complex formation ***(24,25)***. In fact, structural transitions of flexible conformational domains, which can be identified and characterized in detail by measuring single-molecule conformational fluctuation dynamics, are likely common in biomolecular recognition processes ***(9,10)***.

We have applied single-molecule spectroscopy to study the interactions of an intracellular signaling protein Cdc42, a GTPase, with its downstream effector protein Wiskott–Aldrich Syndrome Protein (WASP) ***(10)***. In this work, a WASP fragment that binds only the activated Cdc42 was labeled with a solvatochromic dye to probe the hydrophobic interactions at the protein–protein interface that are significant to Cdc42/WASP signaling recognition. Cdc42 is a monomeric GTP-binding protein (a GTPase) that acts as a molecular switch in signaling pathways to regulate various cellular responses ***(26–28)***. Cdc42 GTPase can be activated by binding to GTP (guanosine 5'-triphosphate), and the activated Cdc42 then binds and activates a series of effector proteins via direct protein–protein interactions. Our single-molecule spectroscopy study revealed static and dynamic inhomogeneous conformational fluctuations of the protein complex involving bound and loosely bound states of the protein–protein interaction complex ***(10)***.

Single-molecule spectroscopy has been applied to dissect protein–DNA interaction dynamics, revealing the conformational fluctuation of molecular noncovalent interactions within protein–DNA complexes in a DNA damage recognition system associated with DNA repair ***(9)***. DNA damage recognition and repair have been one of the most active research areas in molecular biology, structural biology, and biophysics in recent years ***(24,25,29–31)***. It is critical to maintain the integrity of the genome from harmful mutations due to endogenous and exogenous sources of DNA damage. Molecular-level identification and characterization of subtle interactions of protein–DNA complexes provide quantitative information on the dynamic parameters regulating recognition-repair protein–DNA interactions, such as changes in conformations and protein–DNA binding associated rates.

There are primarily two types of DNA damage repair systems, base excision repair (BER) and nucleotide excision repair (NER). BER works mainly on lesions caused by endogenous agents, and NER works mainly on helix-distorting damage caused by environmental mutagens and some lesions occurring as a consequence of oxidative metabolism ***(29–31)***. DNA damage recognition involves a complex mechanism of protein interactions and conformational motions: first, damage recognition protein(s) must recognize and bind to highly variable kinds of DNA lesions; then, the resulting damage recognition complex must signal and recruit additional proteins to the damage site; finally, the recognition proteins must eventually remove themselves from the resulting complex sufficiently to enable DNA repair as additional proteins participate. It is apparent that if such complex processes of dynamic multistep molecular interactions are ensemble-averaged, many critical stochastic conformational changes and protein interaction motions become indistinguishable and averaged out in the measurements. Single-molecule approaches, studying one protein–DNA complex at a time, are the reasonable choice for measuring this complex and inhomogeneous mechanism.

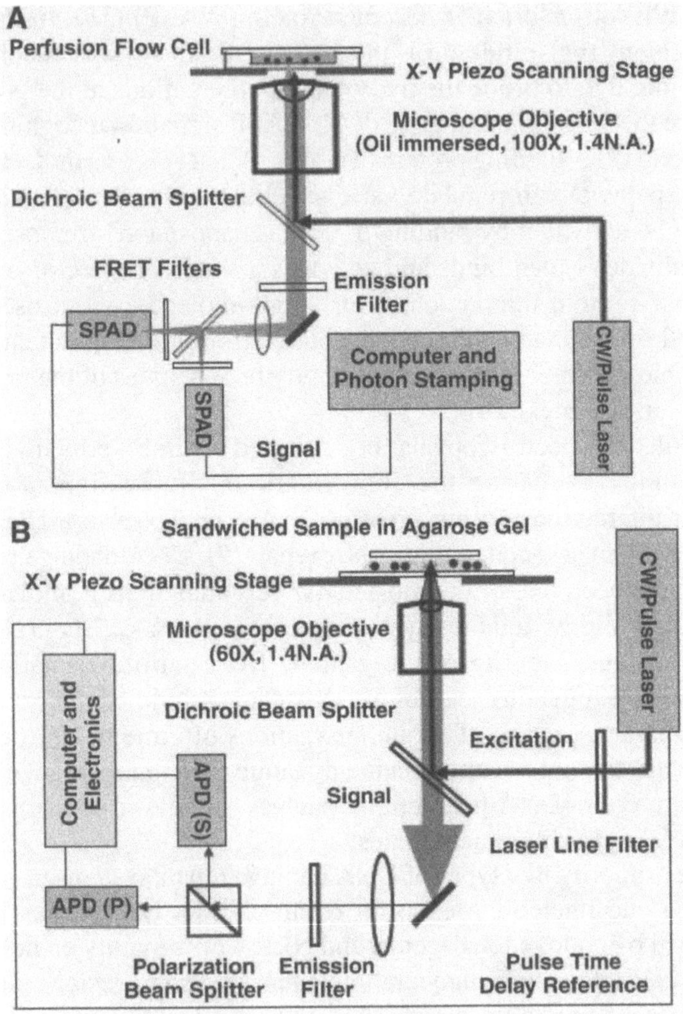

Fig. 1. Instrumentation for single-molecule fluorescence spectroscopy and imaging experiments. Single-molecule measurements were typically performed on an inverted confocal fluorescence microscope (Nikon Diaphot 300 or Zeiss Axiovert 200). The excitation light is, depends on the experiments, from a CW or pulse laser source coupled through fiber optics into the microscope. Individual protein complexes embedded in agarose gel or covalently tethered on glass surface were located through fluorescence imaging by raster-scanning the sample using a computer controlled close-loop x-y electro-piezo scanning stage (Queensgate, UK, or Physik Instrumente, Germany). After obtaining two-dimensional single-molecule fluorescence image, the laser focal spot can park on top of a single molecule to collect time trajectories of fluorescence. Fluorescence photons were directed onto a pair of avalanched photodiodes (APD) for acquiring emission images and time trajectories. *(Continued on next page.)*

2. Materials and Methods

This section contains a discussion of single-molecule spectroscopic methods that are directly related to single-molecule spectroscopy studies on protein–protein and protein–DNA interaction dynamics.

2.1. Single-Molecule Fluorescence Spectral and Intensity Fluctuation Spectroscopy

Single-molecule spectroscopy (**Fig. 1A,B**) is able to record time trajectories of a single-molecule property in real time, and such time trajectories contain detailed dynamic information. Single-molecule trajectory recording has been demonstrated for translational diffusion *(32–34)*, rotational diffusion *(11,35–37)*, spectral fluctuation *(38)*, and conformational fluctuation *(5,6,8a,8b–12)*. Of particular interest has been real-time observation of chemical reactions of biomolecules, including DNA–protein and protein–protein complex formations and conformational changes and protein–DNA *(9)* and protein–protein interactions *(10)*. These approaches remove or characterize the inhomogeneities of complex biological systems by studying one molecule or molecular complex at a time. These trajectories reveal intrinsic single-molecule conformational dynamics and local environmental change dynamics through calculation of first-order and higher-order autocorrelation functions from the trajectories *(9,10)*.

2.2. Single-Pair Fluorescence Resonant Energy Transfer

Single-pair fluorescence resonant energy transfer (spFRET) may occur between two dye molecules at the distance of about 80 Å or less, and their respective emission and absorption spectra overlap to give a Förster energy transfer radius of about 50 Å. In a spFRET event, excited state excitation energy is transferred from a donor to an acceptor. The spFRET can be measured between a single donor-acceptor pair attached to a single protein or a single complex (**Fig. 1A**). The quantum efficiency (E) of spFRET is inversely related to the sixth power of the distance, r (Å), and is sensitive to the relative orientation of the donor and acceptor transition dipoles *(12,14a,14b,40)*. Molecule conformational changes, protein motions, and protein–DNA and protein–protein interactions can be monitored by exciting the donor and monitoring either

---

Fig. 1. *(continued)* (**A**) Single-molecule fluorescence fluctuation spectroscopy and fluorescence resonant energy transfer are measured by detecting fluorescence from either a single probe dye or a pair of donor-acceptor FRET dye probes. (**B**) Single-molecule nanosecond anisotropy and static anisotropy are measured by detecting parallel and perpendicular polarized emission, and the protein rotational moditions, as well as conformational motions can be characterized at a wide time scale from subnanoseconds to minutes.

the donor's or the acceptor's intensity trajectories, or both simultaneously. Site-selectively labeling DNA with a donor and acceptor on either side of the damaged site enables one to observe the binding-induced conformational changes through monitoring donor and/or acceptor fluorescence intensity and lifetime fluctuations due to single-pair FRET. Single-molecule spFRET studies of enzymatic reaction ***(12,14)***, RNA and ribozyme conformational changes ***(14,42,43)***, and a few other important biological processes ***(39–41)*** have been reported in recent years. Because of photobleaching, probe molecule diffusion motions, and spectral fluctuations of single molecules, spFRET is a reliable and capable method for probing distance change dynamics but not reliable for measuring exact distances.

2.3. Single-Molecule Photon Stamping Spectroscopy

Photon-stamping detection of single-molecule fluorescence using single-photon counting has been widely applied and demonstrated recently in a number of laboratories ***(5,11,44,45)***. In our experiments (**Figs. 1B** and **2**), photon stamping can be done in two ways: 1) A continuous-wave (CW) laser is used for measuring the time trajectory of the single-molecule fluorescence intensity fluctuation by detecting each photon and *stamping* (recording) its arrival time. This technique has pushed the single-molecule fluorescence intensity detection to the detection limit so that the time sequence of every detected photon is recorded, and subsequent off-line binning (taking an average of a certain number of channels consecutively) provides the highest time resolution. The time resolution is only limited by the photon flux, excitation saturation, photobleaching, and detector counting rate (15MHz, SPCM-AQR 16 Avalanche Photodiode). 2) A pulsed-laser based approach is used in which each emitted photon is detected and its delay time stamped, Δt, from the pump laser pulse. The delay time is the duration from the time of excitation of the single molecule to the emission by the same molecule, which is a single event measurement of the fluorescence lifetime of this molecule. As a Poisson process, the distribution of the delay time gives an exponential decay, and the statistical mean of the distribution reports the value of the single-molecule fluorescence lifetime.

2.4. Single-Molecule Static and Transient Anisotropy

Single-molecule rotational motions can be probed by using p- and s-polarized two-channel detection (**Fig. 2B**). This measurement was typically conducted using a CW laser, and therefore the time-resolution is submilliseconds to seconds, which is suitable to study molecular rotational jumps and slow molecular conformational motions ***(36,40)***. To probe faster conformational motions, single-molecule time-resolved fluorescence anisotropy has been developed and applied, and this technique has significantly enhanced the

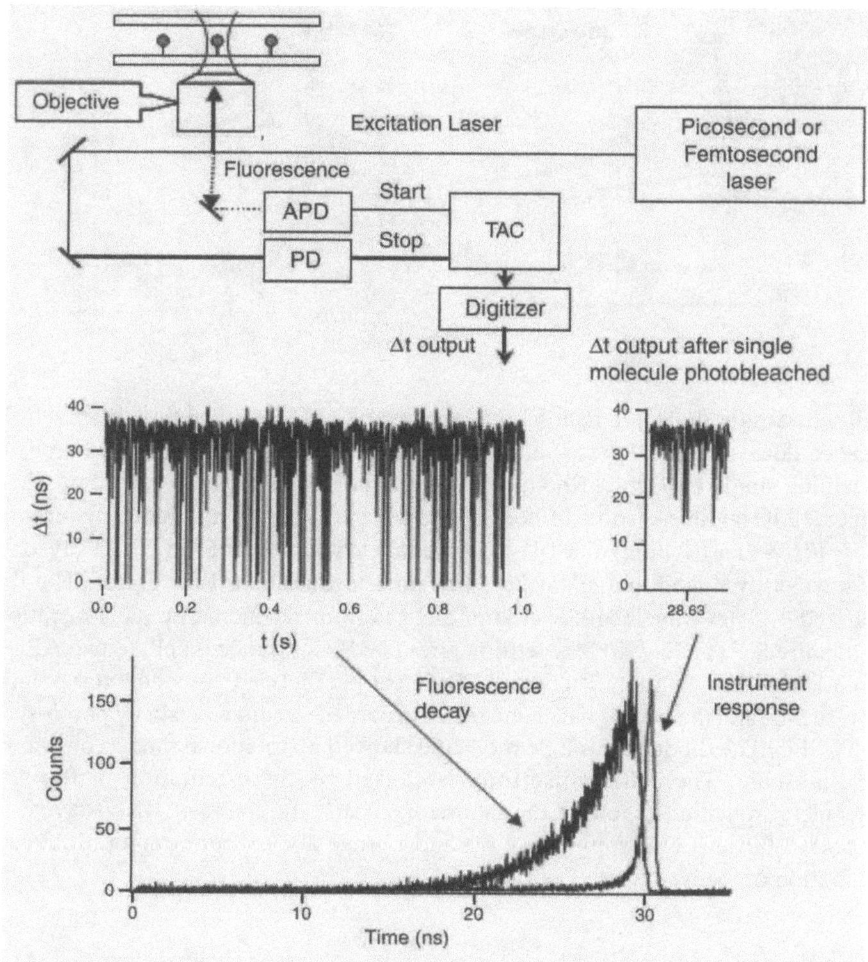

Fig. 2. Diagram of the single-molecule fluorescence photon arriving-and-delay time stamping instrument based on the traditional time-correlated single photon counting (TCSPC) approach. A picosecond or femtosecond laser is used to excite samples and to generate *start* pulses. The sample fluorescence is detected by an APD to produce *stop* pulses. The arriving time and delay time are recorded by a similar technique of TCSPC using a digitizer of our own design to record the arrival time and amplitude of TAC output, which corresponds to the delay time between the excitation laser pulse and the fluorescence photon. The binning of the arrival time gives a fluorescence intensity trajectory, and the histogram of the delay time of each photon gives decay of the emission that is corresponding to a single expo-nential decay as a Poisson process or a nonexponential decay as a nonPoisson process. Using this approach, single-molecule lifetime can be studied by averaging only hundreds of photons instead of fitting the histogram to exponential decay, making it possible to study the fluctuations in the decay time with adequate time resolution ***(10–12)***.

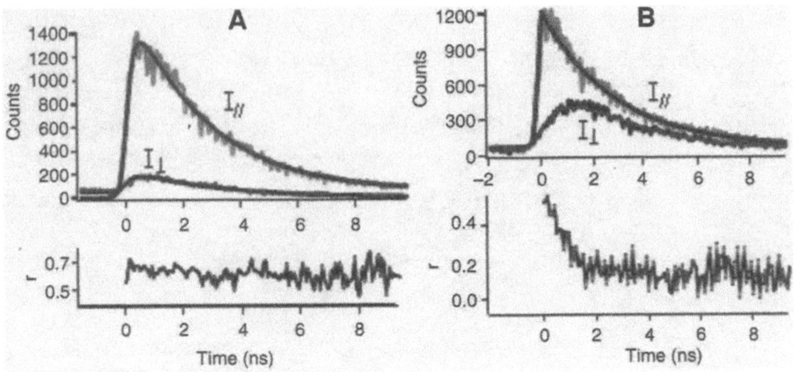

Fig. 3. Single-molecule nanosecond anisotropy ***(11)***. **(A)** The time-resolved fluorescence decays $I_{//}$ and $I_{\perp}$ (top) and anisotropy $r = (I_{//} - I_{\perp})/(I_{//} + 2I_{\perp})$ (bottom) of a rhodamine single molecule. Rhodamine single molecules are spin-coated on PMMA surface. **(B)** The time-resolved fluorescence decays $I_{//}$ and $I_{\perp}$ (top) and anisotropy $r = (I_{//} - I_{\perp})/(I_{//} + 2I_{\perp})$ (bottom) of a T4 lysozyme labeled with Alexa 488. The T4 lysozyme was covalently linked to a silanized glass surface and under PBS buffer. The linear polarized 475 nm wavelength excitation light is from second harmonic generation of Ti:sapphire laser at 75.7 MHz repetition rate. The fluorescence is split to two APDs by a polarizing beam splitter. The two APD detect parallel and perpendicular polarized fluorescence and the two signal channels are recorded simultaneously by photo-stamping TCSPC. The rhodamine single molecule showed no rotational motion on nanosecond timescale. The static anisotropy is caused by the orientation of the single molecule's transition dipole in the laboratory frame. In contrast, T4 lysozyme with Alexa488 showed anisotropy decay at nanoseconds due to the spinning of Alexa488 and motion of T4 lysozyme (Courtesy of Dehong Hu).

single-molecule anisotropy measurement of the protein rotational motions and conformational changes from submilliseconds ***(11)*** to nanoseconds (**Fig. 3**).

Molecule rotational motion rates are dependent upon molecular hydrodynamic shapes and masses that can be changed by conformational changes, molecular interactions, and complex formation and dissociation. Measuring the nanosecond time-dependent anisotropy of a site-specifically labeled dye molecule probes the conformational motions of the labeled domain of the protein ***(11a,11b,46)***. Because the time-scale for dye-molecule spinning is in the subnanosecond range, the nanosecond-to-microsecond protein conformational motions can be differentiated and probed by measuring the time-dependent anisotropy of the dye molecules. Recently, a new dye-labeling technique has provided covalently attached dye-probe molecules. The rotational motions of these molecules are locked on protein matrixes that contain a tetra-cysteine motif, which is genetically inserted within an α-helix region near the amino terminus. The tetra-cysteine motif then reacts with FlAsH, a fluorescein derivative with two As(III) substituents, which fluoresces only after the arsenics bind to cysteine thiols *(47)*. The tight arsenic-binding structure ensures that the fluorescent label has no rotational freedom relative to the protein matrix so that the protein motions can be probed unambigiously without a convolution of fast dye molecule motions *(11b)*.

To be able to measure the single-molecule spectroscopic time trajectories, there are three types of single-molecule sample preparations typically used in a single-molecule spectroscopy measurement (**Fig. 4**). The essential requirements are: 1) confine the translational diffusion within diffraction-limitted laser focus, and 2) maitian the physiological conditions for the single-molecule proteins. The proteins can either be covalently tethered to modified glass surface or confined in agarose gel containing of 99% of buffer solution *(5–12)*.

2.5. Application of Magnetic Tweezers for Detecting Protein–DNA Interactions

Combined with fluorescence measurements, magnetic tweezers have been successfully applied in recent years to single-molecule studies of protein interactions *(48–50)*. With a zero optical background contamination, magnetic tweezers can hold a DNA or a protein molecule in solution while its interactions with another protein are being examined *(48–50)*. In such experiments, the single-molecule DNA or protein is tethered to a glass surface on one end through thiolation and bifunction linker interactions *(51)*, and on the other end tethered to magnetic paramagnetic beads through a biotin-streptavidin linkage. There are different designs of magnetic tweezers *(5,12,52)* and the magnetic field can be turned on and off, high and low, by tuning the electric voltage on the electromagnetic magnets. Sophisticated magnetic tweezers can even rotate and fast-modulate the magnetic field to perform molecular manipulations. The combination of magnetic tweezers and single-molecule fluorescence spectroscopy *(40)* is highly effective in studying single-molecule protein interactions with both DNA and proteins because of its lack of optical contamination, the low-cost of the magnetic tweezers technique, the versatility of its applications, and the biological compatibility of a paramagnetic beads attachment. Although this chapter does not review the details of this approach, the applications of magnetic tweezers will definitely be a relevant and effective approach.

3. Single-Molecule Protein–Protein Interaction Dynamics

3.1. Protein–Protein Interaction System in Cell-Signaling Processes

This section focuses on a protein–protein interaction system, involving a dye-labeled fragment of WASP, CBD (Cdc42 binding domain of WASP),

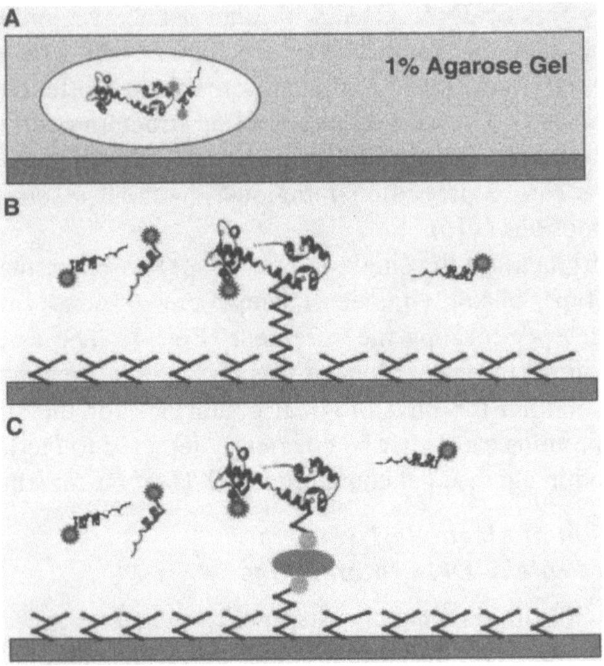

Fig. 4. Sample preparations for studying single-molecule protein-ligand interactions using single-molecule spectroscopy and imaging. **(A)** Covalent tethering of the single-molecule proteins on a hydrocarbon modified glass surface ***(11,12)***. Coverslips were first derivatized with a 10% (v/v) mixture of 3-mercaptopropyl-trimethoxysilane and isobutyltrimethoxysilane (1:10^4 ratio) in DMSO for 24 h. After washing with methanol and PBS buffer, coverslips were incubated in 10 nm of SIAXX (Molecular Probes, S-1668) in PBS buffer pH 7.2 for 1 h. After washing, the coverslips were incubated with 1 nm of protein in 0.2 *M* of carbonate buffer, pH 8.2, for 1 h followed by rinsing with PBS buffer. **(B)** Embedding single protein-ligand complexes at 10^{-10} *M* concentration in agarose gel containing 99% of buffer solution ***(8a,8b–10)***. **(C)** Using avidin-streptavidin linking complex to tether single-molecule proteins on a polyethelyne glycol modified glass surface ***(42,43)***.

which was studied for a single-molecule characterization on its protein–protein binding interactions. CBD is a 13-kDa WASP fragment (residues 201–320) that contains the CRIB (Cdc42/RAC interactive binding) motif (residues 238–251), an N-terminal portion (residues 201–237), and a C-terminal segment (residues 252–320), with dye labeling at residue 271 via a cysteine mutation. The dye, solvatochromic dye I-SO (indolenine-benzothiophen-3-one-1,1-dioxide), exhibits fluorescence that is highly sensitive to solvation energetics and solvent polarity, stimulating severalfold changes in the hydrophobicity of local environments ***(10,26)***.

Ensemble-averaged fluorescence assays were prepared by 1:1 volume-ratio mixing of CBD (300 n*M*) and Cdc42 (1 μ*M*) loaded with GTP or GDP (10 μ*M*) (GTP: guanosine 5'-triphosphate; GDP: guanosine 5'-diphosphate) in buffer solutions (50 m*M* Tris-HCl, pH 7.6, 50 m*M* NaCl, 5 m*M* $MgCl_2$, 1 m*M* DTT). For the single-molecule experiments, Cdc42-CBD protein complexes at n*M* concentration were imbedded in agarose gel (0.5%) (**Fig. 4A**) and sandwiched between two cleaned cover glasses. The formed agarose gel matrix was about 10 to 20 μm in thickness, and the aqueous environment could be maintained for a few hours. In agarose gel, the translational diffusion of the single protein complexes were confined in a physiologically relevant buffer environment within the microscopic laser focal point of approx 300 nm diameter, the rotational motions of the protein complexes within the agarose gel were free, and the interactions between proteins and the agarose gel negligible ***(10)***.

Previous NMR analyses have indicated that the WASP fragments undergo dramatic conformational changes from disordered to ordered tertiary structures in the presence of the GTP-activated Cdc42 ***(53,54)***. Preferential binding of WASP to GTP-activated Cdc42 appears to derive not only from polar and hydrophobic contacts involving highly conserved residues in the CRIB (Cdc42/RAC interactive binding) motif but also from hydrophobic interactions outside the CRIB motif ***(53,55)***. In this single-molecule spectroscopy study, the dye molecule was intended to probe the extra-CRIB hydrophobic interactions significant to Cdc42/WASP recognition. The dye molecule was strategically attached to a site among hydrophobic residues in order to probe relevant Cdc42-CBD interactions while minimally perturbing protein–protein interactions.

3.2. Control Experiments

The fluorescence of the I-SO dye has high sensitivity to changes in solvent hydrophobicity (**Fig. 5A**) ***(10,26)***, increasing by about threefold as the solvent hydrophobicity increases from water to methanol and butanol. The fluorescence lifetime of the solvachromic dye is also found to be highly sensitive to hydrophobicity, changing from approx 200 ps in water to approx 1 ns in more hydrophobic solvents such as methanol.

Upon forming a complex, the protein–protein binding interface often has significantly different hydrophobicity comparing to that of the aqueous environment. The attached I-SO dye, which is highly sensitive to changes in hydrophobicity, probes the accessible active binding site of Cdc42-CBD that forms a hydrophobic interface. The result of a control fluorescence assay experiment (**Fig. 5B**) showed an approximately threefold increase in the activated GTP-loaded Cdc42 that was binding to CBD to form a protein–protein complex ***(10,26,28)***. By comparison, nonactivated GDP-loaded Cdc42 did not significantly change the fluorescence intensity of the dye probe under the same mea-

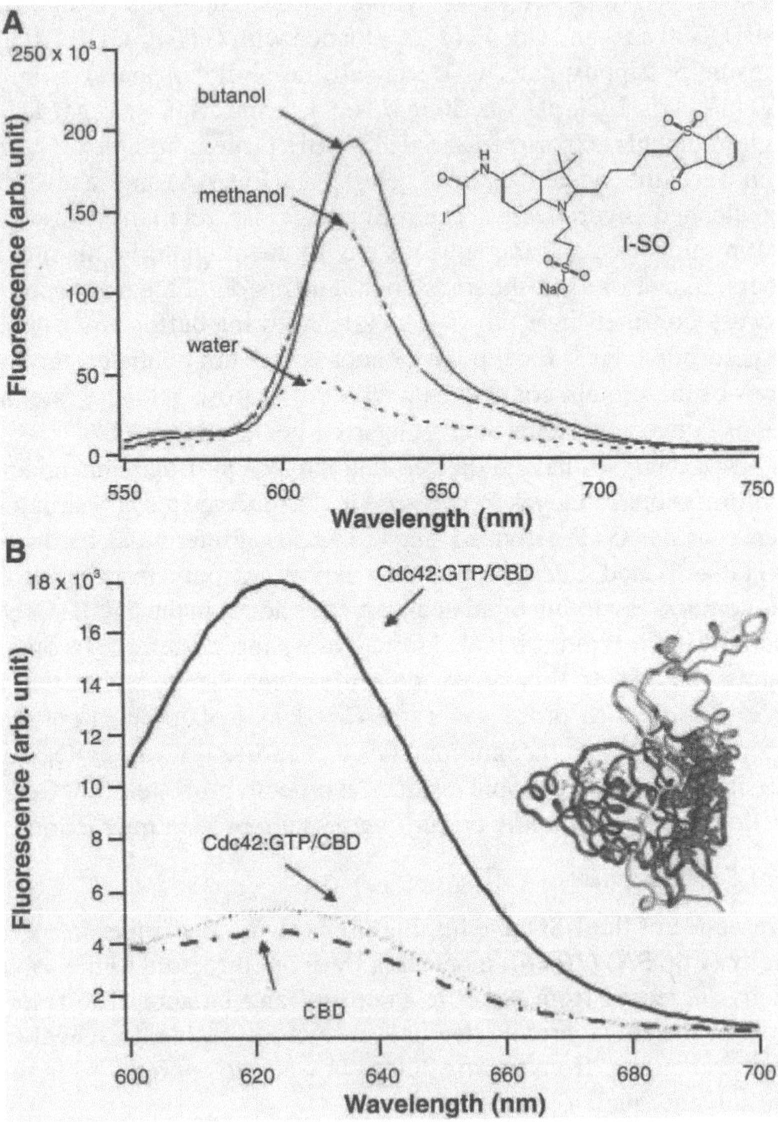

Fig. 5. Protein–protein interaction fluorescence assay and fluorescence characterization of the I-SO dye probe. **(A)** Ensemble-averaged fluorescence emission spectra of dye (I–SO) under the same concentration and excitation intensity in water (dotted curve), methanol (dash-dotted curve), and butanol (solid curve). The fluorescence intensity increases by about factor of 4 as the hydrophobicity increases from water to butanol. **(B)** Ensemble fluorescence assays of Cdc42 with CBD biosensor using fluorescence spectral measurements. Fluorescence spectra were obtained with excitation at 568 nm.

(Continued on the opposite page)

surement conditions because Cdc42, as a GTPase, was not activated by GDP and was incapable of binding to the CBD biosensor *(26,28)*. Furthermore, in control experiments we observed only a low fluorescence intensity for dye-labeled CBD alone in a buffer solution, suggesting that CBD does not provide a hydrophobic environment for the dye molecule. Without the presence of Cdc42 in buffer and agarose gel, the single molecules of dye-labeled CBD alone were not detectable beyond background noise in a single-molecule spectroscopy measurement.

3.3. Single-Molecule Experiments on Protein–Protein Interaction Dynamics

All single-molecule measurements were performed under ambient conditions. We used an inverted confocal fluorescence microscope (Nikon Diaphot 300) *(10)* with an optical-fiber coupled multiple laser system, consistent with a continuous-wave krypton-ion laser (Innova 90C, Coherent), a CW Argone-ion laser (MWK Laser, Riverside, CA), and a Nd:YAG pumped picosecond-pulsed R6G dye laser (Coherent) tunable from 560 nm to 600 nm (**Fig. 1B**). A high numerical aperture objective (Nikon 60×, N.A. = 1.4) was used to focus the linear polarized light near diffraction-limited spot, typically around a 300 nm diameter. A photon-stamping detection system recorded the time trajectories of single-molecule fluorescence photon counting and anisotropy. Individual protein complexes were located by raster-scanning the sample, recording the fluorescence photon counting at each imaging pixel. A computer-controlled closed-loop x-y electropiezo nanometer-resolved scanning stage (Physical Instruments, Germany) was used to register the single-molecule fluorescence images and facilitate the moving of laser focal spots on top of particular single molecules to collect time trajectories. Using single-molecule spectroscopy and fluorescence imaging, we collected fluorescence emission images (**Fig. 6A**)

Fig. 5. *(continued)* The solid curve represents the active GTP-activated Cdc42 forming complexes with dye-labeled CBD; the dotted curve represents the GDP-loaded Cdc42 with dye-labeled CBD; and the dash-dotted curve represents the dye-labeled CBD alone. A nonhydrolyzable GTP analogue, GTP-γ-S, was used to lock Cdc42 in the active conformation. Thus the GTP binding and unbinding process was eliminated from the single-molecule experiments that measured only the activated Cdc42 interacting with CBD. The effects of GTP-γ-S are the same as that of GTP, and the biological relevance and validity of using GTP-γ-S has been well established in literature *(26–28)*. **Inset:** a structure based on MD calculations of protein complex Cdc42 (dark ribbon structure on the left)/WASP (light ribbon structure on the right) with dye attachment (dark ball structure) outside the CRIB motif among hydrophobic residues (purple). (Adapted in part with permission from **ref. *10***. Copyright 2003 American Chemical Society.)

and time trajectories (**Fig. 6B**) from individual Cdc42-CBD complexes immobilized in an agarose gel matrix, as shown in **Fig. 4A**. Emission photons were directed onto one or two silicon avalanche photodiodes (APD) to acquire emission images and time trajectories from single protein complexes. For single-molecule anisotropy experiments, a polarization beam-splitter cube (CVI Laser) was used to separate the two orthogonal polarization components into two APD detectors. Fluorescence time-stamping TCSPC (time-correlated single photon counting) ***(10,11)*** data were collected using a PicoQuant Time-Harp 200 card in a Time-Tagged Time-Resolved (T3R) mode and Becker & Hickl fluorescence lifetime imaging electronics.

Figure 6B shows raw data consisting of pairs of registered times for each detected photon, the sequential arrival time (t) and the delay time (Δt) with respect to the laser excitation reference signal ***(10,11)***. The fluorescence intensity trajectory can be calculated from binning the arrival time (t) (**Fig. 6C**), and a nanosecond fluorescence decay curve can be calculated from the distribution of the delay time (Δt) of the emission photons (**Fig. 6D**). Photon stamping detection was able to record both the fluorescence intensity and the lifetime simultaneously ***(10,11)***.

From the trajectories of single protein complexes, $\{I(t)\}$, we observed a more than twofold variation in fluorescence intensity fluctuations at a time-scale ranging over two orders of magnitude (**Fig. 6C**). Based on the control experiments that indicated that the detectable single-molecule fluorescence comes only from the Cdc42-CBD complex, we attributed the fluorescence intensity fluctuations to conformational fluctuations of the protein complex ***(10)***. The fluorescence intensity fluctuations were further confirmed by additional control experiments to be caused by protein–protein interactions rather than to dye motions, hindered rotation, or rotational jumps of the protein complex or to intrinsic photophysical processes ***(10)***. The dissociation constant of activated Cdc42 and the CBD fragment of WASP has been reported ***(27,28)*** in the range of approx 10^{-2} s^{-1}, suggesting that dissociation of the Cdc42-CBD complex is a rare event during our single-molecule trajectory measurement within a few seconds. A typical statistical method to analyze fluctuation dynamics is to use the autocorrelation function to analyze the rate of the fluctuation ***(1,5,7,10,12)***. The autocorrelation function, $C(t) = < \Delta I(t)\Delta I(0) >/< \Delta I(0)^2 >$, is calculated from a single complex fluorescence intensity trajectory $\{I(t)\}$, where $\Delta I(t) = I(t) - < I(t) >$. **Figure 7A** shows an example of the autocorrelation function of a single-complex intensity trajectory. A typical spike at $t = 0$ is a result of the uncorrelated measurement noise and faster fluctuations beyond the instrument time-resolution. For $t > 0$, the autocorrelation functions of all single-complex trajectories are fit to either single- or bi-exponential decays, $C(t) = A_f \exp(-k_f t) + A_s \exp(-k_s t)$.

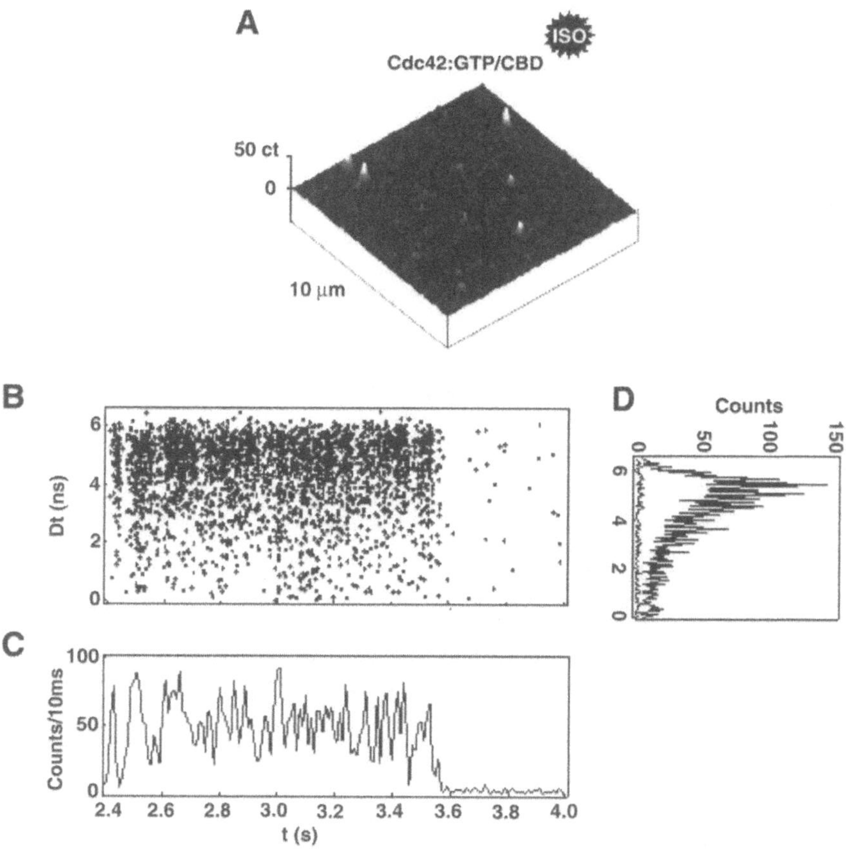

Fig. 6. Single-molecule spectroscopy and imaging study of Cdc42-CBD binding dynamics. **(A)** A single-molecule fluorescence raster-scanning image of GTP-loaded Cdc42 in complex with dye-labeled CBD biosensor. **(B)** An example of the raw data of one-channel single-molecule photon time-stamping TCSPC trajectory. Each dot corresponds to a photon stamped with an arrival time (t; x-axis) and a delay time (Δt; y-axis). **(C)** The fluorescence intensity trajectory is calculated from the histogram of the arrival time (t) with 0.01 s time-bin resolution. The dye molecule was photobleached at approx 3.6 s. **(D)** The nanosecond fluorescence decay curves (top-right plot) are the histograms of the delay time (Δt) of the fluorescence photons ($t < 3.6$ s) and background photons ($t > 3.6$ s). (Adapted with permission from **ref. *10***. Copyright 2003 American Chemical Society.)

Detailed experimental characterizations of the protein–protein interaction conformational fluctuation conclude that the measured conformational fluctuations were spontaneous rather than photo-induced. Moreover, the dynamics of the conformational fluctuations involve bound (B) and loosely-bound (LB)

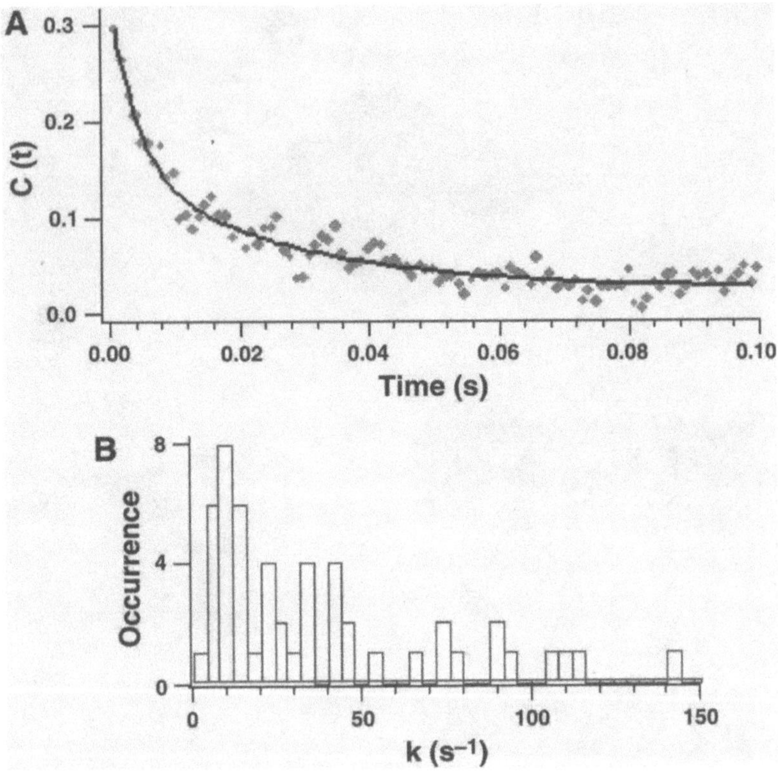

Fig. 7. (**A**) The second-order autocorrelation function, C(t), calculated from a fluorescence intensity trajectory I(t) of a single Cdc42-CBD complex. The solid curve is a bi-exponential fit with decay rates of 250 ± 60 and 45 ± 10 s^{-1}. (**B**) The occurrence histogram of the single-complex Cdc42-CBD conformational fluctuation rates is constructed by using parameters from fitting the autocorrelation functions for 60 individual protein complexes. (Adapted in part with permission from **ref. *10***. Copyright 2003 American Chemical Society.)

states of the protein complex ***(10)***. The LB states are a subset of conformations with deviated nuclear displacements from the bound equilibrium states; they distort the protein–protein interaction interface and the local environment of the dye probe without disrupting the subnanometer long-range interactions ***(10)*** so that the overall protein complex is still associated ***(10)***. Compared with the B state, the LB state gives significantly lower fluorescence intensity as the distorted protein–protein interaction interface probed by I-SO dye becomes more solvent-accessible and hydrophilic. With respect to the ensemble-averaged assay experiment (**Fig. 5B**), the B state corresponds to the high emission-intensity Cdc42-CBD bound equilibrium state where the local environment of the dye probe is more hydrophobic, and the LB state has a hydrophilic local environment of the dye probe resembling the environment of CBD alone. In the measured single-molecule experiments, fluorescence fluctuations at millisecond and subsecond time-scales reflect the Cdc42-CBD conformational changes between the B and LB states.

One of the most interesting observations is that the conformational fluctuation rates are found to be highly inhomogeneous (**Fig. 7B**), which is reflected by a broad distribution of fluctuation rates. Variations of more than two orders of magnitude occur in the conformational fluctuation rates among individual protein complexes. Although it is difficult to identify exactly how many conformational states contribute to the inhomogeneous distribution, at least two subgroups of states are associated with conformational fluctuations at approx 10 s^{-1} and approx 40 s^{-1} (**Fig. 7B**). About 25% of the single-complex fluorescence intensity trajectories demonstrated bi-exponential decays, indicating non-Poisson kinetics ***(1–12)***. The non-Poisson behavior suggests that the Cdc42-CBD interactions have both static and dynamic inhomogeneities. The dynamic inhomogeneity associated with the rate fluctuation, which is beyond the conventional kinetics has been observed for other protein systems ***(5–12,60–62)***.

We have further explored the possible molecular structure of these multiple conformational states (B and LB) using molecular dynamics (MD) simulations. The MD simulations explored the interfacial structures of protein–protein interaction in the dye-labeled Cdc42-CBD complex. A representative complex of Cdc42 (blue) and CBD (yellow) is shown in **Fig. 8A**. The solvatochromic dye was located within the protein interfacial pocket. Among amino acid residues (purple) surrounding the dye molecule (red) were Cdc42, Phe37, and Phe56 residues and WASP Leu263 and Leu267 residues, as shown in **Fig. 8B**. These residues formed important hydrophobic contacts. The MD calculations further showed that the dye molecule probed important hydrophobic interactions outside the conserved binding motif CRIB (white). Various modes of interactive motions, especially those involving the important hydrophobic residues (Cdc42, Phe37, and Phe56 and WASP Leu263 and Leu267), could give rise to multiple conformations between B and LB states while the overall complex is still associated. **Figures 8C,D** demonstrate that the solvent-accessible surface area of the dye molecule undergoes a decrease from approx 377 Å^2 to approx 225 Å^2 upon Cdc42-CBD binding. This result signifies an increasingly hydrophobic environment surrounding the dye molecule upon formation of the protein complex, which is consistent with ensemble fluorescence measurements (**Fig. 5**). It is likely that the B and LB conformational states correspond to different degrees of distortion of the solvent-accessible surface and that the LB states are more solvent-accessible.

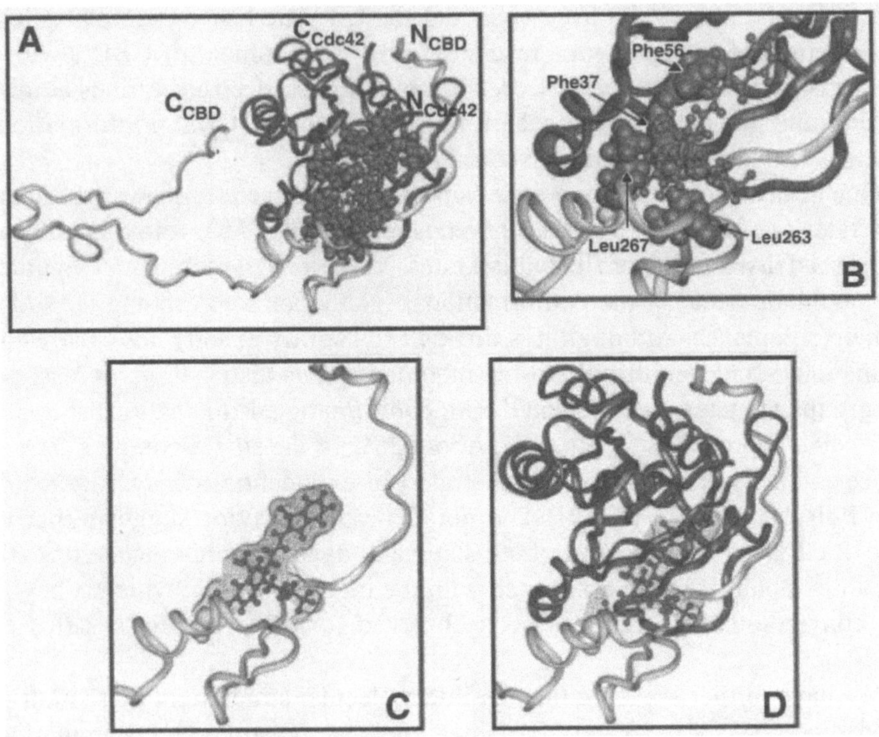

Fig. 8. **(A)** A representative structure of Cdc42 (dark ribbon structure on the right)/ CBD (light ribbon structure) complex based on MD calculations. The CPK renderings illustrate the amino acid residues (dark ball structure) surrounding the dye molecule (ball-stick structure). **(B)** The important hydrophobic interactions involving Cdc42 Phe37 and Phe56 and WASP Leu263 and Leu267 were probed by the dye molecule. **(C)** Solvent-accessible surface (mesh rendering) of the dye molecule prior to binding of CBD to Cdc42; the surface area was approx 377 Å^2. **(D)** Solvent-accessible surface of the dye molecule after Cdc42-CBD binding; the surface area reduced to approx 225 Å^2. (Adapted with permission from **ref. *10***. Copyright 2003 American Chemical Society.)

4. Protein–DNA Interaction Dynamics

4.1. Protein–DNA Interaction System in a DNA Damage-Recognition Process

The second protein interaction system we discussed below is a protein–DNA interaction complex associated with DNA damage recognition. Xeroderma Pigmentosum group-A protein (XPA) is known to be involved in recognizing DNA topographic distortion lesions ***(29)***. Our experiments used recombinant, full-length *Xenopus* XPA and a double-stranded (ds) 55-mer oligonucleotide with fluorescein covalently attached at a 5'-end. As a chemical modification, fluorescein is a *bona fide* lesion for DNA repair ***(9,13)***, and fluorescence quenching is observed upon XPA binding ***(9,13)*** in this system. Fluorescence titration and anisotropy titration of dye-labeled DNA with XPA have proven that the XPA–DNA binding is a one-to-one interaction ***(9,13)***. Single XPA–DNA complexes at n*M* concentration were imbedded in a 20 μm thick, 0.5% agarose gel, which was sandwiched between two glass coverslips ***(9)***.

Nearly two-thirds of active, full-length XPA appears to be relatively unstructured based upon NMR analysis and partial proteolysis ***(24,56,57)***. Although it is noteworthy that XPA involvement with DNA damage recognition is in a primarily flexible domain, static and ensemble-averaged structural analyses have left its dynamic behavior largely unknown. Ensemble-averaged techniques include NMR experiments, time-dependent X-ray spectroscopy, conventional equilibrium or stop-flow fluorescence experiments ***(13)***, spectral hole-burning ***(58)***, and fluorescence line-narrowing spectroscopy ***(59)***. NMR measurements require a relatively long time-scale that cannot effectively detect submillisecond interactions nor provide detailed structural data for proteins of mass greater than about 30 kDa. Ensemble-averaged techniques are hampered by the molecular-level asynchrony and disorder that prevent the variations that control the repair process from being revealed. The resulting conjecture and uncertainty can be better resolved by experiments at single-molecule and single-molecular complex levels.

4.2. Control Experiments

Fluorescein covalently labeled double-strain DNA 50-mer oligonucleotide were used for the DNA damage-recognition assay. The purified recombinant XPA protein recognizes fluorescein as a lesion and binds to it, causing a change in the fluorescence quantum yield of the fluorescein ***(9)***. The changes in the fluorescence intensity of fluorescein are caused by the local environmental hydrophobicity differences before and after XPA binding ***(9,13)***. Depending on the fluoroscein labeling position on the DNA, XPA binding can cause either fluorescence intensity suppression for end-labeled DNA or enhancement for internally labeled DNA. Based on the fluorescence intensity titration, the stopped-flow measurement of fluorescence intensity changes with DNA and XPA mixing, and the fluorescence anisotropy titration measurements ***(13)***, the XPA–DNA complex is a 1:1 interaction complex, and the measured binding constants (Kd) of XPA to ds-oligos are approx 24.4 nm.

4.3. Single-Molecule Experiments on DNA Damage Recognition

Single XPA–DNA complexes at n*M* concentrations were confined and imaged in dilute agarose gels (99% aqueous). The single XPA–DNA complexes were located **(Fig. 9A)** using an inverted fluorescence confocal micro-

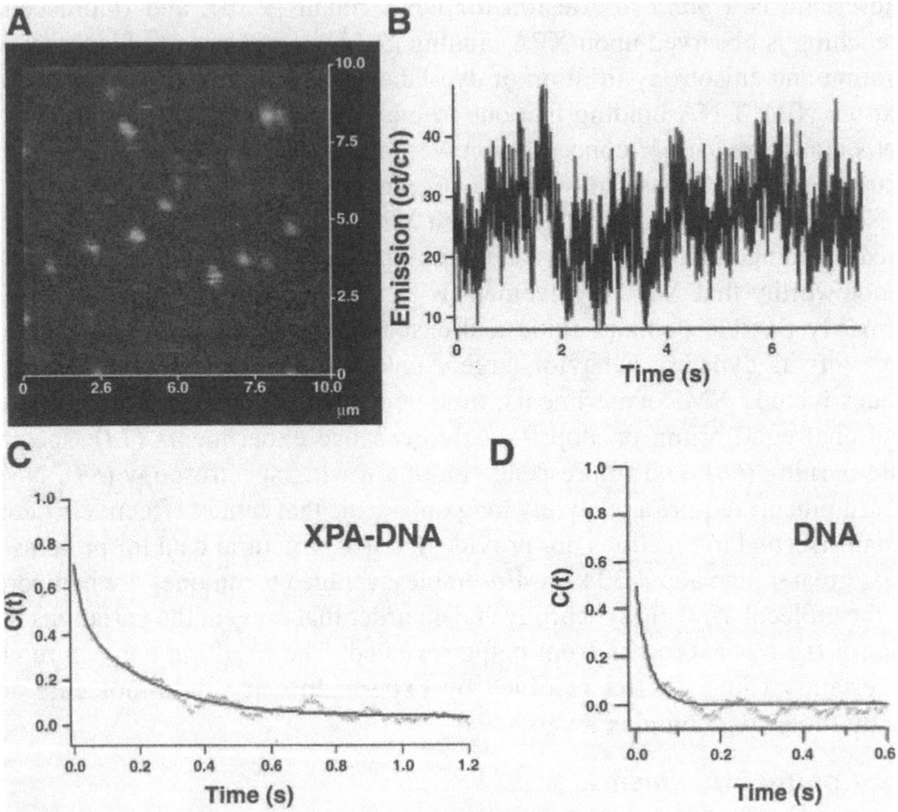

Fig. 9. Single-molecule XPA–DNA conformational fluctuation dynamics. **(A)** Fluorescence image (10 µm × 10 µm) of single XPA–DNA complexes immobilized in a 10 µm-thick film of agarose gel of a 99% buffer solution (pH 7.4). The emission is from the fluorescein labeling on DNA, which mimics DNA damage recognized by XPA proteins. This image was taken in 4 min with an inverted fluorescence microscope by raster-scanning the sample with a focused laser beam of 300 nW at 442 n*M*. Each individual feature is attributed to a single XPA–DNA complex molecule. Additional polarization modulation experiments indicate that the XPA–DNA complex molecules can freely rotate within the agarose gel film although their translational motion is confined. The intensity variation between the molecules is due to different longitudinal positions in the light focus. **(B)** Emission intensity trajectory of a 55mer double-stranded oligonucleotide containing a single fluorescein at its 5'-end interacting with a recombinant, full-length XPA protein. A portion of an emission intensity trajectory recorded at 3 ms per channel exhibits stochastic fluctuating behavior. The emission intensity fluctuation is attributed to the fluorescein-labeled DNA emission quantum efficiency fluctuation due to both XPA–DNA complex conformational and XPA–DNA binding-unbinding fluctuations. **(C)** The autocorrelation function, $C(t)$, calculated from the fluorescence intensity trajectory in **(B)**. *(Continued on next page)*

scope by raster-scanning the sample with a focused diffraction-limited laser beam at 442 nm *(9)*. The laser then was focused on a single complex to collect single-molecule fluorescence intensity trajectories where significant fluorescence intensity fluctuations were recorded (**Fig. 9B**). The fluorescence intensity fluctuations in single XPA–DNA complexes were analyzed using the autocorrelation function, and two different time-scales of the fluctuation dynamics were identified (**Fig. 9C,D**).

To determine whether the two components in the fluctuation dynamics, fast and slow fluctuations at approx 50 s^{-1} and approx 5 s^{-1}, respectively, are spontaneous or photoinduced *(9)*, we recorded single-molecule fluorescence intensity trajectories with excitation rates of 1×10^5 ct/s to 7×10^6 ct/s and found that there was no measurable excitation intensity dependence of the autocorrelation decay rates. Using modulated linear polarized laser excitation light, the single XPA–DNA complexes were shown to freely rotate at a much faster rate (>100 s^{-1}) *(9)* than the observed fluctuation rates. The intensity fluctuations cannot be attributed to rotational and translational motions. Possible DNA hydrodynamic bending motions for this oligonucleotide are most likely at the microsecond time-scale *(9)*. Therefore, the fast and slow rates are attributed to conformational fluctuations of two quasi-independent nuclear coordinates (or sets of collective nuclear coordinates) changing at different rates.

The fast conformational changes were attributed to DNA conformational fluctuation (**Fig. 9D**), and the slow conformational changes are most likely associated with interactive motions of the XPA–DNA complex at the binding site (**Fig. 9C**) *(9)*. These attributions were based on a series of control experiments. We measured the fluorescence intensity trajectories from single DNA molecules in the absence of XPA, and only the fast fluctuations were observed (**Fig. 9D**). They can be a result of either intrinsic conformational changes at the fluorescent site or to the interaction between fluorescein and DNA bases *(9)*. After addition of XPA, more than 70% of the trajectories gave biexponential autocorrelation functions with additional slow decays (**Fig. 9C**). Our rate calculations and agarose-concentration-dependent experiments confirmed that slow fluctuations are not caused by dissociation of the single complexes *(9)*.

Fig. 9. *(Continued)* $C(t)$ shows a non-exponential decay that can be fit to a double exponential with $A_f = 0.3$, $k_f = 45 \pm 3\ s^{-1}$, $A_s = 0.5$, $k_s = 5 \pm 2\ s^{-1}$, where $C(t) = A_f \exp(-k_f t) + A_s \exp(-k_s t)$ for $t < 0$. The solid line is a biexponential fit with fast and slow decay rates of $45 \pm 3\ s^{-1}$ and $5 \pm 2\ s^{-1}$, respectively. **(D)** The autocorrelation function calculated from a fluorescence intensity trajectory of a fluorescein-labeled-DNA molecule alone, without interacting with a XPA protein. The solid line is an exponential fit with decay rate of $65 \pm 5\ s^{-1}$. (Adapted in part with permission from **ref. *9***. Copyright 2001 American Chemical Society.)

The slow intensity fluctuation amplitude change is comparable to that of the bound and unbound states of XPA–DNA observed by conventional spectrofluorimetry. In single-molecule experiments, fluorescence fluctuations reflect conformational changes between the B and LB states while the overall complex is still associated. The LB states were postulated to be a subset of conformations with deviated nuclear displacements from equilibrium. These states partially restore the local environment of the fluorescein, but do not completely lose the subnanometer long-range interactions *(9,19)* between XPA and DNA. The conformational fluctuations associated with the hydrophobicity changes at the XPA–DNA interaction interface are probed by the fluorescence fluctuations of the dye probes. Slow and large-amplitude conformational motions by XPA–DNA complexes are consistent with XPA's ability to recognize a wide variety of DNA lesions *(29–31)*.

We have observed significant inhomogeneities in the conformational fluctuation dynamics of the XPA–DNA recognition complexes under the same physiological condition. **Figure 10A** shows that a seven- to tenfold variation occurs in the rates of the slow interactive conformational motions among the individual complexes *(9)*. The possible origin of the inhomogeneity is likely associated with the existence of different subsets of protein conformations seeking an induced fit to a fluctuating conformation of DNA lesion. XPA–DNAXPA–DNA complexes under their association states involve spontaneous XPA–DNA conformational motions, which are trapped in a free-energy potential minimium, and the rates of slow interactive motions are determined by the rugged free-energy landscape (**Fig. 11**) *(9)* of XPA–DNA interactions like "fly-fishing" motions of stochastically casting onto the target but never lightly bind to it. Conformational dimensionality reduction *(19,21)*, fly-casting driven by induced-fit *(19,22)*, protein interior packing relaxation *(22,23)*, and hydrogen bond formations *(30)* are all likely to be associated with the potential barriers for the trapped interactions. These trapped and slow-fluctuating interactions could be crucial in XPA–DNA interaction dynamics and intrinsically pertinent to the protein–DNA complexes in other DNA damage-recognition processes. Although this model is consistent with current experimental results and with theoretical understanding *(19–23)* on protein interactions further experimental characterizations on both single-molecule and ensemble-averaged levels, as well as molecular dynamics simulations will definitely shed more light on the detailed description of the mechanism.

Our results suggest that interaction dynamics may be essentially fit to a two-state model *(9,60–62)*, assuming that the single-complex intensity fluctuations reflect the state population changes. However, we cannot rule out possible memory effects *(9,20)*, i.e., multiple states associated with rate fluctuations. The XPA–DNA complexes showed conformational fluctuations between

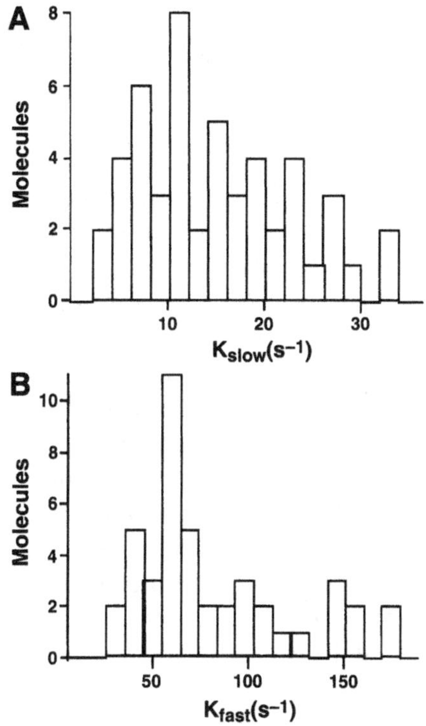

Fig. 10. The rate constant distributions of XPA–DNA interation dynamics. **(A)** Distribution of the slow conformational fluctuation rate constant, which is attributed to XPA–DNA binding–unbinding motions. A tenfold variation of the XPA binding–unbinding motion rate is observed. **(B)** Distribution of the fast conformational fluctuation rate constant, which is attributed to DNA conformational fluctuations. There is no significant inhomogeneity observed in DNA conformational fluctuation. (Adapted in part with permission from **ref. *9***. Copyright 2001 American Chemical Society.)

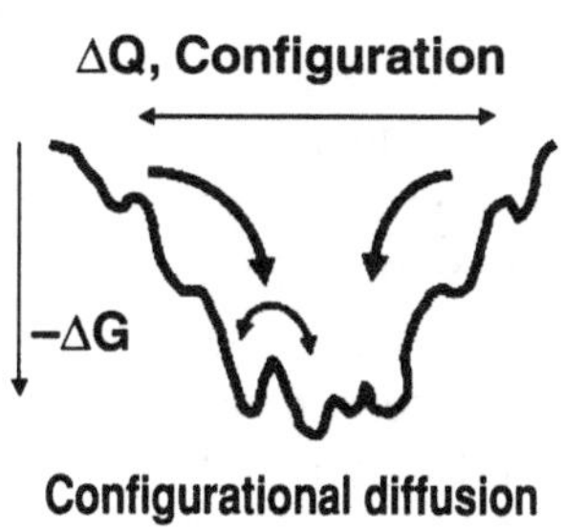

Fig. 11. Conceptual diagram of the free energy potential surface of the fluctuating protein–protein and protein–DNA interactions. The interaction complex is trapped and involves significant conformational fluctuations within the global minimum of the potential surface.

bound and loosely bound states while the overall complex was still associated. The results suggest highly dynamic rather than static protein–DNA interactions in this DNA damage-recognition system.

5. Summary and Perspectives

Single-molecule spectroscopy and imaging reveal detailed protein interaction mechanisms and dynamics that are unobtainable by conventional ensemble-averaged approaches due to the static and dynamic inhomogeneities and nonsynchronizable conformational fluctuations of the protein–DNA and protein–protein interactions. Real-time single-molecule experiments under physiological conditions demonstrate that protein recognition complexes dynamically fluctuate over subsecond to ultrafast time-scales as a result of heterogeneous binding–unbinding interactions while exhibiting a wide variation of motion for both proteins and DNA conformation. Based on the single-molecule studies, it was substantiated that the tertiary structures of the complex are important in controlling the recognition interactions, and large inhomogeneous rates were often observed in these dynamic processes. It is highly promising that the methodologies of single-molecule spectroscopy, real-time optical imaging, and spectroscopy will be widely applicable to understanding the general interaction mechanisms of important protein–ligand interactions.

Characterizing the temporal, spatial, and energetic properties of single-complex interactions amongst proteins and DNA molecules will provide the detailed descriptions and molecular analyses of some of the most important biological processes, including the earliest events in DNA damage recognition, DNA repair, protein sensing, and cell signaling. The new information, will help determine the range of important time and distance parameters for such protein interaction processes. It is now possible to obtain detailed analyses of local environmental effects, such as protein concentrations, pH, ionic strength, temperature, etc., on the protein interactions at the single-molecule level.

More applications and methodological developments will further extend the horizon of single-molecule spectroscopy. Already, it has substantially changed the landscape of the biological and biophysical sciences. New fluorescence probes, photon-detection techniques, electronic control and data-collection techniques, Raman spectroscopy, near-field optics, ultrafast spectroscopy, statistical modeling, and even molecular dynamics simulation will all enhance single-molecule research. Ultimately, protein–protein and protein–DNA interactions will be directly studied under highly complex backgrounds in living cells in real time.

Acknowledgments

The author thanks Xin Tan, Dehong Hu, Erich Vorpagel, and Yu Chen for their crucial contributions to the work discussed here; Klaus Hahn and Eric

Ackerman for collaboration; Steve Colson for helpful discussions; and Gordon Anderson for help with the instrument development. Our work was supported by the Chemical Sciences Division of the Office of Science of the U.S. Department of Energy (DOE), and by the Laboratory Directed Research and Development Program of Pacific Northwest National Laboratory, operated for the U.S. Department of Energy by Battelle Memorial Institute.

References

1. Zwanzig, R. (1990) Rate processes with dynamical disorder. *Acct. Chem. Res.* **23,** 148–152.
2. Wang, J., Wolynes, P. (1995) Intermittency of single molecule reaction dynamics in fluctuating environments *Phys. Rev. Lett.* **74,** 4317–4320.
3. Yang, S. L., Cao, J. S. (2001) Two-event echos in single-molecule kinetics: A signature of conformational fluctuations. *J. Phys. Chem. B* **105,** 6536–6549.
4. Jung, Y. J., Barkai, E., Silbey, R. J. (2002) Current status of single-molecule spectroscopy: Theoretical aspects. *J. Chem. Phys.***117,** 10,980–10,995.
5. Xie, X. S. (2002) Single-molecule approach to dispersed kinetics and dynamic disorder: Probing conformational fluctuation and enzymatic dynamics. *J. Chem. Phys.* **117,** 11,024–11,032.
6. Harms, G. S., Orr, G., Montal, M., Thrall, B. D., Colson, S. D., and Lu, H. P. (2003) Probing conformational changes of gramicidin ion channels by single-molecule patch-clamp fluorescence microscopy. *Biophys. J.* **85,** 1826–1838.
7. Moerner, W. E. and Orrit, M. (1999) Illuminating molecules in condensed matter. *Science* **283** 11,670–11,676.

8a. Lu, H. P., Xun, L., and Xie, X. S. (1998) Single-molecule enzymatic dynamics *Science* **282,** 1877–1882.

8b. Xie, X. S. and Lu, H. P. (1999) Single-molecule enzymology, *J. Biol. Chem.* **274,** 15,967–15,970.

9. Lu, H. P., Iakoucheva, L. M., and Ackerman, E. J. (2001) Single-molecule conformational dynamics of fluctuating noncovalent protein-DNA interactions in DNA damage recogntion, *J. Am. Chem. Soc.* **123,** 9184–9185.
10. Tan, X., Nalbant, P., Toutchkine, A., Hu, D., Vorpagel, E. R., Hahn, K. M., and Lu, H. P. (2004) Single-molecule study of protein–protein interaction dynamics in a cell signaling system. *J. Phys. Chem. B.* **108,** 737–744.

11a. Hu, D. H. and Lu, H. P. (2003) Single-molecule nanosecond anisotropy dynamics of tethered protein motions, *J. Phys. Chem. B* **107,** 618–626.

11b. Tan, X., Hu, D., Squier, T. C., and Lu, H. P. (2004) Probing nanosecond protein motions of calmodulin by single-molecule fluorescence anistropy. *Appl. Phys. Lett.* **83,** 2420–2422.

12. Chen, Y., Hu, D., Vorpegal, E., and Lu, H. P. (2002) Probing of single-molecule T4 lysozyme conformational dynamics by intramolecular fluorescence energy transfer. *J. Phys. Chem. B.* **107,** 7947–7956.
13. Iakoucheva, L. M., Walker, R. M., van Houten, B., and Ackerman, E. J. (2002) Fluorescence equilibrium and stopped-flow kinetic studies on the interactions of

nucleotide excision repair protein XPA and replication protein A with DNA. *Biochemistry* **41,** 131–143.

14a. Zhuang, X., Bartley, L. E., Babcock, H. P., Russell, R., Ha, T., Herschlag, D., and Chu, S. (2000) A Single-molecule study of RNA catalysis and folding. *Science* **288,** 2048–2051.

14b. Zhuang, X., Kim, H., Pereira, M. J. B., Babcock, H. P., Waiter, N. G., and Chu, S. (2002) Correlating structural dynamics and function in single ribozyme molecules. *Science* **296,** 1473–1476.

15. Wennmalm, S., Edman, L., and Rigler, R. (1997) Conformational fluctuations in single DNA molecules. *Proc Natl Acad Sci USA* **94,** 10,641–10,646.

16. Ha, T. Ting, A. Y., Liang, J., et al. (1999) *Proc. Natl. Acad. Sci. USA* **96,** 893–898.

17. Bustamante, C. Guthold, M., Zhu, X., and Yang, G. (1999) Facilitated target location on DNA by individual *Escherichia coli* RNA polymerase molecules observed with the scanning force microscope operating in liquid. *J. Biol. Chem.* **274,** 16,665–16,668.

18. Argaman, M., Golan, R., Thomson, N. H., and Hansma, H. G. (1997) Phase imaging of moving DNA molecules and DNA molecules replicated in the atomic force microscope. *Nucleic Acids Res.* **25,** 4379–4384.

19. Shoemaker, B. A., Portman, J. J., and Wolynes, P. G. (2000) Speeding molecular recognition by using the folding funnel: The fly-casting mechanism. *Proc. Natl. Acad. Sci. USA* **97,** 8868–8873.

20. Frauenfelder, H., Sligar, S. G., and Wolynes, P. G. (1991) The energy landscapes and motions of proteins. *Science* **254,** 1598–1603.

21. Richter P. H. and Eigen, M. (1974) Diffusion controlled reaction rates in spherioidal geometry. application to repressor-operator association and membrane bound enzyme. *Biophys. Chem.* **2,** 255–263.

22. Frieden, C. (1979) Slow transition and hysteretic behavior in enzymes. *Annu. Rev. Biochem.* **48,** 471–489.

23. Koshland, D. E. (1998) Conformational changes: how small is big enough? *Nat. Med.* **4,** xii–xiv.

24. Iakoucheva, L. M., Kimzey, A. L., Masselon, C. D., Bruce, J. E., Garner, E. C., Brown, C. J., Dunker, A. K., Smith, R. D., and Ackerman, E. J. (2001) Identification of intrinsic order and disorder in the DNA repair protein XPA. *Prot. Sci.* **10,** 560–571.

25. Wright, P. E. and Dyson, H. J. (2001) Intrinsically unstructured proteins: re-assessing the protein structure- function paradigm. *J. Mol. Biol.* **293,** 321–331.

26. Toutchkine, A., Kraynov, V., and Hahn, K. (2003) Solvent-sensitive dyes to report protein conformational changes in living cells. *J. Am. Chem. Soc.* **125,** 4132–4145.

27. Hahn, K. and Toutchkine, A. (2002) Live-cell fluorescent biosensors for activated signaling proteins. *Curr. Opin. Cell Biol.* **14,** 167–172.

28. Kraynov, V. S., Chamberlain, C., Bokoch, G. M., Schwartz, M. A., Slabaugh, S., and Hahn, K. M. (2000) Localized Rac activation dynamics visualized in living cells. *Science* **290,** 333–337.

29. Lindahl, T. and Wood, R. D. (1999) Quality control by DNA repair. *Science* **286,** 1897–1905.
30. Spolar, R. S. and Record, Jr., M. T. (1994) Coupling of local folding to site-specific binding of proteins to DNA. *Science* **263,** 777–784.
31. Wakasugi, M. and Sancar, A. (1999) Order of assembly of human DNA repair excision nuclease. *J. Biol. Chem.* **274,** 18,759–18,768.
32. Schmidt, T., Schutz, G. J., Baumgartner, W., Gruber, H. J., and Schindler, H. (1995) Characterization of photophysics and mobility of single molecules in a fluid lipid membrane. *J. Phys. Chem.* **99,** 17,662–17,668.
33. Xu, X. H. and Yeung, E. S. (1997) Direct measurement of single-molecule diffusion and photodecomposition in free solution. *Science* **275,** 1106–1109.
34. Dickson, R. M., Norris, D. J., Tzeng, Y. L., and Moerner, W. E. (1996) Three-dimensional imaging of single molecules solvated in pores of poly(acrylamide) gels. *Science* **274,** 966–969.
35. Ha, T. T. E., Dhemla, D. S., Selvin, P. R., and Weiss, S. (1996) Single molecule dynamics studied by polarization modulation. *Phys. Rev. Lett.* **77,** 3979–3982.
36. Ha, T., Glass, J., Enderle, T., Chemla, D. S., and Weiss, S. (1998) Hindered rotational diffusion and rotational jumps of single molecules. *Phys. Rev. Lett.* **80,** 2093–2096.
37. Noji, H., Yasuda, R., Yoshida, M., and Kinosita, K., Jr. (1997) Direct observation of the rotation of F1-ATPase. *Nature* **386,** 299–302.
38. Lu, H. P. and Xie, X. S. (1997) Single-molecule spectral fluctuations at room temperature. *Nature* **385,** 143–146.
39. Dickson, R. M., Cubitt, A. B., Tsien, R. Y., and Moerner, W. E. (1997) On/off blinking and switching behaviour of single molecules of green fluorescent protein. *Nature* **388,** 355–358.
40. Weiss, S. (1999) Fluorescence spectroscopy of single biomolecules. *Science* **283,** 1676–1683.
41. Jia, Y., Talaga, D. S., Lau, W. L., Lu, H. S. M., Degrado, W. F., and Hochstrasser, R. M. (1999) Folding dynamics of single GCN4 peptides by fluorescence resonant energy transfer confocal microscopy. *Chem. Phys.* **247,** 69–83.
42. Ha, T., Rasnik, I., Cheng, W., Babcock, H. P., Gauss, G. H., Lohman, T. M., and Chu, S, (2002) Initiation and re-initiation of DNA unwinding by the *Escherichia coli* Rep helicase *Nature* **419,** 638–641.
43. Ha, T., Zhuang, X., Kim, H. D., Orr, J. W., Willamson, A. R., and Chu, S. (1999) Ligand-induced conformational changes observed in single RNA molecules. *Proc. Natl. Acad. Sci. USA* **99,** 9077–9082.
44. Eggeling, C., Fries, J. R., Brand, L., Gunther, R., and Seidel, C. A. (1998) Monitoring conformational dynamics of a single molecule by selective fluorescence spectroscopy. *Proc. Natl. Acad. Sci. USA* **95,** 1556–1561.
45. Schaffer, J., Volkmer, A., Eggeling, C., Subramaniam, V., Striker, G., and Seidel, C. A. M. (1999) Identification of single molecules in aqueous solution by time-resolved fluorescence anisotropy. *J. Phys. Chem. A.* **103,** 331–336.

46. Kaim, G., Prummer, M., Sick, B., Zumofen, G., Renn, A., Wild, U. P., and Dimroth, P. (2002) Coupled rotation within single F0F1 enzyme complexes during ATP synthesis or hydrolysis. *FEBS Lett.* **525,** 156–163.
47. Griffin, B. A., Adams, S. R., Tsien, R. Y. (1998) Specific covalent labeling of recombinant protein molecules inside live cells. *Science* **281,** 269–272.
48. Charvin, G., Bensimon, D., and Croquette, V. (2003) Single-molecule study of DNA unlinking by eukaryotic and prokaryotic type-II topoisomerases. *Proc. Natl. Acad. Sci. USA* **100,** 9820–9825.
49. Strick, T. R., Allemand, J.F., Bensimon, D., and Croquette, V. (1998) Behavior of supercoiled DNA. *Biophys. J.* **74,** 2016–2028.
50. Dekker, N. H., Rybenkov, V. V., Duguet, M., Crisona, N. J., Cozzarelli, N. R., Bensimon, D., and Croquette, V. (2002) The mechanism of type IA topoisomerases. *Proc. Natl. Acad. Sci. USA* **99,** 12,126–12,131.
51. Haber, C., Wirtz, D. (2000) Magnetic tweezers for DNA micromanipulation. *Rev. Sci. Inst.* **71,** 4561–4570.
52. Amblard, F., Yurke, B., Pargellis, A., Leibler, S. (1996) A magnetic manipulator for studying local rheology and micromechanical properties of biological systems. *Rev. Sci. Inst.* **67,** 818–827.
53. Abdul-Manan, N., Aghazadeh, B., Liu, G. A., Majumdar, A., Ouerfelli, O., Siminovitch, K. A., Rosen, M. K. (1999) Structure of Cdc42 in complex with the GTPase-binding domain of the 'Wiskott–Aldrich syndrome' protein. *Nature* **399,** 379–383.
54. Rudolph, M. G., Bayer, P., Abo, A., Kuhlmann, J., Vetter, I. R., Wittinghofer, A. (1998) The Cdc42/Rac Interactive Binding Region Motif of the Wiskott Aldrich Syndrome Protein (WASP) Is Necessary but Not Sufficient for Tight Binding to Cdc42 and Structure Formation. *J. Biol. Chem.* **273,** 18,067–18,076.
55. Hoffman, G. R., Cerione, R. A. (2000) Flipping the Switch: Minireview The Structural Basis for Signaling through the CRIB Motif. *Cell* **102,** 403–406.
56. Iakoucheva, L. M. and Dunker, A. K. (2003) Order, disorder, and flexibility: prediction from protein sequence. *Structure* **11,** 1316–1317.
57. Buchko, G. W., Ni, S., Thrall, B. D., and Kennedy, M. A. (1998) Structural features of the minimal DNA binding domain (M98-F219) of human nucleotide excision repair protein XPA. *Nucleic Acids Res.* **26,** 2779–2788.
58. Milanovich, N., Ratsep, M., Reinot, T., Hayes, J. M., and Small, G. H. (1998) Stark hole burning of aluminum phthalocyanine tetrasulfonate in normal and cancer cells. *J. Phys. Chem. B.* **102,** 4265–4268.
59. Devanesan, P., Ariese, F., Jankowiak, R., Small, G. J., Rogan, E. G., and Cavalieri, E. L. (1999) Preparation, isolation, and characterization of dibenzo-[a,1]pyrene diol epoxide-deoxyribonucleoside monophosphate adducts by HPLC and fluorescence line-narrowing spectroscopy. *Chem. Res. Toxicol.* **12,** 789–795.
60. Agmon, N. (2000) Conformational Cycle of a Single Working Enzyme. *J. Phys. Chem.* **104,** 7830–7834.
61. Agmon, N. and Hopfield, J. J. (1983) Transient kinetics of chemical reactions with bounded diffusion perpendicular to the reaction coordinate: Intramolecu-

lar processes with slow conformational changes. *J. Chem. Phys.* **78,** 6947–6959.

62. Schenter, G. K., Lu, H. P., and Xie, X. S. (1999) Statistical analyses and theoretical models of single-molecule enzymatic dynamics. *J. Phys. Chem. A.* **103,** 10,477–10,488.

20

Application of Fluorescence Correlation Spectroscopy to Hapten–Antibody Binding

Theodore L. Hazlett, Qiaoqiao Ruan, and Sergey Y. Tetin

Summary

Two-photon fluorescence correlation spectroscopy 2P–FCS has received a large amount of attention over the past ten years as a technique that can monitor the concentration, the dynamics, and the interactions of molecules with single molecule sensitivity. In this chapter, we explain how 2P–FCS is carried out for a specific ligand-binding problem. We briefly outline considerations for proper instrument design and instrument calibration. General theory of autocorrelation analysis is explained and straightforward equations are given to analyze simple binding data. Specific concerns in the analytical methods related to IgG, such as the presence of two equivalent sites and fractional quenching of the bound hapten–fluorophore conjugate, are explored and equations are described to account for these issues. We apply these equations to data on two antibody–hapten pairs: antidigoxin IgG with fluorescein–digoxin and antidigitoxin IgG with Alexa488–digitoxin. Digoxin and digitoxin are important cardio glycoside drugs, toxic at higher levels, and their blood concentrations must be monitored carefully. Clearly, concentration assays based on IgG rely on accurate knowledge of the hapten–IgG binding strengths. The protocols for measuring and determining the dissociation constants for both IgG–hapten pairs are outlined and discussed.

Key Words: Fluorescence; immunoglobulin G (IgG); autocorrelation; digoxin; digitoxin; fluorescein; FCS; equilibrium constant; dissociation constant; diffusion constant; hapten; fluorescence correlation spectroscopy.

1. Introduction

There are many important facets in association reactions such as the binding strength, the entropic and ethalpic contributions to the binding, the specific bonds involved, and the allosteric play between multiple binding to name a

From: *Methods in Molecular Biology, vol. 305: Protein–Ligand Interactions: Methods and Applications*
Edited by: G. U. Nienhaus © Humana Press Inc., Totowa, NJ

few. It is the goal of this chapter to introduce the reader to fluorescence correlation spectroscopy (FCS) and illustrate its use in examining a biologically relevant binding reaction. FCS is, however, still a technique under development. In fact, novel applications periodically appear in the literature and discussions on data collection and treatment are ongoing. For this reason, it would behoove the researcher to keep a close eye on the advancement of FCS and strive to keep current in the latest thoughts, which may very well influence interpretations based on correlation data.

To illustrate the use of FCS in ligand binding, we will look at the determination of immunoglobulin G (IgG) and hapten binding strengths, critical for many assays within the diagnostics and pharmaceutical industries. IgGs have two equivalent and independent binding sites that can be selected to recognize an amazing array of antigens. In addition to the use of their ability as detection elements, antibody–antigen interactions provide an excellent model system for probing and understanding protein–biomolecule binding interfaces. IgG's malleability have made them ideal subjects for studies on the nature of protein recognition elements. Antibody–hapten interactions have generally high affinities with dissociation constants approaching subnanomolar level, and a complete range of affinities can be isolated for any specific hapten. FCS is an excellent tool for measuring antibody affinities at these levels because of its high sensitivity and ability, in fact, a requirement, to work at low sample concentrations.

1.1. Why Fluorescence Correlation Spectroscopy?

The fact that FCS can be used to efficiently determine a binding stoichiometry does not make FCS the only or best tool for a given system. Other techniques described in this book are equally effective and, depending on the system under study, perhaps more appropriate. Fluorescence methods involving polarization and intensity are commonly used to study ligand binding. However, FCS can be applied in the absence of binding-associated fluorescence intensity or fluorescence anisotropy change, which are not always present. For this reason, linker chemistry is less restricted in the synthesis of labeled conjugates for FCS. In addition, fluorescence polarization-based binding assays are, unlike FCS, restricted by the fluorophore's excited state lifetime that limits the size of the macromolecules that can be examined. A comparison between FCS and fluorescence polarization methodologies in looking at hapten–antibody binding has been addressed in the literature ***(1,2)***.

FCS is primarily a technique for looking at the dynamic properties of molecules ***(3–7)***. Recovering translational diffusion constants in solution and in cells ***(8–12)*** is probably the most common application of FCS, but one also sees measurements on conformational fluctuations in proteins ***(13)***, enzymatic activity ***(14)***, and measurements of flow velocities ***(15,16)***. In addition to dynamics, FCS data also contain information on a sample's molecular concentration and sample's molecular brightness. Molecular brightnesses are best extracted using statistical fluctuation analyses—photon counting histogram (PCH) ***(17,18)*** or the similar fluorescence intensity fluctuation analysis (FIDA) ***(19)***. Though we will not be focused on fluctuation methods in this chapter, comments are made when appropriate.

Whereas specialized equipment can vary the absolute limits, standard FCS instrumentation is constrained to measuring dynamics from the submicrosecond to second times, and is limited in determining specie concentrations from approx 10^{-10} up to 10^{-6} *M*. The accessible concentration range of FCS restricts its use in binding studies to interactions involving dissociation constants between 0.5 n*M* and 200 n*M*. Despite these limitations, FCS has been widely used to look at molecular associations, because this method can often extract information not available through other techniques and be applied where other techniques cannot. For example, the dissociation of dimeric phospholipase A2 in binding monomer and micellar concentrations of lipid analog could never be well addressed by standard fluorescence techniques ***(20)***. Additionally, interactions between biomolecules within cells and their cellular diffusion rates can be successfully studied by FCS ***(8–12)***. Furthermore, cross-correlation FCS can directly assess the interaction between two species within a cell (or in solution), and thus goes one step beyond the standard colocalization studies in cellular fluorescence microscopy ***(21,22)***.

1.2. Fluorescence Correlation Spectroscopy: General Theory

The principle of FCS in measuring diffusion is quite straightforward. The fluorescence emission is measured as a function of time from a small volume within a larger sample. Data collection must be rapid and is usually on the microsecond or submicrosecond timescale. The volume, in FCS using two-photon excitation (2P–FCS), is defined at the focal point of our microscope where the photon density is sufficiently high to result in two-photon excitation of the fluorophores present (**Fig. 1A**). The intensity time series will show peaks in the fluorescence as particles pass through the volume (**Fig. 1B**) from which an autocorrelation curve is then calculated. The data presented in **Fig. 1C** were collected from a dilute sample of very bright particles to better illustrate the technique. The width of the intensity peaks is the diffusion time of the fluorophore-conjugate through our observation volume whereas the frequency of peaks can be related to the average particle number. Knowing the observation volume and shape, these data can be used to calculate the diffusion constant and concentration of the fluorescent specie(s). It should be noted that the autocorrelation curve for a single particle will relate to the random walk of that

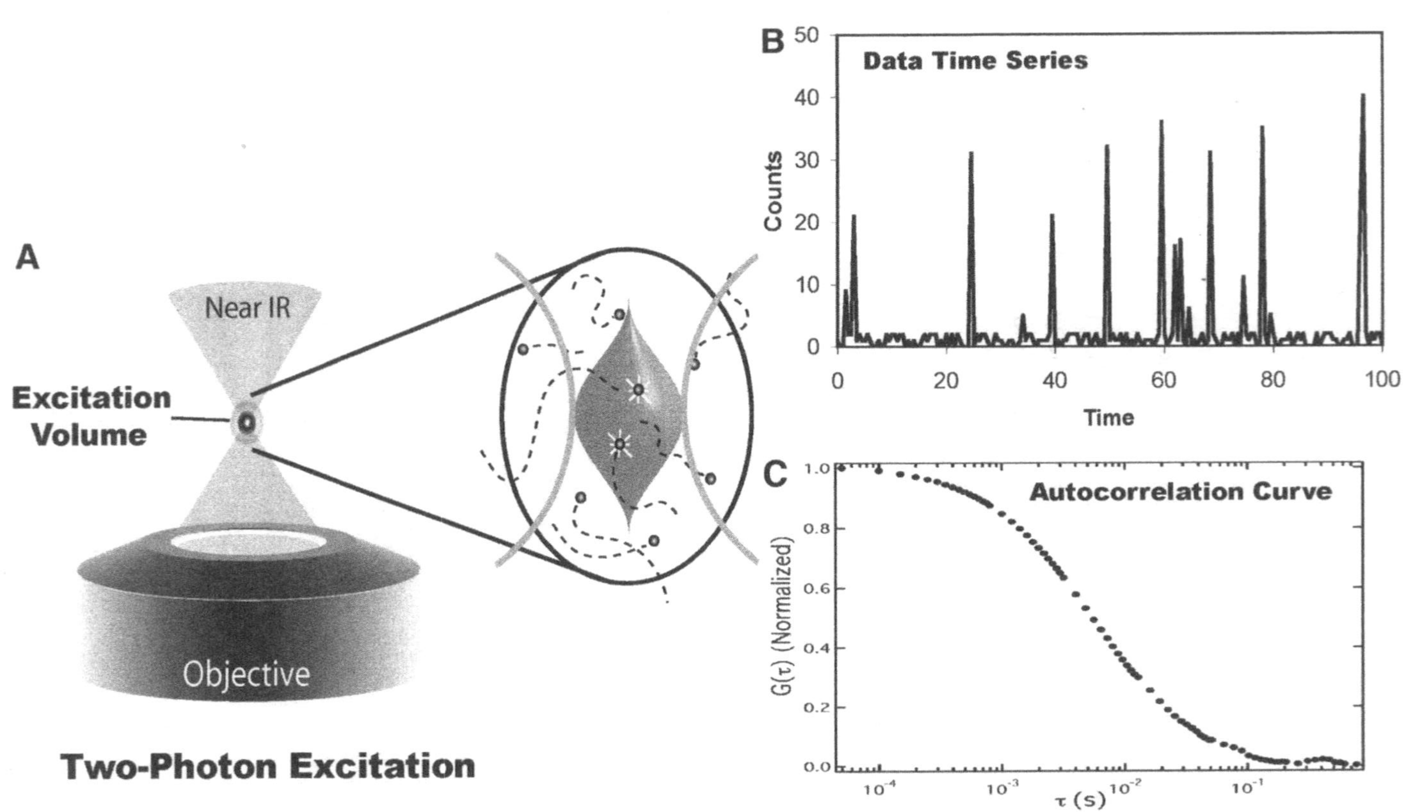

Fig. 1. Part **(A)** illustrates a characterization of a microscope objective focal point showing a three-dimensional Gaussian excitation volume. **(B)**, An FCS intensity time series of a solution containing a low concentration of bright particles is plotted. The intensity trace illustrates photon peaks that are the result of fluorescent particles diffusing through the excitation volume. **(C)**, An example autocorrelation function from large, bright particles randomly diffusing through the excitation volume.

particle through our observation volume and can vary between long and short extremes. One generally does not examine each particle separately, but performs an autocorrelation analysis on a time series representing a collection of individual random walks of a large number of diffusing particles. The autocorrelation function, $G(\tau)$, is used to extract the information from the time series and is given as,

$$G(\tau) = \frac{\langle \delta F(t) \cdot \delta F(t+\tau) \rangle}{\langle F(t) \rangle^2} \tag{1}$$

where the $\langle\ \rangle$ symbols indicates the time average of the enclosed value(s), F is the fluorescence; δF is the difference between the fluorescence at a given time, t, and the average fluorescence; and τ is the time shift variable. The autocorrelation curve presented in **Fig. 1C** shows the characteristic sigmoid shape of this function in a semilog plot.

In both one- and two-photon excitation FCS, it is commonly accepted that the shape of our observation volume formed with high numerical aperture objectives is well approximated by a three-dimensional Gaussian form *(23)*. If we have collected sufficient data to properly sample the volume, then we may impose the following equation which relates the autocorrelation function to the diffusion constant of our particle, D, and an amplitude parameter, $G(0)$, which is an extrapolated value representing the time-zero autocorrelation,

$$G(\tau) = G(0)\left(1+\frac{8D\tau}{w^2}\right)^{-1}\left(1+\frac{8D\tau}{z^2}\right)^{-1/2} \tag{2}$$

where w, is the $1/e^2$ radius of our observation volume waist (x–y axis), and z is the $1/e^2$ radius of the volume length. $G(0)$ is a function of a shape factor, γ, and the average particle number in our observation volume, $\langle N\rangle$ [$G(0) = \gamma/\langle N\rangle$]. The shape factor for a three-dimensional Gaussian shape is 0.3535 *(24)*. Eq. 2 is the form for two-photon excitation. For data collected with one-photon excitation the two eights in **Eq. 2** are simply replaced with fours. There is also another equivalent form of the equation that is often found in the literature which relates the autocorrelation function to the diffusion time (average diffusion time through the observation volume), and a structure parameter which is the ratio of z to w (for addition information *see* **ref. *25***).

In ligand binding there are two species of the ligand present, the bound and free. The extension of **Eq. 2** to multiple species is given below:

$$G(\tau)_{\text{sample}} = \sum_{i=1}^{M} f_i^2 \cdot G(0) \cdot \left(1+\frac{8D_i\tau}{w^2}\right)^{-1}\left(1+\frac{8D_i\tau}{z^2}\right)^{-1/2} \tag{3}$$

where i denotes the separate species and f is the fractional intensity. From **Eq. 3**, it can be seen that the amplitude for the zero time limit, $G(0)_{\text{sample}}$, is a

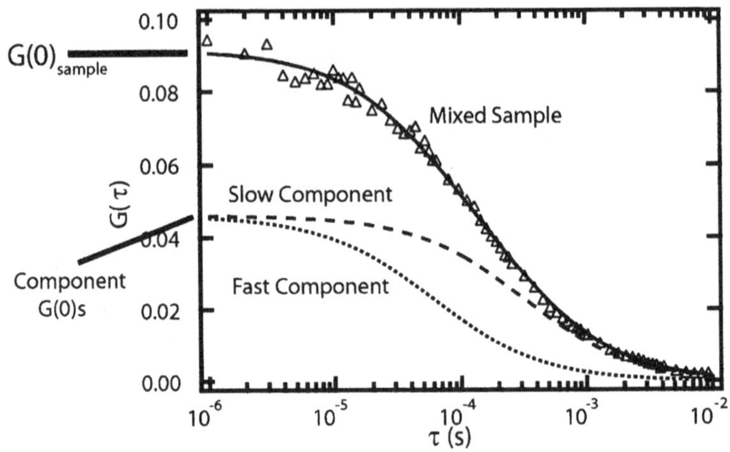

Fig. 2. An autocorrelation data set for a 50:50 mixture of a fast, 280 μm^2s^{-1} and slow 53 μm^2s^{-1} is shown along with a curve representing a fit to **Eq. 3**. The autocorrelation curves for the individual components are also drawn for illustration. The sample $G(0)$ and component $G(0)$s are indicated.

function of the separate specie $G(0)$s multiplied by the square of their associated fractional intensities and is no longer directly related to the reciprocal of the particle number. For this reason, one must take care in interpreting the apparent sample $G(0)$ because it is not always known how many species are present. Bright particles will factor heavily into the autocorrelation curve, and contaminating fluorophores in the buffer will give rise to additional fast moving components. In contrast, uncorrelated contaminating light, such as room light or dim background fluorescence, would not influence the calculated diffusion constant (because sample $G(0)_{uncorrelated} = 0$) but would decrease the curve magnitude through its contribution to the fluorescence and lead to an overestimate of the particle number.

An application of **Eq. 3** on autocorrelation data for a 50% mixture of a fast, $D = 280\ \mu m^2s^{-1}$, and a slow, $D = 53\ \mu m^2s^{-1}$, component with similar brightnesses is given in **Fig. 2**. The data points come from a stochastic simulation, generated using SimFCS, a software program developed at the Laboratory for Fluorescence Dynamics (University of Illinois, Urbana, IL) for data simulation, acquisition, and manipulation. The autocorrelation amplitude at short times (as τ approaches 0) is the sample $G(0)$ while the position of the curve along the time axis defines the diffusion time. Using **Eq. 3**, the data points can be fitted to multiple components extracting both diffusion constants and their separate amplitudes. The curve for the best fit of the data points with a two-component model (solid line) and dashed lines representing curves for the individual components are shown.

1.3. Measuring Binding Reactions With Fluorescence Correlation Spectroscopy

Association reactions between chemicals and between molecules and macromolecules, under equilibrium conditions, can be treated similarly. A simple binding equilibrium can be described as,

$$LS \leftrightarrow L + S \qquad (4)$$

where L and S stand for the free ligand and free binding site concentrations, respectively, while LS is the bound complex. The equilibrium condition can be written as:

$$K_d = \frac{1}{K_a} = \frac{L \cdot S}{LS} \qquad (5)$$

The equilibrium is shown in terms of the dissociation constant, K_d, and the association constant, K_a. Throughout this chapter we will be using the dissociation constant when we discuss binding. One commonly carries out a binding experiment by holding one component constant and making a titration with the companion. Some measurable parameter is then examined to extract the amount of bound and free material that can be used to calculate the binding constant from **Eq. 5**. The only known variables are the total concentrations of the materials and some value, (e.g., intensity, anisotropy, FCS amplitudes) related to the amount of complex formed. If the component being held constant can be kept well below the expected K_d, then the total concentration of the titrant will approximate the free concentration of titrant and **Eq. 6** can be used. From an experimental standpoint, the binding between antibody and hapten is usually monitored using a fluorophore-tagged hapten at constant concentration and subsequent titration of the solution with antibody. This experimental protocol is common in fluorescence experiments and used to avoid excessive emission from fluorescent titrants that must be added to high concentrations. Using then the binding sites (two per [IgG]) as our titrant into a solution of a fluorescently-labeled hapten, the relationship between the fraction of hapten bound, Fb, and the concentration of free sites S is given by,

$$Fb = \frac{m \cdot S}{K_d + S} + c \qquad (6)$$

in which m and c have been added as scaling and offset variables, and Fb is a parameter that is proportional to fraction bound. The fraction bound is typically plotted as a function of free sites. When the concentration of the fluorescent hapten is near or above the K_d, one must use the total concentration of sites (S_t) and total hapten concentration (L_t) to find an apparent K_d. This apparent K_d can then be used to approximate the true fraction bound (from **Eq. 7**), and thus the true concentration of free binding sites (first approximation).

$$Fb = \frac{K_d + S_t + L_t - \left[K_d^2 + 2 \cdot K_d\left(S_t + L_t\right) + S_t\left(S_t - 2 \cdot L_t\right) + L_t^2\right]^{\frac{1}{2}}}{2 \cdot S_t} \qquad (7)$$

The true K_d can be resolved through successive iterations of this procedure. The reader can find further information on experimental strategies within the chapters of this book as well as in numerous reviews ***(26–30)***.

In FCS there are two analytical tools that can be used to determine the free and bound fluorescent species present in a given sample: the autocorrelation analyses described in the last section and photon counting histogram analysis (PCH ***[17,18]***, fluorescence intensity distribution analysis [FIDA] ***[19]***) and an expanded FIDA analysis, fluorescence intensity multiple distributions analysis (FIMDA) ***(31)***, that attempts to extract both diffusion and brightness parameters simultaneously. Autocorrelation analysis of hapten and IgG association, the focus of this chapter, resolves the free and bound species by virtue of the grossly differing diffusion constants of a small fluorescently tagged hapten (molecular weight <1000 Da) and a relatively large IgG molecule (molecular weight = 150,000 Da). Labeling of the hapten allows one to observe a large change in diffusion time between the free and bound species. The diffusion coefficients allow us to separate the contribution of each species to the observed sample autocorrelation data. The information on the concentration of each is derived from the associated $G(0)$s.

Setting τ to zero in **Eq. 3** provides us with a simpler equation relating the sample $G(0)$ to the specie $G(0)$s and associated fractional intensities. We can expand the equation by substituting the separate specie $G(0)$s with $\gamma/\langle N_i \rangle$, and the fractional intensities, f, with the brightnesses, ε, multiplied by the average particle numbers and divided by the average fluorescence,

$$G(0)_{\text{sample}} = f_f^2 G(0)_f + f_b^2 G(0)_b = \frac{\gamma \cdot f_f^2}{\langle N_f \rangle} + \frac{\gamma \cdot f_b^2}{\langle N_b \rangle} = \frac{\gamma \cdot \varepsilon_f^2 \cdot \langle N_f \rangle}{\langle F \rangle^2} + \frac{\gamma \cdot \varepsilon_b^2 \cdot \langle N_b \rangle}{\langle F \rangle^2} \qquad (8)$$

where variables related to free and bound ligand are denoted with a subscript f and b, respectively. We perform this exercise because fitting data with **Eq. 3** determines not the individual specie $G(0)$s but the composite $f^2G(0)$ values. The individual brightnesses must be known to determine the average particle numbers. The brightness of the labeled ligand may be determined using pure samples of known concentrations, and the brightness of the bound fluorescent ligand can be measured at saturation with antibody. Given that the average

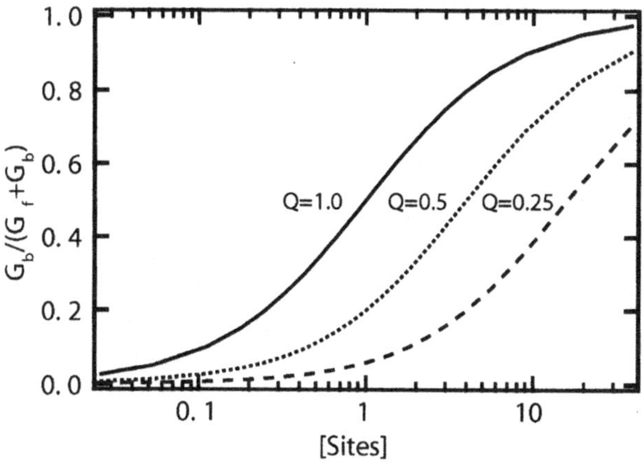

Fig. 3. Simulated equilibrium binding plots for the association of a small fluorescent ligand with a large, single-site, macromolecule. The effect of the brightness ratio, $Q = \varepsilon_1/\varepsilon_2$ on the fractional amplitude ($G_b/(G_f + G_b)$, *see* text) and the curve is shown. The dissociation constant was set equal to 1.0 unit.

fluorescence can be taken from the data trace (also note that: $\langle F\rangle = \varepsilon_f \cdot \langle N_f\rangle + \varepsilon_b \cdot \langle N_b\rangle$), we can calculate the average numbers of each specie directly and then determine the K_d. In the simplest cases when we have a single site, and the bound and free ligand have similar brightnesses, $\varepsilon_a = \varepsilon_b$, we can also calculate the fraction of ligand bound, Fb, from ratios of the $f^2G(0)$ terms,

$$Fb = \frac{f_b^2 G(0)_b}{f_f^2 G(0)_f + f_b^2 G(0)_b} = \frac{G_b}{G_f + G_b} \quad (9)$$

where G_f (amplitude of the fast diffusing component–free hapten) and G_b (amplitude of the slow diffusing component–bound hapten) are our shorthand for the composite $f^2G(0)$ variables that are derived from the analyses using **Eq. 3**. The ligand fraction bound can then be readily calculated and the dissociation constant can be estimated with **Eq. 6**. However, once the brightness of the free and bound hapten differs, the data must be properly weighted. Though one can calculate absolute brightnesses for the bound and free species, and then determine concentrations of the bound and free hapten directly. It is often more convenient to calculate the brightness ratio of these two species. If we define Q as the brightness ratio, $Q = \varepsilon_b/\varepsilon_f$, then **Eq. 9** can be rewritten as,

$$Fb = \frac{G_b}{Q^2 \cdot G_f + G_b} \quad (10)$$

The effect of using **Eq. 9** to calculate fraction bound when $Q < 1$ (quenching) and $K_d = 1$ is illustrated in **Fig. 3**.

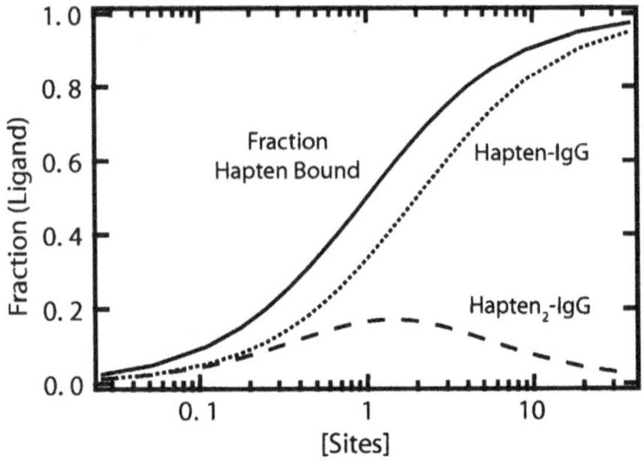

Fig. 4. The fractions of fluorescent hapten free in solution, bound to singly liganded IgG (hapten–IgG) and to doubly liganded IgG (hapten2–IgG) are given as a function of IgG binding site concentration. The simulation assumes a K_d of 1 *unit* and a fluorescent hapten concentration also of 1 *unit*.

Interestingly, the sample $G(0)$ will not be constant when $Q \neq 1$ despite the fact that particle number does not change. An equation relating Fb and Q to the sample $G(0)$ can be written as,

$$G(0)_{\text{sample}} = \frac{\gamma}{\langle N_{\text{total}} \rangle} \left[\frac{(1-F_b)+Q^2 \cdot F_b}{(1-F_b+Q \cdot F_b)^2} \right] \tag{11}$$

Equation 11 is given elsewhere ***(32)*** in a study of another antidigoxin monoclonal antibody. One could use **Eq. 11** in conjunction with **Eq. 6** to fit $G(0)$ as a function of free concentration of IgG sites. However, unless the difference in brightnesses between the states is large enough, the changes observed in $G(0)$ will not be sufficient to precisely determine Fb.

The IgG and hapten binding model has a complication as a result of the presence of two, identical, noninteracting binding sites per protein molecule (also discussed in **ref. *32***). This second site adds an additional specie to the sample: the doubly-bound IgG. Normally, this correction is minor, but is enhanced with labels that become brighter upon binding. Using the experimental protocol where a fluorescence ligand is held constant and IgG is titrated in, we will have sensible concentrations of the doubly-labeled IgG only if our hapten concentration is held near or above the K_d. **Figure 4** shows the extent of the problem when the K_d and total hapten concentration are similar. In many cases, the hapten concentration can be kept low relative to the K_d, but if the binding is very tight this condition is difficult to achieve. This new specie will have the same diffusion constant as the singly-labeled IgG and therefore will be part of the G_b value from our fit, but will have a brightness of twice the singly labeled IgG. We can relate G_b, G_f, and the K_d using **Eq. 6**, **Eq. 8**, and recognizing that the population of doubly-bound IgG is a simple probability. The resulting expression is given in **Eq. 12**.

$$\frac{G_b}{G_f + G_b} = m \cdot \frac{S \cdot Q^2 \cdot (S + K_d + 2 \cdot L_t)}{S \cdot Q^2 \cdot (S + 2 \cdot L_t + K_d) + K_d \cdot (S + L_t + K_d)} + c \qquad (12)$$

where, m is a scaling factor (ideally = 1), c is an offset variable (ideally = 0), Q is our brightness ratio (bound/free, also found in **Eq. 10**), S is the free site concentration, and L_t is the total hapten concentration. This equation can be used to fit the normalized specie amplitudes derived from fitting the FCS autocorrelation data with a two-component model. The true K_d will be resolved accounting for the presence of the doubly labeled specie and a non-unity Q. One should note that the sample $G(0)$ will increase with the concentration of the doubly-bound IgG.

2. Materials

2.1. Reagents

1. Buffers and general chemicals (Sigma-Aldrich, MO).
2. Alexa488 succinimidyl ester, fluorescein, rhodamine 110, and fluorescein isothiocyanate (Molecular Probes, OR).
3. Lysozyme (Sigma-Aldrich, MS).
4. NAP-5 G-25 Sephadex desalting columns and Protein A columns (Amersham Biosciences, Upssala, Sweden).
5. Pierce Protein Assay kit (Pierce, Rockford, IL).
6. Samples were placed in 8-Chamber *Lab-Tek* Cover glass System (Nalge Nunc, IL) for data collection.
7. IgG concentration was measured on a Cary 4BIO spectrophotometer using A278 = 1.45 for the 1 mg/mL solution.
8. Fluorescein-digoxin (M.W. = 732) was synthesized as previously published ***(33)*** and is >97% pure by analytical HPLC.
9. The digitoxin-Alexa488 conjugate was prepared by reacting Alexa488 succinimidyl ester and digitoxigenin amine ***(33)***.
10. Antidigoxin and antidigitoxin IgG antibodies (M.W. approx 150,000) were purified from hyper immune rabbit serum on a Protein-A column.
11. Measurements with IgG were performed in 0.1 *M* NaH_2PO_4 buffer, pH 7.4 at room temperature (*see* **Note 1**).

2.2. Instrumentation

1. Two-photon fluorescence correlation microscope: A simplified diagram of a standard, single channel, two-photon microscope is given in **Fig. 5**. There are several commercially available instruments containing the same set of basic components.

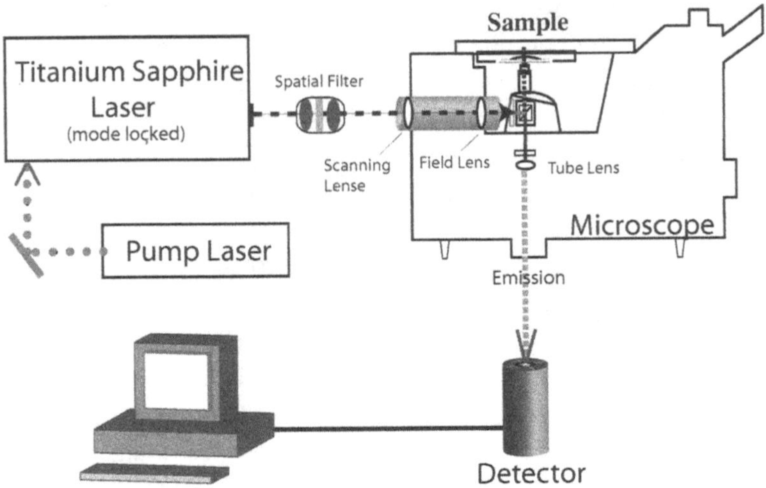

Fig. 5. The general scheme for a single channel, two-photon, FCS microscope is shown.

Data reported here were collected on one of the homebuilt instruments at the Laboratory for Fluorescence Dynamics) or on an Alba FCS (ISS instrument, Campaign, IL) (Abbott Laboratories), both using two-photon excitation. Two-photon excitation requires a mode locked laser (80 MHz repetition rate) with a short pulse (100 fs), high peak power characteristics. Excitation source was either a Spectra Physics Tsunami (Spectra-Physics Lasers Inc., CA) (at Abbott) or a Coherent, Mira-900 (Coherent Laser Division, Santa Clara, CA) Titanium Sapphire laser with similar ranges. The output wavelengths of these lasers are useful for exciting a variety of fluorophores including fluorescein, most rhodamine derivatives, and many of the Alexa dyes, to name a few. The laser beam is directed into the back port of a Nikon Model, model TE300, inverted microscope (Nikon, NY) in the ISS Alba Spectrometer (Abbott Laboratories; ISS Alba) or to a Zeiss Axiovert 135 TV (LFD; homebuilt). One should be certain that the beam is guided with mirrors coated for the near infrared otherwise the average intensity and laser pulse width will suffer.

2. Excitation path components: A long pass glass emission filter, Schott RR68 (blocking light below 700 nm), is routinely placed in the excitation path to eliminate any short wavelength contamination. In 2P–FCS, the excitation beam should be expanded to overfill the back aperture of the chosen objective. The beam should underfill the back aperture for optimal performance in one-photon instruments where the beam qualities and the pinhole size can have particularly dramatic effects on the resulting point-spread function *(23)*. The excitation beam is expanded and collimated with the spatial filter, ULM-TILT Laser Beam Expander (Newport, Fountain Valley, CA), and further with the scanning–field lens pair within the microscope. If the beam is sufficiently wide, the spatial filter may not

be required. Neutral density filters are used in the excitation path to control excitation power. An appropriate dichroic beam splitter, 700DCSPXR, (Chroma Technology Corp.,VT), is used to direct the near infrared beam to the objective and the sample, and then pass the 400 nm to 700 nm emission to the light detector.

3. Microscope objective: The microscope objective is a crucial part of a properly set instrument and must be selected carefully. Objectives with high numerical apertures (NA) are a necessity for efficient and proper focusing. Ideally, objectives for two-photon excitation should be well corrected for color aberrations over a broad wavelength range spanning from the visible to the near infrared. Many objectives are not rated for near infrared wavelengths, and it is better to consult the manufacturers to determine if their objectives are suitable, or the objective should be directly tested *in situ*. Data presented here were collected using a Nikon 60X 1.2 NA, infinity corrected, water objective (Abbott; Alba System) or a Zeiss, 63X Plan-Neofluar 1.25 NA oil objective (LFD; homebuilt) (*see* **Note 2**). For tissue or cell work, infinity corrected water objectives are the preferred choice because they are corrected to show less distortion as the focal point moves away from the cover glass and into an aqueous sample. As a general rule, focusing close to the cover glass surface minimizes distortions in the point-spread function of any objective.
4. Emission optical components: A glass short pass optical filter, BG39 or an E700sp-2pv (Chroma Technology Corp.) are commonly used to reduce any leakage of the excitation light into the emission channels. A blocking of at least 99.9% (OD = 3.0) of wavelengths above 700 nm is a standard requirement. The BG39 filter will begin to significantly light absorption above 610 nm and should not be used for far red emitting dyes. Interference filters with infrared coatings (OD > 3.0 for near IR) can be used for any fluorophores in place of the BG39 or E700sp-2pv.
5. Detectors: Avalanche photodiodes (APD) (model SPCM-AQR-15-Si, Perkin Elmer, Norwalk, CT) and their single photon counting model were used as detectors for the data collected on both systems. APDs are very sensitive and have low dark counts making them ideal detectors for FCS measurements on samples with relatively low signals. APDs, however, are most sensitive in the red wavelength region and are not necessarily the best choice for fluorophores emitting at shorter wavelengths. A wider spectral range can be attained through the use of photomultiplier tubes such as the GaAsP photocathode, series H7421 by Hamamatsu (Shizuoka-ken, Japan). The H7422P-40 PMT of this series is very sensitive and has a broad wavelength range spanning from 300 nm to 700 nm. Photon counting electronics are most common and are required in situations involving single fluorophore or low multifluorophore molecules or particles. Very bright particles can saturate these systems if the electronics and detectors are not fast enough to properly collect the peak photon flux during the particle trajectory. For highly heterogeneous samples, analog detection methods may provide wider dynamic range, and a trivial reduction of the excitation power can minimize losses at peak emissions for bright, homogeneous samples *(34)*.

3. Methods

3.1. Preparation of a Fluorescein-Labeled Lysozyme

1. Lysozyme was dissolved in 50 m*M* borate buffer pH 9.0 to a final concentration of approx 250 μ*M* (3.6 mg/mL) determined using an extinction coefficient for lysozyme of 36,500 $M^{-1}cm^{-1}$ at 280 nm *(35)*.
2. Fluorescein isothiocyanate (FITC) was added in a 1:2 molar ratio to the protein.
3. The sample was thoroughly mixed and allowed to react for 1 h at room temperature slowly rotating on a mixer.
4. The reaction is quenched with the addition of glycine, also in borate buffer, to a final concentration of 10 m*M*.
5. A NAP-5 column, equilibrated with 25 m*M* Tris-HCl, pH 8.0, 100 m*M* NaCl, 2 m*M* NaN3 and 200 m*M* Urea, was then used to remove free and glycine-bound FITC.
6. The trailing protein fractions are discarded to minimize potential contamination by free fluorescein. The concentration of lysozyme was determined using the Pierce Coomassie Protein assay kit following the manufacturers instructions using unlabeled lysozyme as the protein standard.
7. The lysozyme-Fl concentration was calculated using the extinction coefficient for sodium fluorescein at 490 nm of 88,000 M^{-1} cm^{-1} *(36)*. The labeling stoichiometry should be less than 0.20 to insure that only one fluorescein binds per lysozyme molecule.

A standard solution of 10 n*M* lysozyme–Fl solution, with an additional 10 μ*M* of unlabeled lysozyme added, can be made and stored at 4°C, in the dark, and used for several months. The unlabeled lysozyme serves to coat the glass surface of the sample chambers and effectively eliminates loss of lysozyme–Fl to these surfaces during data acquisition (*see* **Note 3**).

3.2. Instrument Calibration

3.2.1. Excitation Volume Determination

1. The FCS instrument was calibrated using either 10 n*M* fluorescein in 10 m*M* NaOH or 35 n*M* rhodamine 110 in water.
2. Stock solutions of higher fluorophore concentrations were diluted immediately before use. The extinction coefficients used for fluorescein and rhodamine 110 were 88,000 $M^{-1}cm^{-1}$ *(36)* and 92,000 $M^{-1}cm^{-1}$ (Molecular Probes, OR), respectively.
3. A 50 uL droplet of the chosen fluorophore was placed centrally in a sample well and the objective was focused into the solution (*see* **Note 4**).
4. With the excitation set to 780 nm, data on 35 n*M* rhodamine 110 in water or on 10 n*M* fluorescein was collected at 100 kHz (or faster) for several minutes.
5. The autocorrelation data were fitted using the ISS software or Globals Unlimited (LFD, University of Illinois, Urban IL) to a single specie model using a three-dimensional Gaussian shape for the excitation volume. The translational diffu-

sion rate for fluorescein and rhodamine was fixed in the analysis to 300 μm^2 s^{-1} *(6)* and the data are fit leaving $G(0)$, the beam waist w, and the z/w ratio variable. If the z/w guess is set too low (less than 1) the analysis may find an incorrect second minimum with an oblate shape for the excitation volume. The beam waist typically is found to be 0.35 μm or slightly wider and the z/w ratio near 4 for the Nikon, 60X, 1.2 NA water objective (*see* **Note 5**). Excitation power must be relatively low when using fluorescein because fluorescein is prone to photobleaching, which can lead to erroneously narrow beam wastes.

The quality of the fit to a standard, single specie sample gives a good indication of the instruments proper alignment and whether or not the laser excitation power has been chosen correctly. Using a well-defined standard on a regular basis helps to accurately judge the instrument's performance level. The primary culprits of poor and erroneous data are: detector afterpulsing *(37)*, optical misalignment, triplet state generation *(38)*, excitation saturation *(39)*, and sample bleaching *(38)*. Afterpulsing can be recognized by the presence of a fast diffusion constant in the data. Afterpulsing is detector-dependent and generally appears below 1 μs. This artifact can be reduced through selection of quiet detectors or eliminated in a two-channel instrument through splitting of the emission signal to two detectors and then cross-correlating the two resulting intensity time series. Optical misalignment is instrument-dependent and can occur for many reasons. There should be a standard protocol within a given laboratory outlining for an alignment procedure, so that it can be quickly and easily checked.

Fluorophore issues such as triplet state formation, fluorophore bleaching, and fluorophore saturation are probe specific. These problems commonly arise from excessive excitation power. Experimentally, one can test for the proper excitation by running a power curve on a particular fluorophore (or conjugate) of interest. By collecting FCS data on a known standard through a range of excitation powers, one can define the optimal settings for a particular system. The probe artifacts generally have signatures in the intensity traces and autocorrelation curves. Triplet state formation is more prevalent in one-photon excitation than in 2P–FCS. It can be recognized by the appearance of a fast diffusing specie (mathematical description is not identical, however) (*see* **Note 6**). The triplet state can be reduced by lowering the excitation power, and can be corrected within the fitting function by adding a triplet state term *(38)*. It is also possible to use triplet state quenchers such as mercaptoethylamine *(2)*, however, this adds an additional component to the sample buffer that may interfere with the processes being studied. Photobleaching can result in a shortening of the average diffusion time. When extreme, this artifact can be quickly identified through the intensity time trace. Minor photobleaching, however, is difficult to detect and it can result in increasing apparent diffusion constants.

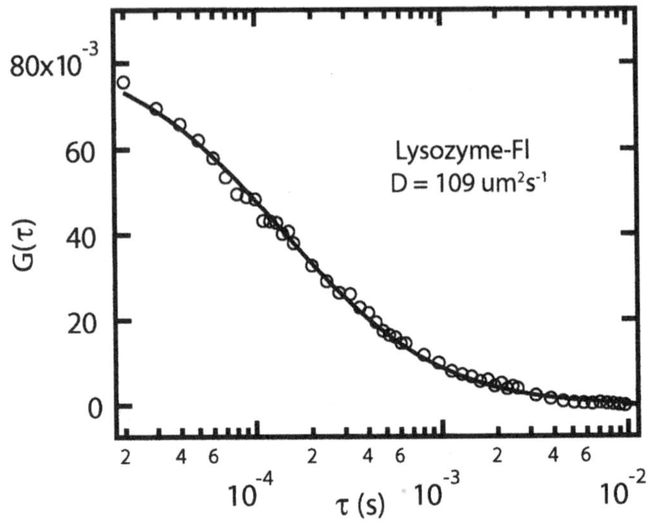

Fig. 6. The autocorrelation curve for lysozyme–Fl (fluorescein–labeled lysozyme) collected at 100 kHz using a 63X Fluar Plan-Neofluar1.25 N.A. oil objective. The solid line represents the best fit to a single diffusion model using a beam waste of 0.35 and *z/w* ratio of 5.

Finally, high excitation powers can distort the autocorrelation curve shapes as a result of the saturation of the fluorophore. The shape of the excitation remains three-dimensional Gaussian, but the emission profile becomes a centrally flattened and is no longer properly described by this form. This effect is caused by a limit of the excitation rate that is an inherent property of a given fluorophore.

3.2.2. Protein Standard Evaluation

The lysozyme–Fl conjugate can be used to validate the instrument calibration performed with the free fluorescent dye (*see* **Subheading 3.2.1.**). Lysozyme has a molecular weight of 14,300 Da and its translational diffusion constant of approx 104 $\mu m^2 s^{-1}$ ***(40)***. One should expect a reasonable agreement between the experimental and the expected values. **Figure 6** shows the autocorrelation data points calculated from an FCS time trace collected on a 10 n*M* lysozyme–Fl sample. The curve represents the best fit to the data for a single species model. The presented data give a proper diffusion constant for lysozyme, 109 $\mu m^2 s^{-1}$ with an error from the fit of 5%. Longer data sets can be collected to sharpen the curve but there is a point of diminished return. The question of how much data to collect is a common one and must be specifically addressed for each system. Estimating errors in FCS data is complex and has been discussed in the literature ***(41,42)***. One common and practical method is to collect a number of data sets on the same sample and use the average and standard deviation of the fitted variables.

3.2.3. Determination of the Equilibrium Binding Constants of an Antidigoxin and Antidigitoxin Rabbit poly-Ab (see **Note 7**)

3.2.3.1. Sample Preparation

1. Samples with serial dilutions of antibodies were prepared in the concentration range from 0.04 n*M* to 80 n*M* and mixed with the equal volume of the fluorescent hapten. The final concentration of each sample contained 1 n*M* hapten, in the case of fluorescein–digoxin, and 6 n*M*, for Alexa488–digitoxin.
2. Samples were equilibrated at room temperature for approx 30 min before being measured on an Alba FCS Spectrometer.

3.2.3.2. Instrument Linearity With Fluorophore–Hapten Conjugate

The linearity of the FCS data for the hapten conjugates, fluorescein–digoxin and Alexa488–digitoxin was verified by experiment. A single component system can be fit with **Eq. 2**. The sample $G(0)$ is proportional to the reciprocal of the average particle number ($\langle N \rangle$) in the excitation volume during the course of the experiment. Autocorrelation files collected over a range of Alexa488–digitoxin concentrations are shown in **Fig. 7A**. The linearity of the $1/G(0)$ values for this hapten conjugate continues over a wide range of concentrations (**Fig. 7B**).

3.2.3.3. Fluorescence Correlation Spectroscopy Measurements on Antidigoxin Antibody

All samples in a given titration series were measured in the same chamber well in order to minimize the experimental noise created by movements of the cover glass and refocusing of the microscope objective. FCS data were collected using an excitation wavelength of 780 nm and a 60X 1.2 NA Nikon water objective. The autocorrelation data were fitted using either Globals Unlimited or the ISS analysis software (*see* **Note 8**). The series of autocorrelation curves were then analyzed globally by linking the diffusion parameters for the free and bound hapten, fluorescein–digoxin (*see* **Note 9**). The fluorescein–digoxin hapten showed a 15% quenching ($Q = 0.85$) upon binding. In the current experiment, we fixed the fast and slow diffusion constants to average values for the free hapten and hapten saturated with IgG, 240 $\mu m^2 s^{-1}$ and 55 $\mu m^2 s^{-1}$, respectively. Three autocorrelation files are given in **Fig. 8A**. The resolved amplitudes of the fast, G_f, and slow, G_b, diffusing components were analyzed for the dissociation constant by using **Eq. 12** (Q set to 0.85) and **Eq. 7** (*see* **Note 10**). The data points and the fit are shown in **Fig. 8B**. The dissociation constant of 4.2 nanomolar is in reasonable agreement with fluorescence aniso-

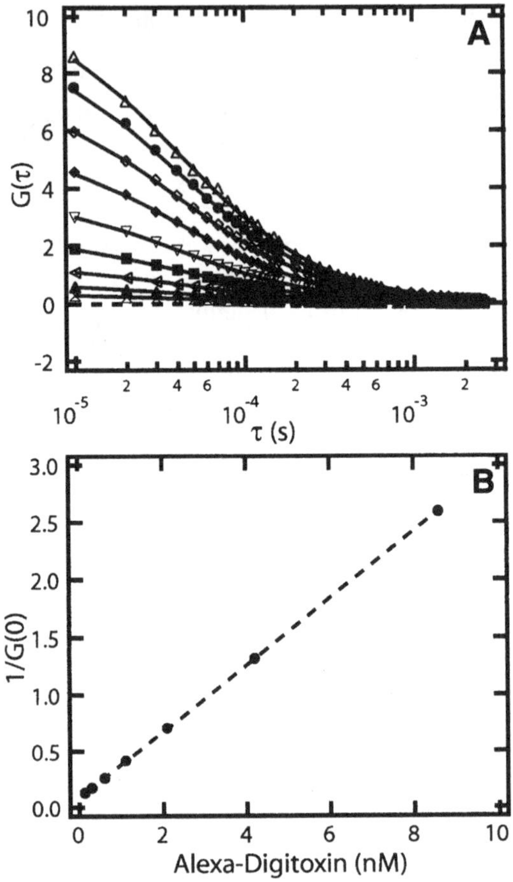

Fig. 7. Autocorrelation curves for Alexa488–Digitoxin at concentrations ranging from 0.02 n*M* to 9.0 n*M* are shown in **(A)**. The plot in **(B)** shows the linear behavior $1/G(0)$ with particle number (as concentration).

tropy measurements on the same samples that gave a K_d of 6.2 nanomolar (*see* **Fig. 8**).

3.2.3.4. Fluorescence Correlation Spectroscopy Measurements on Antidigitoxin Antibody

Using a similar protocol, a series of autocorrelation curves were collected for samples of antidigitoxin IgG and Alexa488–digitoxin. The hapten showed a 53% quenching ($Q = 0.47$) upon binding. Three normalized autocorrelation files are given in **Fig. 9A**. The resolved amplitudes of the fast (free hapten, G_f) and slow (bound hapten G_b), components were determined and used in **Eq. 12** and **Eq. 7**, as above. The data points and the fit are shown in **Fig. 9B**.

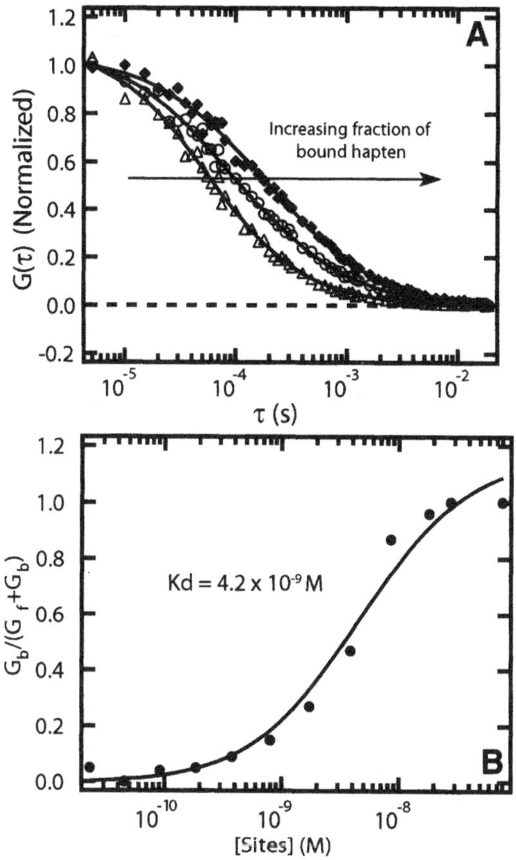

Fig. 8. The dissociation constant of antidigoxin IgG with digoxin–fluorescein was determined using FCS. **(A)** Example autocorrelation files for the data at 0%, 41%, and 74% fraction of ligand bound. **(B)** The autocorrelation fractional amplitude of the slow diffusion component (bound) is plotted as a function of free antibody sites. The curve represents the best fit to the data points (filled circles) by **Eq. 12** using a brightness ratio of 0.85.

The dissociation constant was found to be 4.9 nanomolar. In contrast to the previous case, we observed a significant binding-dependent quenching of the hapten. In addition, the hapten concentration was greater than the K_d. These obstacles result in more severe corrections in the fitted curve. One familiar with standard binding plots (fraction bound vs free sites) would see the K_d at the 0.5 fraction bound point. The plot in **Fig. 9B** is clearly shifted to the higher concentrations. The presence of doubly bound IgG also distorts this plot. The bound form is half as bright and the autocorrelation curve weights the sepa-

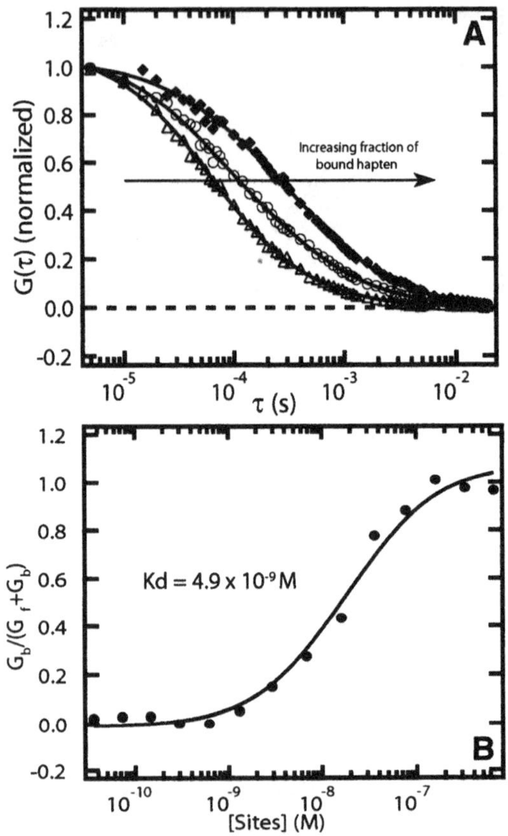

Fig. 9. The dissociation constant of antidigtoxin IgG with Alexa488–digitoxin was determined using FCS. Normalized autocorrelation curves for the 18%, 70%, and 94% fraction of ligand bound are shown in **(A)**. The autocorrelation amplitude fraction of the slow diffusing component is plotted as a function of free antidigitoxin IgG binding sites. The curve represents the best fit to the data points (filled circles) by **Eq. 12** using a brightness ratio of 0.47.

rate amplitudes with the square of this value. **Equation 12** compensates for this issue and reports the corrected K_d.

4. Notes

1. The IgG buffer also contained 0.1% of bovine gamma globulin (BGG) to minimize nonspecific binding of the hapten to the antibody and, importantly, to the walls of the 8-well plate (*see* **Subheading 2.1.**).
2. Despite less than perfect color corrections, we have found that the Zeiss 40X 1.3 NA Fluar objective works reasonably well. By focusing the beam near the cover glass, distortion of the point-spread function of this objective can be minimized.

Also, an Olympus 60X UPlan Apo/IR water objective works very well on our Olympus IX70-based FCS instrument. A note of caution though: a similar 60X Olympus objective designated "PSF," for point-spread function, performs very poorly in two-photon excitation but works well in a one-photon excitation setup (*see* **Subheading 2.1.**).

3. Clearly, other proteins can be used as standards. However, labeling of proteins often results in changes in the extinction coefficient and quantum yield of the labeled probe. For concentration standards a more rigorous characterization of the protein–fluorophore conjugate is required (*see* **Subheading 3.1.**).
4. The difference between the emission intensity in the cover glass and that in the sample is used to aid in setting the focal plane. Often long cover glass are used as samples holders, but one must be aware that samples will begin to dry so data must be collected within a short time (*see* **Subheading 3.2.1.**).
5. The determination of the z-axis is much less precise than the beam waist and the values can vary widely. As one can see from the mathematics (**Eq. 2**), the z-axis is defined in an expression that has much less impact on the autocorrelation curve than the beam waste. In fact, one may fit solution data with a two dimensional Gaussian shape (**Eq. 2** with $z = 0$) with relatively minor systematic fitting errors (*see* **Subheading 3.2.1.**).
6. The triplet state is more severe in one-photon instruments, but can occur with two-photon excitation as well. The autocorrelation curve can be fit with an additional term representing the triplet state (below) *(38)* (*see* **Subheading 3.2.1.**).

$$\left(1+\frac{T}{1-T}\cdot e^{\frac{-\tau}{\tau_T}}\right)$$

7. The present antidigoxin antibody is a separate antibody to the one discussed by Chen et al. *(32)*. The antidigoxin IgG examined quenches the fluorescein–digoxin hapten by less than 15% (*see* **Subheading 3.2.1.**).
8. For a two-component diffusion model, the ISS analysis package fits to the sample $G(0)$, the $\%f1$ and the particle numbers. The fitting algorithm contains the assumption that the brightnesses of all particles are the same (**Eq. 9**), which then allows the average particle number to be solved from the sample $G(0)$ ($G(0) = \gamma / \langle N \rangle$). The specie number fractions are reported through $\%f1$, which is the $G_b/(G_f + G_b)$ expression, if the first component is the slow diffusing specie, (which simplifies to $\langle N_b \rangle / (\langle N_f \rangle + \langle N_b \rangle)$ when brightnesses are the same) (*see* **Subheading 3.2.3.2.**).
9. It is best to globally link common variables across the data sets. This significantly reduces the errors in the resolved variables (*see* **Subheading 3.2.3.2.**).
10. A simplistic iterative procedure was used to solve for K_d. **Equation 12** was used to determine an apparent K_d from the autocorrelation function amplitudes and the concentration of binding sites (total sites not free sites). This K_d was then used in **Eq. 7** to determine the approximate amount of ligand bound from which the concentration of free sites could be determined. **Equation 12** was again used to fit

the data but now with an estimate of the free sites concentration present. More robust algorithms can be implemented but this method is simple to implement and instructional because one can observe the iterative process (*see* **Subheading 3.2.3.2.**).

Acknowledgments

T.L.H. is supported through the National Institutes of Health, grant RR03155. The authors would like to thank Enrico Gratton and David Jameson for their helpful comments and reading of the manuscript.

References

1. Matayoshi, E. D. and Swift, K. M. (2001) Application of FCS to protein–ligand interactions: comparison with fluorescence polarization, in: *Fluorescence Correlation Spectroscopy. Theory and Applications.* (Rigler, R. and Elson, E. S., ed.) Springer-Verlag, Berlin, Germany, pp. 84–98.
2. Tetin, S. Y., Swift, K. M., and Matayoshi, E. D. (2002) Measuring antibody affinity and performing immunoassay at the single molecule level. *Anal. Biochem.* **321,** 183–187.
3. Elson, E. L. and Magde, D. (1974) Fluorescence correlation spectroscopy. I. Conceptual basis and theory. *Biopolymers* **13,** 1–27.
4. Magde, D., Elson, E. L., and Webb, W. W. (1972) Thermodynamic fluctuations in a reacting system—measurement by fluorescence correlation spectroscopy. *Phys. Rev. Let.* **29,** 705–708.
5. Magde, D., Elson, E. L., and Webb, W. W. (1974) Fluorescence correlation spectroscopy. II. An eExperimental realization. *Bioploymers* **13,** 20–61.
6. Thompson, N.L. (1991) Fluorescence correlation spectroscopy, in: *Topics in Fluorescence Spectroscopy. Volume 1*(Lakowicz, J.R., ed.) Plenum Press, New York, pp. 337–378.
7. Thompson, N. L., Lieto, A. M., and Allen, N. W. (2002) Recent advances in fluorescence correlation spectroscopy. *Curr. Opin. Struct. Biol.* **12,** 634–641.
8. Brock, R., Hink, M. A., and Jovin, T. M. (1998) Fluorescence correlation microscopy of cells in the presence of autofluorescence. *Biophys. J.* **75,** 2547–2557.
9. Berland, K. M., So, P. T. C., and Gratton, E. (1995) Two-photon fluorescence correlation spectroscopy: method and application to the intracellular environment. *Biophys. J.* **68,** 694–701.
10. Pramanik, A., Olsson, M. Langel, U., Bartfai, T., and Rigler, R. (2001) Fluorescence correlation spectroscopy detects galanin receptor diversity on insulinoma cells. *Biochemistry* **40,** 10,839–10,845.
11. Schwille, P., Haupts, U., Maiti, S., and Webb, W. W. (1999) Molecular dynamics in living cells observed by fluorescence correlation spectroscopy with one- and two-photon excitation. *Biophys. J.* **77,** 2251–2265.
12. Weiss, M., Hashimoto, H., and Nilsson, T. (2003) Anomalous protein diffusion in living cells as seen by fluorescence correlation spectroscopy. *Biophys. J.* **84,** 4043–4052.

13. Bismuto, E., Gratton, E., and Lamb, D. C. (2001) Dynamics of ANS binding to tuna apomyoglobin measured with fluorescence correlation spectroscopy. *Biophys. J.* **81,** 3510–3521.
14. Rigler, R., Edman, L., Foldes-Papp, Z., and Wennmalm, S. (2001) Fluorescence correlation spectroscopy in single-molecule analysis: enzymatic catalysis at the single molecule level. *Single Mol. Spec.: Nobel Conf. Lects.* **67,** 177–194.
15. Magde, D., Webb, W. W., and Elson, E. L. (1978) Fluorescence correlation spectroscopy. III. Uniform translation and laminar flow. *Biopolymers* **17,** 361–376.
16. Lumma, D., Best, A., Gansen, A., Feuillebois, F., Radler, J. O., and Vinogradova, O.I. (2003) Flow profile near a wall measured by double-focus fluorescence cross-correlation. *Phys. Rev. E.* **6705,** 6313–6318.
17. Muller, J. D., Chen, Y., and Gratton, E. (2000) Resolving heterogeneity on the single molecular level with the photon counting histogram. *Biophysical J.* **76,** 474–486.
18. Muller, J. D., Chen, Y., and Gratton, E. (2001) Photon counting histogram statistics, in: *Fluorescence Correlation Spectroscopy. Theory and Applications* (Rigler, R. and Elson, E. L., eds.) Springer-Verlag, Berlin, Germany, pp. 410–437.
19. Kask, P., Palo, K., Ullmann, D., and Gall, K. (1999) Fluorescence-intensity distribution analysis and its application in biomolecular detection technology. *Proc. Natl. Acad. Sci. USA* **96,** 1379–1376.
20. Sanchez, S. A., Chen, Y., Muller, J. D., Gratton, E., and Hazlett, T. L. (2001) Solution and interface aggregation states of Crotalus *atrox* venom phospholipase A2 by two-photon excitation fluorescence correlation spectroscopy. *Biochemistry* **40,** 6903–6911.
21. Patel, R. C., Kumar, U., Lamb, D. C., Eid, J. S., Rocheville, M., Grant, M., et al. (2002) Ligand binding to somatostatin receptors induces receptor-specific oligomer formation in live cells. *Proc. Natl. Acad. Sci. USA* **99,** 3294–3299.
22. Kim, S. A., Heinze, K. G., Waham, M. N., and Schwille, P. (2004) Intracellular calmodulin availability accessed with two-photon cross-correlation. *Proc. Natl. Acad. Sci. USA* **101,** 105–110.
23. Hess, S. T. and Webb, W. W. (2002) Focal volume optics and experimental artifacts in confocal fluorescence correlation spectroscopy. *Biophys. J.* **83,** 2300–2317.
24. Chen, Y., Muller, J. D., Berland, K. M., and Gratton, E. (1999) Fluorescence correlation spectroscopy. *Methods* **19,** 234–252.
25. Pramanik, A. and Rigler, R. (2001) FCS Analysis of ligand-receptor interactions in living cells, in: *Fluorescence Correlation Spectroscopy. Theory and Applications,* (Rigler, R. and Elson, E. S., eds.) Springer-Verlag, Berlin, Germany, pp. 101–131.
26. Johnson, M. L. (1992) Analysis of ligand-binding data with experimental uncertainties in independent variables, in: *Numerical Computer Methods* (Brand, L. and Johnson, M., eds.). Academic Press, New York, NY, pp. 68–87.
27. Klotz, I. M. and Hunston, D. L. (1984) Mathematical models for ligand-receptor binding. *J. Biol. Chem.* **259**, 10,060–10,062.
28. Tetin, S. Y. and Hazlett, T. L. (2000) Optical spectroscopy in studies of antibody-hapten interactions. *Methods* **20,** 341–361.

29. Winzor, D. J. and Sawyer, W. H. (1995) *Quantitative Characterization of Ligand Binding*. Wiley-Liss, Inc., New York, NY.
30. Weber, G. and Anderson, S. R. (1965) Multiplicity of binding. Range of validity and practical rest of Adair's equation. *Biochemistry* **4,** 1942–1947.
31. Palo, K., Mets, U., Jager, S., Kask, P., and Gall, K. (2000) Fluorescence intensity multiple distributions analysis: concurrent determination of diffusion times and molecular brightness. *Biophys. J.* **79,** 2858–2866.
32. Chen, Y., Muller, J. D., Tetin, S. Y., Tyner, J. D., and Gratton, E. (2000) Probing ligand protein binding equilibria with fluorescence correlation spectroscopy. *Biophys. J.* **79,** 1074–1084.
33. Adamczyk, M. and Grote, J. (1999) Efficient synthesis of 3–aminodigoxigenin and 3-aminodigitoxigenin probes. *Bioorg. Med. Chem. Lett.* **9,** 771–774.
34. Hazlett, T. and Gratton, E. (2004) Photon counting and analog data acquisition in fluorescence correlation spectroscopy: issues of sensitivity and dynamic range. *Biophys. J.* **86,** 157A.
35. Praissman, M. and Rupley, J. A. (1968) Comparison of proteinstructure in the crystal and in solution. 3. Tritium-hydrogen exchange of lysozyme and a lysozyme-saccharide complex. *Biochemistry* **7,** 2446–2450.
36. Klonis, N. and Sawyer, W. H. (2000) Effect of solvent-water mixtures on the prototropic equilibria of fluorescein and on the spectral properties of the monoanion. *Photochem. Photobiol.* **72,** 179–185.
37. Zhou, M., Jin, L., Chen, B., Ding, Y., Ma, H., and Chen, D. (2003) Afterpulsing and its correction in fluorescence correlation spectroscopy experiments. *Appl. Op.* **42,** 4031–4036.
38. Widengren, J., Mets, U., and Rigler, R. (1995) Fluorescence correlation spectroscopy of triplet states in solution: a theoretical and experimental study. *J. Phys. Chem.* **99,** 13,368–13,379.
39. Berland, K. and Shen, G. (2003) Excitation saturation in two-photon fluorescence correlation spectroscopy. *Appl. Op.* **42,** 5566–5576.
40. Cantor, C. R. and Schimmel, P. R. (1980) *Biophysical Chemistry. Part II: Techniques for the Study of Biological Structure and Function.* W. H. Freeman and Company, San Francisco, CA, p. 584.
41. Wohland, T., Rigler, R. , and Vogel, H. (2001) The standard deviation in fluorescence correlation spectroscopy. *Biophys. J.* **80,** 2987–2999.
42. Meseth, U., Wohland, T., Rigler, R., and Vogel, H. (1999) Resolution of fluorescence correlation measurements. *Biophys. J.* **76,** 1619–1631.

21

Atomic Force Microscopy Measurements of Protein–Ligand Interactions on Living Cells

Robert H. Eibl and Vincent T. Moy

Summary

Cell adhesion receptors are expressed on the surface of cells and can mediate binding to other cells and to the extracellular matrix. Here, we describe in detail the use of atomic force microscopy (AFM)-based force spectroscopy for studying cell detachment forces on living leukocytes. With this technique it is now possible to measure force with resolution down to the level of individual molecules. AFM force spectroscopy is particularly well suited for research in cell adhesion, which has relevance in both the medical and life sciences including immunology, cancer and stem cell research, and human pharmacology. Along with its limitations, we herein, describe how the rupture force of a single complex formed between the integrin receptor leukocyte function-associated antigen (LFA)-1, expressed on the surface of a living leukocyte, and immobilized intercellular adhesion molecule-1 (ICAM-1) was measured. With only minor modifications this protocol can be used to study other adhesion receptors on almost any mammalian cell or bacterial system. This protocol is also suitable for studying single-molecule de-adhesion events in cell-free systems as well as between two living cells.

Key Words: Atomic force microscopy (AFM); force spectroscopy; cell adhesion; adhesion receptors; cell adhesion molecule; LFA-1; ICAM-1; leukocyte; rupture force; single molecule measurements.

1. Introduction

Since its introduction approx 20 yr ago *(1)*, atomic force microscopy (AFM) has developed into a highly versatile technique, capable of imaging individual atoms and measuring the interactions of proteins *(2–4)*. Recently, the technique of force spectroscopy was introduced as a method to characterize the

From: *Methods in Molecular Biology, vol. 305: Protein–Ligand Interactions: Methods and Applications*
Edited by: G. U. Nienhaus © Humana Press Inc., Totowa, NJ

dynamic response of individual ligand-receptor complexes to a pulling force *(5–9)*. Here, we provide a detailed description of how single molecule force spectroscopy measurements are carried out in our laboratory. For illustration, we describe our recent characterization of the leukocyte function-associated antigen-1/immobilized intercellular adhesion molecule-1 (LFA-1/ICAM-1) interaction *(10–12)*, which involves the interaction between a live cell and a protein immobilized on a substrate *(13)*. Similar approaches can be used to measure other ligand-receptor interactions *(14)*, including measurements of cell–cell de-adhesion on the molecular level *(15)*.

2. Materials

All chemical reagents and regular lab equipment were purchased from Sigma (St. Louis, MO), R&D Systems (Minneapolis, MN), Becton Dickinson (Franklin Lakes, NJ), and Fisher Scientific (Hampton, NH) if not otherwise specified.

1. AFM unsharpened cantilevers (model no. MLCT-AUHW, Thermomicroscope, Sunnyvale, CA).
2. Biotinylated BSA (Sigma), 0.5 mg/mL in 0.1 *M* sodium bicarbonate, pH 8.6.
3. Streptavidin, 0.5 mg/mL in PBS (10 m*M* phosphate buffer, 150 m*M* NaCl, pH 7.4).
4. Biotinylated concanavalin A (ConA), 0.5 mg/mL in PBS.
5. Bovine serum albumin (BSA), 0.5 mg/mL in PBS.
6. ICAM-1/Fc fusion protein, 250 μg/mL in PBS (R&D Systems).
7. Acetone.
8. UV lamp.
9. LFA-1 expressing cell line (e.g., 3A9).
10. RPMI 1640 medium, supplemented with 10% FBS, 1% glutamine, 50 U/mL penicillin, 50 μg/mL streptomycin.
11. Cell culture dishes, 35 mm (Falcon 351008).
12. HEPES buffer.
13. Blocking monoclonal antibodies (anti-ICAM-1 (e.g., BE29G1) and anti-LFA-1 (e.g., FD441.8)) and control antibodies (e.g., rat IgG).
14. AFM and analysis software.

3. Methods

Our AFM force measurements of ligand-receptor adhesion are carried out using a homebuilt AFM that is equipped with a single axis piezoelectric translator (**Fig. 1** and **Note 1**). To measure the force needed to rupture individual LFA-1/ICAM-1 bonds, we employed an experimental system consisting of a 3A9 cell coupled to the end of an AFM cantilever and immobilized ICAM-1 (**Fig. 2**). Our AFM experiments of ligand-receptor adhesion are divided into eight components:

1. Immobilization of the ligand protein on the Petri dish.
2. Functionalization of the cantilever tip.

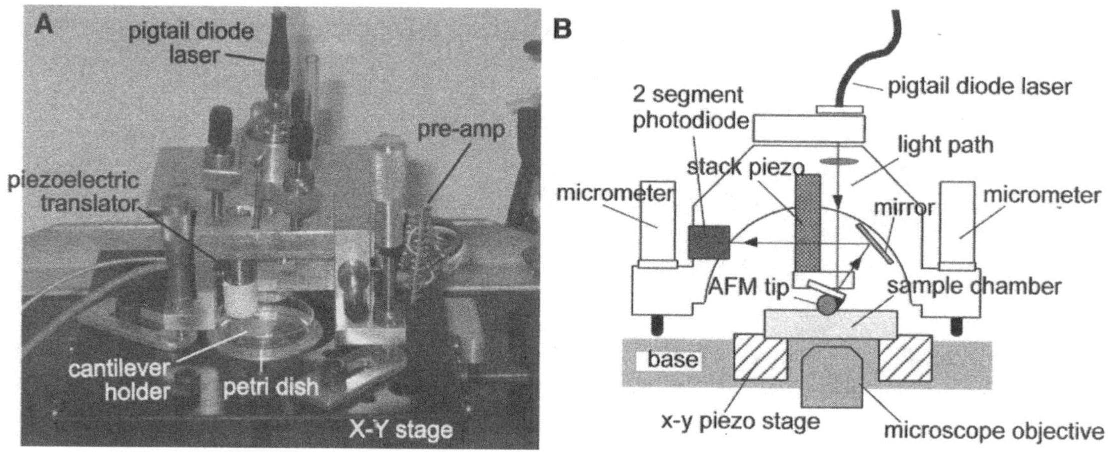

Fig. 1. **(A)** Photograph of a AFM apparatus used to measure the force of biomolecular interactions of live cells. A plexiglass cantilever holder is attached to the end of a piezoelectric translator. The base of the translator is wrapped with Teflon tape to prevent shortening of the piezoelectric element. The AFM rests on a mechanical X–Y stage that permits the cantilever to probe over an area of the sample (i.e., Petri dish). The sample remains stationary during the measurements. **(B)** Schematics of AFM shown in **(A)**. In this design, expansion of the piezoelectric transducer lowers the cantilever onto the sample. The deflection of the cantilever is monitored optically using a pigtail diode laser. The laser light reflected off the cantilever is directed into a 2-segment photodiode detector by an adjustable mirror. A pre-amplifier coupled directly to the photodetector amplifies the signal that is subsequently sent to a computer controlled data acquisition system. An inverted optical microscope system was used in coupling the cell to the cantilever and in the alignment of the laser.

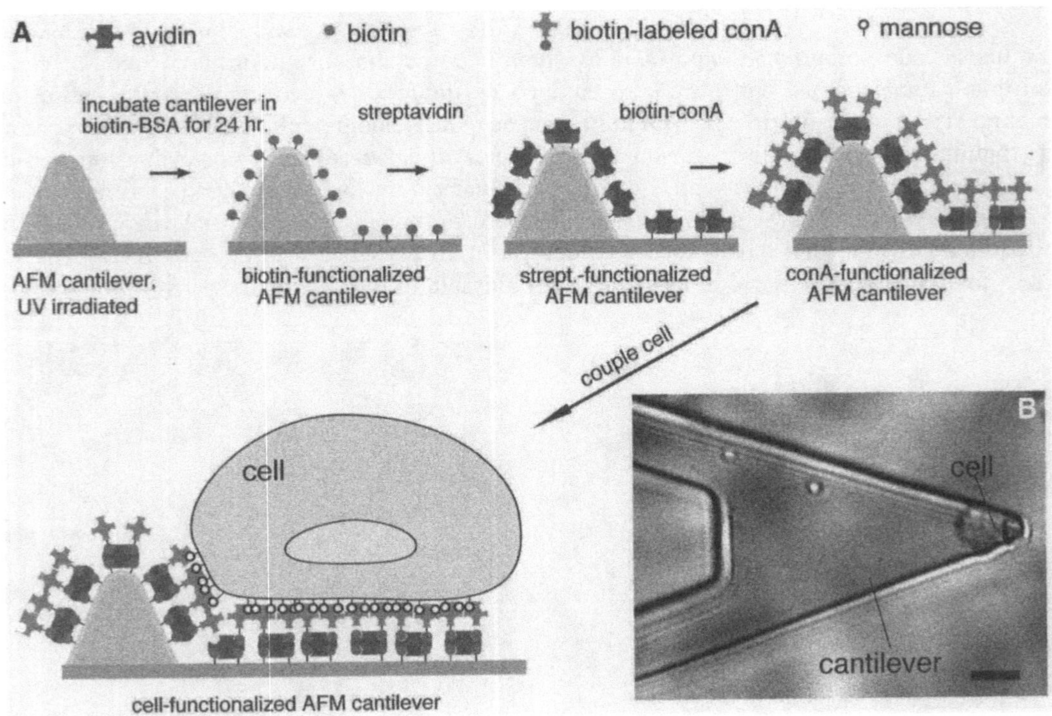

Fig. 2. **(A)** Cell functionalization of AFM cantilever tip. **(B)** Micrograph of a cell attached to the end of a AFM cantilever.

3. AFM setup.
4. Calibration of the cantilever.
5. Immobilization of the cell onto the cantilever.
6. AFM measurement.
7. Specificity control measurements.
8. Analysis of measurements.

In the following sections, we will describe each of these components. In preparing this chapter, we assume that the readers, though may not be familiar with AFM force spectroscopy, have a good understanding of how the AFM operates and of the terminology used in the AFM field.

3.1. Protein Immobilization

For our studies of the LFA-1/ICAM-1 interaction, we used a recombinant ICAM-1/Fc fusion protein (R&D Systems) generated by ligating the D1-D5 of ICAM-1 with the Fc region of human IgG_1. This soluble form of ICAM-1 was adsorbed at four different concentrations on the plastic surface of a Petri dish prior to the AFM measurements as follows:

1. Draw four small circles (3 mm dia.) close to the center of a 35 mm Petri dish.
2. Place 25 μL of ICAM-1/Fc onto the four circles of the Petri dish at concentrations of 2.5, 5, 10, and 20 μg/mL (*see* **Note 2**).
3. Place the Petri dish in a humid chamber (e.g., small container with a damp piece of 3M paper to prevent ICAM-1/Fc from drying.
4. Adsorb for 12–16 h at 4°C.
5. Remove unbound ICAM-1/Fc by washing Petri dish with PBS, three times.
6. Block by flooding Petri dish with BSA, 0.5 mg/mL for 30 min.
7. Wash once with PBS.

3.2. Functionalization of Atomic Force Microscopy Cantilever Tips

To position a cell on the tip of the cantilever, the tip has to be functionalized with a suitable *glue*, which in this case is a molecular sandwich made up of biotinylated BSA, streptavidin, and biotinylated ConA (**Fig. 2A**).

1. Break cantilevers from wafer.
2. Clean cantilever in acetone for 5 min in a glass Petri dish.
3. Irradiate with UV light for 15 min.
4. Wash gently with PBS, three times.
5. Wash with 0.1 *M* $NaHCO_3$, pH 8.6.
6. Coat cantilever with 100 μL biotinylated BSA (0.5 mg/mL) for 12–16 h at 4°C.
7. Wash coated cantilever gently with PBS, three times.
8. Coat cantilever with 100 μL streptavidin (0.5 mg/mL) for 15 min at RT.
9. Wash with PBS, three times.
10. Coat cantilever with 100 μL biotinylated ConA (0.5 mg/mL) for 15 min (*see* **Note 3**).
11. Wash with PBS to remove unbound protein.

3.3. Atomic Force Microscopy Setup

The following protocol describes steps for mounting a functionalized cantilever onto the AFM and connecting the AFM to the control electronics and data acquisition system.

1. Mount clean cantilever holder on the piezoelectric element of the AFM.
2. Mount cantilever on the cantilever holder; check carefully for orientation (8°) and alignment of the cantilever; fix the cantilever to the holder. Avoid scratching the Plexiglas cantilever holder.
3. Place AFM with mounted cantilever over an ICAM-1/Fc coated Petri dish filled with approx 2mL RPMI 1640 medium.
4. Connect pigtail diode laser to AFM; connect AFM to control electronics and data acquisition system.

3.4. Calibration of the Cantilever

Although the manufacturer provides a nominal value for the spring constant of its cantilever, the actual value may vary considerably. Hence, the cantilevers need to be individually calibrated. We use a method based on measuring the cantilever's thermal fluctuation ***(16)***.

1. Locate the cantilever with the inverted light microscope system (*see* **Note 4**).
2. Turn on laser and focus laser spot on the end of the cantilever.
3. Turn on AFM data acquisition system.
4. Adjust the mirror angle so that the reflected laser beam hits the 2-segment photodetector.
5. Optimize the $A+B$ signal, where A and B are the signals from the upper and lower segment of the photodetector, respectively.
6. Adjust mirror to zero the differential signal (i.e., $A–B$) of the photodetector.
7. Adjust laser intensity to achieve an $A+B$ of approx 8V on a scale from 0.0 to 10V.
8. Collect five sample scans at a sampling frequency of 20 kHz to measure the thermal fluctuation of the cantilever. The cantilever should be free in the medium and not less than several μm from the surface of the Petri dish.
9. Lower the cantilever tip manually to the bottom of the Petri dish and collect five force scans. The slope of the approach trace upon contact with the Petri dish surface will be used to calibrate the photodetector (*see* **Note 5**).
10. Analyze measurements to obtain spring constant of the cantilever ***(16)*** (*see* **Note 6**).

3.5. Immobilization of the Cell Onto the Cantilever

After its calibration, the functionalized cantilever is used to capture an unbound single cell from the bottom of the Petri dish onto the very end of the cantilever tip (**Fig. 2**).

1. Manually retract the cantilever and position it over a region of the Petri dish without immobilized ICAM-1/Fc. The cantilever should be >20 μm away from the bottom of the dish.

2. Add 5–20 μL of cells (10^5/mL) to a spot close to the cantilever; wait 2 min to allow cells to settle on the bottom of the dish.
3. Locate a cell with the light microscope and position the end of cantilever above the cell (*see* **Note 7**).
4. Manually lower the cantilever onto the cell. Ideally, the cell should rest directly behind the pyramidal tip of the cantilever (*see* **Note 8**).
5. Allow cantilever a few seconds to bind to the cell.
6. Manually retract the cantilever. The cell should be attached to the cantilever.

3.6. Atomic Force Microscopy Measurement

At the start of the adhesion measurements, the LFA-1 expressing cell on the tip of the cantilever is brought in contact with the immobilized ICAM-1 on the Petri dish. A series of ten force scans are acquired over the each of four regions of the dish that were coated with different concentrations of ICAM-1/Fc. These force scans provide an estimate of the frequency of adhesion at each of the four regions of the dish. An adhesion frequency of 30% is optimal for detecting the unbinding of individual LFA-1/ICAM-1 bonds (*see* **Note 9**).

1. Position the cell-functionalized cantilever over the ICAM-1/Fc coated region of the Petri dish.
2. Collect force scans.
3. Set the frequency of de-adhesion events to approx 30% by adjusting the contact force and contact time (*see* **Note 10**).
4. Collect at least 100 force scans to obtain an accurate estimate of adhesion frequency.
5. Collect force scans with the same settings over an uncoated surface to determine the frequency of nonspecific adhesion.
6. Collect force scans at different loading rates to determine the force spectrum of the LFA-1/ICAM-1 interaction (**Fig. 3** and **Note 11**).

3.7. Specificity Control Measurements

To determine the specificity of the LFA-1/ICAM-1 interaction in the AFM force measurements, we used inhibitory monoclonal antibodies against LFA-1 (i.e., FD441.8) and noninhibitory polyclonal rat IgG.

1. Set AFM to optimal conditions as determined in **Subheading 3.6.** for measuring LFA-1/ICAM-1 interactions.
2. Collect at least 100 force scans.
3. Add function noninhibitory polyclonal rat IgG to cell culture medium to a final concentration of 20 μg/mL.
4. Wait 30 min.
5. Continue force scans with same conditions, collecting at least 100 force scans to determine the adhesion frequency.
6. Repeat experiment with antibodies against LFA-1. Determine adhesion frequency.

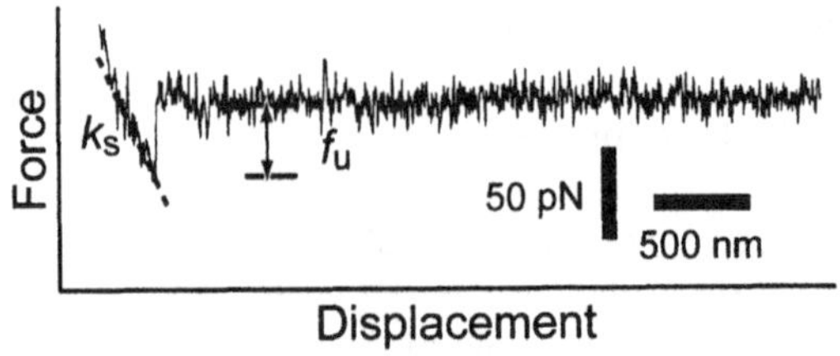

Fig. 3. Force-displacement (retract) traces between LFA-1 and ICAM-1 under conditions of minimal contact. is the rupture force of the LFA-1/ICAM-1 bond. is the system spring constant and was derived from the slope of the force-displacement trace. The cantilever retraction rate of the measurements was 2 µm/s.

3.8. Data Analysis

The analysis of our AFM measurements are carried out using macros written for the IGOR Pro 4.0 software (*see* **Note 12**).

1. Open AFM sample scan files with IGOR Pro software. Determine the spring constant of cantilever (*see* **Note 6**).
2. Open AFM force scan file with IGOR Pro software. Measure the rupture force of the de-adhesion event.
3. Measure the spring constant of the LFA-1/ICAM-1 linkage to determine the loading rate of the measurement (*see* **Note 11**).
4. Add hydrodynamic correction to rupture force values (*see* **Note 13**).
5. Collect rupture force values for different loading rates and plot force histogram.
6. Determine the most probable rupture force as a function of loading rate.

4. Notes

1. Asylum Research (Santa Barbara, CA) manufactures a single-axis force measurement instrument that has been optimized for force spectroscopy measurements.
2. The Petri dish was coated with different concentrations of protein to ensure that one of these concentrations would provide the conditions required for detecting single molecule interaction. If the immobilized protein concentration were too dense, multiple bonds would form upon contact. At the other extreme, if the immobilized protein were too low then the frequency of adhesion would be too low for the experiment to continue. We usually prepare a stock solution of 250 µg/mL protein and use dilutions of 1:10 to 1:160 to determine the optimal concentration for single-receptor binding. We recommend that protein adsorption be allowed to occur overnight at 4°C because this appears to give more reproducible results. It should also be noted that the efficiency of protein adsorption may vary considerably for different proteins and substrates. Moreover, the frequency of adhesion in the AFM measurements will depend on the amount of receptors expressed on the surface of a cell, which may vary from cell to cell of the same culture.

3. Although we have successfully coupled different cell types to the end of AFM cantilevers using the ConA linkage, it may not be the ideal choice for immunological studies because ConA may activate leukocytes. Poly-D-lysine (70–150 kD, 0.1–0.5 mg/mL for 15–30 min, Sigma) may also be used to mediate the attachment of cells to the AFM cantilever. A major drawback of poly-D-lysine is that it often attracts unwanted proteins and debris from the cell culture media that interfere with the laser deflections.
4. Our AFM system includes a simple inverted light microscope that is used to help align the laser of the AFM and couple the cell to the AFM cantilever.
5. Monitor the *A–B* signal while manually lowering the cantilever to determine when the cantilever comes in contact with the substrate.
6. The AFM cantilevers were calibrated using the observed thermally induced vibration of the cantilever ***(16)***. Because each vibration mode receives thermal energy commensurate to one degree of freedom, $k_BT/2$, the measured variance of the deflection $\langle x^2 \rangle$ was used to calculate the spring constant according to the relation: $k_BT/2 = C\langle x^2 \rangle/2$. To separate deflections belonging to the basic (and predominant) mode of vibration from other deflections or noise in the recording system, the power spectral density of the temperature-induced deflections was determined, and only the spectral component corresponding to the basal mode of vibration was used to estimate the spring constant.
7. Although it is possible to transfer a cell coupled to the cantilever from one Petri dish to another, it is much simpler to couple a cell to the cantilever within the same Petri dish used for the measurements. In this protocol, we couple the cell to the cantilever over areas free of immobilized ligand, and then position the cell-functionalized cantilever over a region with the ligand for the AFM force measurements.
8. Although it is not absolutely essential to place the cell directly behind the pyramidal tip of the cantilever, the contact between the cell and the pyramidal tip stabilizes the coupling of the cell to the cantilever. It should be noted that the height of the pyramidal tip is about 4 μm. Hence, tipless cantilevers are required for experiments involving cells less than 5 μm in diameter ***(13)***.
9. An adhesion frequency of <30% in the force measurements ensured that there is a >83% probability that the adhesion event is mediated by a single ligand-receptor bond ***(11)***.
10. Our typical settings include an approach rate of 1 μm/s, a contact duration of 10 msec, and a compression force of 100 pN.
11. We were able to acquire measurements at loading rates (r_f) between 20 and 50,000 pN/s ***(11)***. This was achieved by varying the retraction rate of the cantilever (v) from 0.1 to 15 μm/s in conjunction with the variations in the local compliance of the cell, which allowed for the effective spring constant of the cell–cantilever combination (k_s) to have a range of values between 0.1 and 5 mN/m (i.e., $r_f = k_s \times v$).
12. We use customized software for collecting the force measurement and a customized version of IGOR Pro 4.0 (Lake Oswego, OR) for the force analysis. The commercially available AFMs come with their own software for data acquisition and analysis.

13. At fast cantilever retraction speeds (>1 µm/s), the hydrodynamic drag on the cantilever resulted in smaller forces recorded than were actually applied to rupture the LFA-1/ICAM-1 complex. To correct for the hydrodynamic force exerted on the cantilever, we determined the damping coefficient of the cantilever in the culture medium by retracting the cantilever at different speeds *(11)*.

Acknowledgments

We thank Dr. M. Benoit for providing his excellent expertise in cell-based AFM technology. The scientific contributions of Drs. X. Zhang, F. Li, E. Wojcikiewicz, and A. Chen are gratefully acknowledged. We thank D. Bogorin and M. Abdulreda for helpful discussions and C. Freites for excellent technical support.

References

1. Binnig, G., Quate, C. F., and Gerber, C. (1986) Atomic force microscope. *Phys. Rev. Lett.* **56,** 930–933.
2. Radmacher, M., Tillmann, R. W., Fritz, M., Gaub, H. E. (1992) From molecules to cells: imaging soft samples with the atomic force microscope. *Science* **257,** 1900–1905.
3. Lee, G. U., Kidwell, D. A., and Colton, R. J. (1994) Sensing discrete streptavidin-biotin interactions with AFM. *Langmuir* **10,** 354–361.
4. Moy, V. T., Florin, E.-L., and Gaub, H. E. (1994) Adhesive forces between ligand and receptor measured by AFM. *Colloids and Surfaces* **93,** 343–348.
5. Evans, E., and Ritchie, K. (1997) Dynamic strength of molecular adhesion bonds. *Biophys. J.* **72,** 1541–1555.
6. Merkel, R., Nassoy, P., Leung, A., Ritchie, K., and Evans, E. (1999) Energy landscapes of receptor-ligand bonds explored with dynamic force spectroscopy. *Nature* **397,** 50–53.
7. Yuan, C., Chen, A., Kolb, P., and Moy, V. T. (2000) Dynamic strength of individual ligand-receptor complexes measured by atomic force microscopy. *Biochemistry* **39,** 10,219–10,223.
8. Florin, E. L., Moy, V. T., and Gaub, H. E. (1994) Adhesion forces between individual ligand-receptor pairs. *Science* **264,** 415–417.
9. Hinterdorfer, P. (2002) Molecular recognition studies using the atomic force microscope. *Meth. Cell Biol.* **68,** 115–139.
10. Springer, T. A. (1994) Traffic signals for lymphocyte recirculation and leukocyte emigration: the multistep paradigm. *Cell* **76,** 301–314.
11. Zhang, X. H., Wojcikiewicz, E., and Moy, V. T. (2002) Force spectroscopy of the leukocyte function-associated antigen-1 (LFA-1)/intercellular adhesion molecule-1 (ICAM-1) interaction. *Biophys. J.* **83,** 2270–2279.
12. Wojcikiewicz, E. P., Zhang, X., Chen, A., and Moy, V. T. (2003) Contributions of molecular binding events and cellular compliance to the modulation of leukocyte adhesion. *J. Cell Sci.* **116,** 2531–2539.

13. Benoit, M., Gabriel, D., Gerisch, G., and Gaub, H. E. (2000) Discrete interactions in cell adhesion measured by single-molecule force spectroscopy. *Nat. Cell Biol.* **2,** 313–317.
14. Li, F., Redick, S. D., Erickson, H. P., and Moy, V. T. (2003) Force measurements of the alpha(5)beta(1) integrin-fibronectin interaction. *Biophys. J.* **84,** 1252–1262.
15. Zhang, X., Chen, A., DeLeon, D., Li, H., Noiri, E., Moy, V. T. and Goligorsky, M. S. (2003) Atomic force microscopy measurement of leukocyte-endothelial interaction. *Am. J. Physiol.* **286,** H359–H367.
16. Hutter, J. L. and Bechhoefer, J. (1993) Calibration of atomic-force microscope tips. *Rev. Sci. Instrum.* **64,** 1868–1873.

22

Computer Simulation of Protein–Ligand Interactions

Challenges and Applications

Sergio A. Hassan, Luis Gracia, Geetha Vasudevan, and Peter J. Steinbach

Summary

The accurate modeling of protein–ligand interactions, like any prediction of macromolecular structure, requires an energy function of sufficient detail to account for all relevant interactions and a conformational search method that can reliably find the energetically favorable conformations of a heterogeneous system. Both of these prerequisites represent daunting challenges. Consequently, the routine docking of small molecules or peptides to proteins in their correct binding modes, and the reliable ranking of binding affinities remain unsolved problems. Nonetheless, computational techniques are continually evolving so as to broaden the range of feasible applications, and the accuracy of predictions and theoretical approaches can often be of great help in guiding and interpreting experiments. We discuss the energetics of protein–ligand systems and survey conformational searching techniques. We illustrate how molecular modeling of a protein–ligand complex sheds light on the observed resistance of a mutant dihydrofolate reductase to the antibiotic trimethoprim. In another example, we show that relaxation of side chains in different crystal structures of the same complex, benzamidine bound to trypsin, is needed to draw sensible conclusions from the calculations. The results of these relatively simple conformational searches underscore the importance of incorporating protein flexibility in simulations of protein–ligand interactions, even in the context of relatively rigid binding pockets.

Key Words: Molecular mechanics; molecular dynamics; Monte Carlo simulation; conformational searching; protein–ligand interaction; implicit solvent; free energy; binding affinity.

From: *Methods in Molecular Biology, vol. 305: Protein–Ligand Interactions: Methods and Applications*
Edited by: G. U. Nienhaus © Humana Press Inc., Totowa, NJ

1. Introduction

Success in the computational study of protein–ligand interactions depends critically on both sampling techniques and the energy function employed. Much of the methodological development in the last decade has focused on the improvement and design of sampling techniques, e.g., docking algorithms ***(1–4)***, and recently increasing attention has been paid to the energy function used in such calculations ***(5–20)***. The challenges posed by energetics and sampling are, in fact, intimately linked because the estimation of free energies (i.e., enthalpy and entropy contributions) requires thorough sampling of a high-dimensional energy landscape. We present our view of the difficulties that must be resolved before protein–ligand interactions can be reliably quantified to accuracies on the order of thermal energy or within the range of experimental errors. In particular, we focus on the role of the solvent and on implicit solvent models, and survey some of the most commonly used conformational sampling techniques. We report two simulation studies that demonstrate that the incorporation of protein flexibility can be critical in understanding experimental observations of protein–ligand systems. First, Monte Carlo-minimization (MCM) was used to explore the interactions of the antibiotic trimethoprim with a mutant dihydrofolate reductase, and to provide an atomic-level explanation for the observed drug resistance of the mutant. Second, the change in enthalpy was estimated for the binding of benzamidine to trypsin, using multiple structures reported in the literature for this complex. An initial calculation from each structure yielded differences in binding enthalpy of several kcal/mol. However, following relaxation of side chains in the binding pocket by a simulated annealing/Monte Carlo search, the binding enthalpies derived from each complex converged to the same value.

1.1. Protein–Ligand Thermodynamics

From a macroscopic or thermodynamic point of view, a system in equilibrium at constant temperature T and pressure P has minimal Gibbs free energy, $G = H - TS$ where H is the enthalpy and S the entropy of the system ***(21–23)***. Thus, chemical reactions spontaneously occur in the direction that minimizes G. Consider the binding of ligand L to protein receptor P in solution in the simple reaction: $\mathrm{P} + \mathrm{L} \rightleftarrows \mathrm{PL}$. The binding free energy, or change in G upon ligand binding, is given by $\Delta G = G_{\mathrm{PL}} - G_{\mathrm{P+L}} = G_{\mathrm{PL}} - G_{\mathrm{P}} - G_{\mathrm{L}} = \Delta H - T\Delta S$, where ΔH and ΔS are the change in enthalpy and entropy of the system, respectively. The equilibrium constant, or binding affinity, K_a, is defined as $a_{PL}/ (a_P\, a_L)$, where a_x is the (dimensionless) activity of species x, which for ideal-dilute solutions equals the numerical value of the molar concentration [x] of x, i.e., K_a=[PL]/[P][L]. K_a is related to the free energy of binding through $\Delta G = -RT \ln K_a$, where R is the gas constant (energies given in units of kcal/mol).

From a microscopic or statistical mechanical point of view, the enthalpy and entropy of a system for which the pressure P, temperature T, and number of particles N are kept constant are given respectively by $H = RT^2 \partial \ln Z/\partial T$ and $S = RT\, \partial \ln Z/\partial T + R \ln Z$, where the derivatives of the isothermal–isobaric partition function Z with respect to temperature are taken at constant N and P ***(21–23)***; Z is given by:

$$Z = P_N \int_0^{\infty} Z_N \exp(-PV/RT)\, d(PV/RT) \tag{1}$$

where

$$P_N = \left[\prod_{i=1}^{m} N_j! \Lambda_j^{3N_j} \right]^{-1}$$

originates from an integration over the momenta **p** of all the particles in the system, V is the system volume, and m is the number of molecular species in the system. In the present context, $m = 3$ because the system in equilibrium contains protein, ligand, and protein–ligand complex. N_j is the number of molecules of each species in equilibrium, and, $\Lambda_j = h/\sqrt{2\pi M_j kT}$, where M_j is the molecular mass of the jth species, h and k are Planck's and Boltzmann's constants, respectively. The configurational integral Z_N is given by $Z_N = \int \exp[-V(\mathbf{q})/RT]d\mathbf{q}$, where **q** denotes the set of (generalized) coordinates of all molecules in the system and $V(\mathbf{q})$ is the potential term in the Hamiltonian $\mathcal{H}$ that describes the dynamics of the system. Note that **Eq. 1** assumes that $\mathcal{H}$ does not mix the momenta **p** and coordinates **q**, i.e., it is formally separable in the form $\Sigma \mathbf{p}^2 + \Sigma V(\mathbf{q})$. If a solution for Z_N was available, the conditions of thermodynamic equilibrium (minimal G) would determine the activities or the molar concentrations of each species in the system. Thus, an expression for the affinity K_a would be obtained from first principles, i.e., from the fundamental interactions in the system. However, the configurational integral cannot be solved easily, except in very simple cases, such as an ideal gas or in particle systems interacting with simplified potentials $V(\mathbf{q})$.

The ideal-gas approximation allows estimation of the entropy and enthalpy changes caused by the loss of translational and rotational degrees of freedom of the protein and ligand upon binding ***(24)***. For an ideal gas, changes in translational and rotational enthalpy can be readily estimated and depend only on the temperature. The estimation of translational (S_t) and rotational (S_r) entropy loss can be more difficult. S_t is described by the Sackur–Tetrode equation, which depends on the concentration of the molecules in the medium but not on structural details. By constrast, S_r depends on the shape and mass distribution of the interacting molecules around their respective principal axes but not on concentrations. To obtain more realistic results, corrections to this ideal-gas approach are usually done by introducing the effects of the aqueous solvent on the translational/rotational motion of the molecules based on physical arguments. For example, small molecules do not translate freely in the solvent but *jump* from one position to the next, and *librate* instead of rotating freely in the medium. Molecular mechanics (MM) simulations have been used to calculate these effects (e.g., *see* **ref. *25*** and references therein). An ideal-gas approximation to $V(\mathbf{q})$, however, ignores the explicit interactions between all protein and ligand atoms, the internal degrees of freedom of each molecule, and the effects of the solvent. These effects are the most important in the majority of cases (particularly to calculate relative free energies of binding of similar molecules), and we describe in this chapter techniques developed to calculate the thermodynamic properties of realistic protein–ligand systems.

The free energy of association ΔG has been estimated computationally at various levels of physical rigor. Phenomenological approaches, such as those representing ΔG as an *ad hoc* scoring function that relies on adjustable, optimized parameters ***(26–31)*** have been implemented in a variety of computer software (Dock, Ludi, FlexX, Gold, Glide, ChemScore, Dscore, Gscore, PMF, etc). More rigorous (but far less practical) approaches aim to calculate ΔG (ΔH and ΔS, and the different terms into which each can be partitioned) from first principles (*see* review in **ref. *32***), mostly using molecular mechanics force fields and special sampling methodologies, such as free energy perturbation ***(33)*** and thermodynamic integration ***(34)*** methods, or some recent developments based on nonequlibrium statistical mechanics ***(35)***.

Accurate prediction of the binding affinity is critical for structure-based drug design. A number of sophisticated docking methodologies have been developed in recent years that may well evolve into a reliable tool for the prediction of binding modes. It is probable that the theoretical description of the energetics will be the rate-limiting barrier to the routine prediction of ligand-binding modes. Although high-resolution X-ray crystallography and nuclear magnetic resonance (NMR) techniques, along with calorimetric methods, are essential tools for structure-based docking prediction, theoretical estimation of binding affinity from structural considerations alone is not straightforward. Ligand binding is a very complex process involving a number of interrelated physical effects that are difficult to predict and quantify. Widely recognized as important are hydrogen bonding, and electrostatic and hydrophobic interactions, which in turn, are affected by the presence of electrolytes and cosolvents. In addition, protein–ligand interactions can be affected by changes upon binding in the ionization states of key interacting groups in both the ligand and protein, by entropy–enthalpy compensation, by changes in isomerization states of the ligand, and by individual water molecules (and ions) in or near the binding pocket. Therefore, the synergistic interaction between computer simulations and experimental studies (e.g., **ref. *35a***) is likely to be the most promising and fruitful approach to rational ligand design in the future. An example of the difficulties in predicting theoretically the relative free energy of congeneric compounds has been recently reported ***(36)***. In this study, a combination of thermodynamics (isothermal titration calorimetry (ITC); *see* Chapter 1 of this volume) and crystallographic studies were performed for a series of small ligands binding to the well-known proteins trypsin and thrombin, and some of the effects we have mentioned were analyzed. Finally, it is important to recognize the intrinsic limitations of the force field used, both for the calculations of binding properties, e.g., using molecular-mechanics force fields, and for structure determination, e.g., by X-ray crystallography ***(37)***, two areas that continue to evolve.

1.2. Classical Potential Energy Function

Simulations of biomolecular interactions can only be as good as the energy function employed. We restrict our discussion to classical simulation techniques, or MM simulations (*see* **Note 1**). Whereas the use of quantum mechanical methods and polarizable force fields in the study of protein–ligand interactions will become increasingly important, these promising approaches are not yet predominant.

In classical simulations the system configuration is specified by the coordinates of all N atoms, $\{\mathbf{r}_i\}_{i=1,N}$. All physical interactions between atoms of the system are approximated in terms of the potential-energy function, $V(\{\mathbf{r}_i\})$. Because the force on an atom is simply the negative gradient of its potential energy ($F = -\nabla V$), the description of the system energetics is often referred to as a force field.

The potential energy used in classical simulations is generally of the form $V = V_{bond} + V_{nonbond}$. The bonded components, V_{bond}, arise from the electron density as determined by covalent interactions. Typically, harmonic functions are used to approximate the energetics of bond stretching, angle bending (involving three consecutively bonded atoms), and the oscillation of so-called improper dihedrals that restrain one atom to be in or out of the plane defined by the three atoms to which it is bonded. Periodic functions describe the energetics of rotations about single bonds.

The partitioning of the nonbond interactions is less arbitrary because different physical processes contribute in more definite ways. First, $V_{nonbond}$ accounts for the repulsion of atoms at short range and the van der Waals attraction between oscillating electronic clouds in neighboring atoms. We refer to the sum of these effects as V_{vdW}. It is computationally expedient to use a Lennard–Jones parameterization of V_{vdW} that sums a repulsive r_{ij}^{-12} term with an attractive r_{ij}^{-6} term, where r_{ij} is the separation between atoms i and j. Other forms of V_{vdW} have been proposed and shown to produce promising results ***(38)***. For simplicity, it is assumed that V_{vdW} interactions can be summed pairwise.

Second, $V_{nonbond}$ quantifies electrostatic interactions, typically through Coulomb's law, whereby the interaction between charges q_i and q_j is proportional to $q_i q_j/(\varepsilon r_{ij})$, where ε is the dielectric coefficient used to characterize the medium between the two charged atoms. In explicit-solvent simulations, it is conventional to use $\varepsilon = 1$ (*see* **Note 2**). Implicit solvent models are discussed in detail later.

Although hydrogen bonding interactions in vacuum are now commonly described through the parameterization of the van der Waals and Coulomb interactions, some energy functions include an additional, explicit hydrogen-bond term in $V_{nonbond}$. As discussed later, other effects, some of thermodynamic origin such as hydrophobic interactions, might also be accounted for in implicit-solvent representations.

The parameterization of a ligand molecule in terms of its atomic partial charges and van der Waals radii is a crucial prerequisite to simulating its binding to a protein receptor using molecular mechanics. Historically, force field development has focused on amino acids and nucleotides, and ligand charges have often been estimated, as needed, using standard quantum mechanical procedures. An increasing number of more automated solutions are becoming available. For example, the Merck molecular force field (MMFF) ***(38)*** was designed with protein–ligand applications in mind to provide charges for novel covalent topologies as are encountered in small molecules. An algorithm to automatically assign charges has been recently developed that reproduces standard force field charges quite well ***(39)***.

Soft-core potentials are modifications to $V(\{\mathbf{r}_i\})$ designed to improve sampling in condensed phases such as in protein cores and ligand-binding pockets. These modified potentials are used to reduce the nonbond forces (and energies) at very short range while preserving the energetics at long range. Forces that would otherwise diverge are made finite so that atoms have a greater chance of getting past or going through each other. The use of soft-core potentials and molecular dynamics simulation has been shown to improve efficiency in flexible-ligand docking studies ***(40,41)***.

1.3. Conformational Searching

The energy function determines an energy landscape in a high-dimensional space with a very large number of local minima. This hypersurface must be searched efficiently for conformations of the protein–ligand system that are thermally accessible in equilibrium. The following methods sample the energy landscape in different ways, and the form of the force field used should be considered when refining the conformational search.

The conformational change induced by *energy minimization* is usually very modest but of considerable practical importance. Energy minimization simply seeks the bottom of the energy well occupied by the initial conformation. Energy minimization should be done prior to dynamics simulation to reduce local strains (large forces) before they can detrimentally distort the system configuration.

Molecular dynamics (MD) simulation is the numerical integration of Newton's second law of motion. The net force on each atom is calculated as the negative gradient of its potential energy and used to accelerate the atom in the direction of the force. MD is most commonly used to explore conformations when all atoms, including solvent, are explicitly simulated.

Langevin dynamics (LD) simulation is essentially MD modified for use with an implicit environment. In addition to the net force on an atom arising from its interactions with all other atoms explicitly simulated, two forces are added to mimic the random kicks and viscous drag that would result from interactions with the implicit (e.g., solvent) atoms, if they were present. The magnitude of these two forces are related according to the fluctuation-dissipation theorem and governed by a collision frequency chosen to correspond to dynamics in the implicit environment at a given temperature.

Unlike the methods described earlier, in its purest form, Monte Carlo (MC) simulation does not utilize the forces acting on atoms. The energy of the system is evaluated following an imposed, random conformational change and compared to the energy of the prior conformation. If the energy has decreased the new conformation is accepted as the state of the system. If not, the energy increase ΔE is cast in terms of a probability via the Boltzmann factor, $\exp(-\Delta E/kT)$, which in turn is compared to a random number generated uniformly in the range from 0 to 1. If the Boltzmann factor exceeds the random number, the new conformation is accepted; otherwise it is rejected and the state and energy of the system remain unchanged. In this way, the conformations sampled (accepted) conform to the canonical ensemble at the temperature T. The practical challenge is to impose conformational changes that represent significant jumps across the energy landscape but are modest enough to be accepted frequently.

Monte Carlo-minimization (MCM) *(42,43)* is a hybrid approach that employs relatively large conformational changes immediately followed by energy minimization to generate the trial structures to be accepted or rejected based on Boltzmann probabilities. Because MCM does not satisfy detailed balance, the distribution of accepted conformations does not formally correspond to a thermodynamic (e.g., canonical) ensemble. Although the minimization precludes a sampling of the entire conformational space, MCM may converge to lower energies than traditional MC. Consider a conformation strained by steric overlap or short-range electrostatic repulsion. Upon energy minimization, this strain can be dramatically reduced with only a slight change in atomic coordinates. Thus, when searching conformations in condensed media such as proteins, the likelihood of accepting trial conformations can be much higher with MCM than with MC, easily justifying the added expense per step due to the energy minimization. MCM was recently used to find the lowest-energy conformations of three peptides of known structures starting from an extended conformation and relaxing to near-native folds ***(43a)***.

Simulated annealing is a method of temperature modulation used together with another search protocol (e.g., MD, LD, MC, MCM) to enhance the chance of surmounting barriers and ultimately settling into a low energy conformation. Generally, the system is heated to promote barrier crossing and then cooled more slowly in hopes of reaching low energies.

1.4. Effects of the Solvent

The aqueous solvent is an important part of any protein–ligand system, as it influences the structure, dynamics, and function of biologically active macromolecules ***(44–49)***. In many cases, the role of explicit water–protein interactions can be elucidated with the help of high resolution X-ray crystallography ***(50,51)***. At ligand binding sites, individual water molecules can stabilize ligands and control binding affinity (e.g., *see* **refs. *52–54***). In the bulk solvent, the polarization, reorganization, and orientation of water molecules make it difficult to characterize bulk-solvent electrostatic effects quantitatively. Thus, it is desirable and conceptually simple to use an explicit representation of the solvent in molecular dynamics simulations, where all these effects are automatically accounted for in the potential energy function, limited by the theoretical model used to represent individual water molecules. However, many problems in computational biophysics and computational chemistry are difficult to treat with this explicit representation, given the large number of solvent molecules that must be included for meaningful results. Considerable computer time is required in such calculations to simulate the dynamics of the solvent, in addition to the protein of interest. The enormous computational requirements still preclude dynamics simulations of sufficient length to address many important biological processes, which can easily occur on timescales in the millisecond range and beyond. Use of explicit water in MC simulations of protein systems severely limits the spatial extent of the random moves that are to be accepted with reasonable probability.

Consequently, the development of implicit-solvent models (ISMs) that allow deletion of the explicit water while retaining its effects has a history that parallels the development of force fields and MM simulation techniques. When solvation effects are treated implicitly, the energy function used represents a potential of mean force, W, that includes the free energy of the solvent and the solute potential energy ***(55)***. Although we emphasize throughout this chapter the difficulty in accurately calculating and ranking binding affinities, the development of a reliable ISM could prove very useful in less ambitious calculations, such as when very large libraries of molecules are screened to select a manageable number of promising lead compounds for drug candidates.

The bulk solvent modulates the electrostatics of the protein–ligand system in two ways ***(56,57)***, through the self-energy defined by the interaction of a charge distribution with the polarization it induces in the bulk solvent and through the shielding of the electrostatic interaction between all pairs of atoms. Often, calculations of noncovalent ligand binding have neglected the self-energy component, using only a simple Coulombic interaction with a constant or linearly distance-dependent dielectric function to characterize the electrostatic protein–ligand interactions. The process of desolvation involves these important bulk-solvation effects as well as the competition with individual water molecules for hydrogen bonds. Not only are these solvation effects readily motivated on physical grounds, their importance in practical applications has been clearly demonstrated ***(57–62)***. Recently, for example, failure to consider ligand solvation/desolvation effects was shown to lead to incorrect results for known inhibitors of a number of proteins ***(5,14,63,64)***. Moreover, inclusion of ligand solvation in docking calculations was shown to change the relative ranking of compounds in database screening ***(14)***.

Therefore, it is clear that the desolvation of a ligand upon binding, especially to protein binding pockets with both polar and nonpolar regions, must be accounted for in the energy function used to estimate energies of the conformations. The force field used needs to quantify the preference of charged and/or polar ligands for polar environments and of nonpolar molecules for more hydrophobic pockets. Moreover, the relative magnitude of self-energy and interaction terms is important, as has been demonstrated in stringent tests of protein electrostatics ***(58–62)***. An imbalance in the solvation and interaction components is expected to affect the characterization of specificity, binding mode, and affinity.

Although simple models of solvation were used recently to avoid prohibitive computing time, the results obtained were a marked improvement over those obtained using a force field without any solvation term at all ***(10,12, 65,66)***. This observation encourages the development and use of more realistic models of the nonbonded interactions. Whereas simple models may be suitable for a fast, preliminary screening to discriminate between likely and very unlikely ligand inhibitors, more sophisticated models are needed for a detailed analysis that includes the calculation of binding free energies (absolute and relative) and the evaluation of specificity. Specificity can be of primary practical importance since enhancing the specificity of a drug for a given protein reduces its potential toxicity and the side effects resulting from its association with similar binding sites of a related protein family. The similarity in binding sites of different protein targets may reflect not only the overall three dimensional structure, but also local reactivity properties such as electrostatics, hydrophobicity, hydrogen-bonding pattern, relative location of ionizable groups, electronic polarization, local structural flexibility (which is known to contribute to local dielectric properties of the protein), etc. These functionally important properties are difficult to characterize, but efforts in this direction have been reported ***(67–70)*** and applied to the inverse protein-folding problem ***(71,72)***, that is to say the design of a protein sequence that leads to desired structural properties.

1.5. Implicit Solvent Models

In addition to bulk-solvent electrostatic effects, the removal of solvent eliminates or dramatically perturbs other important effects and properties of the system ***(73)***. Solvent molecules affect hydrophobic interactions, directionality of solute-solvent hydrogen bonding, pressure, viscosity, cooperative effects, and intrasolute interactions. Also important are effects of the granularity of the solvent around exposed side chains, solvent density fluctuations, and electrostriction around highly charged groups. Furthermore, an ISM should be used in combination with appropriate simulation techniques (e.g., LD) to account for some of the dynamical effects mentioned above. Also, the use of an implicit solvent should not preclude the introduction of explicit water molecules when they are deemed important, as when studying ligand binding. However, little has been done to incorporate explicit solvent molecules or ions in the context of an ISM. In the absence of the ligand, the binding pocket may contain a few water molecules that upon ligand binding might or might not be resolvated to the bulk. In either case, their enthalpic and entropic contributions to the change in free energy have to be accounted for to obtain accurate results. Moreover, because most experimental systems, in vivo or in vitro, typically contain salt on the order of $1M$, the effects of electrolytes have to be included. Accounting for salt is a difficult task and very few advances have been made beyond solving the nonlinear Poisson–Boltzmann (PB) equation numerically (see following) to quantify electrostatic changes. However, electrolytes are known to dramatically affect both electrostatic ***(22,74,75)*** and hydrophobic ***(76–79)*** interactions and are critical in ligand–protein interactions. Therefore, a continuum–electrostatics model for use with a classical force field is only one important part of a complete ISM.

1.5.1. Electrostatics

Poisson–Boltzmann Equation: The Poisson equation, $-\nabla \cdot [\varepsilon(\mathbf{r})\nabla\phi(\mathbf{r})] = \rho(\mathbf{r})$, is one of the basic equations in electrostatics ***(80)***. Here ϕ is the electrostatic potential at position $\mathbf{r}$, ρ is the explicit charge density ($\rho_s + \rho_{\text{ions}}$), and ε is the dielectric function (assumed here to be a scalar quantity). If m ionic species coexist with the solute's charge distribution ρ_s, they are distributed in equilibrium according to a Boltzmann distribution. If the potential of mean force experienced by an ion is approximated as the ion's electrostatic potential energy $q\phi$, we obtain the nonlinear PB equation:

$$-\nabla \cdot [\varepsilon(\mathbf{r}) \nabla \phi(\mathbf{r})] = \rho_s(\mathbf{r}) + \sum_{i=1}^{m} q_i \eta_i \exp[-q_i \phi(\mathbf{r}) / kT] \tag{2}$$

where q_i and η_i are the charge and concentration of the ith ionic species. At sufficiently low ionic concentrations, the Taylor expansion of the exponential can be truncated after its linear term ***(21,22)***, and the linear PB equation is obtained in the form $-\nabla \cdot [\varepsilon(\mathbf{r})\nabla\phi(\mathbf{r})] + \varepsilon(\mathbf{r})\kappa^2\phi(\mathbf{r}) = \rho_s(\mathbf{r})$, where the ionic solution has been assumed to be electrically neutral, i.e., $\Sigma q_i \eta_i = 0$. The Debye–Hückel screening length κ is given by $\kappa^2 = 2Ie^2/[kT\ \varepsilon(\mathbf{r})]$, where e is the electron charge and I is the ionic strength of the system: $I = \Sigma \eta_i (q_i/e)^2/2$.

The PB equation, both in its linear (Debye–Hückel approximation) and nonlinear form, has been applied in many areas of physics, and its usefulness in studying protein electrostatics is well documented ***(74,81,82)***. Although analytical solutions exist only for very simple geometries (e.g., spherical or cylindrical), numerical solutions of these equations have allowed the PB formalism to be applied to proteins of arbitrary size and shape, including the calculation of electrostatic effects in protein–ligand binding. Software packages such as UHBD ***(83,84)*** are routinely used to solve these equations for a variety of systems.

Of all the commonly used ISMs, PB solvers represent the most rigorous way to include salt effects in macromolecular applications. However, the user must define the dielectric properties of the system. Because these properties are not known *a priori*, and depend on local properties within the macromolecules, this requirement can be problematic ***(57,62,85,86)***. Typically, the protein-solvent system is represented simply by an external medium (water) characterized by a single, high dielectric constant ε_w, and an internal medium (protein) with a single, low dielectric constant ε_i, with an interfacial boundary separating the media. This boundary is generally defined as a molecular surface or a solvent-accessible surface and expressed in terms of atomic radii for protein atoms. Changes in any of these quantities may significantly affect the numerical value of the property being calculated. A smooth solute/solvent interface can improve the results. In addition, real systems display local dipole moments and even higher-order terms in a multipole expansion. PB calculations that account for the point dipoles produce different numerical solutions than those using point charges alone ***(86)***. Thus, caution should be exercised when PB solvers are applied to macromolecules and specific conclusions are drawn from numerical results. As has been shown and discussed, the value chosen for ε_i and the definition and location of the solute-solvent boundary are arbitrary ***(62,85–96)***, and very different values have been invoked for the internal dielectric constant in order to reproduce experimental data ***(85,88,92–94)***. Nonetheless, solutions to the PB equation are considered to be the *gold standard* for protein electrostatics, as long as common sense and caution are applied when comparing the numerical values obtained to quantities measured in real systems.

Because PB calculations are computationally expensive, simpler implicit-solvent models have been developed to quantify the electrostatics of solvated macromolecules. Two conceptually different approaches have produced a number of models that evaluate solvation energies of the form $E = E_{\text{int}} + E_{\text{sol}}$, where E_{int} is the shielded Coulomb interaction energy and E_{sol} includes the self-energy. The first approach is to propose both a phenomenological functional form for E and a partition into the two terms (E_{int} and E_{sol}) that is expected to capture the two individual effects. Such models include statistical energy functions ***(28,65)***, surface area models ***(97,98)*** and generalized Born (GB) formulations ***(99)***. The second class of approach seeks a functional form for E derived under basic physical assumptions that lead naturally to a partition into E_{int} and E_{sol} and a functional form for each. Models of this type include the screened Coulomb potentials–implicit solvent model (SCP–ISM) ***(56,57,100)***, the effective energy function model (EEF1) ***(55)***, and semimicroscopic approaches such as the Protein Dipoles Langevin Dipoles model (PDLD) ***(101,102)***. Models built from fundamental theory are advantageous in that their inherent approximations and range of applicability are readily understandable and their refinement can be pursued accordingly. It is also likely to facilitate the incorporation of nonpolar interactions between solute and solvent and salt effects en route to a complete and general ISM. To varying degrees, both approaches ultimately require the fitting of adjustable parameters to experimental data and/or simulation results obtained using a supposedly more reliable theoretical model (e.g., PB or explicit–solvent simulations).

Generalized Born (GB) Models: Among the phenomenological ISMs available, GB models are the most commonly used. Most GB variants adopt the functional form, proposed by Still et al. ***(99)*** for the energy contribution of the dielectric, E_{pol}, for a system of N point charges $\{q_i\}$, i.e.,

$$E_{\text{pol}} = \frac{1}{2}\left(\frac{1}{\varepsilon_s} - \frac{1}{\varepsilon_i}\right)\sum_{i,j}^{N} \frac{q_i q_j}{\sqrt{r_{ij}^2 + \alpha_{ij}^2 \exp\left(-r_{ij}^2 / 4\alpha_{ij}^2\right)}} \tag{3}$$

where $a_{ij} = (a_i a_j)^{1/2}$, a_i is the conformation-dependent Born radius of atom i, ε_i, and ε_s are the dielectric constants of the protein and of the solvent, respectively, and r_{ij} is the distance separating atoms i and j. This functional form was chosen from among several alternatives based on the requirement that E_{pol} take certain values in the limits $r_{ij} \to 0$ and $r_{ij} \to \infty$. Upon fitting the Born radii to experimental hydration energies, Still et al. were able to reproduce the hydration energies measured for a number of small organic molecules. Subsequently, **Eq. 3** has been applied to larger systems, from small peptides to proteins and nucleic acids. While many GB models have been reported, most differ only in the calculation of the a_i and/or approaches to speed up the calculation ***(103–109)***. Incorporation of GB models in simulation packages has allowed an economical evaluation of self-energies, representing a step beyond the overly simplistic use of a linearly distance-dependent dielectric coefficient in place of explicit solvent. Attractive results have been obtained with various GB models in many applications. However, a unique GB model has not yet been shown to be general enough for application to biomolecules, and so different approaches to design new GB models are still being reported ***(109)***.

The Screened Coulomb Potential Implicit Solvent Model (SCP–ISM): Based on the Lorentz–Debye–Sack theory of polar/polarizable liquids ***(110–117)***, the continuum electrostatics model in the SCP–ISM is derived from basic electrostatic theory of polar solvation ***(56,57,73,100)***. In contrast to other approaches, it does not assume a boundary separating a low-dielectric protein interior from a high-dielectric solvent. Because dielectric properties in an aqueous medium vary over dimensions comparable to the size of biomolecules, no precise boundary can, in principle, be defined. Instead, the system (e.g., protein, ligand, and solvent) is characterized by a smooth dielectric function that permeates all space, reaching bulk-solvent dielectric values only far (e.g., 7–10 Å) from the solutes ***(56)***. This smooth variation of the dielectric function and the absence of molecular boundaries are characteristics of the model that may ultimately prove to be important when studying processes that occur at the surface of proteins, such as ligand binding. It is worth noting that macroscopic models that assume a dielectric discontinuity at the solute/solvent interface should address the induced forces that arise from the discontinuity. Whereas PB developers have recognized and explicitly accounted for these effects ***(81,118,119)***, other implicit models that assume this purely macroscopic view have not addressed the issue.

The functional form of the electrostatic energy E and the solvation energy have been formally derived under reasonable physical assumptions using a standard thermodynamic path ***(56,57)***. The electrostatic energy for the SCP–ISM is given by

$$E = \frac{1}{2}\sum_{i \neq j}^{N} \frac{q_i q_j}{D(r_{ij}) r_{ij}} + \frac{1}{2}\sum_{i=j}^{N} \frac{q_i^2}{R_{i,B}}\left[\frac{1}{D(R_{i,B})} - 1\right], \tag{4}$$

where $D(r)$ is the screening function that characterizes all the shielding mechanisms in the system and has a nonlinear, sigmoidal distance dependence, and $R_{i,B}$ is the effective Born radius of atom *i*. Note that **Eq. 4** reduces to Coulomb's law for a system in vacuum [$D(r) = 1$] and to the Born expression for the electrostatic contribution to the solvation free energy of a single ion in a homogeneous dielectric [$D(r) = \varepsilon$ =constant]. Note also that the inhomogeneity of the dielectric (i.e., spatial variation of *D*) governs both the electrostatic interactions and the self-energies. Born radii of atoms in the solvated system are defined in terms of the size of the cavity formed by the atoms in their local environment. The screening function used in the SCP–ISM is given by $D(r) = (1 + D_s)/[1 + k \exp(-\alpha r)] - 1$, where $k = (D_s - 1)/2$ and $D_s = \varepsilon_s =$ 78.39 (the approximate bulk dielectric constant value of water at 25°C) (*see* **Note 3**). Thus, *D* varies smoothly from 1 to D_s at a rate determined by the parameters α. The self-energy of atom *i* is screened as controlled by $\alpha = \alpha_i$, whereas the interaction between atoms *i* and *j* is screened using $\alpha = (\alpha_i \alpha_j)^{1/2}$. The relation between *D* and the experimentally measurable ε follows from the definition of the electric potential ϕ, $\mathbf{E}(\mathbf{r}) = -\nabla\phi(\mathbf{r})$, and the definition of the screening function *D* for a point charge *q*, $\phi(\mathbf{r}) = q/[D(\mathbf{r})r]$ (*see* **Note 4**). As discussed earlier ***(56,57)***, many solvation effects can be embodied in the quantities α that control the screening, in particular solvent-exclusion effects and salt effects.

1.5.2. Hydrogen Bonding

Polar groups buried inside proteins are common despite the enthalpic cost of removing a hydrophilic group from water. If the buried group forms internal hydrogen bonds (HBs) with the protein, the energetic penalty associated with desolvation can be overcome ***(120,121)***, and in many cases, the system enthalpy is lowered ***(120)***. Thus, there is a delicate balance between the electrostatic component to macroscopic solvation and HB interactions. A strong HB can stabilize a buried polar group even when the macroscopic solvation is correctly represented, while a weak HB might not fully compensate for the solvation penalty ***(56,100,122,123)***. Clearly, an imbalance in the relative strength of these two effects will degrade the characterization of ligand-binding modes and the estimation of affinities.

The solvation/desolvation energy associated with a HB-forming group is modulated not only by the dielectric properties of the bulk solvent (i.e., as a result of reorientation and polarization of water dipoles), but also by the formation of HBs with the solvent molecules. Because of this local competition for hydrogen-bonding partners, a HB between two fully solvated groups is less stabilizing than one between two fully buried groups. If one of these groups is in the ligand and the other is in the protein, the competion with explicit waters must be included in the ISM ***(73,100,122,123)***. In classical force fields, HB interactions are usually parameterized in terms of attractive electrostatics and repulsive Lennard–Jones interactions. These parameters are largely based on quantum mechanical calculations of small systems in vacuum, not in bulk solvent. The development of an electrostatic description other than Coulomb's law with $\varepsilon = 1$ (e.g., E of **Eq. 4** in the SCP–ISM) must be accompanied by a recalibration of HBs to account for these local interactions with the solvent. Because both a plain dielectric screening and an approximation based on Born radii involve macroscopic quantities derived from bulk properties of polar liquids, neither should be assumed to describe HB interactions quantitatively ***(123a)***.

1.5.3. Hydrophobic Interactions

Hydrophobic interactions affect the stabilization and dynamics of macromolecules ***(124,125)*** and make important contributions to protein–protein and protein–ligand binding. The relative importance of hydrogen bonding and hydrophobic interactions in the binding of small ligands to proteins has been discussed ***(126)***. These interactions are also involved in self-assembly phenomena such as the formation of membranes and micelles ***(127–129)***, and are thought to be important in early protein-folding events ***(130)***. There is an increasing body of experimental evidence that hydrophobic interactions can be linked to protein recognition and specificity ***(126,131–133)***. The cold denaturation and high-pressure unfolding of proteins seem to be related to hydrophobic interactions ***(124,125,134–137)***. Hydrophobic effects have not yet been fully implemented in a general ISM for use in MM simulations, although recent attempts have been reported ***(138)***. Traditionally, they have been described by a solvent accessible surface area (SASA) approach where the proportionality constant is related to the solute/solvent macroscopic surface tension ***(139,140)***. This proportionality can be derived from calculations of free energy of cavity formation, and accounts reasonably well for the loss of entropy of water molecules around the solute ***(139)***. However, it has been shown that hydrophobicity is a rather complex phenomenon, especially for small and medium-sized solutes such as amino acids and small organic molecules ***(141)***, i.e., at the size scale of protein-binding ligands and drugs in general. Recent studies have suggested that in this size regime there is no unique constant of proportionality between hydration free energy and SASA ***(141)***. Corrections have been proposed, for example, by including the curvature of the solute ***(140,142,143)***. A more sophisticated, semi-empirical model for the calculation of absolute free energies of small molecules was developed ***(139)*** where the authors showed convincingly that a simple SASA-based approach would be unable to rationalize the differences in free energy of a number of molecules (e.g., the difference between 1 propanol vs ethanol [obtained by a nonpolar substitution], or 1-butanamine vs dimethylamine).

Statistical Mechanical Basis of the Hydrophobic Interaction: Ligand design will benefit substantially from efforts to implement a more sophisticated description of hydrophobic effects. The formulation of a more fundamental framework based on statistical mechanics has a long history ***(144–149)***. Recently, progress has been made in understanding the hydrophobic effect, based on a theoretical description of water density near hydrophobic solutes ***(141,150,151)***. This approach has formally shown that hydrophobic interactions have a different physical origin depending on solute size, as had been suggested by Stillinger in the context of a modification to scaled-particle theory ***(148)***.

1. *Small length scales:* From a thermodynamic point of view, hydrophobicity is characterized by a temperature-dependent balance between enthalpic and entropic contributions. Water molecules are excluded from the region occupied by a small solute but can reorganize around it to preserve the HB network. The loss of entropy associated with this rearrangement determines the low solubility of small species in water. Simulations have shown that the stable conformation of *n*-butane in solution is the gauche conformer, owing to the reorganization of surrounding water molecules to decrease the total free energy of the system ***(151a)***.
2. *Large length scales:* Water molecules cannot rearrange around large solutes so as to preserve the network of hydrogen bonds. In this case, a depletion of water density at the interface between the solute and the solvent is induced, leading to partial drying or dewetting. This depletion of solvent around solutes can eventually induce association of two large nonpolar surfaces caused by the pressure imbalance. The strong attraction can be measured in surface force experiments even at large solute separation, for example, >10 nm ***(152–154)***. This effect was observed in MD simulations of two large nonpolar surfaces immersed in explicit water at constant pressure ***(155)***: at a critical distance, the solvent between the two solutes was excluded, leading to attraction of the plates. Partial dewetting appears to dominate the hydrophobic interactions between large surfaces.

The sizes of biomolecules, and in particular, of ligands and drug-like organic molecules, fall in the range between the two regimes, where the physical nature of the hydrophobic effect changes. Therefore, an ISM intended to be of general applicability in molecular biophysics should describe the two limits. In particular, at small length scales an accurate description of hydrophobicity is needed to calculate ligand-binding free energies and affinities and to study the specificity of peptides and small organic molecules.

Clearly, the development of an ISM that includes all the physical effects of the solvent would have a lasting impact on the field of computational biophysics and macromolecular simulation, in particular, on the study of protein–ligand interactions. Much work remains to be done, but the combination of a reliable ISM and a modest number of explicit water molecules and ions with appropriate conformational sampling techniques may represent the most promising approach to solve biophysical problems using computer simulations. These methods may prove essential for the efficient calculation of free energies and the rational design of drugs.

1.6. Estimating Free Energies From Computer Simulation

The components of the entropy of a system in the NPT (isothermal-isobaric) ensemble were discussed in **Subheading 1.1.** The calculation of entropy changes upon ligand binding is an area of theoretical and practical interest that can be addressed using computer simulations ***(32)***. Entropy changes affect absolute and relative free energies and are relevant in attempts to predict how to improve ligand affinity and specificity. Entropy changes associated with ligand binding can be divided into several terms, each amenable to different computer techniques ***(25,156–165)***: 1) solvent contribution caused by restructuring and reorganization of water molecules mainly around hydrophobic, but also around hydrophilic groups; 2) configurational entropy as a result of the rotation about single bonds in the ligand and in protein side chains in the binding pocket, and backbone conformational entropy in flexible binding pockets; 3) vibrational entropy arising from other internal degrees of freedom of both the protein and ligand; and 4) loss of S_r and S_t discussed in **Subheading 1.1.**

1.6.1. Free Energy Perturbation and Thermodynamic Integration

Free Energy Perturbation (FEP): Consider a system of particles in a particular statistical ensemble (e.g., constant NPT), and assume that its Hamiltonian $\mathcal{H}_s(\mathbf{p},\mathbf{q})$ depends on an explicit parameter s. This parameter represents some sort of perturbation to the system, perhaps the application of an external electric field, a constraint restricting the system to a certain subset $\mathbf{q}_c$ of coordinates, or even the nonphysical transformation of a molecule M_1 in the system into another molecule M_2. Again, **p** and **q** are the sets of generalized momenta and coordinates of all the particles in the system, respectively. A *microstate* of the system is defined by one particular point (**p**,**q**) in the phase space, whereas a *state* is defined by the parameter s of the Hamiltonian. The free-energy difference between two states $s = 1$ and $s = 2$ of a classical system is given by ***(166)***:

$$\Delta A_{1\to 2} = -RT\ln\left\langle \exp\left\{-\left[\mathcal{H}_2(\mathbf{p},\mathbf{q}) - \mathcal{H}_1(\mathbf{p},\mathbf{q})\right]/kT\right\}\right\rangle_1 \tag{5}$$

where A is the free energy of the system (e.g., Gibbs, G, in an isothermal–isobaric ensemble; Helmholtz, F, in a canonical ensemble, etc.), and $\langle \ldots \rangle$ is the statistical average at equilibrium in state $s = 1$. **Equation 5** follows immediately from basic statistical mechanics. For example, in the canonical ensemble, the probability of the system to be in a given microstate is given by $\mathcal{P} = Z^{-1}\exp(-\mathcal{H}/kT)$, and $\langle \exp[-(\mathcal{H}_2 - \mathcal{H}_1)/kT]\rangle_1 = Z_1^{-1}\int \exp[-(\mathcal{H}_2 - \mathcal{H}_1)/kT]\exp(-\mathcal{H}_1/kT)\,d\mathbf{p}d\mathbf{q} = Z_2/Z_1$, where $Z_x \equiv \int \exp(-\mathcal{H}_x/kT)d\mathbf{p}d\mathbf{q}$ is the partition function in state x.

Since $A_x \equiv -RT \ln Z_x$, **Eq. 5** follows. In other ensembles, the appropriate probability $\mathcal{P}$ and partition function Z must be used, but the reasoning is similar, i.e., in the isothermal–isobaric ensemble, a factor $\exp(-PV/kT)$ and an integral over the volumes V has to be introduced in $\mathcal{P}$ and Z, respectively. **Equation 5** is the basis of the method known as FEP to calculate free-energy differences ***(167)***.

In practice, the difference $\Delta A_{1\rightarrow 2}$ is calculated from a simulation (e.g., MD or MC) as

$$\exp\left(\Delta A_{1\rightarrow 2} / kT\right) \approx \frac{1}{N}\sum_{i=1}^{N} \exp\left[-\left\{\left[\mathcal{H}_2(\Omega_i) - \mathcal{H}_1(\Omega_i)\right] / kT\right\}\right] \tag{6}$$

where N is the number of microstates sampled, Ω_i denotes the element $(\mathbf{p}_i,\mathbf{q}_i)$ in the phase space sampled in the ith step, and the approximation becomes an equality in the limit $N \rightarrow \infty$. In practice, sufficient sampling can be difficult to achieve, especially when states 1 and 2 are fairly different. Because the distribution of microstates being sampled corresponds to state 1, extensive conformational sampling is needed to gather enough information about the system in state 2, unless the states are very similar. To enhance convergence, methods such as the umbrella sampling ***(168)*** and the overlapping-distributions method ***(169)*** have been developed. Because FEP is so computationally intensive, poorly designed simulation protocols or insufficient sampling can render the estimated free energies useless.

To illustrate these concepts, two early uses of FEP to calculate relative binding free energies of enzyme inhibitors are now discussed. In the first example, the difference in binding free energy is calculated for phosphonamidate and phosphonate binding to the enzyme thermolysin ***(170)***. The second involves the ligands benzamidine and p-fluorobenzamidine binding to trypsin ***(171)***.

Example 1: The two molecules Cbz–Gly–(NH)–Leu–Leu and Cbz–Gly–(O)–Leu–Leu differ only by the presence of an oxygen atom in the place of an amide group ***(170)***. The empirical energy function used in a molecular mechanics calculation is expressed so that by changing a parameter s from 0 to 1, the energy function goes from that of the first molecule (with NH) to that of the second molecule (with O). To evaluate the difference in binding free energies, $\Delta\Delta G_{12} = \Delta G_1 - \Delta G_2$, simulations must be performed for the enzyme-inhibitor complex and for the free inhibitor as state 1 is transformed to state 2. Because changing NH to O involves a rearrangement of electron density, the partial charges of the inhibitors used in the empirical force field were recalculated. This was done using *ab initio* quantum mechanics with the STO–3G* basis set. Then, changing the parameter s in the FEP method has to change the charges from those of the first molecule to those of the second.

To improve convergence s was changed from 0 to 1 using 20 windows, each of length $ds = 0.05$. This *window-sampling* method was used to calculate the difference $dG(s) = G(s + ds) - G(s)$ over the 20 intervals, and summing the 20 values to obtain the desired free energy change: $\Delta G = \sum dG(s)$. By changing s from 0 to 1 over 20 windows, convergence is improved by increasing the overlap of the microstate distributions, as discussed above. In each window, an explicit–solvent MD simulation was carried out in the NPT ensemble for the solvated complex and for the solvated ligand. The average in **Eq. 6** was calculated for each window, yielding a converged value for $dG(s)$. Summing all the energy differences $dG(s)$, a value of $\Delta\Delta G_{12} = 4.21 \pm 0.54$ kcal/mol was obtained, comparable to the experimental value of 4.1 kcal/mol. These calculations were carried out with the AMBER simulation package ***(172)***.

Example 2: The relative affinity of two benzamidine inhibitors of trypsin was calculated using MD simulations performed with the GROMOS program ***(173)*** in the canonical ensemble (constant NVT) ***(171)***. Thus, the relevant energy is the Helmholtz free energy F. The difference in binding energy between the two inhibitors was estimated as $\Delta\Delta F = 2.1$ kcal/mol, lower than the experimentally determined quantity of 3.8 kcal/mol. The discrepancy may reflect insufficient sampling and convergence problems, as noted by the authors, or it might result from force field limitations. Remember, classical force fields represent approximations that, while successful in many instances, cannot be expected to describe all experiments quantitatively. However, the calculations did show, at least qualitatively, that benzamidine binds to trypsin more strongly than does p-fluorobenzamidine, an important conclusion for the design of new drugs.

Thermodynamic Integration (TI): Thermodynamic integration is another technique commonly used to calculate free energies of binding. Given the identity ***(174)***,

$$\frac{\partial A_s}{\partial s} = \left\langle \frac{\partial \mathcal{H}_s}{\partial s} \right\rangle_s \tag{7}$$

the binding free energy is given by $\Delta A = \int_0^1 ds \partial A / \partial s = \int_0^1 ds \langle \partial \mathcal{H}_s / \partial s \rangle_s$. In practice, the interval $s = [0,1]$ is divided into m intervals, and $\langle \partial \mathcal{H}_s / \partial s \rangle_s$ is evaluated from a MC or MD simulation in each of the states s. The value ΔA is obtained by numerical integration of these average values over all values of s.

1.7. Ligand Docking

Any computational-based approach for ligand design must be able to predict the conformation of a bound ligand and its binding affinity. Clearly, the prediction of the correct geometry is a necessary, but insufficient, condition for the calculation of binding affinities and characterizing specificity. The positioning of a ligand in the binding pocket is known as docking, and the possible ways the ligand binds to the target are called binding modes.

1.7.1 Effects of Protein Dynamics on Protein–Ligand Interactions

The conformation of protein side chains can exert control on ligand binding through hydrogen bonding, hydrophobic packing, and electrostatic interactions. Side-chain and main-chain relaxations are not always given sufficient consideration in rapid calculations of protein–ligand interactions, even though local properties (e.g., electrostatic potential) can change dramatically as a result of modest protein relaxation. This sensitivity can cause problems when a single dielectric constant is used to describe the protein interior ***(86,175)***.

A body of theoretical and experimental evidence indicates that the folded state of a protein is made up of many different conformational substates that are similar in energy ***(176–180)***. Any of these substates could represent the optimal target for a particular ligand. The binding mode, characterized by the ligand's position and orientation in the binding site, could be affected by this multiplicity ***(181–183)***. The implication for ligand binding and drug design may be that a single protein structure (obtained from NMR or X-ray spectroscopy) is only useful to identify ligands for one or a few particular substates. Whenever possible, multiple protein substates should be considered in the search for ligands and their binding modes. It has been suggested that NMR structures can mimic the set of conformational substates accessible to a protein in solution and should be used whenever these structures are available ***(181)***. Such a strategy was recently employed ***(183)*** in which different X-ray crystal structures of the same protein were used as targets that were complexed with different ligands; each protein structure was assumed to correspond to a different substate. Yet other studies have shown that pharmacophore models created with multiple protein structures, compared more favorably with known inhibitors than models created with a single structure ***(184)***. All of this work underscores the difficulties and limitations of docking ligands to a rigid binding pocket. A sampling technique used to search for the preferred binding mode of a ligand should also generate the possible substates of the protein, even though the conformational search grows more difficult.

Assume that a target protein's structure and its binding site for a given ligand are known and the binding pocket has been identified. Suppose the goal is to rank different ligands according to their affinities (e.g., as in the FEP examples above) and to propose modifications to a particular ligand to enhance its binding affinity and/or its specificity. As already noted, this calls for an accurate energy function and a robust sampling technique. In addition, upon ligand binding, the pocket environment might change substantially such that the pKa values of important ionizable groups are shifted. These shifts could alter protonation and significantly affect the interaction of protein and ligand (e.g., *see* **ref. *36***). Yet, these changes cannot be rationalized solely in energetic terms; entropy changes might oppose the energetic changes (known as enthalpy–

entropy compensation) *(185)*, resulting in a small overall gain/loss of binding free energy, thus leaving the affinity virtually unchanged. Improving the binding free energy by as little as 3–7 kcal/mol may correspond to a sufficiently enhanced affinity for a given drug candidate. However, the prediction of enthalpy–entropy compensation is a difficult task in computational biophysics. Both the enthalpic and entropic contributions to the association process can be measured independently using calorimetric methods (e.g., ITC, *see* Chapter 1 of this volume). Such experiments have shown that the addition of a hydrogen bond between protein and ligand, while stabilizing the complex enthalpically, can actually produce a tighter complex that is entropically less favorable, resulting in a decrease in binding free energy (e.g., *see* **ref. *36***). Ultimately, the accurate prediction of such enthalpy–entropy compensation that may be necessary for a definitive ranking of plausible ligands may require an exhaustive *ab initio* calculation to evaluate the changes of vibrational entropy of the ligand upon binding.

2. Materials

There are many molecular modeling software tools available, some freely available and some commercial. A number of programs were used to obtain the results reported in this section. GeneMine *(186)* was used for homology modeling, RasMol for molecular visualization, and MolScript *(187)* and Raster3D *(188)* for molecular rendering. Conformational searching and energetic evaluation was done using CHARMM *(189)*, and PB calculations were performed with UHBD. Partial charges were calculated using Gaussian 98 *(190)*. Some of the routines used in this work, such as those to bias side-chain moves in MCM calculations, were written in-house, but similar conformational searching can be performed with other programs, such as ICM, AMBER, CHARMM, GROMOS, and Macromodel. We note that a commonly used and freely available (for noncommercial use) docking program, Autodock *(191–193)*, makes use of a screened electrostatics term with a sigmoidal dielectric function akin to the SCP–ISM described previously. An alternative homology-modeling program called NEST has been used successfully in recent, blind structure-prediction tests *(194)*. Molecular modeling calculations are generally performed on one or more workstations, and graphics software is used for visualization.

3. Methods

3.1. Example 1: Trimethoprim Binding to Dihydrofolate Reductase

Next, we apply MCM to the study of interactions of an antibiotic, trimethoprim (TMP), with the protein dihydrofolate reductase (DFHR) from the bacterium *Streptococcus pneumoniae*. A single-point mutation, Ile100Leu, in mutants of the TMP-susceptible R6 strain of DHFR–*S. pneumoniae* has been shown to confer resistance to TMP *(195)*. This mutation is thought to adversely affect hydrogen bonding between the DHFR and TMP.

Here, we report the modeling of the R6 DHFR and the Ile100Leu mutant, both with TMP bound. First, the SegMod algorithm *(195a)* implemented in the program GeneMine was used to build a model of DHFR–*S. pneumoniae* based on its sequence homology to DHFR–*E. coli* and the crystal structure of DHFR–*E. coli* with TMP bound *(196)*. The *S. pneumoniae* and *E. coli* proteins are similar in composition, particularly near the TMP binding site. For 57 of the 168 (33.9%) α carbons in the model, the nearest α carbon in the template structure is from an identical residue. For modeled residues with any atom within 5, 4, and 3 Å of any TMP atom, the fraction with identical nearest neighbors in the template is 50.0, 57.1, and 58.3%, respectively.

In the context of this DHFR–TMP study, we now outline the steps performed in a *typical* macromolecular simulation. Some details reflect this application and the use of the CHARMM program, but essentially all simulations require consideration of the same basic issues.

1. Define the system in terms of the atoms to be simulated explicitly and the parameterized force field to be used to evaluate the system energy. Thoughtful consideration of protonation states (e.g., for histidines) is required at this first step, because in typical simulations using classical mechanics, protonation states are initially assumed and then left unchanged throughout the study. Here, we generated three *segments*, with SEGIDs STRP, TMP, and WAT for use with the Merck Molecular Force Field (MMFF).
2. Assign initial coordinates. Here, heavy-atom coordinates were obtained from a homology model derived from an X-ray crystal structure. When not available from experiment, add hydrogen atoms to the system and assign their coordinates, as called for by the system description employed (all-atom vs polar-hydrogen force field). The HBUILD procedure is often used with CHARMM force fields to place hydrogens. With the MMFF, hydrogens can be initially placed near the bonded heavy atom and relaxed by energy minimization and/or some other conformational search, holding all heavy atoms fixed. In the current application, prior to modeling the effects of the mutation at position 100, the binding pocket in the homology model was refined. Atoms involved in TMP binding to DHFR–*S. pneumoniae* (TMP, I8, I100, the nearby water, and the COO atoms of E30) were relaxed during energy minimization so as to more closely adopt the conformation of the corresponding atoms in the crystal structure of the *E. coli* complex. This refinement was done using the CONS HARM RELATIVE restraint in CHARMM.
3. Perform conformational search (e.g., MD, LD, MC, MCM, etc.). One hundred independent MCM simulations were done of the DHFR–TMP complex for both R6 DHFR–*S. pneumoniae* and the I100L mutant. The MMFF force field was used as implemented in the program CHARMM. Each step in the simulations involved the random change in side-chain conformation for residue 100, followed

by energy minimization during which atoms distant from the binding pocket were fixed, but nearby DHFR, TMP, and water atoms were free to relax. A linearly distance-dependent dielectric (*rdie*) was used to approximate electrostatic forces that were *shift*ed (monotonically) to 0 at separations beyond (*ctofnb*) 12 Å. The CHARMM commands used to constrain the system and evaluate its energy were (*see* **Note 5**):

```
define nearby sele .byres. ( segid tmp .around. 4.0 ) end
define bonded sele .bonded. nearby end
define nearwat sele segid wat .and. ( segid tmp .around. 10.0 ) end
define free sele .byres. ( nearby .or. bonded .or. nearwat ) show end
cons fix sele .not. free end
energy rdie inbfrq –1 cutnb 14.0 ctofnb 12.0 ctonnb 8.0 shift vshift
```

The MCM procedure was implemented using the CHARMM scripting language and can be summarized as:

Minimize the energy of the DHFR–TMP model in its initial conformation. Call the energy-minimized structure $\{\mathbf{r}_i\}$ and its potential energy V.

Repeat …

Randomly change the χ_1 and χ_2 dihedral angles of residue 100. Perform 5 steps of steepest descents and 150 steps of adopted-basis Newton–Raphson minimization, with χ_1 and χ_2 restrained to the trial values (and distant atoms fixed). Call the resultant structure $\{\mathbf{r}_i'\}$ and its minimized potential energy V'.

If $V' < V$, then

Accept the change. That is, replace $\{\mathbf{r}_i\}$ with $\{\mathbf{r}_i'\}$ and V with V'.

Else

Generate a random number, RAND, uniformly distributed between 0 and 1, and compare it to the Boltzmann factor $p = e^{-(V'-V)/RT}$ with 600 K $\leq T \leq$ 900 K.

If ($p >$ RAND) then

Accept the change. Replace $\{\mathbf{r}_i\}$ with $\{\mathbf{r}_i\}$ and V with V'.

Else

Reject the change (retain $\{\mathbf{r}_i\}$ and V).

End if

End if

… Until 500 structural modifications have been attempted or 100 consecutive changes have been rejected.

The random changes in side-chain dihedral angles were biased to favor conformations observed in protein crystal structures using the parameters of reference ***(197)***. This biasing is demonstrated in **Fig. 1**.

The lowest-energy structure obtained in each of the MCM runs was energy minimized to an rms gradient of 0.001 kcal/mol/Å. The 100 structures obtained in this way for the I100L complex were sorted by potential energy, and characterized in terms of the hydrogen-bond distance between the carbonyl O atom of

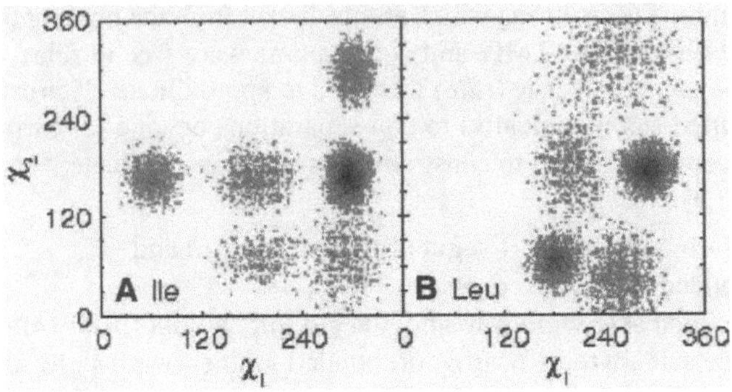

Fig. 1. Conformations observed in 10,000 biased samples of the χ_1 and χ_2 angles of Ile (**A**) and Leu (**B**)

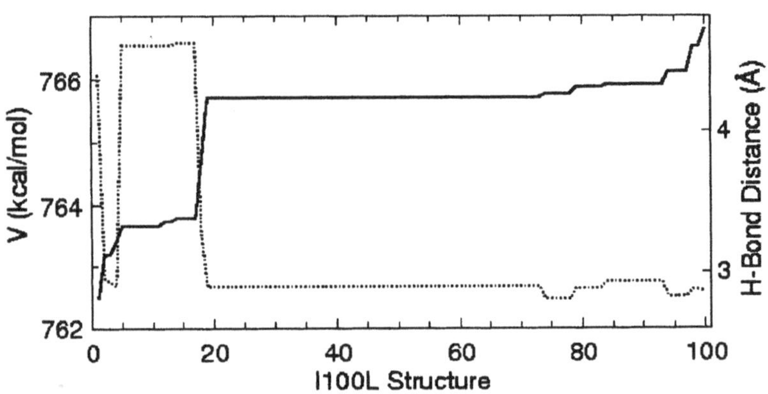

Fig. 2. Lowest-energy structure found in each of 100 MCM simulations of the I100L-TMP complex, sorted in order of increasing energy (solid) The distance between the carbonyl O atom of Leu 100 and the 4-amino N atom of TMP is also plotted (dotted).

Leu 100 and the 4-amino N atom of TMP (**Fig. 2**). This hydrogen bond is broken (i.e., > 4.3 Å) in 14 of the 100 predictions and these 14 structures are among the 17 of lowest energy. By contrast, each of the 100 simulations of the R6 DHFR–TMP complex produced the same lowest-energy structure in which the hydrogen bond was intact. The lowest-energy structures found for the two DHFR–TMP complexes are superimposed in **Fig. 3**.

Thus, this rather simple conformational search, in which the ligand remained docked throughout, provides an atomic level description of the clinically

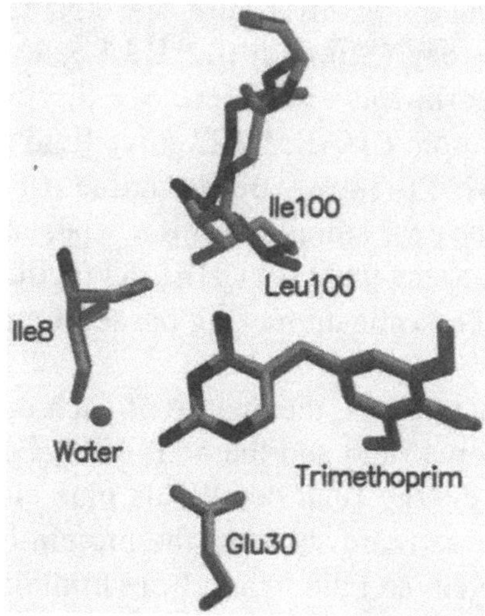

Fig.3. Binding pockets of the lowest-energy structures found for TMP bound to the R6 DHFR (light carbon atoms) and I100L (dark carbons) mutant. For clarity, hydrogen atoms are not shown. Relaxation of the main chain at G101 and G102 results in a broken hydrogen bond between DHFR and TMP in the I100L mutant.

observed TMP resistance. The packing of a leucine side chain in the space vacated by an isoleucine side chain, results in a relaxation of the main chain and a broken hydrogen bond between DHFR and TMP. This application illustrates the need to account for protein flexibility and main-chain reorganization when accounting for the effects of mutation. This observation is consistent with the importance of protein flexibility in ranking the binding affinity of different ligands.

3.2. Example 2: Benzamidine–Trypsin Complex

We consider another relatively simple protein–ligand study, for which the availability of seven different crystal structures of the same complex allows us to circumvent the difficult problem of docking. The ligand, benzamidine, is a small molecule with little conformational flexibility. In addition to the seven crystal structures of benzamidine complexed with bovine trypsin (PDB codes: 1bty, 1c1n, 1ce5, 1j8a, 1tio, 2tio, 3ptb), extensive thermodynamic data are available, and both enthalpic and entropic contributions are known ***(198)***. Thus, the estimated binding energies can be compared to experimental enthalpies. This example illustrates the need for a pre-processing protein relaxation, even in a binding pocket considered to be *rigid* as in most serine proteases ***(199)***. The calculations reported here represent a preliminary step prior to a quantitive comparison of theoretical binding free energies, e.g., calculated by using free-energy perturbation theory (**Subheading 1.6.1.**), with binding enthalpies obtained from ITC experiments *(36)*. Here, we discuss two sets of calculations, one using the all-atom CHARMM22 force field with the SCP–ISM and another using the UHBD PB solver. Benzamidine parameters were based on the CHARMM parameters for atoms of similar type, and atomic charges were calculated using the Gaussian 98 *(190)* CHELPG facility at a HF 6-31G* level of theory after an initial optimization using the semi-empirical AM1 Hamiltonian (*see* **Note 6**).

In the first set of calculations, the energy of each complex was minimized using the CHARMM force field and the SCP–ISM. Following minimization, the effective nonbond energy (van der Waals plus electrostatics), W_{PL}, was evaluated. The ligand was removed from the protein-binding site, and effective energies of the protein and the ligand were minimized separately to relax their conformations in isolation. The nonbond energies W_P and W_L were then evaluated for the protein and ligand, respectively, and the binding energy was approximated as $\Delta W = W_{PL} - W_P - W_L$ (*see* **Note 7**).

In the second set, the energy was minimized using a linear distance-dependent dielectric and the CHARMM program, prior to the PB calculations using UHBD. These PB calculations were performed at zero ionic strength (i.e., Poisson equation) using $\varepsilon_w = 78$ and $\varepsilon_i = 2$, and a solute/solvent boundary equal to the solvent accessible surface defined with a probe radius of 1.4 Å. A grid of 150 × 150 × 150 cells was used, together with two-step focusing: a coarse-graining step using a cell length of 1.5 Å, followed by a fine-graining step with cell length of 0.4 Å.

In all calculations, the water molecules and coligands were removed, even though the importance of water molecules in the binding of ligands to serine proteases, including trypsin, has been discussed previously *(200)*. In each of the seven crystal structures, five water molecules were found to mediate the interaction between the protein and ligand. Although these waters should be taken into account when calculating binding affinities, here their contribution is expected to be the same in each of the structures, and they were ignored (*see* **Note 8**).

Four different energy-minimization protocols were employed, but the results obtained were not qualitatively affected (*see* **Note 9**). The differences in binding energies shown in the first two columns of **Table 1** reflect the different conformations of some side chains in the binding pocket that interact with the ligand. The conformational heterogeneity might result from differences in crystallization conditions and/or structure-determination/refinement protocols. For example, in some structures Gln192 interacts directly with the benzamidine ring, but in others, it points away from the ligand.

Table 1
Binding Energy Calculated for the Seven Complexes of Trypsin and Benzamidine With Known Coordinates

Complex PDB code	UHBD[a]	CHARMM[a]	CHARMM MC[a]
1bty	–10.7	–12.7	–16.6
1cln	–13.3	–14.3	–16.8
1ce5	–15.3	–15.9	–15.8
1j8a	–12.6	–13.9	–14.9
1tio	–17.4	1.3	–16.2
2tio	–12.5	–11.1	–15.3
3ptb	–13.1	–39.8	–14.8

[a]Energies in kcal/mol.

First and second columns: binding energies calculated using UHBD and CHARMM without protein side-chain relaxation; third column: binding energies obtained after MC relaxation of binding-pocket side chains and Boltzmann averaging (*see* text).

In an attempt to reduce the variation in the binding-energy estimates, a simulated-annealing/MC simulation was performed to relax the protein side chains and to create a Boltzmann distribution from which an average binding energy could be calculated. This simulation was performed with an in-house program that was implemented into the CHARMM code (usersb.src) and used with the SCP–ISM. Protein side chains within 5 Å of any ligand atom were moved during the MC simulation while the ligand remained in place. The temperature was reduced from 3000 to 300 K using a logarithmic schedule with ten steps, with 5×10^5 MC trial moves performed at each temperature. Accepted moves at 300 K were saved to a file for analysis. Following energy minimization, the binding energy of the *i*th accepted structure was calculated as $\Delta W_i = W_{PL,i} - W_{P,i} - W_{L,i}$. The binding energy was then obtained for each complex by averaging ΔW_i over the N structures accepted at 300 K: $\langle \Delta W \rangle = N^{-1} \Sigma \Delta W_i$. Convergence of the MC search was confirmed from the average values, shown in the third column of **Table 1**. The MC results demonstrate that this limited conformational sampling was sufficient to shift the distribution of side-chain conformations in each complex to similar regions in the conformational space, thus minimizing the uncertainties in the estimated binding energies. Following the MC search, the binding energies obtained from the seven different crystal structures converge to the same value, within the statistical errors estimated from the variance of the distributions. However, it is worth keeping in mind that use of another energy function may well have produced different conformational distributions and hence different binding energies. Following this preprocessing of side-chain conformations, any of the seven structures could be used as the initial conformation in a more sophisticated calculation of binding affinities.

4. Notes

Throughout this chapter, we have stressed the considerable challenges that must be met in order to use atomistic modeling for the reliable prediction of ligand-binding modes and the ranking of ligands by affinity. In general, the computer simulation of macromolecules cannot be done in an automated, *black-box* fashion. Each application is different, requiring a carefully planned protocol and close monitoring by the simulator to safeguard the reliability of the results.

1. Clearly, the description of protein–ligand interactions in terms of Newtonian physics is suspect whenever electronic degrees of freedom and/or high-frequency motions are important. However, in many cases, calculations can be designed so that the errors incurred cancel, at least partially, as when relative changes in free energy $\Delta\Delta G$ are calculated. An introductory discussion of quantum mechanical effects in relation to the classical approximations commonly used to simulate macromolecular systems is given elsewhere ***(201)***.
2. To reduce the computer time spent evaluating $V_{nonbond}$, spherical-cutoff schemes can be used to ignore the van der Waals and electrostatic interaction between atoms separated by distances greater than a specified cutoff radius. To avoid the introduction of deep, artificial minima in the energy landscape, the particular cutoff methods used should be chosen so that the respective *forces* are trimmed monotonically to zero ***(202)***. It has been shown that the *shift* function implemented in CHARMM is not the best choice for explicit–solvent simulations ($\varepsilon = 1$). Better cutoff options have been developed, such as the *force shift* and *force switch*, that can be used to reduce electrostatic forces monotonically ***(202)***. For the simulation of fully solvated systems, the use of periodic boundary conditions and Ewald–summation methods to calculate the infinite-range electrostatics within the periodically replicated system ***(203,204)*** have become the standard.
3. Physical intuition suggests that the screening function D is sigmoidal, and both theory and experiment indicate this to be true ***(56)***. Many sigmoidal functions have been proposed to damp electrostatic interactions in macromolecules. The functional form used in the SCP–ISM is referred to as Boltzmann sigmoidal in mathematical statistics. This choice is appealing because it can fit the screening function obtained for experimental data of simple liquids like water, acetone, acetamide, etc. Also, the resultant errors are small compared to dielectric profiles obtained from basic theory of polar liquids (namely, the LDS theory; *see* text). Moreover, the Boltzmann sigmoid is an exact solution of a simple first-order differential equation with sigmoidal solutions, which makes it suitable for algebraic manipulations ***(56,57)***.

4. Details of the derivation of **Eq. 4**, an expression for the solvation energy in the SCP–ISM, and a review of electrostatics in polar liquids can be found online in http://cmm.cit.nih.gov/~mago and references therein.
5. Because the goal of this calculation was essentially the optimal packing of a hydrophobic leucine side chain in a hydrophobic pocket, it was possible to avoid many of the problems associated with the energetics of solvation and the sampling of ligand conformations in a typical ligand-docking study. In addition, much of the system was held fixed so that any strain present in the homology model could not distort the protein conformation markedly. The linear dielectric function (*rdie*) used is a rather poor description of electrostatic shielding. In addition, no solvation term was used (*see* **Subheading 1.5.1.**). Therefore, this crude continuum electrostatic approach is not recommended for applications where polar effects are expected to be more predominant.
6. The results obtained with any empirical force field clearly depend on the parameters used. Throughout the text, we have cautioned that quantitative results can change with force-field details or with the inputs to PB solvers. Because the SCP–ISM was parameterized for routine use with peptides and proteins, parameters have to be derived for all applications involving general organic molecules other than amino acids. Before carrying out the calculations reported in this example, parameters of benzamidine were derived by analogy with other atom types in proteins. Formally, however, parameters should be derived as described previously (*see* text) for side-chain analogs, i.e., by fitting to the solvation free energy of the molecule determined experimentally and by careful reevaluation of the resulting HB strengths ***(123a)***.
7. Generally, the conformations of the isolated ligand and protein in solution will not be those in the complex, and so the separated molecules were energy-minimized to mimic the relaxation to their noncomplexed conformations. Although common, this simplistic approach should be revisited in more sophisticated calculations.
8. Just as for other small molecules (*see* **Note 6**), water molecules used with an implicit–solvent model necessitate an initial parameterization. Experimental water hydration energies can be used to determine the overall damping of electrostatics, but protein–water HB interactions have to be calibrated appropriately ***(123a)***.
9. The different energy-minimization protocols did not affect the qualitative result that all binding energies converged to similar values (the only goal of these calculations). However, the numerical values obtained differed by several kcal/mol depending on whether a complete energy minimization or a partial minimization (to eliminate large steric clashes) was performed. This observation is to be expected given the simplified (although quite common) estimation of ΔW used here and is independent of the choice of continuum model (e.g., SCP–ISM or PB).

Acknowledgments

We thank David Matthews for providing the coordinates of TMP bound to DHFR–*E. coli* and Jacob Donkersloot, Jerry Keith, Andreas Pikis, and Peter

Munson for stimulating discussions regarding the DHFR–TMP calculations. LG was supported by NIH grant P01 GM66531. No endorsement by the U.S. Government should be inferred from the mention of trade names, software packages, commercial products, or organizations.

References

1. Schneidman-Duhovny, D., Nussinov, R., and Wolfson, H. J. (2004) Predicting molecular interactions in silico: II. Protein–protein and protein–drug docking. *Curr. Med. Chem.* **11(1),** 91–107.
2. Brooijmans, N. and Kuntz, I. D. (2003) Molecular recognition and docking algorithms. *Annu. Rev. Biophys. Biomol. Struct.* **32,** 335–373.
3. Taylor, R. D., Jewsbury, P. J., and Essex, J. W. (2002) A review of protein-small molecule docking methods. *J. Comp. Aid. Mol. Des.* **16(3),** 151–166.
4. Lengauer, T. and Rarey, M. (1996) Computational methods for biomolecular docking. *Curr. Op. Struct. Biol.* **6(3),** 402–406.
5. Perez, C. and Ortiz, A. R. (2001) Evaluation of docking functions for protein–Ligand docking. *J. Med. Chem.* **44,** 3768.
6. Roche, O., Kiyama, R., and Brooks, C. L. (2001) Ligand–Protein database: linking protein–ligand complex structures to binding data. *J. Med. Chem.* **44(22),** 3592–3598.
7. Rarey, M. and Lengauer, T. (2000) A recursive algorithm for efficient combinatorial library docking. *Perspect. Drug Discov. Des.* **20(1),** 63–81.
8. Morelli, X., Dolla, A., Czjzek, M., Palma, P. N., Blasco, F., Krippahl, L., Moura, J. J. G., and Guerlesquin, F. (2000) Heteronuclear NMR and soft docking: an experimental approach for a structural model of the cytochrome c(553)-ferredoxin complex. *Biochemistry* **39,** 2530–2537.
9. Hoffmann, D., Kramer, B., Washio, T., Steinmetzer, T., Rarey, M., and Lengauer, T. (1999) Two-stage method for protein–ligand docking. *J. Med. Chem.* **42(21),** 4422–4433.
10. Budin, N., Majeux, N., Tenette-Souaille, C., and Caflisch, A. (2001) Structure-based ligand design by a build-up approach and genetic algorithm search in conformational space. *J. Comp. Chem.* **22,** 1956–1970.
11. Thormann, M. and Pons, M. (2001) Massive docking of flexible ligands using environmental niches in parallelized genetic algorithms. *J. Comp. Chem.* **22,** 1971–1982.
12. Majeux, N., Scarsi, M., Apostolakis, J., Ehrhardt, C., and Caflisch, A. (1999) Exhaustive docking of molecular fragments with electrostatic solvation. *Proteins* **37,** 88–105.
13. Teng, M. K., Smolyar, A., Tse, A. G., Liu, J. H., Liu, J., Hussey, R. E., Nathenson, S. G., Chang, H. C., Reinherz, E. L., and Wang, J. H. (1998) Identification of a common docking topology with substantial variation among different TCR-peptide-MHC complexes. *Curr. Biol.* **8,** 409–12.
14. Shoichet, B. K., Leach, A. R., and Kuntz, I. D. (1999) Ligand solvation in molecular docking. *Proteins* **34,** 4–16.

15. Schafferhans, A. and Klebe, G. (2001) Docking ligands onto binding site representations derived from proteins built by homology modelling. *J. Mol. Biol.* **307,** 407–427.
16. Sandak, B., Wolfson, H. J., and Nussinov, R. (1998) Flexible docking allowing induced fit in proteins: insights from an open to closed conformational isomers. *Proteins* **32,** 159–174.
17. Makino, S. and Kuntz, I. D. (1997) Automated flexible ligand docking method and its application for databese search. *J. Comp. Chem.* **18,** 1812–1825.
18. Lybrand, T. P. (1995) Ligand–protein docking and rational drug design. *Curr. Opin. Struct. Biol.* **5,** 224–228.
19. Lorber, D. M. and Shoichet, B. K. (1998) Flexible ligand docking using conformational ensembles. *Protein Sci.* **7,** 938–950.
20. Mandell, J. G., Roberts, V. A., Pique, M. E., Kotlovyi, V., Mitchell, J. C., Nelson, E., Tsigelny, I., and Ten Eyck, L. F. (2001) Protein docking using continuum electrostatics and geometric fit. *Protein Eng.* **14,** 105–113.
21. Pathria, R. K. (1972) *Statistical Mechanics. International Series in Natural Philosophy. Vol. 45.* Pergamon Press.
22. McQuarrie, D. A.(1976) *Statistical Mechanics.* Harper & Row, New York, NY.
23. Hill, T. L. (1986) *An Introduction to Statistical Thermodynamics.* Dover, New York, NY.
24. Bruce Yu, Y., Privalov, P. L., and Hodges, R. S. (2001) Contribution of translational and rotational motions to molecular association in aqueous solution. *Biophys. J.* **81,** 1632–1642.
25. Siebert, X. and Amzel, L. M. (2004) Loss of translational entropy in molecular associations. *Proteins* **54,** 104–115.
26. Bohm, H. J. (1994) The development of a simple empirical scoring function to estimate the binding constant for a protein–ligand complex of known three-dimensional structure. *J. Comp. Aid. Mol. Des.* **8,** 243–256.
27. Meng, C., Shoichet, B. K., and Kuntz, I. D. (1992) Automated docking with grid-based energy evaluation. *J. Comp. Chem.* **13,** 505–524.
28. Eldridge, M. D., Murray, C. W., Auton, T. R., Paolinine, G. V., and Mee, R. P. (1997) Empirical scoring functions .1. The development of a fast empirical scoring function to estimate the binding affinity of ligands in receptor complexes. *J. Comp. Aid. Mol. Des.* **11,** 425–445.
29. Muegge, I. and Martin, Y. C. (1999) A general and fast scoring function for protein–ligand interactions: A simplified potential approach. *J. Med. Chem.* **42,** 791–804.
30. Jones, G., Willett, P., and Glen, R. C. (1995) Molecular recognition of receptor-sites using a genetic algorithm with a description of solvation. *J. Mol. Biol.* **245,** 43–53.
31. Rarey, M., Kramer, B., Lengauer, T., and Klebe, G. (1996) A fast flexible docking method using an incremental construction algorithm. *J. Mol. Biol.* **261,** 470–489.
32. Gilson, M. K., Given, J. A., Bush, B. L., and McCammon, J. A. (1997) The statistical-thermodynamic basis for computation of binding affinities: A Critical Review. *Biophys. J.* **72,** 1047–1069.

33. Singh, U. C., Brown, F. K., Bash, P. A., and Kollman, P. A. (1987) An approach to the application of free-energy perturbation–methods using molecular-dynamics–applications to the transformations of Ch3oh- Ch3ch3, H3o+- Nh4+, glycine- alanine, and alanine- phenylalanine in aqueous-solution and to H3o+ (H2o)3- Nh4+(H2o)3 in the gas-phase. *J. Am. Chem. Soc.* **109,** 1607–1614.
34. Straatsma, T. P. and Berendsen, H. J. C. (1988) Free-energy of iIonic hydration—analysis of a thermodynamic integration technique to evaluate free-energy differences by molecular-dynamics simulations. *J. Chem. Phys.* **89,** 5876–5886.
35. Jarzynski, C. (1997) Nonequilibrium equality for free energy differences. *Phys. Rev. Lett.* **78,** 2690–2693.
35a. Sharma, P., Steinbach, P. J., Sharma, M., Amin, N. D., Barchi, J. J., and Pant, H. C. (1999) Identification of substrate binding site of cyclin-dependent kinase 5. *J. Biol. Chem.* **274,** 9600–9606.
36. Dullweber, F., Stubbs, M. T., Musil, D., Sturzebecher, J., and Klebe, G. (2001) Factorising ligand affinity: A combined thermodynamic and crystallographic study of trypsin and thrombin inhibition. *J. Mol. Biol.* **313,** 593–614.
37. Ryde, U. and Nilsson, K. (2003) Quantum chemistry can locally improve protein crystal structures. *J. Am. Chem. Soc.* **125,** 14,232–14,233.
38. Halgren, T. A. (1996) Merck molecular force field .1. Basis, form, scope, parameterization, and performance of MMFF94. *J. Comput. Chem.* **17,** 490–519.
39. Gilson, M. K., Gilson, H. S. R., and Potter, M. J. (2003) Fast assignment of accurate partial atomic charges: An electronegativity equalization method that accounts for alternate resonance forms. *J. Chem. Inf. Comput. Sci.* **43,** 1982–1997.
40. Vieth, M., Hirst, J. D., Kolinski, A., and Brooks, C. L. (1998) Assessing energy functions for flexible docking. *J. Comput. Chem.* **19,** 1612–1622.
41. Wu, G. S., Robertson, D. H., Brooks, C. L., and Vieth, M. (2003) Detailed analysis of grid-based molecular docking: a case study of CDOCKER—a CHARMm-based MD docking algorithm. *J. Comput. Chem.* **24,** 1549–1562.
42. Li, Z. Q. and Scheraga, H. A. (1987) Monte-Carlo-minimization approach to the multiple-minima problem in protein folding. *Proc. Natl. Acad. Sci. USA* **84(19),** 6611–6615.
43. Abagyan, R. A. and Totrov, M. (1999) Ab initio folding of peptides by the optimal-bias Monte Carlo minimization procedure. *J. Comp. Phys.* **151,** 402–421.
43a. Steinbach, P. J. (2004) Exploring peptide energy landscapes: a test of force fields and implicit solvent models. *Proteins* **57,** 665–677.
44. Steinbach, P. J. and Brooks, B. R. (1993) Protein hydration elucidated by molecular-dynamics simulation. *Proc. Natl. Acad. Sci. USA* **90(19),** 9135–9139.
45. Steinbach, P. J. and Brooks, B. R. (1996) Hydrated myoglobin's anharmonic fluctuations are not primarily due to dihedral transitions. *Proc. Natl. Acad. Sci. USA* **93,** 55–59.
46. Luecke, H., Schobert, B., Richter, H.-T., Cartailler, J.-P., and Lanyi, J. K. (1999) Structural changes in bacteriorhodopsin during ion transport at 2 angstrom resolution. *Science* **286,** 255–260.

47. Bryant, R. G. (1996) The dynamics of water–protein interactions. *Annu. Rev. Biophys. Biomol. Struct.* **25,** 29–53.
48. Ooi, T. (1994) Thermodynamics of protein-folding. Effects of hydration and electrostatic interactions. *Adv. Biophys.* **30,** 105–154.
49. Ben-Naim, A. (1980) *Hydrophobic Interactions.* Plenum Press, New York, NY.
50. Teeter, M. M. (1991) Water–protein interactions: theory and experiment. *Annu. Rev. Biophys. Biophys. Chem.* **20,** 577–600.
51. Otting, G., Liepinsh, E., and Wütrich, K. (1991) Protein hydration in aqueous solution. *Science* **254,** 974–980.
52. Bailly, C., Chessari, G., Carrasco, C., Joubert, A., Mann, J., Wilson, W. D., and Neidle, S. (2003) Sequence-specific minor groove binding by bis-benzimidazoles: water molecules in ligand recognition. *Nucleic Acids Res.* **31,** 1514–1524.
53. Poornima, C. S. and Dean, P. M. (1995) Hydration in drug design .1. Multiple hydrogen-bonding features of water molecules in mediating protein–ligand interactions. *J. Comp. Aid. Mol. Des.* **9,** 500–512.
54. Poornima, C. S. and Dean, P. M. (1995) Hydration in drug design .3. Conserved water molecules at the ligand-binding sites of homologous proteins. *J. Comp. Aid. Mol. Des.* **9(6),** 521–531.
55. Lazaridis, T. and Karplus, M. (1999) Effective energy function for proteins in solution. *Proteins* **35,** 133–152.
56. Hassan, S. A., Guarnieri, F., and Mehler, E. L. (2000) A general treatment of solvent effects based on screened Coulomb potentials. *J. Phys. Chem. B.* **104,** 6478–6489.
57. Hassan, S. A. and Mehler, E. L. (2002) A critical analysis of continuum electrostatics: the screened Coulomb potential-implicit solvent model and the study of the alanine dipeptide and discrimination of misfolded structures of proteins. *Proteins* **47,** 45–61.
58. Warshel, A. and Aqvist, J. (1991) Electrostatic energy and macromolecular function. *Annu. Rev. Biophys. Biophys. Chem.* **20,** 267–298.
59. Mehler, E. L. and Warshel, A. (2000) Comment on "a fast and simple method to calculate pProtonation states in proteins." *Proteins* **40,** 1–3.
60. Kassner, R. J. (1972) Theoretical model for effects of local nonpolar heme environments on redox potentials in cytochromes. *J. Am. Chem. Soc.* **95,** 2674.
61. Eisenman, G. and Horn, R. (1983) Ionic selectivity revisited. The role of kinetic and equilibrium processes in ion permeation through channels. *J. Membr. Biol.* **76,** 197.
62. Warshel, A. and Papazyan, A. (1998) Electrostatic effects in macromolecules: fundamental concepts and practical modeling. *Curr. Opin. Struct. Biol.* **8,** 211–217.
63. Arora, N. and Bashford, D. (2001) Solvation energy density occlusion approximation for evaluation of desolvation penalties in biomolecular interactions. *Proteins* **43,** 12–27.
64. Camacho, C. J., Weng, Z. P., Vajda, S., and DeLisi, C. (1999) Free energy landscapes of encounter complexes in protein–protein association. *Biophys. J.* **76,** 1166–1178.

65. Muegge, I. (2000) A knowledge-based scoring function for protein–ligand interactions: probing the reference state. *Perspect. Drug Discov.* **20,** 99–114.
66. Palma, P. N., Krippahl, L., Wampler, J. E., and Moura, J. J. G. (2000) BiGGER: a new (soft) docking algorithm for predicting protein interactions. *Proteins* **39,** 372–384.
67. Varnai, P. and Warshel, A. (2000) Computer simulation studies of the catalytic mechanism of human aldose reductase. *J. Am. Chem. Soc.* **122,** 3849–3860.
68. Luo, N., Mehler, E., and Osman, R. (1999) Specificity and catalysis of uracil DNA glycosylase. A molecular dynamics study of reactant and product complexes with DNA. *Biochemistry* **38,** 9209–9220.
69. Mehler, E. L., Fuxreiter, M., Simon, I., and Garcia-Moreno E., B. (2002) The role of hydrophobic microenvironment in modulating pKa dhifts in proteins. *Proteins* **48,** 283.
70. Mehler, E. L. and Guarnieri, F. (1999) A self-consistent, microenvironment modulated screened Coulomb potential approximation to calculate pH dependent electrostatic effects in proteins. *Biophys. J.* **77,** 3–22.
71. Bowie, J. U., Luthy, R., and Eisenberg, D. (1991) A method to identify protein sequences that fold into a known three-dimensional structure. *Science* **253,** 164–170.
72. Kleiger, G., Beamer, L. J., Grothe, R., Mallick, P., and Eisenberg, D. (2000) The 1.7Å crystal structure of BPI: a study of how two dissimilar amino acid sequences can adopt the same fold. *J. Mol. Biol.* **299,** 1019–1034.
73. Hassan, S. A., Mehler, E. L., Zhang, D., and Weinstein, H. (2003) Molecular dynamics simulations of peptides and proteins with a continuum electrostatic model based on screened Coulomb potentials. *Proteins* **51,** 109–125.
74. Honig, B. and Nicholls, A. (1995) Classical electrostatics in biology and chemistry. *Science* **268,** 1144–1149.
75. Fogolari, F., Zuccato, P., Esposito, G., and Viglino, P. (1999) Biomolecular electrostatics with the linearized Poisson–Boltzmann equation. *Biophys. J.* **76,** 1.
76. Smith, P. E. (1999) Computer simulation of cosolvent effects on hydrophobic hydration. *J. Phys. Chem. B.* **103,** 525.
77. Mancera, R. L. (1999) Influence of salt on hydrophobic effects: a molecular dynamics study using the modified hydration-shell hydrogen-bond model. *J. Phys. Chem. B.* **103,** 3774.
78. Kokkoli, E. and Zukoski, C. F. (1998) Interactions between hydrophobic self-assembled monolayers. Effect of salt and the chemical potential of water on adhesion. *Langmuir.* **14,** 1189.
79. Christenson, H. K., Claesson, P. M., and Parker, J. L. (1992) Hydrophobic attraction. A reexamination of electrolyte effects. *J. Phys. Chem.* **96,** 6725.
80. Jackson, J. D. (1975) *Classical Electrodynamics*. 2nd Ed. Wiley.
81. Gilson, M. K., Davis, M. E., Luty, B. A., and McCammon, J. A. (1993) Computation of electrostatic forces on solvated molecules using the Poisson–Boltzmann equation. *J. Phys. Chem.* **97,** 3591–3600.
82. Sharp, K. A. (1995) Polyelectrolyte electrostatics. Salt dependence, entropic, and enthalpic contributions to free energy in the nonlinear Poisson–Boltzmann model. *Biopolymers* **36,** 227.

83. Davis, M. E., Madura, J. D., Luty, B. A., and McCammon, J. A. (1991) Electrostatics and diffusion of molecules in solution: simulations with the University of Houston Brownian Dynamics program. *Comp. Phys. Comm.* **62,** 187–197.
84. Madura, J. D., Briggs, J. M., Wade, R. C., Davis, M. E., Luty, B. A., Ilin, A., Antosiewics, J., Gilson, M. K., Bagheri, B., Scott, L. R., and McCammon, J. A. (1995) Electrostatics and diffusion of molecules in solution: simulations with the University of Houston Brownian dynamics program. *Comp. Phys. Commun.* **91,** 57–95.
85. Garcia-Moreno, B., Dwyer, J. J., Gittis, A. G., Lattman, E. E., Spencer, D. S., and Stites, W. E. (1997) Experimental measurement of the effective dielectric in the hydrophobic core of a protein. *Biophys. Chem.* **64,** 211–224.
86. Sham, Y. Y., Muegge, I., and Warshel, A. (1998) The effect of protein relaxation on charge-charge interactions and dielectric constants of proteins. *Biophys. J.* **74,** 1744–1753.
87. Scarsi, M., Apostolakis, J., and Caflisch, A. (1997) Continuum electrostatic energies of macromolecules in aqueous solutions. *J. Phys. Chem. B.* **101,** 8098–8106.
88. Antosiewicz, J., McCammon, J. A., and Gilson, M. K. (1994) Prediction of pH-dependent properties of proteins. *J. Mol. Biol.* **238,** 415–436.
89. Pollock, E. L., Alder, B. J., and Pratt, L. R. (1980) Relation between the local field at large distances from a charge or dipole and the dielectric constant. *Proc. Natl. Acad. Sci. USA* **77,** 49–51.
90. Sham, Y. Y., Chu, Z. T., and Warshel, A. (1997) Consistent calculations of pKa's of ionizable residues in proteins: semi-microscopic and microscopic approaches. *J. Phys. Chem. B.* **101,** 4458–4472.
91. Takashima, S. and Schwan, H. P. (1965) Dielectric dispersion of crystalline powders of amino acids, peptides and proteins. *J. Phys. Chem.* **69,** 4176–4182.
92. Gilson, M. K. and Honig, B. H. (1986) The dielectric-constant of a folded protein. *Biopolymers* **25(11),** 2097–2119.
93. Oda, Y., Yamazaki, T., Nagayama, K., Kanaya, S., Kuroda, Y., and Nakamura, H. (1994) Individual ionization constants of all the carboxyl groups in ribonuclease HI from *Escherichia coli* determined by NMR. *Biochemistry* **33,** 5275–5284.
94. Nakamura, H., Sakamoto, T., and Wada, A. (1988) A theoretical-study of the dielectric-constant of protein. *Protein Eng.* **2,** 177–183.
95. King, G., Lee, F. S., and Warshel, A. (1991) Microscopic simulations of macroscopic dielectric constants of solvated proteins. *J. Chem. Phys.* **91,** 3647–3661.
96. Smith, P. E., Brunne, R. M., Mark, A. E., and Vangunsteren, W. F. (1993) Dielectric-properties of trypsin-inhibitor and lysozyme calculated from molecular-dynamics simulations. *J. Phys. Chem.* **97,** 2009–2014.
97. Wesson, L. and Eisenberg, D. (1992) Atomic solvation parameters applied to molecular dynamics of proteins in solution. *Protein Sci.* **1(2),** 227–235.
98. Ooi, T., Oobatake, M., Némethy, G., and Scheraga, H. A. (1987) Accessible surface areas as a measure of the thermodynamic parameters of hydration of peptides. *Proc. Natl. Acad. Sci. USA* **84,** 3086–3090.

99. Still, W. C., Tempczyk, A., Hawley, R. C., and Hendrickson, T. (1990) Semi-analytical treatment of solvation for molecular mechanics and dynamics. *J. Am. Chem. Soc.* **112,** 6127–6129.
100. Hassan, S. A., Guarnieri, F., and Mehler, E. L. (2000) Characterization of hydrogen bonding in a continuum solvent model. *J. Phys. Chem. B.* **104,** 6490–6498.
101. Warshel, A. and Russell, S. T. (1984) Calculation of electrostatic interactions in biological systems and in solutions. *Quart. Rev. Biophys.* **17,** 283–422.
102. Russell, S. T. and A., W. (1985) Calculations of electrostatic energies in proteins. *J. Mol. Biol.* **185,** 389–404.
103. Qiu, D., Shenkin, P. S., Hollinger, F. P., and Still, W. C. (1997) The GB/SA continuum model for solvation. A fast analytical method for the calculation of approximate Born radii. *J. Phys. Chem. B.* **101,** 3005–3014.
104. Schaefer, M. and Karplus, M. (1996) A comprehensive analytical treatment of continuum electrostatics. *J. Phys. Chem.* **100,** 1578–1599.
105. Ghosh, A., Rapp, C., and Friesner, R. A. (1998) Generalized Born model based on a surface area formulation. *J. Phys. Chem. B.* **102,** 10,983.
106. Dominy, B. N. and Brooks, I., C.L. (1999) Development of a generalized Born model parametrization for proteins and nucleic acids. *J. Phys. Chem. B.* **103,** 3765–3773.
107. Onufriev, A., Bashford, D., and Case, D. A. (2000) Modification of the generalized Born model suitable for macromolecules. *J. Phys. Chem. B.* **104,** 3712–3720.
108. Zhu, J. A., Shi, Y. Y., and Liu, H. Y. (2002) Parametrization of a generalized Born/solvent-accessible surface area model and applications to the simulation of protein dynamics. *J. Phys. Chem. B.* **106,** 4844.
109. Lee, M. S., Salsbury, F. R., and Brooks, C. L. (2002) Novel generalized Born methods. *J. Chem. Phys.* **116,** 10,606–10,614.
110. Lorentz, H. A. (1952) *Theory of Electrons.* Dover, New York, NY.
111. Debye, P. (1929) *Polar Molecules.* Dover, New York, NY.
112. Debye, P. and Pauling, L. (1925) The inter-ionic attraction theory of ionized solutes. IV. The influence of variation of dielectric constant on the limiting law for small concentrations. *J. Am. Chem. Soc.* **47,** 2129–2134.
113. Sack, V. H. (1926) The dielectric constant of electrolytes. *Phys. Z.* **27,** 206–208.
114. Sack, V. H. (1927) The dielectric constants of solutions of electrolytes at small concentrations. *Phys. Z.* **28,** 199–210.
115. Bucher, M. and Porter, T. L. (1986) Analysis of the Born model for hydration of ions. *J. Phys. Chem.* **90,** 3406–3411.
116. Ehrenson, S. (1989) Continuum radial dielectric functions for ion and dipole solution systems. *J. Comp. Chem.* **10,** 77–93.
117. Mehler, E. L. (1996) The Lorentz-Debye-Sack theory and dielectric screening of electrostatic effects in proteins and nucleic acids, in *Molecular Electrostatic Potential: Concepts and Applications.* (Murray, J. S. and Sen, K., eds.) Elsevier Science, Amsterdam, The Netherlands, pp. 371–405.
118. Davis, M. E. and McCammon, J. A. (1990) Calculating electrostatic forces from grid-calculated potentials. *J. Comp. Chem.* **11,** 401.

119. Sharp, K. (1991) Incorporating solvent and ionic screening into molecular dynamics using the finite-difference Poisson–Boltzman method. *J. Comp. Chem.* **12,** 454–468.
120. Myers, J. K. and Pace, C. N. (1996) Hydrogen bonding stabilizes globular proteins. *Biophys. J.* **71**, 2033–2039.
121. Pace, C. N. (2001) Polar group burial contributes more to protein stability than nonpolar group burial. *Biochemistry* **40,** 310–313.
122. Hassan, S. A. and Mehler, E. L. (2001) A general screened Coulomb potential based implicit solvent model: calculation of secondary structure of small peptides. *Int. J. Quant. Chem.* **83,** 193–202.
123. Hassan, S. A., Mehler, E. L., and Weinstein, H. (2002) Structure calculations of protein segments connecting domains with defined secondary structure: A simulated annealing Monte Carlo combimed with biased scaled collective variables technique, in *Lecture Notes in Computational Science and Engineering.* (Hark, K. and Schlick, T., eds.) Springer Verlag, New York, NY, pp. 197–231.

123a. Hassan, S. A. (2004) Intermolecular potentials of mean force of amino acids side chain interactions in aqueous medium. *J. Phys. Chem. B* **50,** 19,501–19,509.

124. Privalov, P. L. and Makhatadze, G. I. (1993) Contribution of hydration to protein folding thermodynamics. II.The entropy and Gibbs energy of hydration. *J. Mol. Biol.* **232,** 660.
125. Makhatadze, G. I. and Privalov, P. L. (1993) Contribution of hydration to protein folding thermodynamics. I. The enthalpy of hydration. *J. Mol. Biol.* **232,** 639.
126. Davis, A. M. and Teague, S. J. (1999) Hydrogen bonding, hydrophobic interactions, and failure of the rigid receptor hypothesis. *Angew. Chem, Int. Ed.* **38,** 736–749.
127. Tanford, C. (1973) *The Hydrophobic Effect. Formation of Micelles and Biological Membranes.* Wiley-Interscience, New York, NY.
128. Fink, A. L. (1998) Protein aggregation: folding aggregates, inclusion bodies and amyloid. *Folding Des.* **3,** R9.
129. Cheng, Y.-K. and Rossky, P. J. (1998) Surface topography dependence of biomolecular hydrophobic hydration. *Nature* **392,** 696.
130. Choe, S. E., Li, L., Matsudaira, P. T., and Wagner, G. (2000) Differential stabilization of two hydrophobic cores in the transition state of the villin 14T folding reaction. *J. Mol. Biol.* **304,** 99.
131. Hadi, M. Z., Ginalski, K., Nguyen, L. H., and Wilson III, D. M. (2002) Determinant in nuclease specificity of ape1 and ape2, human homogues of scherichia coli exonuclease III. *J. Mol. Biol.* **316,** 853.
132. Starich, M. R., et al. (1998) The solution structure of the Leu22–>Val mutant AREA DNA binding domain complexed with a TGATAG core element defines a role for hydrophobic packing in the determination of specificity. *J. Mol. Biol.* **277,** 621.
133. Shiba, T. Takatsu, H., Nogi, T., et al. (2002) Structural basis for recognition of acidic-cluster dileucine sequence by GGA1. *Nature* **415,** 937–941.

134. Peng, X. and Jonas, J. (1994) High-pressure NMR study of the dissociation of arc repressor. *Biochemistry* **33,** 8323.
135. Nelson, C. J., LaConte, M. J., and Bowler, B. E. (2001) Direct detection of heat and cold denaturation for partial unfolding of a protein. *J. Am. Chem. Soc.* **123,** 7453.
136. Zhang, J., Peng, X., Jonas, A., and Jonas, J. (1995) NMR study of the cold, heat, and pressure unfolding of ribonuclease A. *Biochemistry* **34,** 8631.
137. Hummer, G., Garde, S., Garcia, A. E., Paulaitis, M. E., and Pratt, L. R. (1998) The pressure dependence of hydrophobic interactions is consistent with the observed pressure denaturation of proteins. *Proc. Nat. Acad. Sci. (USA)* **95,** 1552.
138. Wagner, F. and Simonson, T. (1999) Implicit solvent models: combining an analytical formulation of continuum electrostatics with simple models of the hydrophobic effect. *J. Comp. Chem.* **20,** 322–335.
139. Cramer, C. J. and Truhlar, D. G. (1992) An SCF solvation model for the hydrophobic effect and absolute free energies of aqueous solvation. *Science.* **256,** 213–217.
140. Simonson, T. and Brunger, A. T. (1994) Solvation free energies estimated from macroscopic continuum theory: an accuracy assessment. *J. Phys. Chem.* **98,** 4683–4694.
141. Lum, K., Chandler, D., and Weeks, J. D. (1999) Hydrophobicity at small and large length scales. *J. Phys. Chem. B.* **103,** 4570.
142. Reiss, H. (1965) A acaled particle methods in the statistical thermodynamics of fluids. *Adv. Chem. Phys.* **9,** 1.
143. Tolman, R. C. (1949) The effect of droplet size on surface tension. *J. Chem. Phys.* **17,** 333.
144. Pratt, E. A. and Chandler, D. (1980) Effect of solute-solvent attractive forces on hydrophobic correlations. *J. Chem. Phys.* **73,** 3434.
145. Pratt, L. R. and Chandler, D. (1977) Theory of hydrophobic effect. *J. Chem. Phys.* **67,** 3683.
146. Reiss, H., Frisch, H. L., and Lebowitz, J. L. (1959) Statistical mechanics of rigid spheres. *J. Chem. Phys.* **31,** 369.
147. Pierotti, R. A. (1963) Solubility of gases in liquids. *J. Phys. Chem.* **67,** 1840.
148. Stillinger, F. H. (1973) *J. Solution Chem.* **2,** 141.
149. Chandler, D. (1987) *Introduction to Modern Statistical Mechanics.* Oxford University Press, New York, NY.
150. Huang, D. M. and Chandler, D. (2002) The hydrophobic effect and the influence of solute-solvent attractions. *J. Phys. Chem. B.* **106,** 2047.
151. Weeks, J. D., Katsov, K., and Vollmayr, K. (1998) Roles of repulsive and attractive forces in determining the structure of nonuniform liquids: generalized mean field theory. *Phys. Rev. Lett.* **81,** 4400.
151a. Jorgensen, W. L. (1982) Quantum and statistical mechanical studies of liquids. Monte Carlo simulation of n-butane in water. Conformational evidence for hydrophobic effect. *J. Chem. Phys.* **77,** 5757–5765.

152. Pashley, R. M., McGuiggan, P. M., Ninham, B. W., and Al., E. (1985) Attractive forces between uncharged hydrophobic surfaces. Direct measurements in aqueous solution. *Science* **229,** 1088.
153. Christenson, H. K. and Claesson, P. M. (1988) Cavitation and the interaction between macroscopic hydrophobic surfaces. *Science* **239,** 390.
154. Christenson, H. K. (1992) *Modern Approaches to Wettability: Theory and Applications* (Schrader, M.E. and Loeb, G. eds.) Plenum, New York, NY.
155. Wallqvist, A. and Berne, B. J. (1995) Computer simulation of hydrophobic hydration forces on stacked plates at short range. *J. Phys. Chem.* **99,** 2893.
156. Karplus, M. and McCammon, J. A. (2002) Molecular dynamics simulations of biomolecules. *Nature Struct. Biol.* **9,** 646–652.
157. Andricioaei, I. and Karplus, M. (2001) On the calculation of entropy from covariance matrices of the atomic fluctuations. *J. Chem. Phys.* **115,** 6289–6292.
158. Levy, R. M., Karplus, M., Kushick, J., and Perahia, D. (1984) Evaluation of the configurational entropy for proteins—application to molecular-dynamics simulations of an alpha-helix. *Macromolecules* **17,** 1370–1374.
159. Brady, J. and Karplus, M. (1985) Configurational entropy of the alanine dipeptide in vacuum and in solution: a molecular dynamics study. *J. A. Chem. Soc.* **107,** 6103–6105.
160. Eriksson, M. A. and Nilsson, L. (1995) Structure, thermodynamics and cooperativity of the glucocorticoid receptor DNA-binding domain in complex with different response elements. Molecular dynamics simulation and free energy perturbation studies. *J. Mol. Biol.* **253,** 453–72.
161. van der Vegt, N. F. A. and van Gunsteren, W. F. (2004) Entropic contributions in cosolvent binding to hydrophobic solutes in water. *J. Phys. Chem. B* **108,** 1056–1064.
162. Swanson, J. M. J., Henchman, R. H., and McCammon, J. A. (2004) Revisiting free energy calculations: a theoretical connection to MM/PBSA and direct calculation of the association free energy. *Biophys. J.* **86,** 67–74.
163. Amzel, L. M. (1997) Loss of translational entropy in binding, folding, and catalysis. *Proteins.* **28,** 144–149.
164. Daquino, J. A., Gomez, J., Hilser, V. J., Lee, K. H., Amzel, L. M., and Freire, E. (1996) The magnitude of the backbone conformational entropy change in protein folding. *Proteins* **25,** 143–156.
165. Lee, K. H., Xie, D., Freire, E., and Amzel, L. M. (1994) Estimation of changes in side-chain configurational entropy in binding and folding—general-methods and application to helix formation. *Proteins* **20,** 68–84.
166. Zwanzig, R. W. (1954) High-temperature equation of state by a perturbation method. I. Nonpolar gases. *J. Chem. Phys.* **22,** 1420.
167. Kollman, P. A. (1993) Free energy calculations: Applications to chemical and biochemical phenomena. *Chem. Rev.* **93,** 2395–2417.
168. Torrie, G. M. and Valleau, J. P. (1977) Nonphysical sampling distributions in Monte Carlo free energy estimation:Umbrella sampling. *J. Comput. Phys.* **23,** 187–189.
169. Bennett, C. H. (1976) Efficient estimation of free energy differences from Monte Carlo data. *J. Comput. Phys.* **22,** 245–268.

170. Bash, P. A., Singh, U. C., Brown, F. K., Langridge, R., and Kollman, P. (1987) Calculation of the relative change in binding free energy of a protein-inhibitor complex. *Science* **235,** 574–576.
171. Wong, C. F. and McCammon, J. A. (1986) Dynamics and design of enzymes and inhibitors. *J. Am. Chem. Soc.* **108,** 3830–3832.
172. Weiner, S., Kollman, P., Case, D., Singh, U. C., Ghio, C., Alagona, G., Profeta, S. J., and Weiner, P. (1984) A new force field for molecular mechanical simulation of nucleic acids and proteins. *J. Am. Chem. Soc.* **106,** 765–784.
173. van Gunsteren, W. F. (1987) *GROMOS.* Groningen Molecular Simulation Computer Program Package, University of Groningen. The Netherlands.
174. Kirkwood, J. G. (1935) Statistical mechanics of fluid mixtures. *J. Chem. Phys.* **3,** 300–313.
175. Olson, M. A. and Reinke, L. T. (2000) Modeling implicit reorganization in continuum descriptions of protein-protein interactions. *Proteins* **38,** 115–119.
176. Frauenfelder, H. (1995) Complexity in proteins. *Nature Struct. Biol.* **2,** 821–823.
177. Frauenfelder, H., Sligar, S. G., and Wolynes, P. G. (1991) The energy landscapes and motions of proteins. *Science.* **254,** 1598–1603.
178. Noguti, T. and Go, N. (1989) Structural basis of hierarchical multiple substates of a protein. *Proteins* **5,** 97–103.
179. Elber, R. and Karplus, M. (1987) Multiple conformational states of proteins: A molecular dynamics analysis of myoglobin. *Science* **235,** 318–321.
180. Leeson, D. T. and Wiersma, D. A. (1995) Looking into the energy landscape of myoglobin. *Nature Struct. Biol.* **2,** 848–851.
181. Carlson, H. A. and McCammon, A. (2000) Accomodating protein flexibility in computational drug design. *Mol. Pharmacol.* **57,** 213–218.
182. Gane, P. J. and Dean, P. M. (2000) Recent advances in structure-based drug design. *Curr. Opin. Struct. Biol.* **10,** 401–404.
183. Bouzida, D., Rejto, P. A., Arthurs, S., Colson, A. B., Freer, S. T., Gehlhaar, D. K., Larson, V., Luty, B. A., Rose, P. W., and Verkhivker, G. M. (1999) Computer simulations of ligand–protein binding with ensembles of protein conformations: A Monte Carlo study of HIV-1 protease binding energy landscapes. *Int. J. Quantum Chem.* **72,** 73–84.
184. Carlson, H. A., Masukawa, K. M., and McCammon, A. (1999) Method for including the dynamic fluctuations of a protein in a computer-aided drug design. *J. Phys. Chem. A.* **103,** 10,213–10,219.
185. Gallicchio, E., Kubo, M. M., and Levy, R. M. (1998) Entropy-enthalpy compensation in solvation and ligand binding revisited. *J. Am. Chem. Soc.* **120,** 4526–4527.
186. Lee, C. and Irizarry, K. (2001) The GeneMine System for genome/proteome annotation and collaborative data mining. *IBM Systems Journal.* **50,** 592–603.
187. Kraulis, P. J. (1991) MOLSCRIPT: A program to produce both detailed and schematic plots of protein structures. *J. Appl. Cryst.* **24,** 946–950.
188. Merritt, E. A. and Bacon, D. J. (1997) Raster3D: photorealistic molecular graphics, in *Methods in Enzymology*, Vol. 277, Macromolecular Crystallography, Pt.

B., (Carter, C. W., Jr. and Sweet, R. M., eds.) Academic Press, San Diego, CA, pp. 505–524.

189. Brooks, B. R., Bruccoleri, R. E., Olafson, B. D., States, D. J., Swaminathan, S., and Karplus, M. (1983) CHARMM: A program for macromolecular energy, minimization and dynamics calculations. *J. Comput. Chem.* **4,** 187–217.
190. Frisch, M. J., Trucks, G. W., Schlegel, H. B., Scuseria, G. E., Robb, M. A., Cheeseman, J. R., et al. (2001) *Gaussian 98*. Gaussian, Pittsburgh, PA.
191. Goodsell, D. S. and Olson, A. J. (1990) Automated docking of substrates to proteins by simulated annealing. *Proteins* **8,** 195.
192. Goodsell, D. S., Morris, G. M., and Olson, A. J. (1996) Automated docking of flexible ligands: applications of AutoDock. *J. Mol. Recognit.* **9,** 1.
193. Morris, G. M., Goodsell, D. S., Huey, R., and Olson, A. J. (1996) Distributed automated docking of flexible ligands to proteins: parallel applications of AutoDock 2.4. *J. Comp. Aid. Mol. Des.* **10,** 293.
194. Petrey, D., Xiang, Z. X., Tang, C. L., Xie, L., Gimpelev, M., Mitros, T., Soto, C. S., Goldsmith-Fischman, S., Kernytsky, A., Schlessinger, A., Koh, I. Y. Y., Alexov, E., and Honig, B. (2003) Using multiple structure alignments, fast model building, and energetic analysis in fold recognition and homology modeling. *Proteins* **53,** 430–435.
195. Pikis, A., Donkersloot, J. A., Rodriguez, W. J., and Keith, J. M. (1998) A conservative amino acid mutation in the chromosome-encoded dihydrofolate reductase confers trimethoprim resistance in *Streptococcus pneumoniae*. *J. Infect. Dis.* **178(3),** 700–706.

195a. Levitt, M. (1992) Accurate modeling of protein conformation by automatic segment matching. *J. Mol. Biol.* **226,** 507–533.

196. Matthews, D. A., Bolin, J. T., Burridge, J. M., Filman, D. J., Volz, K. W., Kaufman, B. T., Beddell, C. R., Champness, J. N., Stammers, D. K., and Kraut, J. (1985) Refined crystal-structures of *Escherichia-coli* and chicken liver dihydrofolate-reductase containing bound trimethoprim. *J. Biol. Chem.* **260,** 381–391.
197. Abagyan, R. and Totrov, M. (1994) Biased probability Monte-Carlo conformational searches and electrostatic calculations for peptides and proteins. *J. Mol. Biol.* **235,** 983–1002.
198. Talhout, R. and Engberts, J. B. (2001) Thermodynamic analysis of binding of p-substituted benzamidines to trypsin. *Eur. J. Biochem.* **268,** 1554–1560.
199. Bachovchin, W. W. (2001) Contributions of NMR spectroscopy to the study of hydrogen bonds in serine protease active sites. *Magnet. Res. Chem.* **39,** S199–S213.
200. Sanschagrin, P. C. and Kuhn, L. A. (1998) Cluster analysis of consensus water sites in thrombin and trypsin shows conservation between serine proteases and contributions to ligand specificity. *Protein Sci.* **7,** 2054–2064.
201. Steinbach, P. J. (1998) Introduction to Macromolecular Simulation, in *Biophysics Textbook On-line*, (Bloomfield, V., ed.), Biophysical Society: Bethesda, MD. Website: http://www.biophysics.org/btol/.

202. Steinbach, P. J. and Brooks, B. R. (1994) New spherical-cutoff methods for long-range forces in macromolecular simulation. *J. Comput. Chem.* **15(7),** 667–683.

203. Darden, T., York, D., and Pedersen, L. (1993) Particle mesh Eewald—an N.Log(N) method for ewald sums in large systems. *J. Chem. Phys.* **98(12),** 10,089–10,092.

204. Sagui, C. and Darden, T. A. (1999) Molecular dynamics simulations of biomolecules: Long-range electrostatic effects. *Ann. Rev. Biophys. Biomol. Struct.* **28,** 155–179.

23

Force Probe Molecular Dynamics Simulations

Helmut Grubmüller

Summary

Many proteins are molecular *nano-machines*, which perform their biological function via well-coordinated structural transitions. Often, these motions occur on much slower time scales than those accessible to conventional molecular dynamics techniques, which are limited to submicrosecond time scales by current computer technology. This is also true for ligand binding and unbinding reactions. Force probe simulations (or steered molecular dynamics) provide a powerful means to overcome this limitation, and thus to get insight into the atomistic mechanisms that underlie biological functions such as ligand binding. This chapter provides a basic introduction into this method. It further sketches a simple nonequilibrium statistical mechanics treatment that shows how to relate the results of force probe simulations to atomic force microscopy (AFM) or optical tweezer experiments. As an example, enforced unbinding simulations of streptavidin/biotin complexes are detailed.

Key Words: Molecular dynamics simulation; protein dynamics; force probe simulation; steered molecular dynamics; ligand/receptor unbinding; enforced protein unfolding; rupture force calculation; unfolding forces; rescaling of loading rates; nonequilibrium statistical mechanics.

1. Introduction

Chapter 22 provided an excellent introduction into the molecular dynamics (MD) simulation method on biomolecules, which is also the basis for the force probe simulation technique *(1)* (also known as steered molecular dynamics *[2]*) described herein. Therefore, the basics of MD simulations are not repeated here in detail, but only briefly sketched. We strongly recommend reading Chapter 22 in combination with this chapter.

From: *Methods in Molecular Biology, vol. 305: Protein–Ligand Interactions: Methods and Applications*
Edited by: G. U. Nienhaus © Humana Press Inc., Totowa, NJ

A characteristic feature of protein dynamics is the occurrence of collective structural transitions, so-called conformational transitions *(3)*, which are often crucial for ligand binding and for protein function in general. These differ from the thermal high-frequency vibrations, which fall into the picosecond range, in that they occur on much slower time scales from nanoseconds to seconds. The opening and closing of ion channel proteins in nerve cells is an example for such a functional conformational transition. It is characterized by millisecond rates and can be observed via an electric current through the channel by patch-clamp measurements *(4–6)* on single proteins. Other examples are so-called allosteric effects (e.g., in hemoglobin), which are realized by conformational transitions and induced fit motions upon ligand binding, for example, of an antigen to an antibody. Last but no least, we discuss the elementary steps of motor proteins for muscle contraction, which also represent conformational transitions.

Functional processes in proteins occur on a large variety of different time scales, and so does protein dynamics and conformational dynamics that covers a wide hierarchy of dynamical processes: the high frequency part of the thermal fluctuations of single atoms around their average position is characterized by reciprocal frequencies of several ten to several hundred femtoseconds; collective motions of smaller groups of atoms range up to several tens of picoseconds. Larger structural changes occur on a hierarchy of time scales ranging from nanoseconds to hours. There is a convincing amount of experimental and theoretical evidence for *hierarchical substates* *(3)* for protein conformations, which are linked by conformational transitions on a wide range of time scales *(7)*. These conformational dynamics apparently are a feature specific to proteins and other systems of comparable complexity *(8)*.

The highly ordered, but heterogeneous structure of proteins in combination with the broad range of nonseparating time scales in protein dynamics renders the application of established concepts of many body physics extremely difficult; work in this direction, therefore, is a challenging field in theoretical physics.

Currently, the explicit, atomically resolved molecular dynamics simulation of the protein of interest, as described in Chapter 22, is still to be considered the only reliable theoretical description. The lack of more elegant treatments has a high price: such protein dynamics simulations are computationally extremely demanding and, therefore with presently available hardware restricted to relatively small simulation systems (several 100,000 atoms) and, most severely, to short time scales clearly below microseconds. Nevertheless, the method is now widely and successfully used for a large number of biological processes that fall into this category, and has yielded a large number of correct predictions and insights into protein function at the atomic level *(9)*.

1.1. Protein Dynamics Simulations

Molecular dynamics simulations compute the motion of every single atom within the simulation system (e.g., a protein solvated in physiological solvent, **Fig. 1**, left), which is determined by the interaction forces (**Fig. 1**, right) between all the n atoms of the system ***(10)***. Typically, these forces are described by a *force field* $V(\mathbf{x}_1,\ldots,\mathbf{x}_n)$ which serves to compute a trajectory $\{\mathbf{x}_i(t_j)\}_{i=1\ldots n;\, j=0\ldots N}$ via the (classical) Newtonian equations of motion. This trajectory specifies the position $\mathbf{x}_i$ of each single atom i in the system within a (discretized) period of time, $t_0 = 0, t_1 = \Delta t, t_2 = 2\Delta t, \ldots, t_N = N\Delta t$, for N so-called integration steps. From such a trajectory one can subsequently calculate the observables of interest, often via averages.

For medium-sized systems of about 100,000 atoms, currently simulations of several 10 ns duration are feasible at justifiable computational cost ***(11–15)***. One can estimate from this number that a further increase of computer power by a factor of about 10^4 will be required to cover most biochemical elementary reactions such as enzymatic catalysis or the transport of an ion across the membrane by ion pump proteins. For protein-folding simulations, a 10^6- to 10^8-fold increase would be required. Assuming that the annual increase by a factor of 1.6 observed for the past 35 yr continues into the future, these aims could be reached within 20 to 40 yr. Today, nearly 30 yr after the first molecular dynamics simulation of a protein ***(16,17)***, and despite the impressive 10^5-fold increase of computer power since then, the limited system size and, particularly, limited simulation lengths are the main obstacle in the attempt to derive from first principles physics laws the biochemical processes in proteins that keep us alive ***(18–21)***.

1.2. Force Fields for Protein Dynamics

The forces between the atoms of a solvated macromolecule (**Fig. 1**, left) are diverse (**Fig. 1**, inset, right). Chemical binding forces, symbolized by springs, enforce equilibrium distances or angles between chemically bound atoms (small arrows). Pauli repulsion forces (dark arrows) prevent atoms from moving through each other. Long-ranged interactions, particularly electrostatic forces (light arrows) between partially charged atoms (δ^+, δ^-) contribute substantially to the stability of protein structures and dominate the slow conformational dynamics. Hydrogen bonds, for example, are mainly of electrostatic origin and only to a lesser extent a quantum-mechanical effect, and contribute significantly to the stability of α-helices and β-sheets ***(22)***.

All these forces (and several more, which are not discussed here) determine the three-dimensional structure of a protein as well as the motion of each single atom, and therefore have to be considered in protein dynamics simulations. As mentioned before, this is achieved via the force field $V(\mathbf{x}_1,\ldots,\mathbf{x}_n)$, which describes the influence of the electronic dynamics on the motion of the nuclei. Most force

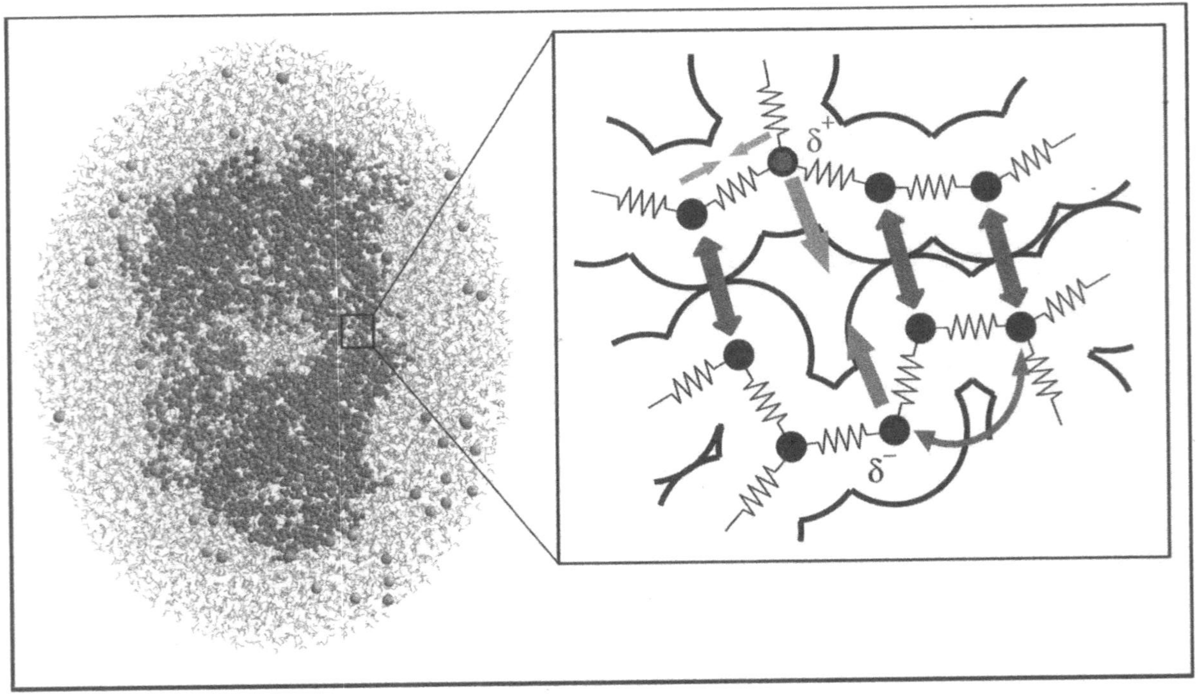

Fig. 1. **Left:** A typical protein dynamics simulation system comprises a protein (dark gray atoms) solvated in water (light gray; only the O–H-bonds are shown). Salt ions (sparse light gray atoms) have been added to the aqueous solvent. **Right:** The inset shows part of the system, together with selected interatomic forces (for example, chemical binding forces, Coulomb forces between atoms that carry partial charges, Pauli repulsion, van der Waals attraction) that determine the molecular motion.

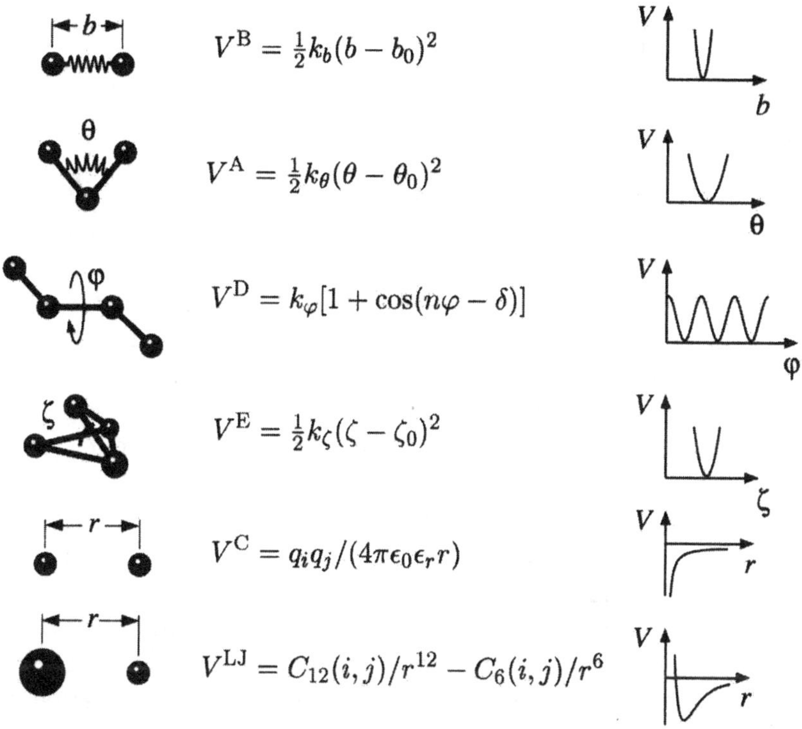

Fig. 2. Interaction contributions to a typical force field. Bond stretch vibrations are described by a harmonic potential V^B, the minimum of which is at the equilibrium distance b_0 between the two atoms connected by chemical bond i (*see* **Eq. 1**; the indices i, j, etc., are not shown in the figure). Bond angles and out-of-plane angles are also described by harmonic potential terms, V^A and V^E where Θ_0 and ζ_0 denote the respective equilibrium angles. Dihedral twists are subjected to a periodic potential V^D; the respective force coefficients are denoted by k's with appropriate indices. Nonbonded forces are described by Coulomb interactions, V^C, and Lennard–Jones potentials, $V^{LJ} = V^P + V^{vdW}$, where the latter includes the Pauli repulsion, $V^P \sim r^{12}$, and the van der Waals interaction, $V^{vdW} \sim -r^6$. (Adapted with permission from **ref. *23***.)

fields are composed of empirically motivated interaction terms. Construction, parameterization, and testing of a force field is a challenging task, and substantial efforts will be required to further improve their accuracy.

Since the 1970s, several force fields for biomolecules and, specifically, for proteins and DNA/RNA have been developed. Well known and widely used are, for example, CHARMm, Gromos96, AMBER, and OPLS. **Figure 2** shows interaction terms from which these force fields are typically composed. They describe interactions that arise from covalent chemical bonds as well as from noncovalent interactions,

$$V = \sum_{\substack{\text{bonds}\\ i}} V_i^{\mathrm{B}} + \sum_{\substack{\text{bonds}\\ \text{angles } j}} V_j^{\mathrm{A}} + \sum_{\substack{\text{dihedral}\\ \text{angles } k}} V_k^{\mathrm{D}} + \sum_{\substack{\text{extraplanar}\\ \text{angles } l}} V_l^{\mathrm{E}} + \sum_{\substack{\text{atom pairs}\\ r,s}} \left(V_{r,s}^{\mathrm{C}} + V_{r,s}^{\mathrm{P}} + V_{r,s}^{\mathrm{vdW}} \right) \tag{1}$$

Covalent forces arise from changes of the length of chemical bonds (B) and of the angle between two bonds (A), from twisting chemical bonds (dihedral, D), and deviations of aromatic carbon atoms from their in-plane positions (extraplanar, E). As noncovalent interactions the Coulomb interaction (C), the Pauli repulsion (P) and the van der Waals interaction (vdW) are included; the latter two are conveniently described by a Lennard–Jones potential (LJ).

Many important details of protein dynamics simulations can only partly be described and discussed at the level of this chapter. These include the treatment of the system boundaries or, alternatively, their periodic continuation; the *freezing* of fast bond vibrations to mimic their quantum-mechanical character; the proper placement of salt ions in the vicinity of the protein; the treatment of protonable residues; simulations in canonical thermodynamic ensembles via appropriate coupling to heat and pressure baths, the treatment of nonpolar hydrogen atoms through *compound atoms*; pros and cons of explicit vs implicit treatments of hydrogen bonds; the set-up of a simulation system using structures derived from X-ray crystallography (*see* **Note 1**) and the problem of sufficiently long equilibration of the system (*see* **Note 3**); the implicit (or, in part, explicit) description of electronic polarizability; the efficient computation of the long-ranged Coulomb forces and the parallelization of the respective algorithms; as well as the proper choice of numerical integrators for the solution of the Newtonian equations of motion. Excellent review articles have been published on these topics *(24–26)*. For a full in-depth treatment, the reader is referred to the books of Gunsteren, Weiner, and Wilkinson *(27)*.

1.3. Force Probe Simulations

In recent years, we have seen dramatic improvements in single-molecule experiments, particularly atomic force microscopy (AFM) methods *(28,29)*, described in Chapter 21, which motivated the force probe simulation technique described in this section.

In single-molecule force probe experiments *(29,30)*, the cantilever of the AFM microscope is used as a piconewton force sensor (**Fig. 3B**). The cantilever can be positioned to subnanometer accuracy, and its deflection—measuring the applied force—is detected via a laser beam with equally high precision.

Figure 3A illustrates the experiment, in which a ligand (here, biotin, light gray) is forced by the cantilever to leave the specific binding pocket of the receptor (streptavidin, dark gray). A polymer linker connects the proteins (with

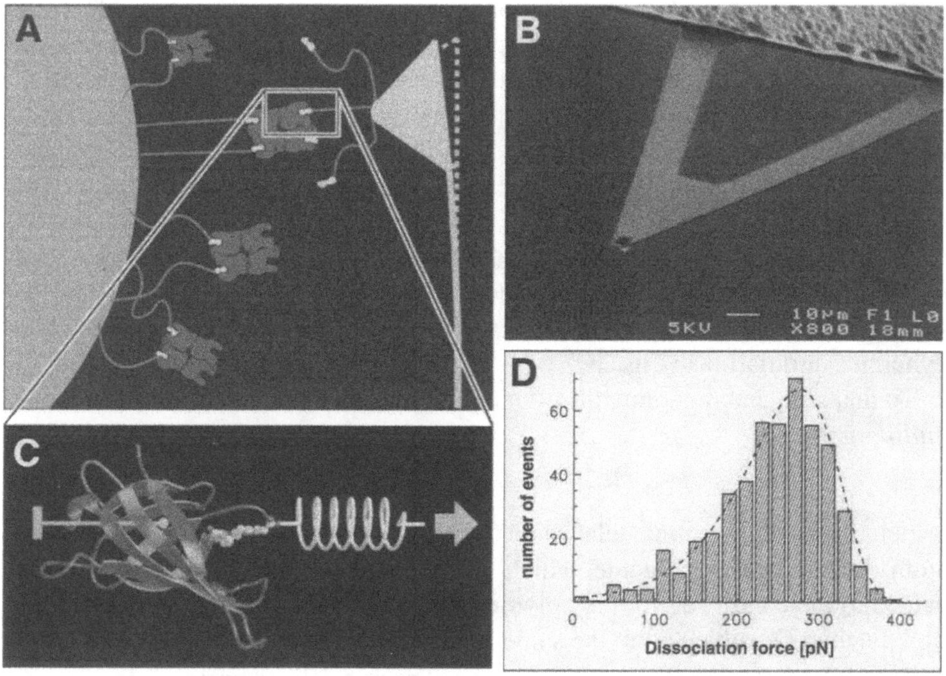

Fig. 3. **(A)** Principle of single molecule AFM experiments; **(B)** electron microscopy picture of an AFM cantilever (picture kindly provided by Nanoscope); **(C)** force probe simulation of a single molecule AFM experiment; **(D)** typical force histogram obtained from a series of single molecule unbinding experiments and fit to theory (dashed).

empty binding pockets) with a surface (left), as well as several biotin molecules with the tip of the cantilever (right). As the cantilever approaches the surface several biotin/streptavidin complexes form, which dissociate one after the other upon subsequent retraction of the cantilever. Occasionally, one *single* complex remains intact until the very end of the experiment, in which case the force required to dissociate this last complex can be measured from the jump of the deflection of the cantilever to zero.

By repeating the experiment several hundred times one obtains a histogram of dissociation forces (**Fig. 3D**). Its maximum denotes the most probable dissociation force, in the shown example about 270 pN. This is the force that the noncovalent binding interaction between ligand and receptor can withstand at the time scale set by the experiment (typically milliseconds to seconds), and thus a measure for the binding strength (but *see* **Note 6**).

In a similar manner, molecular forces have been measured recently for a number of systems. One example is the force generated by single motor proteins *(31)* which is used to drive muscle contraction or to transport intracellular vesicles along filaments. Another one is the force generated by DNA polymerase upon transcription along the DNA primer *(32)*. (For the application of torques rather than linear forces, *see* **Note 4**; for nondirected forces, *see* **refs. *33,34*.**)

However, in those experiments, the underlying atomistic dynamics and interactions that generate the measured forces cannot be observed, which is the main motivation to simulate these experiments by means of atomistic protein dynamics simulations (**Fig. 3C**).

To that aim, and modeling the effect of the cantilever, an additional harmonic *pulling potential*,

$$V_{\text{cant}}(\mathbf{R}_i, t) = \frac{1}{2}k\left[\left(\mathbf{R}_i - \mathbf{R}_i^0\right)\cdot\hat{\mathbf{n}} - vt\right]^2 \tag{2}$$

is included within the molecular force field *(1)*, such that it acts on that particular atom i of the ligand molecule, which in the real experiment is covalently connected to the cantilever via the polymer linker. This ensures that in the simulation the atom is subjected to the same force as in the experiment. In **Fig. 3**, this pulling potential is symbolized by a spring. The normalized vector $\hat{\mathbf{n}}$ denotes the direction of pulling, and $\mathbf{R}_i^0$ the position of atom i at the start of the simulation. As can be seen from **Eq. 2**, the minimum $z_{\text{cant}} = vt$ of the pulling potential (i.e., the equilibrium position of atom i) is subsequently moved with constant velocity v away from the binding pocket (arrow).

To avoid drift of the protein, the center of mass of the protein (consisting of n_p atoms with positions $\mathbf{R}_i$, $i = 1...n_p$) is kept in place at $\mathbf{R}_{\text{fix}}$ by a second harmonic (but stationary) potential,

$$V_{\text{fix}}\left(\mathbf{R}_1,\ldots,\mathbf{R}_{n_p}\right) = \frac{1}{2}k_{\text{fix}}\left[\mathbf{R}_{\text{fix}} - \frac{1}{n_p}\sum_{i=1}^{n_p}\mathbf{R}_i\right]^2 \tag{3}$$

with suitable spring coefficient k_{fix}. This ensures that the protein is free to undergo internal motions, such as induced-fit motions upon ligand unbinding, and that it can adopt to the pulling direction by rotations—as in the real experiment—but prevents translational motions.

As an example, **Fig. 4** shows a detailed model of streptavidin/biotin dissociation derived from force probe simulations (*see* **Note 5** for a caveat), highlighting the sequence and type of stretching and subsequent rupture of single noncovalent bonds between the ligand and the receptor *(1,2)*. These complex sequence of localized rupture events give rise to a complex *force profile* (**Fig. 5**), that is the exerted force plotted during the simulation,

$$F_{\text{pull}} = \hat{\mathbf{n}}\cdot\nabla V_{\text{cant}}(\mathbf{R}_i, t) = k\left[\left(\mathbf{R}_i - \mathbf{R}_i^0\right)\cdot\hat{\mathbf{n}} - vt\right]\hat{\mathbf{n}} \tag{4}$$

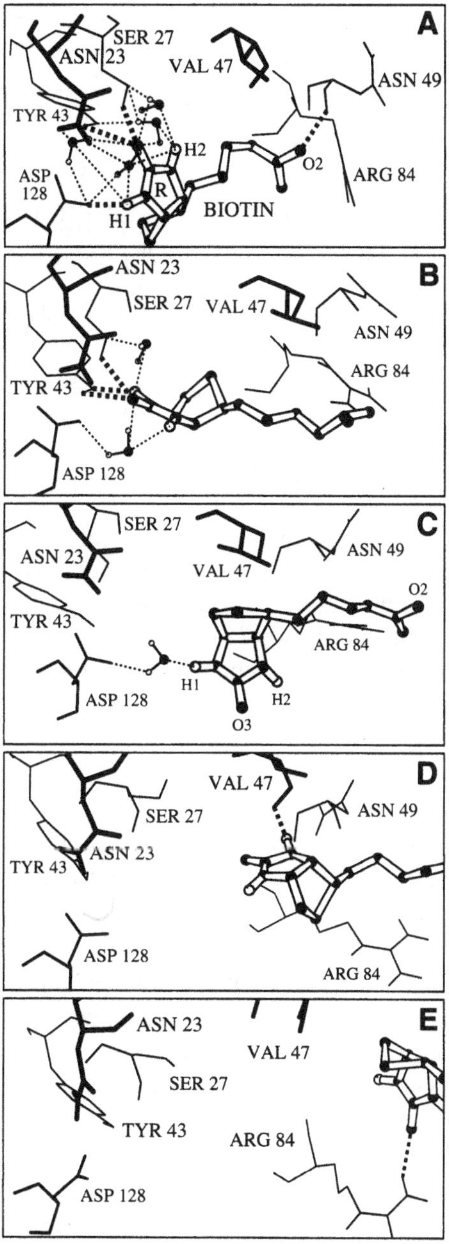

Fig. 4. Snapshots of enforced dissociation of a biotin molecule (white, thick contours) from the streptavidin binding pocket (only the few relevant residues of the binding pocket are shown as stick-models). Hydrogen bonds are shown as bold dashed lines, and few of the many water bridges with thin, dotted lines. The simulation length is one nanosecond; during this time the biotin molecule is moved by about one nanometer. (Reprinted with permission from **ref. *1*.**)

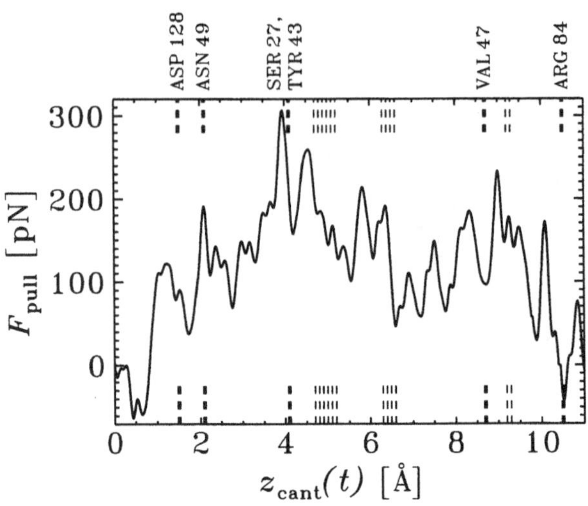

Fig. 5. Force exerted onto the biotin ligand during enforced dissociation (*force profile, see* **Note 2**). A large number of local force maxima can be seen, each of which can be attributed to the rupture of single noncovalent bonds like hydrogen bonds or water bridges. The bold dashed lines at the top denote the respective residues, which are also shown in **Fig. 4**. The thin dashed lines denote rupture of water bridges. (Adapted with permission from **ref. *1*.**)

These force probe AFM simulations of enforced dissociation also allow one to determine dissociation forces from the maximum of the force profile, and to compare them with those measured in AFM experiments. In such comparisons, one has to carefully take into account the fact that the AFM measurements and the force probe simulations are carried out at quite different time scales, namely milliseconds vs nanoseconds. Therefore, in order to compare the respective dissociation forces, the computed force has to be rescaled to the measured ones, for example, for the simplest case using **Eq. 14**, as shown in **Fig. 6** and described later.

This type of force probe simulations, also called *steered molecular dynamics*, is now widely used, for example, to elucidate at the atomic level the structural changes that are induced by mechanically unfolding proteins like titin *(38–44)* or by stretching various other polymers *(45–47)*, and to explain the measured forces in terms of intramolecular interactions. In this context, it is worth pointing out that force profiles contain information on the underlying free energy landscape that governs unbinding or unfolding; indeed, information on this landscape can be obtained from force profiles *(48–50)*. For more detailed information, we refer the reader to **refs. *37,51,52*.**

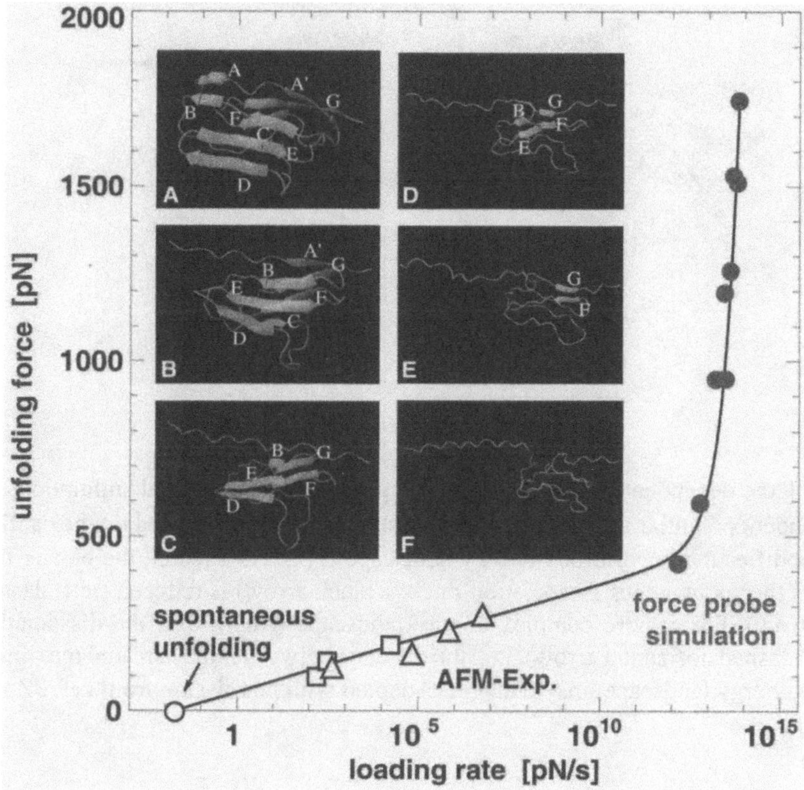

Fig. 6. Measured (□ *[35]* and △ *[36]*) and calculated (●) unfolding forces of single titin molecules as a function of the loading rate *kv*. The solid line shows the fit of **Eq. 14** to the computed unfolding forces and the spontaneous unfolding rate (○). As can be seen, the experimental values are predicted very well. The six insets show snapshots of the protein during simulated unfolding. (Adapted with permission from **ref.** ***37***.)

1.4. Simple Nonequilibrium Statistical Mechanics of Enforced Dissociation

As we have seen in the previous example, both measured and computed dissociation or unfolding forces vary with the time scale at which the process is enforced to occur ***(1,53,54)***. Usually, the forces increase with shorter time scales, that is, increasing loading rates. For thermodynamic equilibrium, such time scale dependency should not appear, hence the underlying processes must be assumed to proceed far from equilibrium. This observation motivated the development of nonequilibrium theories of enforced unbinding reactions ***(2,53–57)***. One of these shall be sketched here.

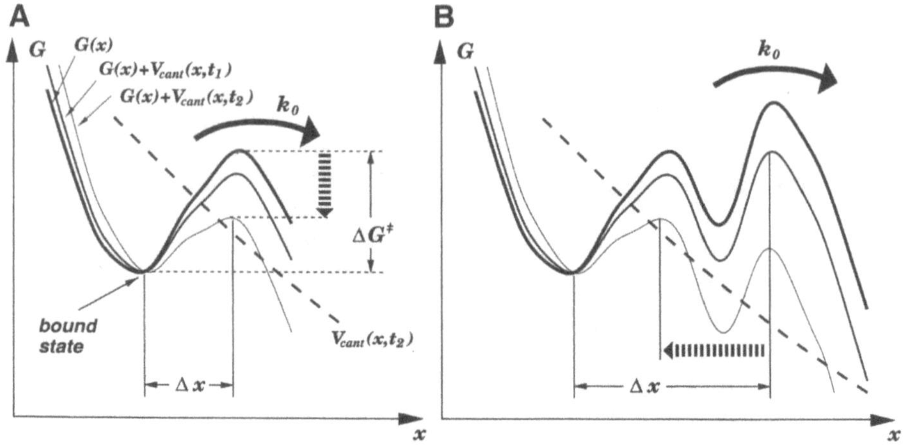

Fig. 7. Time dependent free energy in force probe experiments and simulations. A time-dependent potential $V_{\text{cant}} = (x,t)$, **Eq. 2**, describes the force exerted by the cantilever and modifies the unperturbed energy landscape $G(x)$. As a result, the barrier that determines the spontaneous dissociation rate k_0 (black arrow) is reduced (left: dashed vertical arrow). For a more complex energy landscape (right), also the dissociation length Δx (dashed horizontal arrow), i.e., the distance between minimum and maximum of the free energy landscape, may change. (Adapted with permission from **ref. *37***.)

We assume an effective free energy $G(x)$ along a (one-dimensional) reaction coordinate x (**Fig. 7**, solid bold line), for example, the distance between the center of mass of the ligand and the one of the receptor, which governs the dissociation reaction.

As sketched in the two panels, the bound state (left minimum) is separated from the dissociated state (right) by a more or less structured energy barrier (global maximum). The height $\Delta G^{\ddagger}$ of this barrier determines the rate coefficient k_0 for thermally activated spontaneous dissociation,

$$k_0 = \omega_0 \exp(-\beta \Delta G^{\ddagger}) \tag{5}$$

Here, $\beta = 1/(k_B T)$ denotes the reciprocal thermal energy and ω_0 the attempt frequency or Kramers' prefactor of the bound state ***(58–60)***. For the above streptavidin/biotin dissociation, for example, the reciprocal rate coefficient is several months, whereas it can be as short as milliseconds for antibody–antigen dissociation. Note, that if unbinding is enforced at a time scale that is slower than the reciprocal rate coefficient no dissociation force is measured, because in this case the system has fallen apart already before any significant force has been applied.

As for the force probe simulations described previously, the potential exerted by the cantilever is assumed harmonic and moving with velocity v,

$$V_{\text{cant}}(x,t) = \frac{1}{2}k(x - vt)^2 \tag{6}$$

(dashed curve in **Fig. 7**), where k is the effective spring coefficient. Here, we restrict our discussion to the case of small spring coefficients, that is, soft springs, which covers most realistic situations. Therefore, $V_{\text{cant}}(x,t)$ appears in **Fig. 7** as a nearly straight line, the slope of which increases linearly with time. Because the total potential $G(x) + V_{\text{cant}}(x,t)$ now becomes time dependent, so does the barrier height $\Delta G^{\ddagger}$ and hence, the dissociation coefficient k_0. This effect is demonstrated by the thin, solid lines. As can be seen, the barrier height decreases with time. Dissociation occurs when the barrier has become small enough to be overcome thermally activated at the time scale $\tau = \Delta G^{\ddagger}/(kv\Delta x)$ set by the loading rate kv.

For simple and localized barriers (left picture), we will neglect the slight shift of the position of the minimum that also occurs. In this case, a simple two-state model *(53)* is appropriate, which neglects the details of $G(x)$ and assumes a linear decrease of the barrier height with time,

$$\Delta G^{\ddagger}(t) = \Delta G^{\ddagger} - kvt\Delta x \tag{7}$$

This implies a time dependent dissociation coefficient (*see* **Eq. 5**),

$$k_0(t) = \omega_0 e^{-\beta(\Delta G^{\ddagger} - kvt\Delta x)} \tag{8}$$

which holds as long as $\Delta G^{\ddagger} - kvt\Delta x$ is larger than $k_B T$ and back reactions can be neglected. In analogy to the treatment of radioactive decay, the flux across the barrier decreases the probability $P(t)$ to find the system in the bound state at time t,

$$\frac{dP(t)}{dt} = -P(t)k_0(t) = -P(t)\omega_0 e^{-\beta(\Delta G^{\ddagger} - kvt\Delta x)} \tag{9}$$

with $P(t = 0) = 1$. Solution of **Eq. 9**,

$$P(t) = \exp\left[\frac{\omega_0}{\beta kv\Delta x} e^{-\beta\Delta G^{\ddagger}} \left(1 - e^{\beta kvt\Delta x}\right)\right] \tag{10}$$

yields a distribution $p(t_D)dt_D = -\left.\frac{dP(t)}{dt}\right|_{t_D} dt_D$ of dissociation times t_D,

$$p(t_D)dt_D = \omega_0 e^{-\beta(\Delta G^{\ddagger} - kvt_D\Delta x)} \exp\left[\frac{\omega_0}{\beta kv\Delta x} e^{-\beta\Delta G^{\ddagger}} \left(1 - e^{\beta kvt_D\Delta x}\right)\right] dt_D \tag{11}$$

and dissociation forces $F_D = kvt_D$, respectively,

$$p(F_D)dF_D = \omega_0 e^{-\beta(\Delta G^{\ddagger} - F_D\Delta x)} \exp\left[\frac{\omega_0}{\beta kv\Delta x} e^{-\beta\Delta G^{\ddagger}} \left(1 - e^{\beta F_D\Delta x}\right)\right] dF_D \tag{12}$$

As the main result, the maximum ($dP(F_D)/dF_D = 0$) denoting the most probable dissociation force, reads

$$F_{\max}(v) = \frac{\Delta G^{\ddagger}}{\Delta x} + \frac{1}{\beta \Delta x} \ln \frac{\beta k v \Delta x}{\omega_0} = \frac{1}{\beta \Delta x} \ln \frac{\beta k v \Delta x}{k_0} \tag{13}$$

(where **Eq. 5** has been used) and thus increases logarithmically with the loading rate kv.

This result links the rupture length Δx to the slope of the loading rate dependent dissociation force, and thus provides the basis for dynamic force spectroscopy (**Fig. 8A**). The larger Δx, the faster the energy barrier is decreased (*see* **Fig. 7**), and the stronger is the effect of the loading rate kv on the dissociation time t_D and dissociation force F_D.

Furthermore, this result serves to rescale dissociation forces from force probe simulations to the much slower time scales of the AFM experiments. For the very fast MD time scales, frictional forces can become relevant (which are negligible in the AFM experiments), and in the simplest approach ***(61)*** can be heuristically included into **Eq. 13**,

$$F_{\max}(v) = \gamma v + \frac{1}{\beta \Delta x} \ln \frac{\beta k v \Delta x}{k_0} \tag{14}$$

where γ is the effective friction coefficient. A more rigorous treatment is given, for example, in **refs. *2,54***.

For more complex energy landscapes as shown in **Fig. 7** (right), the simple two-state model is not applicable. For the two barriers shown, there is a critical pulling force for which the left maximum becomes the global one at the time of dissociation (*see* the thin curve in **Fig 7**), and hence the rupture length Δx jumps to smaller values. Accordingly, the logarithmic slope of the dissociation force increases as shown in **Fig. 8B**. (For a more general treatment, the reader is referred to **refs. *56,57***.)

2. Materials

Most molecular dynamics packages can be used, as listed in Chapter 22. Good force probe implementations are in GROMACS ***(62)***, NAMD ***(63)***, and EGO ***(64)***. Small to medium-sized simulations can be run on Linux boxes; for large scale simulations we recommend Linux clusters (*Beowulf*) or special parallel machines such as the IBM SP Series.

3. Methods

This section lists the steps required to carry out a ligand-unbinding force probe simulation.

1. Select protein–ligand complex and *obtain pdb-structure* from the Protein Data Bank ***(65)***, http://www.rcsb.org/pdb/, if available. If not, find a homologous struc-

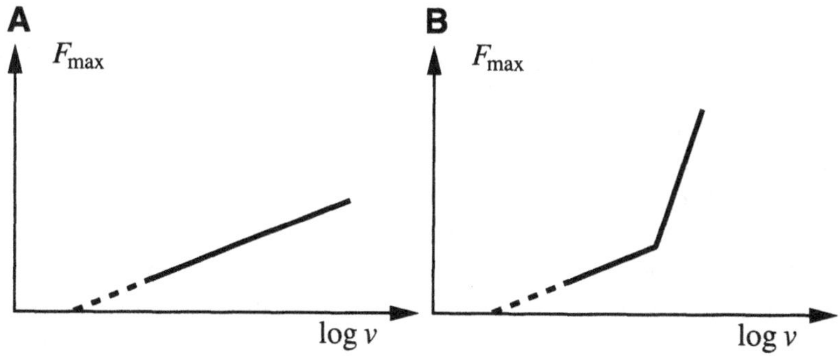

Fig. 8: Dynamic force spectra show the increase of the dissociation force F_{max} with pulling velocity v of the cantilever or, equivalently, loading rate kv. **(A)** For the case of a well-localized energy barrier in $G(x)$, **Fig. 7** (left), the simple two-state model described in the text predicts a logarithmic increase. **(B)** For more complex energy landscapes such as the two barriers sketched in **Fig. 7** (right), the dynamic force spectrum falls into two regimes. For small v, the larger barrier in $G(x)$ dominates, and the resulting large Δx yields a slow increase of the spectrum with v. For large v, however, the (originally) smaller barrier to the left dominates at the time t_D of dissociation, resulting in a steeper logarithmic increase of F_{max} with v.

ture and try homology modeling. This is a risky business, however, since the accuracy of homology models is generally questionable, and the atomic details that are crucial for the ligand-receptor interactions may be wrong. If the ligand is not present in the protein structure, try docking approaches; the same caveats apply here, however, as for homology modeling. Carefully study the Protein Data Bank entry: is the resolution sufficient (i.e., generally better than 2.5 Å)? Are there missing residues that need to be added by modeling? Have special compounds been added to facilitate crystallization that need to be removed? Is the ligand in the crystal the biologically relevant one, or is it an analog that needs to be changed?

2. For the protein, the *force field* is typically implemented in the MD package (*see* **Note 1**, however). For the ligand, often not. In this case, the chemical force field parameters can possibly be derived from the structure or from chemically related amino acid motifs. More challenging are the partial charges, which sensitively depend on the chemical environment. Often, it will be necessary to carry out quantum mechanical calculations to get sufficiently accurate charges (*see* Chapter 24 on density functional (DFT) calculations).
3. Construct the *native environment* for the protein, typically a water box with solvated salt ions for soluble proteins or a solvated lipid bilayer for membrane proteins. There are several tools available for this task, typically included with MD packages, or stand-alone, for example, SOLVATE ***(66)***. The latter takes particu-

lar care of the thermodynamically correct placement of the ions according to a Debye–Hückel distribution. As the ion diffusion is rather slow at the MD time scale, correct placement can be crucial for the system to get close to thermal equilibrium. Pay particular attention to the selected system size: in contrast to conventional MD simulations, the solvation box may have to be chosen a bit larger to leave room for the ligand to move away from the protein without hitting the system boundaries or, via periodic boundary conditions, the protein again from the other side. This is particularly true if, rather than ligand unbinding, protein unfolding is studied with force probe simulations. In this case, step-wise increase of the box size as the protein unfolds, or, alternatively repeated truncation of already unfolded segments can help to keep the simulation system size reasonably small, and thus save computer time.

4. *Minimize and equilibrate* the simulation system as in conventional MD simulations (*see* Chapter 22). To check whether the system is sufficiently equilibrated, just monitoring energies is typically insufficient. Therefore, always monitor equilibration via the root mean square (rms) deviation from the start structure (*see* **Note 3**).
5. Once a satisfactory simulation *start system* has been obtained, *force probe simulations* can be carried out, typically each starting from this equilibrated *start structure*. Crucial parameters are:
   a. *Pull selection*: Which atom or atoms should be subjected to the pulling potential? If a single molecule AFM experiment is to be simulated, a natural choice is that particular atom of the ligand molecule, to which in the AFM experiment the cantilever is linked via a polymer (if known). If the chemistry in the experiment is not known, as a general rule, try various choices and check whether, and to what extent, these affect the results. Do not forget to avoid that the receptor protein is dragged together with the ligand, for example, by immobilization of the protein's center of mass *(see* **Eq. 3**). Do not immobilize each protein atom separately, as this would suppress conformational motions that are often crucial for binding-unbinding of a ligand.
   b. *Pull direction*: In the AFM experiment, the ligand-receptor complex has sufficient time to adapt, by rotation, to the pulling direction set by the AFM. Because of Stokes' friction, such adaptation takes quite a while at the MD time scale, and thus should be avoided by carefully selecting the correct pulling direction right away, typically by visual inspection. If in doubt, do a *prepulling* phase and monitor rigid-body adaptation motions of the system. Start the production runs only after the adaptation motions have stopped.
   c. *Stiffness of the cantilever spring*: This choice depends on the questions to be addressed. If an AFM experiment is to be simulated, choose the effective spring coefficient of the experimental set-up. Note, that just using the spring coefficient of the cantilever may be inadequate, as also other elements may contribute to the effective spring that acts onto the ligand-receptor complex. Typically, this will be the entropic elasticity of the linker polymer(s), which can, for example, be described to sufficient accuracy by the *worm like chain* model ***(67)***. If emphasis is on the unbinding pathway, a soft spring ($k < 0.01$

N/m) can be adequate, as it leaves the ligand the freedom to move as soon as the force has become large enough to overcome the next energy barrier. If, rather the force profile along the unbinding pathway is of interest, a stiff spring (>1 N/m) is required to obtain sufficient spatial resolution.

d. *Pulling velocity or loading rate*: The simple rule is: as slow as possible, given the computational limitations. In fact, the pulling simulations will necessarily, almost always be carried out too fast, that is to say at too short time scales. This is particularly true if AFM or optical tweezers experiments are simulated: whereas those are carried out at a milliseconds to seconds time scale, the simulations fall into the nanoseconds scale, i.e., are many orders of magnitude too fast. To alleviate this discrepancy, one should carry out several force probe simulations with different pulling velocities, and study the variation of the calculated rupture force in order to estimate if the experimental value falls onto this curve (see the previous Theory Section). To change this sometimes a bit risky (unstable) extrapolation into an interpolation, a good trick is to also consider the independently measured spontaneous unbinding rate (k_{off}), which corresponds to vanishing rupture force.

6. Considerable effort has to be spent in *analyzing the obtained force probe simulations* in order to provide a causal picture of the unbinding process and to relate the observed forces to interatomic interactions (*see* **Note 5**). Monitoring interactions like individual hydrogen bonds, water bridges, hydrophobic contacts, salt bridges, electrostatics, or van der Waals contacts, as well as water accessible surface or larger structural rearrangements during unbinding are typical ways to gain microscopic insight. Be careful when relating forces to interaction energies (**Note 6**). No general rules on what observables should be analyzed can be given, though, and one often has to resort to biophysical and biological intuition. Not sufficiently appreciated particularly in the physics community is the fact that from pure inspection of movies (generated from the trajectories), much insight can be gained. In particular, movies often are an invaluable tool to develop ideas for subsequent quantitative analyses.

4. Notes

1. *Histidines and S-bonds*. One cannot overemphasize the necessity to carefully check protonation states—particularly of histidines (possibly, if in doubt, with Poisson–Boltzmann calculations)—and the correct placement of disulfide bonds. Errors here can hide frustratingly long, even until after publication!
2. *Smoothing of force profiles*. Particularly when using stiff pulling potentials V_{cant} the *raw* force profiles as obtained from **Eq. 4** contain both the unbinding forces of interest, but also high frequency contributions arising from the thermal fluctuation of the atom subjected to V_{cant}. Typically, these force fluctuations are much larger than the underlying unbinding forces, and therefore have to be filtered, for example, with Gaussian filters (i.e., by a convolution of the raw force profile with a Gaussian function $\exp(-t^2/2\tau^2)$ of width τ). Good results have been obtained for τ determined from the resonance frequency of V_{cant}, i.e., $\tau = (m/k)^{1/2}$, where m is

the (total) mass of the atom or atoms subjected to V_{cant}. Note that the choice of τ affects the maximum of the force profile, and hence the calculated dissociation force F_{max}. This slight arbitrariness entails an uncertainty for F_{max}.

3. *Equilibration and rms deviation.* Take care when checking if a simulation system is sufficiently equilibrated ***(68)***. Besides the fact that it is often not quite clearly explained what *sufficiently* means here (strictly, it means that the results should not change if the system is equilibrated even much longer, but this does not help much in practice), there is generally no watertight proof that a system is indeed sufficiently equilibrated (*see* also **refs. *69–73***). Thus, do not uncritically interpret any short-time plateau in the root mean square (rms) deviation from the start structure as *system is equilibrated.* If in doubt, extend the equilibration period to see whether the rms really stays constant. Today, equilibration phases shorter than 1 ns are generally considered insufficient for medium-sized proteins. Note, also that larger rms deviations can originate both from systematic drift away from the start structure (i.e., system is not equilibrated) or from large fluctuations (in which case the system can, but need not, be in equilibrium). Careful analysis, for example, by excluding parts of the protein that show large fluctuations from the rms calculation, is mandatory here. Also helpful are residue-based rms-plots that help to trace where a large rms value comes from. Note finally that an rms of, for example, 2.5 Å may be small for a large protein but unacceptably large for a small, rigid protein.
4. *Exerting torque.* In recent experiments, utilizing magnetic beads ***(74)*** torque has been applied to single DNA molecules rather than the linear forces discussed in the main text. This process can be simulated by subjecting *n* selected atoms (with positions $\mathbf{x}_i$) onto which the torque shall act, to a rotating potential V_{torque} ***(13)***,

$$V_{torque} = \frac{1}{2} k \sum_{i=1}^{n} \left[\Omega(\omega t)\, \mathbf{x}_i (t = 0) - \mathbf{x}_i (t) \right]^2 \tag{15}$$

The rotation matrix $\Omega(\omega t)$ has to be specified such that it moves the minima of V_{torque} along concentric circles around the desired rotation axis, and the spring coefficient k defines the size of the allowed fluctuations around the equilibrium positions of the atoms. The torques $\Theta_i(t)$ exerted on the individual atoms are readily computed from the deflection of the actual atomic positions $\mathbf{x}_i(t)$ from the respective minima,

$$\Theta_i (t) = \Omega(\omega t) \mathbf{x}_i (t = 0) \times k \left[\Omega(\omega t) \mathbf{x}_i (t = 0) - \mathbf{x}_i (t) \right] \tag{16}$$

and the total torque exerted on the molecule,

$$\Theta(t) = \sum_{i=1}^{n} \Theta_i (t) \tag{17}$$

is just the sum of the individual torques. This approach has, for example, been used in MD simulations to mimic the torque exerted by the proton–motive force onto the central stalk (γ-subunit) of F_1-ATP synthase, and thereby to study the mechano-chemical energy conversion involved in ATP synthesis ***(13)***.

5. *Anecdotal events.* The huge computational amount of computer time that is required to carry out just a single force probe simulation at sufficiently low loading rate leads into the temptation to extract information on unbinding reaction pathways from just one trajectory. However, in this case one cannot tell which features of the observed unbinding event are reproducible, and which are purely anecdotal. Therefore, with limited computational resources consider carrying out, instead of one very extended simulation, several at somewhat higher pulling speed.
6. *Forces differ from free energies mainly in two respects.* First, the size of forces and that of free energies are, generally, uncorrelated. Without further support, large forces do not necessarily mean large free energy differences (e.g., between bound and unbound state) and vice versa. If distributed over large distances, as in nonlocal or multiple interactions along an extended reaction path (e.g., protein folding), small forces can build up large free energy differences. On the other hand, a very stiff bond does not need much energy to generate large rupture forces. Second, free energy differences ΔG are thermodynamic state functions, and therefore depend only on start and end state but not on the chosen reaction pathway. This is not true for forces, which do depend on the particular reaction pathway. Therefore, forces yield information on pathways whereas binding free energies do not. One consequence is that when analyzing the effect of point mutations, forces are not additive: whereas the binding free energy changes as a result of a double mutation is to a good approximation simply the sum of the binding free energy change of the two single-point mutations, this is not true for the respective dissociation forces *(75)*.

Acknowledgments

The author would like to thank Berthold Heymann for his contributions which are presented here as an example. The author would also like to thank Frauke Gräter and Gunnar F. Schröder for carefully reading the manuscript.

References

1. Grubmüller, H., Heymann, B., and Tavan, P. (1996) Ligand binding: molecular mechanics calculation of the streptavidin-biotin rupture force. *Science* **271,** 997–999.
2. Izrailev, S., Stepaniants, S., Balsera, M., Oono, Y., and Schulten, K. (1997) Molecular dynamics study of unbinding of the avidin-biotin complex. *Biophys. J.* **72,** 1568–1581.
3. Ansari, A., Berendzen, J., Browne, S. F., Frauenfelder, H., Iben, I. E. T., Sauke, T. B., Shyamsunder, E., and Young, R. D. (1985) Protein states and protein quakes. *Proc. Natl. Acad. Sci. USA* **82,** 5000–5004.
4. Neher, E. and Sakmann, B. (1976) Single-channel currents recorded from membrane of denervated frog muscle fibres. *Nature* **260,** 799–802.
5. Sakmann, B. and Neher, E., eds. (1995) *Single-Channel Recording*. 2nd Ed., Plenum Press.

6. Yellen, G. (2002) The voltage-gated potassium channels and their relatives. *Nature* **419,** 35–42.
7. Ansari, A., Berendzen, J., Braunstein, D., Cowen, B. R., Frauenfelder, H., Hong, M. K., Iben, I. E. T., Johnson, J. B., Ormos, P., Sauke, T. B., Scholl, R., Schulte, A., Steinbach, P. J., Vittitow, J., and Young, R. D. (1987) Rebinding and relaxation in the myoglobin pocket. *Biophys. Chem.* **26,** 337–355.
8. Elber, R. and Karplus, M. (1987) Multiple conformational states of proteins: a molecular dynamics analysis of myoglobin. *Science* **235,** 318–321.
9. Berendsen, H. J. C. (1996) Bio-molecular dynamics comes of age. *Science* **271,** 954–955.
10. Fermi, E., Pasta, J., and Ulam, S. (1955) *Studies in Nonlinear Problems, I. Los Alamos Report 1940*, Los Alamos, LA .
11. de Groot, B. L. and Grubmüller, H. (2001) Water permeation across biological membranes: Mechanism and dynamics of aquaporin-1 and GlpF. *Science* **294,** 2353–2357.
12. Marrink, S. J., Lindahl, E., Edholm, O., and Mark, A. E. (2001) Simulation of the spontaneous aggregation of phospholipids into bilayers. *J. Am. Chem. Soc.* **123,** 8638–8639.
13. Böckmann, R. and Grubmüller, H. (2002) Nanoseconds molecular dynamics simulation of primary mechanical energy transfer steps in F_1-ATP synthase. *Nature Struct. Biol.* **9,** 198–202.
14. Colombo, G., Roccatano, D., and Mark, A. E. (2002) Folding and stability of the three-stranded beta-sheet peptide betanova: insights from molecular dynamics simulations. *Proteins* **46,** 380–392.
15. Aksimentiev, A., Balabin, I. A., Fillingame, R. H., and Schulten, K. (2004) Insights into the molecular mechanism of rotation in the F_o sector of ATP synthase. *Biophys. J.* **86,** 1332–1344.
16. McCammon, J. A., Gelin, B. R., and Karplus, M. (1977) Dynamics of folded proteins. *Nature* **267,** 585–590.
17. van Gunsteren, W. F. and Berendsen, H. J. C. (1977) Algorithms for macromolecular dynamics and constraint dynamics. *Molec. Phys.* **34,** 1311–1327.
18. Berendsen, H. J. C. (2001) Reality simulation—observe while it happens. *Science* **294,** 2304–2305.
19. van Gunsteren, W. F., Bakowies, D., Bürgi, R., Chandrasekhar, I., Christen, M., Daura, X., Gee, P., Glättli, A., Hansson, T., Oostenbrink, C., Peter, C., Pitera, J., Schuler, L., Soares, T., and Yu, H. (2001) Molecular dynamics simulation of biomolecular systems. *Chimia* **55,** 856–860.
20. T. Hansson, C. O. and van Gunsteren, W. (2002) Molecular dynamics simulations. *Curr. Opin. Struct. Biol.* **12,** 190–196.
21. Karplus, M. and McCammon, J. A. (2002) Molecular dynamics simulations of biomolecules. *Nature Struct. Biol.* **9,** 646–652.
22. Pauling, L., Corey, R. B., and Branson, H. R. (1951) The structure of proteins: two hydrogen-bonded helical configurations of polypeptide chain. *Proc. Natl. Acad. Sci. USA* **37,** 205.

23. Schröder, G. F. (2000) *Molekulardynamiksimulation der Flexibilität und Fluoreszenzanisotropie eines an ein Protein gebundenen Farbstoffs*. Diploma Thesis, Universität Göttingen, Germany.
24. van Gunsteren, W. F. and Berendsen, H. J. C. (1990) Computer simulation of molecular dynamics: methodology, applications, and perspectives in chemistry. *Angew. Chem. Int. Ed.* **29,** 992–1023.
25. Parrinello, M. (2000) Simulating complex systems without adjustable parameters. *IEEE Comput. Sci. Eng.* **2,** 22–27.
26. Tuckerman, M. E. and Martyna, G. J. (2000) Understanding modern molecular dynamics: techniques and applications. *J. Phys. Chem. B.* **104,** 159–178.
27. van Gunsteren, W. F., Weiner, P. K., and Wilkinson, A. J., eds. (1989–1997) *Computer Simulation of Biomolecular Systems: Theoretical and Experimiental Applications*, vol. 1–3, Escom, Leiden, The Netherlands.
28. Binnig, G., Quate, C. F., and Gerber, C. H. (1986) Atomic force microscope. Phys. Rev. Lett. **56,** 930–933.
29. Florin, E.-L., Moy, V. T., and Gaub, H. E. (1994) Adhesion forces between individual ligand-receptor pairs. *Science* **264,** 415–417.
30. Lee, G. U., Kidwell, D. A., and Colton, R. J. (1994) Sensing discrete streptavidin-biotin interactions with atomic force microscopy. *Langmuir* **10,** 354–357.
31. Finer, J. T., Simmons, R. M., and Spudich, J. A. (1994) Single myosin molecule mechanics: piconewton forces and nanometer steps. *Nature* **368,** 113–119.
32. Yin, H., Wang, M. D., Svoboda, K., Landick, R., Block, S. M., and Gelles, J. (1995) Transcription against an applied force. *Science* **270,** 1653–1657.
33. Grubmüller, H. (1995) Predicting slow structural transitions in macromolecular systems: conformational flooding. *Phys. Rev. E.* **52,** 2893.
34. Müller, E. M., de Meijere, A., and Grubmüller, H. (2002) Predicting unimolecular chemical reactions: Chemical flooding. *Biophys. J.* **116,** 897–905.
35. Rief, M., Gautel, M., Oesterhelt, F., Fernandez, J. M., and Gaub, H. E. (1997) Reversible unfolding of individual titin immunoglobulin domains by AFM. *Science* **276,** 1109–1112.
36. Carrion-Vazquez, M., Oberhauser, A. F., Fowler, S. B., Marszalek, P. E., Broedel, S. E., Clarke, J., and Fernandez, J. M. (1999) Mechanical and chemical unfolding of a single protein: A comparison. *Proc. Natl. Acad. Sci. USA* **96,** 3694–3699.
37. Rief, M. and Grubmüller, H. (2002) Force spectroscopy of single biomolecules. *Chem. Phys. Chem.* **3,** 255–261.
38. Lu, H., Isralewitz, B., Krammer, A., Vogel, V., and Schulten, K. (1998) Unfolding of titin immunoglobulin domains by steered molecular dynamics simulation. *Biophys. J.* **75,** 662–671.
39. Marszalek, P., Lu, H., Li, H., Carrion-Vazquez, M., Oberhauser, A. F., Schulten, K., and Fernandez, J. M. (1999) Mechanical unfolding intermediates in titin modules. *Nature* **402,** 100–103.
40. Paci, E. and Karplus, M. (1999) Forced unfolding of fibronectin type 3 modules: an analysis by biased molecular dynamics simulations. *J. Molec. Biol.* **288,** 441–459.

41. Lu, H. and Schulten, K. (2000) The key event in force-induced unfolding of titin's immunoglubin domains. *Biophys. J.* **79,** 51–65.
42. Best, R. B., Li, B., Steward, A., Daggett, V., and Clarke, J. (2001) Can non-mechanical proteins withstand force? Stretching barnase by atomic force microscopy and molecular dynamics simulation. *Biophys. J.* **81,** 2344–2356.
43. Paci, E., Caflisch, A., Plückthun, A., and Karplus, M. (2001) Forces and energetics of hapten-antibody dissociation: A biased molecular dynamics simulation study. *J. Molec. Biol.* **314,** 589–605.
44. Fowlere, S. B., Best, R. B., Herrera, J. L. T., Rutherford, T. J., Steward, A., Paci, E., Karplus, M., and Clarke, J. (2002) Mechanical unfolding of a titin Ig domain: Structure of unfolding intermediate revealed by combining AFM, molecular dynamics simulations, NMR and protein engineering. *J. Mol. Biol.* **322,** 841–849.
45. Rief, M., Oesterhelt, F., Heymann, B., and Gaub, H. E. (1997) Single molecule force spectroscopy reveals conformational change in polysaccharides. *Science* **275,** 1295–1297.
46. MacKerell, A. D. and Lee, G. U. (1999) Structure, force, and energy of a double-stranded DNA oligonucleotide under tensile loads. *Europ. Biophys. J.* **28,** 415–426.
47. Heymann, B. and Grubmüller, H. (1999) Elastic properties of poly(ethylene-glycol) studied by molecular dynamics stretching simulations. *Chem. Phys. Lett.* **307,** 425–432.
48. Gullingsrud, J., Braun, R., and Schulten, K. (1999) Reconstructing potentials of mean force through time series analysis of steered molecular dynamics simulations. *J. Comp. Phys.* **151,** 190–211.
49. Park, S., Khalili-Araghi, F., Tajkhorshid, E., and Schulten, K. (2003) Free energy calculation from steered molecular dynamics simulations using Jarzynski's equality. *J. Chem. Phys.* **119,** 3559–3566.
50. Park, S. and Schulten, K. (2004) Calculating potentials of mean force from steered molecular dynamics simulations. *J. Chem. Phys.* **120,** 5946–5961.
51. Isralewitz, B., Gao, M., and Schulten, K. (2001) Steered molecular dynamics and mechanical functions of proteins. *Curr. Opin. Struct. Biol.* **11,** 224–230.
52. Norberg, J. and Nilsson, L. (2003) Advances in biomolecular simulations: methodology and recent applications. Q. Rev. Biophys. **36,** 257–306.
53. Bell, G. I. (1978) Models for specific adhesion of cells to cells. *Science* **200,** 618–627.
54. Evans, E. and Ritchie, K. (1997) Dynamic strength of molecular adhesion bonds. *Biophys. J.* **72,** 1541–1555.
55. Seifert, U. (2000) Rupture of multiple parallel molecular bonds under dynamic loading. *Phys. Rev. Lett.* **84,** 2750–2753.
56. Heymann, B. and Grubmüller, H. (2000) Dynamic force spectroscopy of molecular adhesion bonds. *Phys. Rev. Lett.* **84,** 6126–6129.
57. Hummer, G. and Szabo, A. (2003) Kinetics from nonequilibrium single-molecule pulling experiments. *Biophys. J.* **85,** 5–15.
58. Eyring, H. (1935) The activated complex in chemical reactions. *J. Chem. Phys.* **3,** 107–115.
59. Kramers, H. A. (1940) Brownian motion in a field of force and the diffusion model of chemical reactions. *Physica (Utrecht)* **VII,** 284–304.

60. Hänggi, P., Talkner, P., and Borkovec, M. (1990) Reaction-rate theory: fifty years after Kramers. *Rev. Mod. Phys.* **62,** 251–341.
61. Heymann, B. and Grubmüller, H. (1999) AN02/DNP unbinding forces studied by molecular dynamics AFM simulations. *Chem. Phys. Lett.* **303,** 1–9.
62. Lindahl, E., Hess, B., and van der Spoel, D. (2001) GROMACS 3.0: a package for molecular simulation and trajectory analysis. *J. Mol. Model.* **7,** 306–317.
63. Nelson, M., Humphrey, W., Gursoy, A., Dalke, A., Kalé, L., Skeel, R. D., and Schulten, K. (1995) NAMD–a parallel, object-oriented molecular dynamics program. Technical report, Beckman Institute.
64. Eichinger, M., Heller, H., and Grubmüller, H. (2000) EGO—An efficient molecular dynamics program and its application to protein dynamics simulations, in: *Workshop on Molecular Dynamics on Parallel Computers, John von Neumann Institute for Computing (NIC) Research Centre Jülich, Germany, 8–10 February 1999,* (Esser, R., Grassberger, P., Grotendorst, J., and Lewerenz, M., eds.), pp. 154–174, World Scientific, Singapore 912805.
65. Berman, H. M., Westbrook, J., Feng, Z., Gilliland, G., Bhat, T. N., Weissig, H., Shindyalov, I. N., and Bourne, P. E. (2000) The protein data bank. *Nucleic Acids Res.* **28,** 235–242.
66. Grubmüller, H. (1996) Solvate: a program to create atomic solvent models. http://www.mpibpc.gwdg.de/abteilungen/070/solvate.html.
67. Bustamante, C., Marko, J. F., Siggia, E. D., and Smith, S. (1994) Entropic elasticity of lambda-phage DNA. *Science* **265,** 1599–1600.
68. Hess, B. (2000) Similarities between principal components of protein dynamics and random diffusion. *Phys. Rev. E.* **62,** 8438–8448.
69. Daura, X., Jaun, B., Seebach, D., van Gunsteren, W. F., and Mark, A. E. (1998) Reversible peptide folding in solution by molecular dynamics simulation. *J. Molec. Biol.* **280,** 925–932.
70. Daura, X., van Gunsteren, W. F., and Mark, A. E. (1999) Folding-unfolding thermodynamics of a β-heptapeptide from eqilibrium simulations. *Proteins: Structure, Function, and Genetics* **34,** 269–280.
71. Hamprecht, F. A., Peter, C., Daura, X., Thiel, W., and van Gunsteren, W. F. (2001) A strategy for analysis of (molecular) equilibrium simulations: configuration space density estimation, clustering, and visualization. *J. Chem. Phys.* **114,** 2079–2089.
72. de Groot, B. L., Daura, X., Mark, A. E., and Grubmüller, H. (2001) Essential dynamics of reversible peptide folding: Memory-free conformational dynamics governed by internal hydrogen bonds. *J. Molec. Biol.* **309,** 299–313.
73. L.J. Smith, X. D., and van Gunsteren, W. (2002) Assessing equilibration and convergence in biomolecular simulations. *Proteins* **48,** 487–496.
74. Bryant, Z., Stone, M. D., Gore, J., Smith, S. B., Cozzarelli, N. R., and Bustamante, C. (2003) Structural transitions and elasticity from torque measurements on DNA. *Nature* **424,** 338–341.
75. Heymann, B. and Grubmüller, H. (2001) Molecular dynamics force probe simulations of antibody/antigen unbinding: entropic control and non-additivity of unbinding forces. *Biophys. J.* **81,** 1295–1313.

24

Study of Ligand–Protein Interactions by Means of Density Functional Theory and First-Principles Molecular Dynamics

Carme Rovira

Summary

Density Functional Theory (DFT) is a promising technique to study protein–ligand interactions from an atomistic–electronic point of view. It provides information on the electronic rearrangements upon ligand binding, the structure and the relative energy of the ligand in the binding pocket, among other properties. In addition, DFT-based techniques such as first-principles molecular dynamics (FPMD) (e.g., the Car–Parrinello [CP] method) are used to simulate the short-time dynamics of ligand–protein interactions. These techniques are emerging as a useful tool to decipher complex protein–ligand interactions in which chemical bonds are formed and/or broken during the binding process.

In this chapter, the basis of DFT, its limitations, and current developments of the theory are discussed, focusing on its applications in the area of ligand–protein interactions. The performance of the method is illustrated with three examples in which the ligand binding process induces changes in the spin state or in the protonation state of the active species. The first two examples deal with the binding of oxygen to the active center of myoglobin, whereas the third one describes the binding of a formic acid inhibitor in the active center of catalase.

Key Words: Density functional theory; molecular dynamics; Car–Parrinello molecular dynamics; *ab initio* molecular dynamics; protein–ligand interactions; oxyheme; catalases.

1. Introduction

Ligand–protein interactions are at the basis of many fundamental biological processes such as enzymatic reactions and molecular recognition. In many cases, the interaction of a ligand with a protein involves complex electronic

From: *Methods in Molecular Biology, vol. 305: Protein–Ligand Interactions: Methods and Applications*
Edited by: G. U. Nienhaus © Humana Press Inc., Totowa, NJ

reorganizations of the protein target and/or the ligand. These interactions, in which the bonding pattern changes qualitatively in the course of the binding process, are the focus of this chapter. For instance, when oxygen binds to myoglobin (Mb), the active center changes its electronic configuration from a high-spin (i.e., maximum number of unpaired electrons) to low-spin state, at the same time that the bond between the iron atom and the oxygen ligand develops *(1,2)*. Reactive processes occurring in the active site of enzymes are also complex processes. For instance, the decomposition of the superoxide radical (O_2^-) into hydrogen peroxide and oxygen by superoxide dismutases involves changes in the coordination state of the active species as well as in their oxidation states *(3)*. Deciphering these processes from an electronic point of view is necessary for both understanding the mechanisms behind the ligand-binding interactions, and for the design of small molecules able to affect the biological function of the protein.

The way to approach complex ligand–protein interactions from a theoretical point of view usually rely on quantum–chemistry techniques, where the electronic variables are considered as active degrees of freedom. Among all quantum chemistry techniques, Density Functional Theory (DFT) is often the method of choice because of its good relation between accuracy and computational cost. In a few words, DFT provides a way to obtain the electron density and the ground state energy of a polyatomic system given its atomic coordinates *(4,5)*. Most programs based on DFT are capable to search for energy minima and compute several molecular properties such as atomic charges, multipole moments, vibrational frequencies, and spectroscopic constants. DFT is also the basis of first principles molecular dynamics (FPMD) techniques such as the Car–Parrinello (CP) method, in which the molecules evolve in real time and finite temperature under the instantaneous ground state of the electron cloud *(6)*. Since the electronic density changes during the simulation, polarization effects are described in a natural way, as well as changes in the bonding pattern of the atoms (e.g., bond breaking and bond-forming processes).

However, DFT methods cannot be applied to all possible biological problems. Just as any theoretical approach, they have limitations. One of them is that multiplet states are not as well defined as in multiconfigurational *ab initio* methods, and that the theory is only strictly valid for nondegenerate ground states *(7,8)*. Nevertheless, recent developments such as Restricted Open Shell Kohn–Sham (ROKS) *(7,9,10)* and Time Dependent Density Functional Theory (TDDFT) *(11)* permit the study of excited-state processes such as photochemical reactions. Another limitation of DFT is the poor description of very weak interactions (e.g., van der Waals). This is not a limitation of the theory itself, but a consequence of the commonly used approximations (local density approximation [LDA] and generalized gradient corrections [GGA], which will be described in **Subheadings 1.2.2.** and **1.2.3.**). Several schemes are being developed to overcome this limitation ***(12)***, but they are still far from being applicable to realistic systems. Currently, the only practical way to solve this problem is to add an empirical term to the DFT energy ***(13)***. Other limitations that one faces when using DFT are the small size of the systems that can be investigated (up to ≈ 150 atoms in routine works, although benchmark calculations on systems of ≈ 1200 atoms, such as DNA in solution, have been performed) ***(14)*** and the short times that can be simulated in FPMD (up to tenths of picoseconds, depending on the physics of the system). Several techniques aimed to improve these aspects have been developed, such as hybrid quantum mechanics/molecular mechanics (QM/MM) methods (*see* for instance **refs. *15*** and ***16***), linear scaling methods, and schemes that allow overcoming energy barriers ***(18)***. Moreover, the fast development of computer power will most likely triplicate the size of the systems that can be investigated and the simulation times in just a few years.

1.1. Examples of Applications of DFT in the Area of Ligand–Protein Interactions

The range of applications of DFT to protein–ligand interactions has expanded considerably in recent years as more powerful computers are available and more efficient DFT-based programs have been developed. The scope of this chapter is not to carry out an exhaustive review of the growing literature on this field (*see* **refs. *19–22*** for recent reviews of DFT applications to problems of biological interest), but rather to mention some illustrative examples and provide detailed information on how to address the study of ligand–protein interactions using DFT. Our concept of *ligand* will be restricted to molecules or ions that bind to a protein and affect its function, i.e., we will be dealing with ligands being conceptually separated from the protein. The interaction of metal ions with its protein host ***(21,23)*** or the interaction of cofactors with a protein ***(24)*** will not be considered here.

An intense area of work in the field of protein–ligand interactions is the study of substrate–enzyme interactions. The DFT approach to these problems typically relies on a model system consisting of a fragment of the protein that includes the chemically active residues close to the binding region. For instance, the binding of oxygen to the active site of methane monooxygenase, the enzyme responsible for the conversion of methane into methanol, was modeled using a 36-atom fragment involving the protein residues covalently linked to the two metal atoms of the active site ***(25)***. These are two histidines and four glutamic acid residues. The model was further simplified by replacing the bound His and Glu by NH_2 and carboxylate groups, respectively. Another example is the study of the reaction mechanism of HIV-1 protease by means of DFT-based molecular dynamics (MD) (Car–Parrinello). Using a model consisting of 60 atoms it was shown how the protein brings the substrate into a suitable geometrical position that favors a concerted proton transfer among the catalytic residues and the substrate *(26)*.

The binding and permeation of ions in membrane proteins is starting to be studied by means of DFT. In this respect, calculations on a fragment of the selectivity filter of the potassium channel KcsA provided insight into details of the coordination chemistry and polarization induced by the K^+ ion during the permeation process *(27)*. The structure and dynamics of proton diffusion through a polyglycine analog of the gramicidin A ion channel was also studied by means of DFT-based MD (Car–Parrinello) *(28)*. The simulation showed that transient $H_5O_2^+$ species are formed during the diffusion process. Protonated water species in the interior of other transmembrane proton punps have also been recently modeled *(29)*.

One classical example of ligand binding is the interaction of molecular oxygen with the hemeproteins hemoglobin and myoglobin *(1,2)*. A longstanding problem in this area concerns the origin of the protein discrimination for CO, i.e., the reason why the CO/O_2 affinity ratio is lower in the protein with respect to synthetic heme models. For many years, it was assumed that the protein weakens the heme–CO bond by distorting it with respect to its linear optimum structure *(1,30)*. In fact, most X-ray studies reported a distorted Fe–CO bond in the protein, but a linear bond in synthetic analogues *(31)*. However, DFT calculations demonstrated that the Fe–CO bond is very robust and that the energetic cost of small distortions is marginal *(32–35)*. Together with spectroscopic studies *(36)* and critical revisions of the X-ray data **(30)**, the steric hypothesis was excluded from being responsible of the protein discrimination for CO. The binding and dynamics of the heme–O_2 interaction and its consequences on the ^{17}O nuclear magnetic resonance (NMR) chemical shift tensors have also been analyzed with DFT, showing that the bound ligand undergoes a rotational motion around the iron–ligand bond in the picosecond timescale *(37,38)*. Another relevant problem in this area is the mechanism of ligand recombination in the heme pocket. Recent calculations showed that the interplay of spin states in this process determines the ligand-rebinding rate *(39)*. The binding of ligands in other hemeproteins such as catalases, peroxidases, guanylate cyclase, and cytochromes is also actively investigated with DFT *(40–42)*. For instance, a recent study on the interaction of formic acid with Heliobacter Pylori catalase provides insight into the conformation of the bound ligand when it is trapped between the catalytic residues *(42)*. Practical details of this problem, as well as that of the binding of O_2 to myoglobin, will be described in **Subheading 3.3.**

Although the fragment approximation is very popular, some problems cannot be properly addressed without the explicit treatment of the protein environment. For instance, subtle variations in the stretching frequencies of the ligands bound to Mb depend on changes on the protonation state of protein residues far from the active center ***(43)***. Mixed QM/MM techniques based on DFT are emerging as a very useful tool to study biological problems without having to rely on a model fragment ***(44)***. In the mixed QM/MM approach, the system is partitioned into a chemically active region, which is treated with DFT (e.g., the ligand and the active center) and the rest of the protein, which is treated with empirical potentials. The most delicate issue of QM/MM methods is the treatment of the coupling between the QM and MM regions ***(45,46)***. Several methods have been proposed to saturate the QM region, such as the link–atom approach or the use of monovalent pseudopotentials. None of these methods can fully recover the properties of the original bond and their success is only guaranteed when the QM/MM frontier is far enough from the chemically active site. Nevertheless, several problems in the field of ligand–protein interactions have been successfully solved using QM/MM techniques (*see* for instance **refs. *47–49***).

In this chapter, we will illustrate the performance of DFT in describing protein–ligand interactions. Three examples have been chosen in which small models are used to investigate electronic aspects of a ligand–protein interaction and its picosecond dynamics. The first two examples deal with the binding of oxygen to the active center of myoglobin (the first example focuses on the binding process, while the second one deals with the dynamics of the bound ligand). The third example describes the binding of a formic acid inhibitor in catalase. These problems will be analyzed by means of DFT calculations at fixed structure (i.e., single-point calculations) and DFT-based molecular dynamics within the Car–Parrinello approach. Some parts of the procedure are common to the three examples, and therefore will be described in detail only for example 1. We will refer to the CPMD program for the practical details of the calculations ***(50)***, but similar calculations can be performed with a number of other DFT programs (*see* **Note 1**).

1.2. Density Functional Theory

1.2.1. Basic Equations

DFT provides a framework to obtain the total energy of a polyatomic system given their atomic coordinates. The development of DFT in the area of computational chemistry dates from the mid 1960s when Hohenberg and Kohn ***(51)*** demonstrated that the ground-state energy of a system of interacting electrons subject to an external potential $V(\vec{r})$ is a unique functional of the electron density, i.e. and it can be obtained by minimizing the energy functional with respect to the density, $E = E\,[\rho(\vec{r}\,)]$

$$E^{DFT} = \min_{\rho(\vec{r})} E\left[\rho(\vec{r})\right] \tag{1}$$

Later, Kohn and Sham *(52)* demonstrated that there is an equivalence between the electronic density of this system (our *real system*) and that of a *model system* of noninteracting electrons which are subjected to an effective potential, V_{eff}. This provided a way to solve the problem of finding the density of the many-electron interacting system, via obtaining the electron density of the noninteracting system. This density can be expressed in terms of single-electron orbitals $\psi_i(\vec{r}\,)$, known as Kohn–Sham (KS) orbitals,

$$\rho(\vec{r}) = 2\sum_{i}^{occ.} \left|\psi_i(\vec{r})\right|^2 \tag{2}$$

where the sum extends over the occupied single-particle orbitals (here, we restrict to the most simple situation in which all orbitals are doubly occupied). Because of the relation **Eq. 2**, the energy functional can be either expressed in terms of the density (**Eq. 1**) or the single-electron orbitals,

$$E^{DFT} = \min_{\{\psi_i\}} E^{KS}\left[\{\psi_i(\vec{r})\},\{\vec{R}_N\}\right] \tag{3}$$

The energy functional **(Eq. 1)** can be written as:

$$E^{KS} = 2\sum_{i}^{occ.} \int \psi_i^*(\vec{r})\left(-\frac{\nabla^2}{2}\right)\psi_i(\vec{r})\,d\vec{r} + \int V(\vec{r})\rho(\vec{r})\,d\vec{r} + \frac{1}{2}\int \frac{\rho(\vec{r})\rho(\vec{r}\,')}{|\vec{r}-\vec{r}\,'|}\,d\vec{r}\,d\vec{r}\,' + E_{xc}\left[\rho(\vec{r})\right] \tag{4}$$

The first term in the right-hand side of this expression is the kinetic energy of the noninteracting electrons. The second term corresponds to the interaction of the electrons with the nuclear charges and $V(\vec{r}\,)$ is the potential as a result of the nuclei. In case only valence electrons are explicitly considered in the calculation, $V(\vec{r}\,)$ would be a pseudopotential. The third term corresponds to the classical Coulomb interaction of a density distribution ρ. The fourth term, $E_{xc}\,[\rho(\vec{r}\,)]$, is a functional of the density that accounts for the remaining contributions to the electron–electron interaction.

1.2.2. Local Density Approximation

All terms in **Eq. 4** can be calculated exactly, except $E_{xc}\,[\rho(\vec{r}\,)]$ for which DFT does not provide an explicit form. The theory only demonstrates that a universal expression for it exists, $E_{xc}\left[\rho(\vec{r})\right] = \int \rho(\vec{r})\varepsilon_{xc}\left[\rho(\vec{r})\right]d\vec{r}$. Usually, $E_{xc}\,[\rho(\vec{r}\,)]$ is taken as the exchange and correlation energy of a uniform electron gas, which is precisely known. This is the basis of the so-called local density approximation (LDA). In this approximation it is assumed that the exchange and correlation energy of an electron at a point depends on the density at that point instead of the density at all points in the space ***(53)***. One of the main drawbacks of LDA is that van der Waals interactions, which originate from correlated motions of electrons caused by Coulomb interactions between distant atoms, cannot be properly described. Therefore, special care should be taken when addressing problems in which van der Waals interactions might play a relevant role, such as stacking interactions between π-systems and the diffusion of ligands in purely hydrophobic cavities ***(54)***.

An extension of the LDA to unrestricted cases or open-shells systems (i.e., electronic configurations in which electrons are not paired) leads to the local spin-density approximation (LSD). In this case, not only the total density ρ, but also the electron densities of the electrons with spin α and β (ρ_α and ρ_β, respectively) are employed in the formulation ***(53)***. For instance, the exchange-correlation energy is expressed as

$$E_{xc}^{LSD}\left[\rho_\alpha(\vec{r}),\rho_\beta(\vec{r})\right]=\int\rho(\vec{r})\varepsilon_{xc}\left[\rho_\alpha(\vec{r}),\rho_\beta(\vec{r})\right]d\vec{r}$$

A useful property to describe where α and β electrons are localized in a given system (a molecule, a molecule–ligand complex, solid, etc.) is the distribution of the spin density, i.e., the difference $\rho_\alpha(\vec{r}) - \rho_\beta(\vec{r})$. For a system in which all electrons are paired (e.g., a closed-shell system) the spin density is zero at all points in space. However, any system with unpaired electrons will show regions of nonvanishing spin density. The integral of the spin density over all space (i.e., $\int_{\vec{r}}\left[\rho_\alpha(\vec{r})-\rho_\beta(\vec{r})\right]d\vec{r}$) gives the total number of unpaired electrons (i.e., zero for a singlet state, one for a doublet state, two for a triplet state, etc.).

1.2.3. Generalized Gradient Approximation

The accuracy provided by the local (spin) density approximation is not enough for most applications in chemistry and biology. One of its main drawbacks is that bond distances and binding energies can have large errors that appear in a nonsystematic way. This represents a serious problem for the study of ligand–protein interactions.

A step forward with respect to LDA are the so-called generalized gradient approximation GGA. This approach is based on using not only the density, but also the gradient of the density $\nabla\rho(\vec{r})$ in the functional expression ***(55)*** in order to account for the nonhomogeneity of the true electron density. The functional can be generically written as,

$$E_{xc}^{GGA}\left[\rho_\alpha,\rho_\beta\right]=\int f\left(\rho_\alpha,\rho_\beta,\nabla\rho_\alpha,\nabla\rho_\beta\right)d\vec{r}$$

Several forms for the explicit dependence of the integrand *f* on the densities and their gradients have been proposed, including semiempirical functionals that contain parameters that have been calibrated against reference values, usually using experimental data. In practice, E_{xc}^{GGA} is usually split in two terms corresponding to its exchange and correlation contributions (i.e., $E_{xc}^{GGA} = E_{x}^{GGA} + E_{c}^{GGA}$) and separate forms for each term is provided. Among the most popular GGA exchange and correlation functionals used in biological applications are the ones denoted as BP86 (exchange part by Becke *[55]* and correlation by Perdew *[56]*), BLYP (combination of Becke exchange and correlation developed by Li, Yang and Parr *[57]*), PBE (developed by Perdew, Burke and Ernzerhof in 1996 *[58]*). Hybrid functionals, which includes to some extent *exact exchange energy* in the functional expression, are also widely used. One of the most popular is B3LYP (exact exchange developed by Becke *[59]*, combined with the LYP correlation functional).

The use of the GGA approximation improves considerably the description of bonding (and especially hydrogen bonding) with respect to pure LDA with a very low additional computational cost. The description of weak van der Waals interactions, however, remains problematic. Most of the applications of DFT to systems of biological interest use the GGA approximation (*see* **refs. *19–21*** for recent reviews).

1.2.4. Kohn–Sham Equations

The single electron orbitals $\psi_i(\vec{r})$ of **Eq. 2** and **Eq. 3** can be obtained by solving the following single-particle equations known as Kohn–Sham equations ***(52)***,

$$\left(\underbrace{-\frac{\nabla^2}{2} + V(\vec{r}) + \int d\vec{r}' \frac{\rho(\vec{r})}{|(\vec{r}) - r'|} + V_{xc}(\vec{r})}_{V_{eff}} \right) \psi_i(\vec{r}) = \varepsilon_i \psi_i(\vec{r}) \tag{5}$$

where ε_i are the eigenvalues of the matrix of Lagrange multipliers and are called the Kohn–Sham eigenvalues or Kohn–Sham orbital energies. $V_{xc}(\vec{r})$ is the exchange-correlation potential,

$$V_{xc}(\vec{r}) = \frac{\delta E_{xc}[\rho(\vec{r})]}{\delta \rho(\vec{r})}$$

The KS equations can be solved iteratively given an initial guess for the set of single electron orbitals $[\psi_i(\vec{r})]$. Alternatively, the total energy (**Eq. 3**) can be minimized with respect to the $V_{xc}(\vec{r})$ using gradient search techniques ***(60)***.

In summary, DFT provides a framework to find the total energy of a many-electron interacting system by means of solving the one-electron equations of a model noninteracting system that shares the same density. Based on the generalized gradient corrections approximation and choosing a suitable exchange-correlation potential, many problems of chemistry, physics, and biology can be addressed (*see* **refs.** ***7,19–22*** for recent reviews).

1.2.5. Basis Sets

To solve numerically the KS equations, the KS orbitals are expanded in a basis set,

$$\psi_i(\vec{r}) = \sum_j c_j \, \phi_{ij}(\vec{r}) \tag{6}$$

This expansion should in principle extend to infinity, but it is generally truncated so that only a limited set of basis functions is used. In the chemistry community, Gaussian functions are very popular:

$$\psi_i(\vec{r}) = \sum_j c_j \, e^{-\alpha_j r^2} \tag{7}$$

Several notations are used to specify a particular set of atomic Gaussian functions, such as Pople's split valence basis sets [e.g., 3–21G, 6–31G* or 6–311++G(2d,2p)] or the *correlation-consistent polarized valence N-zeta* basis sets of Dunning (*N* = double, triple, etc., e.g., cc-pVDZ, cc-pVTZ), among others ***(61)***. These notations usually depend on the number of functions representing every atomic angular momentum and the spread of the gaussian function, which is given by the exponent value α (large/small α values result in compact/diffuse functions).

In the physics community, plane waves (PW) are commonly used to expand the KS orbitals,

$$\psi_i(\vec{r}) = \frac{1}{\Omega^{1/2}} \sum_{G}^{G_{\max}} c_G \, e^{i\vec{G}\cdot\vec{G}} \tag{8}$$

where Ω is the volume of the cell and G is the plane wave momentum. *PW* basis sets are denoted by an energy value E_{cut}, which is related to the maximum G value of the *PW* expansion, $G_{\max}$). The number of plane waves N_{PW} can be approximated as

$$N_{PW} \approx \frac{\Omega}{6\pi^2} E_{\text{cut}}^{3/2}$$

*PW*s are not centered at the atoms but extend throughout all the space. In order to reduce the large number of *PW*s necessary to achieve a reliable description of the KS orbitals, the effect of the core electrons is usually described with pseudopotentials acting only on the valence electrons ***(62)***.

The accuracy of a DFT calculation does not depend on the type of basis set used (Gaussian, *PW*, or another), provided that the expansion is complete enough to describe the relevant properties of the system under investigation.

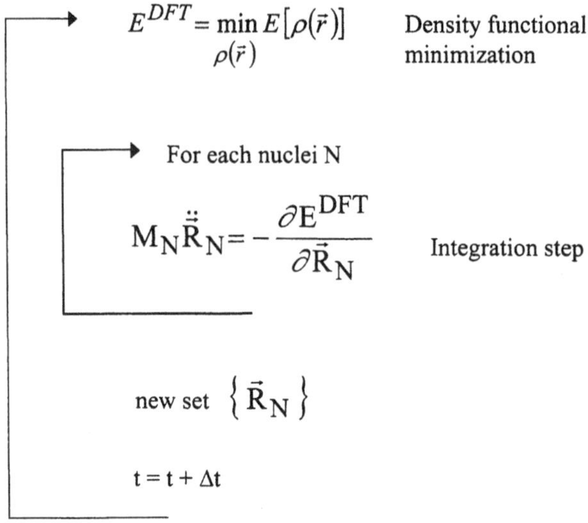

Fig. 1. Schematic diagram of a first-principles molecular dynamics simulation (FPMD).

1.2.6. First-Principles Molecular Dynamics

FPMD is a powerful technique for the study of protein–ligand interactions at an atomic–electronic level ***(63)***. It can be viewed as a series of DFT calculations (*see* **Note 2**) at different instants of time, each one for a different set of atomic positions $\{ \vec{R}_N \}$. These atomic positions are related by the Newton's equations of motion (e.o.m.),

$$M_N \ddot{\vec{R}}_N = -\frac{\partial E_{el}}{\partial \vec{R}_N} \tag{9}$$

which can be derived from the Lagrangian:

$$\mathcal{L} = E_N^{kin} - E_{el} \tag{10}$$

where
$$E_N^{kin} = \sum_N \frac{1}{2} M_N \dot{\vec{R}}_N^2$$

is the kinetic energy of the nuclei, M_N and $\vec{R}_N$ are nuclear masses and positions, respectively, and the electronic energy E_{el} is their potential energy, i.e., E^{DFT}, given by **Eq. 3** (we are assuming throughout this section that the Born–Oppenheimer approximation holds, i.e., the electrons are moving in the field of fixed nuclei).

The basic FPMD procedure consists in repeating two main steps: 1) For a given set of atomic coordinates $\{ \vec{R}_N \}$, find the total energy E^{DFT}. 2) Solve Newton's equations of motion (**Eq. 9**). This procedure is illustrated in **Fig. 1**.

The basic difference between FPMD and classical MD lies in the way the interatomic energy is obtained. In standard MD (*see* Chapter 23) E_{el} is computed from a parametrized energy expression that depends on the structural properties of our system (atomic positions, bond distances, angles, ...) as variables. Instead, in FPMD the interatomic energy (i.e., E^{DFT}) is obtained from quantum mechanics and depends on the atomic positions and the electron density. Nevertheless, the integration of Newton equations of motion (**Eq. 8**) to update the atomic positions at each time instant is performed using similar techniques ***(64)*** as in standard MD.

A very elegant and efficient approach to perform FPMD was introduced by Car and Parrinello in 1985 ***(6)***. Rather than solving **Eqs. 1** and **8** separately, the authors introduced a generalized fully classical Lagrangian for both electrons and nuclei,

$$\mathcal{L} = E_N^{kin} + E_{el}^{kin} - E^{KS} + \Sigma_{ij} \Lambda_{ij} \left[\int d\vec{r} \psi_j^*(\vec{r}) \psi_j(\vec{r}) - \delta_{ij} \right] \tag{11}$$

where $E_{el}^{kin} = \sum \mu \int d\vec{r} |\dot{\psi}_i(\vec{r})|^2$ is a *fictitious* classical kinetic energy term associated with the electronic subsystem [$\psi_i(\vec{r})$], μ is is a parameter that controls the timescale of the electronic motion and Λ_{ij} are Lagrangian multipliers that impose the orthonormality constraints between the orbitals. E_{el} is the electronic energy (i.e., the KS energy, given by **Eq. 4**). The total energy of the CP Lagrangian is given by $E_{tot}^{CP} = E_{el}^{kin} + E_N^{kin} + E^{KS}$ and it is a constant of motion. The corresponding equations of motion are,

$$\mu \ddot{\psi}_i = \frac{\delta E^{KS}}{\delta \psi_i^*} + \Sigma_j \Lambda_{ij} \psi_j(\vec{r}) \tag{12a}$$

$$M_N \ddot{\vec{R}}_N = \frac{\partial E^{KS}}{\partial \vec{R}_N} \tag{12b}$$

The integration of the coupled **Eqs. 12a** and **12b** provides the time evolution of not only the atomic positions [$\vec{R}_N(t)$] but also the KS orbitals [$\psi_i(\vec{r},t)$]. In practice, the orbitals are expanded in a basis set (*see* **Subheading 1.2.5.**) and what is obtained from the integration is the value of the expansion coefficients at each time instant.

Therefore, in a CP simulation both electrons and nuclei are evolved simultaneously. It can be demonstrated that, provided that the electrons are initially in the ground state, they will follow adiabatically the nuclear motion, remaining very close to the instantaneous ground state ***(65)***. From this point of view, the CP method is a procedure to describe computationally what occurs in reality, which is that electrons follow the nuclear motion ***(63)***. In a CP simulation, the electronic energy only needs to be calculated at the beginning of the simulation, and the KS orbitals evolve, following the nuclear motion, as the simulation proceeds.

The electronic energy obtained at a given instantaneous structure { $\vec{R}_N$ } generally differs slightly from the exact DFT energy. However, if the energy exchange between the electronic and nuclear subsystems is small, the trajectory generated will be identical to the one obtained in a standard FPMD simulation *(63)*. This decoupling of the two subsystems can be achieved by a suitable choice of the fictitious electronic mass μ (*see* **Note 3**). As the time needed for energy equipartition between electrons and nuclei is larger than physical nuclear relaxation times, meaningful statistical averages can be obtained from the trajectories (*see* **refs.** ***63***, ***66***, and ***67*** for reviews of the CP method).

2. Materials

1. Protein structures: The structures of oxymyoglobin and catalase with bound formate can be taken from the PDB database: http://www.rcsb.org/pdb (entry codes 1A6M and 1QWM, respectively).
2. DFT program: The Car–Parrinello Molecular Dynamics (CPMD) program and the utility *cpmd2cube* can be downloaded from http://www.cpmd.org.
3. Additional software: A visualization software such as Visual Molecular Dynamics (VMD) (http://www.ks.uiuc.edu/Research/vmd) is needed, as well as a graphics software (e.g., GRACE, http://plasma-gate.weizmann.ac.il/Grace).
4. Hardware: The CPMD program runs in a number of different platforms and operating systems (Linux PC, SGI, DEC, IBM-SP4, etc.). The examples described here have been tested in a single processor Linux Pentium IV, as well as in a SGI Origin R12000 with a minimum of 1.2 GB RAM. The duration of the simulations depends on the number of processors used (*see* **Note 4**).
5. Pseudopotentials: A pseudopotential library is available at the following URL address: http://www.unizh.ch/pci/ (Jürg Hutter).

3. Methods

The procedures described for each example consist of 1) the construction of a computational model, 2) the building of the input files, 3) the calculation procedure, 4) the data collection and analysis, and 5) the discussion of the results.

Input files for all calculations described here are available at the following URL address: http://www.pcb.ub.es/sqpbio/downloads.

3.1. Example 1: Binding of O_2 to the Myoglobin Active Center

The first example illustrates the complexation of O_2 to myoglobin **(Fig. 2)**, starting from a situation in which the unbound ligand is close to the heme active center. In this situation, the interaction between the ligand and the Fe^{2+} atom becomes the most relevant, and the computational model can be reduced to an oxygen molecule interacting with an iron–porphyrin (FeP). The calculation will involve a structural relaxation until the O_2 ligand binds to the Fe atom.

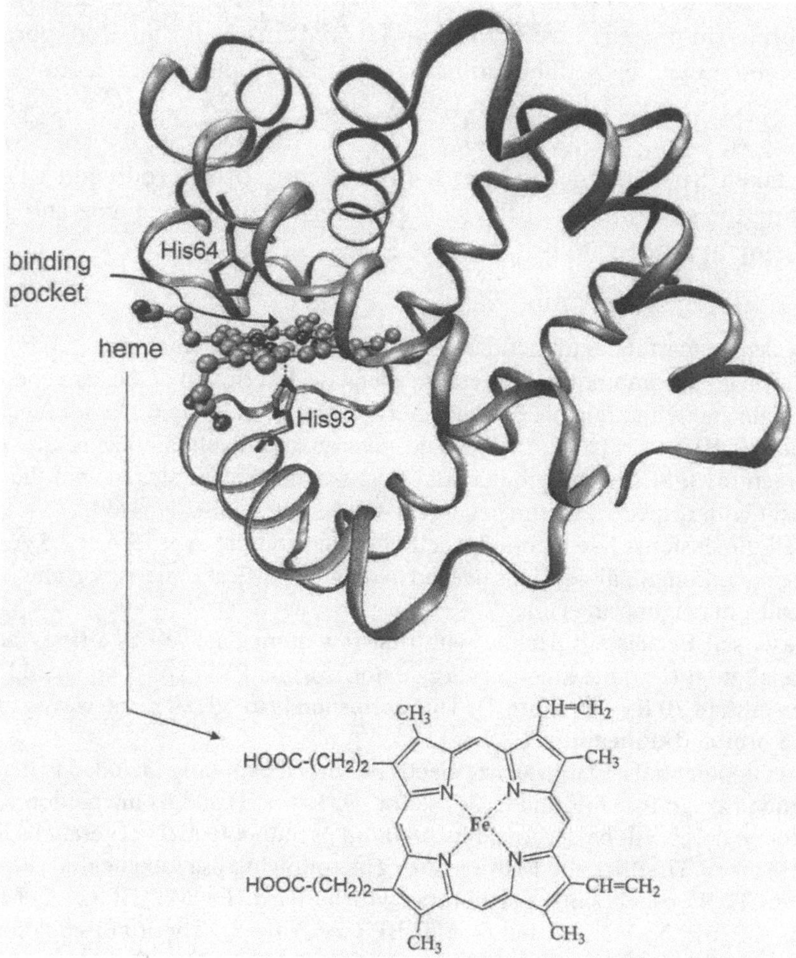

Fig. 2. Structure of myoglobin showing the ligand binding pocket. The active center (heme) is represented in ball and stick. The two histidines close to the heme (His93 in the proximal side and His64 in the distal side) are represented in sticks. Only His93 is covalently bound to the iron atom.

The electronic structure of the final FeP–O_2 complex will be analyzed, in relation to the properties of the unbound species.

3.1.1. Model Building

The starting configuration will consist of an oxygen molecule separated by 3Å from the iron atom (the distance is measured from the center of mass of the oxygen molecule) and with the O–O axis in a parallel plane with respect to the iron–porphyrin plane (**Fig. 3A**). The initial coordinates for the iron–porphyrin and oxygen molecules will be built from their DFT optimized structures in the gas phase (O–O = 1.23 Å, Fe–N = 1.98 Å, N–C^{α} = 1.39Å, C^{α}–C^{β} = 1.44 Å, C^{β}–C^{β} = 1.36 Å, C^{α}–C^{γ} = 1.38 Å, C–H = 1.09 Å). Alternatively, the FeP structure can be taken from the most recent X-ray structure of Mb (pdb entry 1A6M), extracting the coordinates of the iron–porphyrin part of the heme and adding the missing hydrogen atoms (*see* **Note 5**).

3.1.2. Building of the Input Files

1. Exchange–correlation functional: The use of the BP86 functional ***(55,56)*** represents a good compromise between an adequate description of the energy ordering of spin states and reliable binding energy values (*see* **Note 6**). Other functionals such as BLYP, B3LYP, PBE would give quantitatively similar results for the structures ***(68)*** and the ground spin state (an open-shell singlet) but the agreement with respect to binding energies will be just qualitative ***(69)***.
2. Cell dimensions: Use a periodic cell of dimensions 15 Å × 15 Å × 7.5 Å. These are the minimum dimensions needed to avoid significant interaction among molecules in neighboring cells.
3. Basis set: Because of dealing with first-row atoms, as well as a first transition metal atom (Fe), the kinetic energy cutoff for the PW expansion should be not lower than 70 Ry (*see* **Note 7**). This corresponds to 56329 plane waves for each KS orbital (**Subheading 1.2.5.**).
4. Pseudopotentials: Only valence electrons will be explicitly included in the calculation (2s, 2p for O, N and C, 3d, 4s for Fe, 1s for H) and its interaction with the atomic cores will be described by *ab initio* pseudopotentials generated using the scheme of Troullier and Martins ***(70)***. The following pseudopotential files (from the CPMD pseudopotential library) will be used: Fe_MT_BP.cc, C_MT_BP, H_MT_BP, N_MT_BP and O_MT_BP (*see* **Note 8**). The iron pseudopotential should include the nonlinear core correction ***(71)*** (*see* **Note 9**)
5. Scaling of the nuclear velocities: An annealing factor between 0.900 and 0.997 to scale the nuclear velocities at each MD step is sufficient to bring the system towards its minimum energy structure.
6. Timestep and μ: Use a time step of 6 atomic units (a.u.) and set the fictitious mass of the CP Lagrangian to 700 a.u. For an energy minimization, these two parameters are not critical because the total energy does not need to be conserved. Similar values in the range μ = 500–1000 a.u., Δt < 6 a.u. would lead to the same optimized structure. The only difference being the total number of MD steps and the temperature variations during the simulation procedure.
7. Spin: Use the LSD with a spin multiplicity of 1 (M = *1*) for FeP–O_2. The spin multiplicity is defined as 2S+1, where S is the spin quantum number (S = 0 in this case). Setting M = *1* allows to describe both the starting configuration (in which both O_2 and FeP have triplet ground states) and the final FeP–O_2 complex which has an open-shell singlet ground state. In order to compute the ligand binding

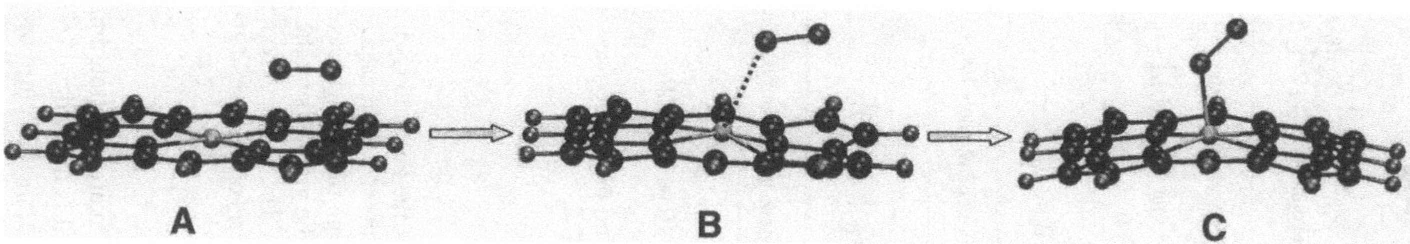

Fig. 3. Selected steps of the optimization of the FeP-O_2 complex. (**A**) initial structure, (**B**) structure after 50 fs of annealed MD, and (**C**) final optimized structure.

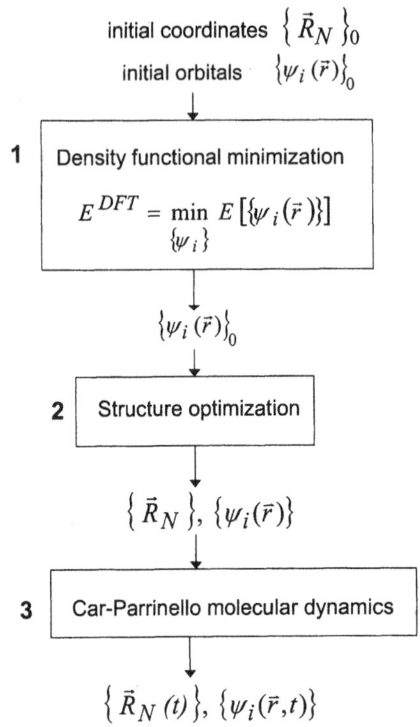

Fig. 4. Main steps of the calculation procedure used in examples 1 (only **steps 1** and **2**), 2, and 3.

energy, separate calculations for the FeP and O_2 fragments should also be performed, considering a triplet state ($M = 3$) in each case.

8. Saved data: In order to avoid large file storage, save the trajectory file (xyz format) every 10 MD steps. Print the gradients on the atoms every 100 steps for a better control of the simulation. The simulation can be stopped when both the maximum component and the norm of the atomic gradients are lower than 1E–04 a.u. (i.e., 0.019 eV . Å^{-1}).

3.1.3. Calculation Procedure

The basic steps of the calculation are illustrated in **Fig. 4** (**steps 1** and **2**). In the first step, the density functional is minimized (*see* **Note 10**). Once the system is in its electronic ground state, the structure optimization can be initiated (*see* **Note 11**). One way to do it is by means of a molecular dynamics simulation with annealing of the nuclear velocities, following the guidelines of **Subheading 3.1.2.** (**step 5**). The simulation should be allowed to run for a total

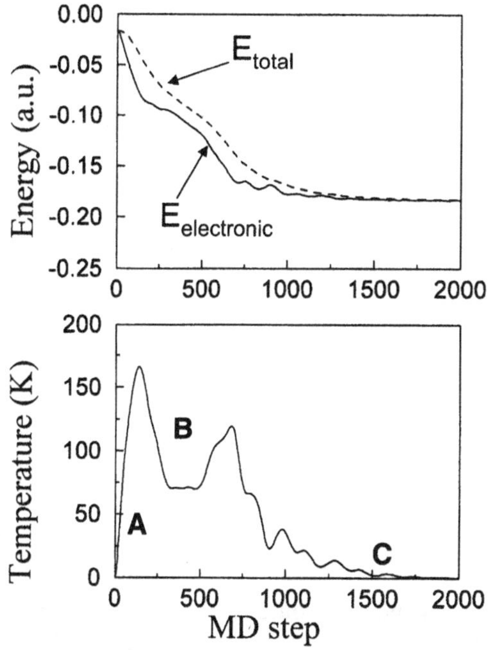

Fig. 5. Variation of the total energy, the KS energy (E^{DFT}) and the nuclear temperature during the annealed MD procedure used to optimize the molecular structure of the heme-ligand complex (FeP–O_2).

time of about 300 fs (i.e. ≈ 2000 MD steps). In an eight processor SGI-Origin R12000 this calculation needs 35 h CPU time.

To compute the binding energy of the Fe–O_2 bond, two additional calculations are needed, one for FeP and another one for O_2, both in their optimized structure (*see* **Subheading 3.1.1.** for the coordinates to be used) and triplet spin state (i.e., defining a spin multiplicity of 3 in the input file). The simulation box should be the same in all cases (*see* **Subheading 3.1.2., step 2**).

3.1.4. Data Collection and Analysis

1. *Energy and temperature evolution:* Plot the temperature, the KS energy and the total energy as a function of time (*see* **Note 12**). Check that the total energy decreases until it converges to a minimum value. As shown in **Fig. 5**, the temperature initially increases (because the structure is far from the minimum) reaching a maximum of ≈ 170 K at ≈ 200 MD steps. After ≈ 700 MD steps (140 fs) the temperature decreases smoothly until it reaches 0 K. At this point the atomic gradients are very small (root mean square [r.m.s.] ≤ 10E–03 a.u.).

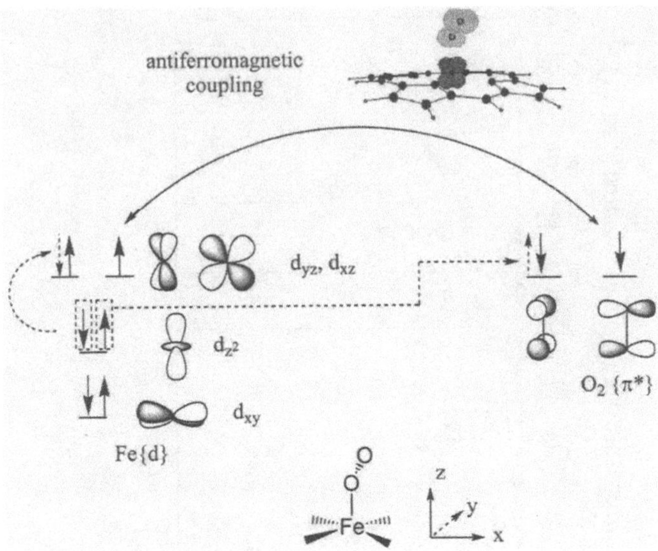

Fig. 6. Qualitative picture of the electronic rearrangements taking place upon O_2 binding to the iron-porphyrin that lead to the birradical state. Only the iron d-orbitals and the π* orbitals of the O_2 molecule are shown. The top-right picture is the spin-density distribution of the $FePO_2$ complex. This distribution reflects the antiferromagnetic coupling between one electron in the Fe-d_{xz} orbital orbital and one electron in a ligand π* orbital.

2. *Structure changes*: Visualization of the trajectory file shows that the ligand starts approaching the iron atom until it binds to it in an angular orientation. Simultaneously, the Fe atom moves out of the porphyrin plane and the iron–porphyrin curves (**Fig. 3**). The final optimized structure is defined by: Fe–O = 1.74 Å, O–O = 1.28 Å, < Fe–O–O = 123º, Fe–N = 2.02 Å, 1.99 Å.
3. *Binding energy:* To compute the binding energy of the Fe–O_2 bond, subtract the energy of the FeP–O_2 complex from the sum of the energies of the isolated fragments (the O_2 ligand and the FeP). The result of this calculation is 9 kcal/mol (*see* **Note 13**).
4. *Spin density:* Plot the spin density distribution at the final optimized structure (*see* **Note 14**). Two surfaces of opposite spin centered in the Fe atom and in the O_2 molecule are observed.
5. *Bonding analysis:* In order to analyze the nature of the Fe–O_2 bond, it is useful to plot the higher occupied spin-orbitals of the FeP(O_2) complex, and classifying them as being either of Fe, O_2, or porphyrin character. This analysis gives a total number of five d-electrons for the iron atom, while it has six in an isolated heme (Fe^{2+}). This is because one electron has been transferred from a Fe d-orbital to a π* orbital of the O_2 molecule. In fact, there has been a more complex electron rearrangement (**Fig. 6**): one of the two electrons being originally in the iron dz^2

Table 1
Computed Structure and Binding Energy (DE) of the Heme–Oxygen Bond in the FeP–O_2 and FeP(Im)–O_2 Models

	FeP-O_2 open-shell	FeP(Im)-O_2 open-shell	FeP(Im)-O_2 closed-shell	exp. (i)	exp. (ii)
Fe-O	1.74	1.77	1.74	1.75 / 1.89	1.81
O-O	1.28	1.30	1.29	1.2 / < 1.22	1.24
< Fe-O-O	123	121	122	< 129 / 131	122
Fe-N	2.02 – 1.99	2.02 – 2.01	2.02 – 2.01	1.98 / 1.99	2.01
Fe-N_ε	—	2.08	2.08	2.07 / 2.12	2.06
DE	–9.0	–15.0	–13.5	–14.3 ± 0.5[a]	–15.3 to –21.0[b]

Distances are in angstroms, angles in degrees, and energies in kcal/mol. The experimental values correspond to the X-ray structures of the oxyheme models Fe(T_{piv}P)(1-MeIm)(O_2) *(70)* and Fe(T_{piv}P)(2-MeIm)(O_2) *(71)* (i) and oxymyoglobin (ii).
[a] Enthalpy of O_2 binding to FeT_{piv}P(1,2-Me_2Im) in toluene solution, from **ref. *76*.**
[b] Enthalpy of O_2 binding to MbO_2, from **ref.** *77*.

orbital moves to a d_π orbital (d_{xz} or d_{yz}), while the other is transferred to a π^* orbital of the O_2 molecule. The resulting unpaired electrons on Fe and O_2 are coupled antiferromagnetically, i.e., Fe $\{d_\pi\}^\uparrow$– O_2 $\{\pi^*\}^\downarrow$.

3.1.5. Discussion

The above results show that a simple FeP–O_2 model already reproduces the main features of the Fe–O_2 bond in MbO_2 and HbO_2, as well as synthetic analogues **(Table 1)** *(72–75)*. This suggest that the essential structure of the FeO_2 bond is not influenced by the heme pocket. In contrast, the Fe–O bond strength (9 kcal/mol) is far from the experimental estimates (between –14.3 and –21.0 kcal/mol) *(76,77)*. We will see in the next example that the missing proximal His residue **(Fig. 2)** is the main determinant of this discrepancy *(32)*.

The electronic structure of the FeP–O_2 complex is particularly interesting: despite the $M = 1$ multiplicity, it is an open shell structure, as evidenced by the spin density distribution (**Fig. 6**, inset). The vanishing integrated spin density (*see* **Subheading 1.2.2.**) is the result of the antiferromagnetic coupling of two regions of opposite spin, centered on the Fe and on the oxygen molecule. The integrated spin density in each of these two regions is approximately one electron, located in the d_{xz} and π^* orbitals of the Fe atom and the O_2 molecule, respectively. This result is not unexpected, given the open shell nature of the interacting molecules (both the iron–porphyrin and the oxygen ligand have triplet ground states, i.e., $M = 3$) and the relatively weak bond between them.

Precisely on the basis of those two considerations, the antiferromagnetism of heme was already proposed by Weiss back in the 1960s *(78)*. The Weiss picture, which describes the bonding as $Fe^{III}–O_2^-$, has been competing for many years with the picture proposed by Pauling *(79)*, based on a $Fe^{II}–O_2$ scheme. From the experimental point of view this issue is still controversial. Several spectroscopic measurements in hemeproteins and synthetic models (Mössbauer *[80]* and optical spectra) have been interpreted in terms of the Weiss description, whereas NMR data strongly points to a diamagnetic state, with no unpaired spins, over a wide range of temperatures (*see* **ref.** ***81*** for an overview). On the theoretical side, this issue is also controversial. Until 1997, all theoretical studies on this problem were done at a fixed structure using mainly semiempirical and HF methods, although some studies based on CI and CASSCF methods using small active spaces were also performed (*see* for instance **refs.** ***82*** and ***83***). All these studies concluded on a closed-shell ground state for oxyheme (i.e., with no unpaired spins). Later on, DFT calculations optimizing the molecular structure gave evidence for an open-shell singlet ground state *(33,84)*. As illustrated in the previous section (point 5) the DFT analysis of the FeO_2 bond leads to a $Fe^{III}–O_2^-$ description, thus supporting Weiss model.

3.2. Example 2: Dynamics of the Bound O_2

In this example, *ab initio* molecular dynamics simulation of the bound O_2 will be performed in order to explore the conformational flexibility of the ligand and its preferred orientations at room temperature.

3.2.1. Model Building

The model of example 1 will be extended with an imidazole molecule bonded to the iron atom in order to obtain a more quantitative description of the heme–ligand bond. The axial imidazole mimics the effect of the *proximal histidine* residue (His93 in **Fig. 2**). We anticipate here that, even though this residue does not change the characteristic structure and electron distribution of the heme–ligand bond, it affects significantly its strength.

The starting coordinates for the FeP(Im)–O_2 system can be taken from the final structure of FeP–O_2 (**Subheading 3.1.4.**) adding an imidazole ring coordinated to the Fe atom via the N_ε atom, at a distance of 2.06 Å. This is the Fe–N_ε distance reported in the most recent X-ray structure of Mb (pdb entry 1A6M) *(72)*. The imidazole internal coordinates can be taken as: C–C = 1.37 Å, N–C = 1.38 Å, C–H = 1.08 Å, N–H = 1–02 Å. For the sake of simplicity, here we will consider only the case in which the ligand does not interact with distal residues. The intrinsic dynamics of the ligand will be analyzed and possible effects of the protein environment will be discussed.

3.2.2. Building of the Input Files

3.2.2.1. Structure Optimization

1. Follow the description of **Subheading 3.1.2.** to build the input for the structure optimization.
2. Perform separate calculations for the closed-shell and the open-shell singlet states (i.e., using the LSD approximation in the latter).

3.2.2.2. Molecular Dynamics

1. Because the potential energy surface for both singlet states turns out to be very similar *(36)* (*see* later), the molecular dynamics simulation will be performed on the closed-shell surface. This reduces considerably the computational cost of the calculation and avoids possible problems of surface crossing (*see* **Note 15**). The coordinates and KS orbitals of the optimized structure of the closed-shell singlet will be used to start the MD simulation.
2. Use an integration time step (Δt) of 0.12 fs (5 a.u.), with the fictitious electronic mass of the CP Lagrangian set to 700 a.u. (*see* **Note 3**).
3. Set the mass of the hydrogen atoms to 2 a.m.u. This allows using a larger time step for integrating the equations of motion (by decreasing the value of the largest nuclear frequency, in our case the C–H stretching) and still increase μ.
4. Use a periodic supercell of dimensions 16 Å x 16 Å x 20 Å.
5. Fix the porphyrin carbon atoms connecting the four pyrrole rings. This will partially account for the sterical restrictions introduced by the chemical groups attached to the porphyrin in the real system (i.e., the protein and synthetic analogs).
6. In order to avoid large file storage, save the trajectory file every 10 MD steps. Save also the trajectory file in xyz format.
7. Start with a temperature of 630 K and let the system to evolve freely. After $\approx$ 500 MD steps the temperature will oscillate around 300 K (*see* **Note 16**).

3.2.3. Calculation Procedure

The main steps of the calculation procedure are illustrated in **Fig. 4**.

1. Minimize the density functional to obtain the energy and electron density at the initial atomic positions.
2. Starting from the previous electron density, optimize the structure. About 1000 MD steps of annealed MD would be needed to relax completely the structure. During this procedure, it is useful to decrease the annealing factor progressively (e.g., + 0.05 every 200 steps).
3. Starting from the coordinates and electron density of the closed-shell state, perform a molecular dynamics simulation for at least 10–12 ps (30,000 – 40,000 MD steps). The first 1.5 ps would be taken as equilibration time and not used for the analysis.

3.2.4. Data Collection and Analysis

1. Visualize the final optimized structures and list the structural parameters, in comparison with the model used in the previous example **(Table 1).** Check that the

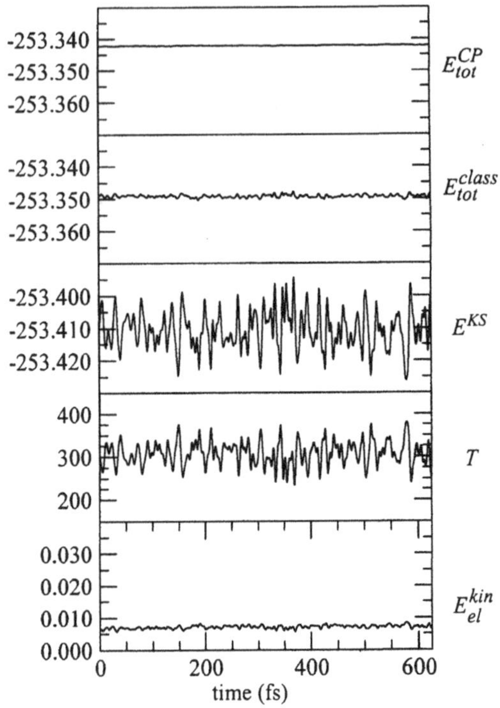

Fig. 7. Variation of the nuclear temperature and the different energy components (*see* **Note 12**) during a 600 fs window of the molecular dynamics run.

optimized structure and total energy are very similar regardless of the type of singlet state. It is also useful to compare the energies for different orientations of the O_2 ligand and the imidazole molecule. The energy difference between both singlet states is less than 1.5 kcal/mol.

2. Plot the electronic kinetic energy (E_{el}^{kin}), the temperature (T), the KS energy (E^{KS}), the total energy (E_{tot}^{CP}), and the classical energy (as a function of time in the MD simulation (*see* **Note 12**). Check that the total energy is constant and that the electronic kinetic energy remains constant and very small **(Fig. 7)**. The temperature should fluctuate around an average value of ≈ 300 K.
3. Visualize the trajectory of the MD simulation. Oscillatory rotations of the ligand around the Fe–O bond can be observed. During these rotations, the projection of the O–O bond on the average porphyrin plane remains within one porphyrin quadrant for a few picoseconds (≈ 2–4 ps), but it eventually moves to a different quadrant.
4. As a way to characterize these transitions, monitor the cosine of the N_{Im}–Fe–O–O dihedral angle. This angle defines the orientation of the O_2 with respect to the porphyrin plane as depicted in **Fig. 8**.

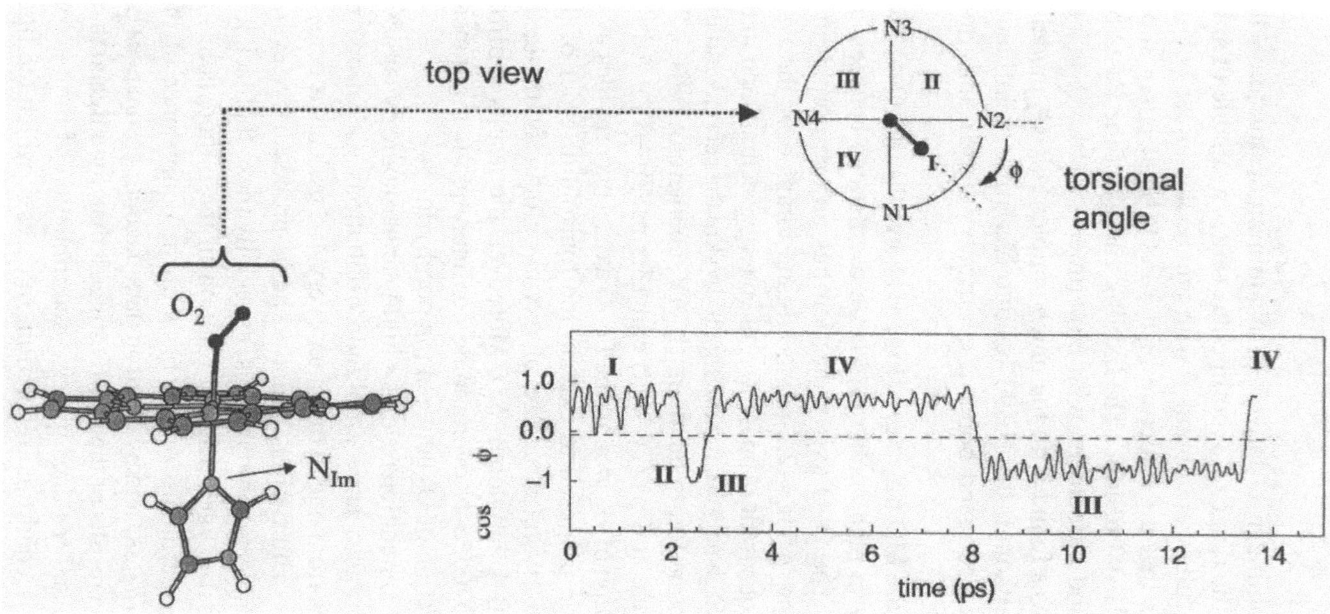

Fig. 8. Time evolution of the cosine of the N_{Im}–Fe–O–O dihedral angle in the FeP(Im)–O_2 complex. This torsional angle defines the position of the O–O bond projected on the porphyrin plane (top picture). The four quadrants of the porphyrin are labeled as I–IV.

3.2.5 Discussion

The structure optimization converges to a minimum structure whose parameters fall within the range of experimental values available (**Table 1**). The fact that similar results were obtained with the much simpler FeP–O_2 model (**Table 1**) suggests that the structure of the FeO_2 bond is not substantially influenced by the His93 residue. The binding energy of the Fe–O bond (–15 kcal/mol) is in good agreement with experiments (–14.3 ± 0.5 kcal/mol) ***(76,77)***, unlike it was found with the much simpler FeP–O_2 model (–9 kcal/mol). Therefore, although the His93 residue has a minor influence on the structure of the heme–ligand bond, it contributes significantly to its bond strength.

The analysis of the MD trajectory (**Fig. 8**) shows that the O–O axis projection on the porphyrin plane undergoes large oscillations between the Fe–N_1 and Fe–N_2 bonds. The O–O axis projection lies on one porphyrin quadrant during the first 2 ps. After ≈ 2.2 ps, the ligand jumps over the Fe–N_2 bond toward the second quadrant. Apparently, the energy accumulated in the Fe–O_2 rotational mode is high enough for the ligand to skip the second and third quadrants and end up in the fourth quadrant (I→IV, counterclockwise). Two more transitions take place at 8 ps (IV→III) and 13.5 ps (III→IV). All transitions take place via rotation of O_2 around the Fe–O axis. This provides evidence for the dynamic motion proposed to explain the fourfold disorder of the Fe–O_2 bond found in the crystal structure of oxymyoglobin synthetic analogues ***(31,74,75)*** and in β–hemoglobin ***(85)***. Moreover, the results confirm the hypothesis that the O–O/Fe–N overlapping configuration is the transition state for the dynamic motion of O_2 between the porphyrin ***(86)***.

In summary, the simulation reveals a highly anharmonic dynamics of the O_2 ligand, which undergoes large amplitude oscillations within one porphyrin quadrant and jumps from one to the other every 4–6 ps. This is consistent with the highly dynamic nature of O_2 bound to heme proposed by several experiments in proteins and synthetic models, especially those that lack a hydrogen bond at the terminal oxygen. Ligand rotation in these models has been evidenced by NMR experiments on the basis of the equivalence of the pyrrole proton resonances ***(87,88)*** and by electron paramagnetic resonance (EPR) measurements ***(89)***. The results of the FPMD suggest that, for a ligand that does not interact with the distal residues, precise determination of the rate of rotation would require picosecond time resolution. In cases in which the ligand interacts with distal residues (e.g., His64), it is expected that one of the porphyrin quadrants will be preferred and the rate of rotation would be considerably lower. However, the essential features of the dynamics such as the heme–ligand rotation mechanism would remain unchanged.

3.3. Example 3: Binding of Formic Acid to Catalase

The enzyme catalase protects organisms against reactive oxygen species through its degradation of hydrogen peroxide to water and oxygen ***(90)***. Formic acid is a common inhibitor of catalases. Upon binding to the enzyme, it impedes the formation of the first reaction intermediate (the so-called *compound I* which can be detected spectroscopically), thus blocking the enzymatic process. Recently, the first crystal structure of catalase with bound formic acid was reported, showing several inhibitor molecules inside the protein, one of them being near the active center ***(42)*** **(Fig. 9)**. However, because the positions of the hydrogen atoms are not available, the protonation state of the ligand (i.e., whether it is present as formic acid or formate anion) cannot be elucidated. Furthermore, the ligand shows a different orientation in the two independent protein subunits. In subunit A it is located half-way between heme and the Asn129 and His56 catalytic residues, while in subunit B it is only close to Asn129 and His56. The fact that there are two different positions for the ligand raised the question whether two different protonation states contribute to the observed densities.

In this example, DFT will be used to get insight into the orientation of the ligand in the binding pocket. We will focus on one of the possible scenarios (the simplest one) in which the ligand is present as formic acid and it interacts only with the Asn129 and His56 catalytic residues.

3.3.1. Model Building

1. Extract the coordinates of His56, Asn129, FMT701 (or FMT702), Ser95, Thr96, and HOH5 from the pdb structure (1QWM) **(Fig. 9A)**. Use either subunit A (FMT701) or B (FMT702).
2. Simplify the residue structures: substitute His by methylimidazole, Asn by methylamide, Thr by formaldehyde and Ser by methanol.
3. Add the missing hydrogen atoms. For the formic acid ligand, add the hydrogen atom to the closest oxygen to the N_ε atom of His56 **(Fig. 9B)**.

3.3.2. Building of the Input Files

3.3.2.1. Structure Optimization

In order to release the strain of the crystal structure, a structural relaxation should be initially performed, following the guidelines of **Subheadings 3.1.2.** and **3.1.3.** As a first approximation, all heteroatoms (C, N, O), except those of the formic acid ligand, will be kept fixed in order to partially account for the steric restraints introduced by the protein.

3.3.2.2. Molecular Dynamics

1. Follow the guidelines of **Subheading 3.2.2.2.** (**steps 3**, **6**, and **7**).

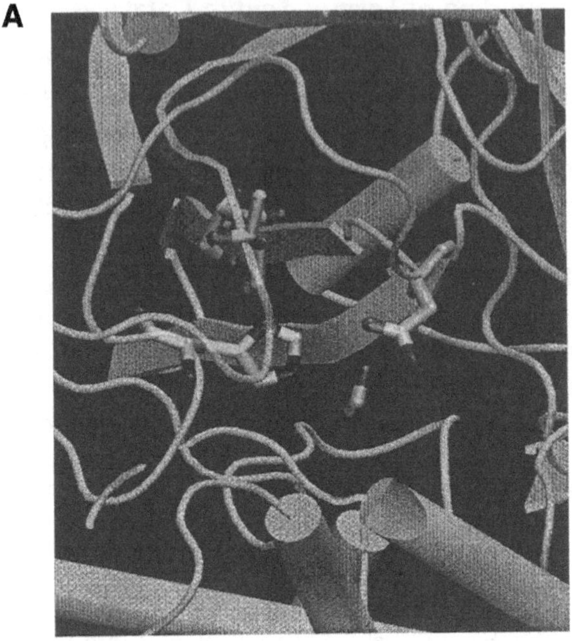

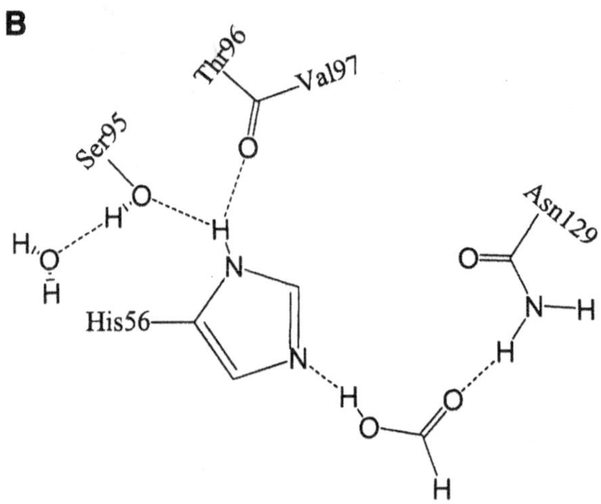

Fig. 9. **(A)** Structure of Heliobacter Pylori catalase. The His56, Asn129, Ser95, Thr96 residues and the formate ligand are highlighted. **(B)** Computational model used for the FPMD simulation.

2. Use an integration time step of 0.12 fs (5 a.u.), with the fictitious electronic mass of the CP Lagrangian set to 700 a.u.
3. Consider a periodic supercell of dimensions 15 Å × 15 Å × 12 Å.

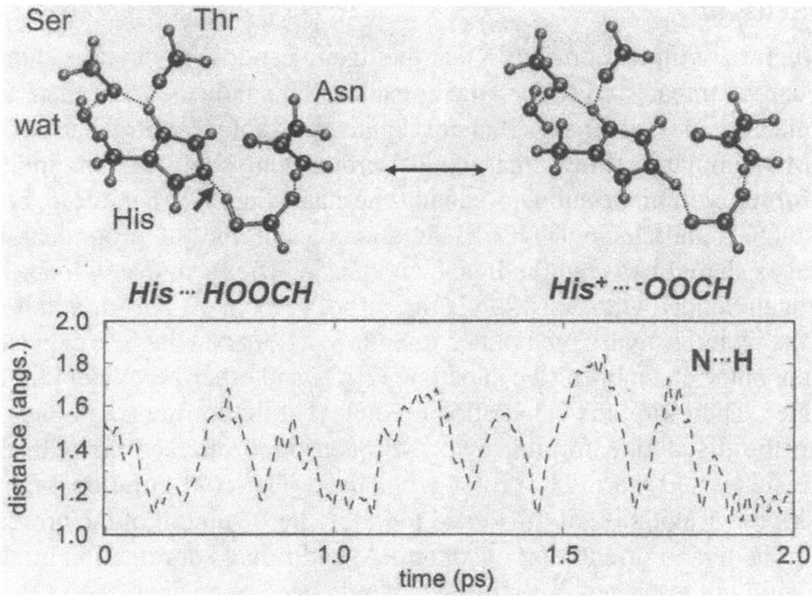

Fig. 10. Distance variation between the imidazole N_ε atom of His56 (unprotonated for the neutral residue) and the hydroxyl proton of the formic acid ligand (small arrow) during the FPMD simulation.

3.3.3. Calculation Procedure

1. In order to optimize the structure, run a MD simulation with annealing and/or quenching of the nuclear velocities (e.g., starting with an annealing factor of 0.800, raise it to 0.998 in steps of ≈ 300 MD steps). Quench the nuclear velocities after ≈ 500 MD steps.
2. Starting with the previous coordinates and electron density, run the MD simulation for at least two picoseconds (i.e., 16,667 MD steps). The first picosecond will be taken as equilibration time and not used in the analysis.

3.3.4. Data Collection and Analysis

1. Control the structure optimization: energy lowering, nuclear gradients, structure and temperature evolution. Whether the formic acid ligand retains its hydroxyl proton or transfers it to the imidazole of His56 depends on the annealing factor used and the number of times the nuclear velocities are quenched.
2. Check the energy variations during the MD simulation: total energy conservation, constant and small electronic kinetic energy, average target temperature of ≈ 300 K. Monitor the distance between the imidazole N_ε atom and the carboxyl proton (**Fig. 10**).

3.3.5. Discussion

The structure optimization shows that the ligand can lose the hydroxyl proton, as it can be transferred to the His56 residue. This indicates that there are two minima very close in energy that are separated by a low energy barrier. In fact, the MD simulation shows that the OH proton jumps between the imidazole and formic acid in the subpicosecond timescale **(Fig. 10)**. Therefore, both $His56^{+}\cdots{}^{-}OOCH$ and $His56\cdots HOOCH$ situations occur and the proton can be considered as shared between the ligand and His56. The formic acid ligand is also hydrogen bonded with Asn129 via the carboxyl oxygen. This orientation, in which the ligand is hydrogen bonded to both Asn129 and His56, is expected to block the entry channel for the substrate (H_2O_2 and other peroxides) to the active center. Therefore, the calculation predicts that the formic acid ligand is trapped in the distal side forming two hydrogen bond interactions with the catalytic residues and sharing its proton with His56. This configuration is compatible with the orientation of the ligand found in the B subunit of the protein. To explain the ligand orientation of subunit A, additional coordination modes of formic acid in the active center of HPC would need to be analyzed ***(42)***.

Note added in proof: After the completion of this chapter, a work describing the development of two new exchange-correlation functionals, named as XLYP and X3LYP, appeared ***(91)***. These functionals are found to describe reasonably well purely dispersive interactions among molecules (i.e., van der Waals interactions), unlike all previously used functionals. Therefore, this opens a promising line to investigate weak protein–ligand interactions by means of DFT.

4. Notes

1. Some currently available DFT computer programs are the following (those marked with an asterisk have ab initio molecular dynamics capabilities):

Program name	Web page	Type of basis set functions used
GAUSSIAN	www.gaussian.com	gaussians
ADF	www.scm.com	slaters
JAGUAR	www.schroedinger.com	gaussians
DeMON		numerical atom-centered functions
TURBOMOLE	www.turbomole.com	gaussians
CPMD*	www.cpmd.org	plane waves
CASTEP*	www.tcm.phy.cam.ac.uk/castep/	plane waves
PAW*	www.pt.tu-clausthal.de/~paw/	plane waves/atomic orbitals
SIESTA*	www.siesta.es	numerical atom-centered functions
VASP*	www.cms.mpi.univie.ac.at	plane waves
ABINIT*	www.abinit.org	plane waves

2. The method is also named as *ab initio* molecular dynamics (AIMD) in many publications. Nevertheless, there is a general consensus that DFT is not an *ab initio* method because most xc funcionals currently used contain external parameters. Other notations found in the literature are Car-Parrinello MD, Hellmann–Feynmann dynamics, quantum chemical MD, on-the-fly MD, direct MD, potential-free MD or quantum MD ***(63)***. Note, that in principle FPMD can be formulated in terms of any electronic structure method for the calculation of the total energy (DFT, HF, MP2, CASSCF, CI, etc.). Nevertheless, only DFT-based MD has been proven to be successful for realistic applications.
3. To save computer time a large timestep should be used. However, using a too large timestep leads to a bad integration of the equations of motion and nonconservation of the total energy. In a standard MD simulation, the optimum timestep depends on the physics of the system. The maximum timestep is determined by the highest vibrational frequency. In most biological molecules this is the C–H stretching, whose period is on the order of 10 fs. Because a vibrational period must be split into at least 8–10 parts to ensure a satisfactory integration of the e.o.m., the integration timestep should therefore be not larger than 0.5–1 fs. In a CP simulation, the highest frequencies correspond to the electronic degrees of freedom. Thus, the timestep needs to be smaller than in standard MD simulations (typical values are in the range 0.1–0.3 fs). On the other hand, the maximum electronic frequency can be shifted by varying $\mu \left(\omega_e^{\max} = \sqrt{E_{cut}/\mu}\right)$. Therefore, in a CP simulation, the optimum timestep depends sensitively on the μ used. To choose a value for μ, it is necessary to remember that this parameter decouples the electronic spectrum from the nuclear one, because it controls the separation between the minimum electronic frequency and the maximum nuclear frequency. The smallest electronic frequency, $\omega_e^{\min}$, is of the order of $\sqrt{E_{gap}/\mu}$, where E_{gap} is the HOMO–LUMO gap of the system. Because μ is a disposable parameter, one can always dynamically decouple electrons and ions for a system with $E_{\text{gap}} > 0$. Typical values for systems with a relatively large gap are μ = 500–1500 a.u. together with a timestep Δt = 5–10 a.u. (0.12–0.24 fs). Before starting the simulation, it is useful to performing short time test simulations in order to find optimum values for μ and Δt. A practical recipe is the following: (a) Start with a conservative timestep (e.g., 3 a.u.) and vary μ (e.g., 500, 600 a.u., ...) until you find the largest value that ensures adiabaticity (i.e., no drift of E_{el}^{kin} is observed). (b) Using this value of μ, increase the timestep and use the largest possible that still conserves the total energy.
4. In order to better use the computational time, it is useful to exploit the parallel capabilities of the CPMD code. For instance, in the case of example 1, one MD step takes 6.7 min in an SGI Origin–R12000 using one processor. Using 2,4, and 8 processors it takes 3.4 min, 1.8 min, and 0.85 min, respectively.
5. Hydrogen atoms can be easily added using several builder software packages such as CERIUS (http://www.msi.com). In order to avoid the large cpu time consumption in optimizing the X–H distances, it is recommended to start with typical DFT-optimized distances: C(sp^3)–H = 1.10 Å, C(sp^2)–H = 1.08 – 1.09 Å, N–H = 1.02 Å.

6. If gradient corrections are not used the Fe–ligand bonds are underestimated by 3% (they become 0.04–0.06 Å too short) and the binding energy is overestimated by more than 100% ***(84)***.
7. Because first-row elements do not have core *p* electrons, their *p* pseudopotentials are very strong and a large number of PWs is necessary to converge the calculation. The same argument holds for the *d* pseudopotentials in the first-series transition metals.
8. In the case of Fe, the *d* pseudopotential should be used as nonlocal and the *s* as local. Within CPMD, this is controlled with the keywords "LMAX=D, LOC=S, SKIP=P" (*see* manual). In all other cases, the pseudopotential function corresponding to the highest/lowest angular momentum should be taken as local/nonlocal. It is recommended to use the Kleinman–Bylander formula ***(92)*** to treat the nonlocal part of the pseudopotential.
9. When one of the core states overlap with the valence states (e.g., in the case of the Fe atom in which the semicore 2p states overlap with the valence 3d states), the pseudopotential approximation fails. However, by treating the nonlinear parts of E_{xc} explicitly it is possible to recover the results that would be obtained with a larger valence configuration. This is the basis of the so-called nonlinear core correction (NLCC) ***(71)***. In this particular case, the Fe–O bond becomes 0.15 Å too long if the NLCC is not used.
10. Because of the presence of many low-lying *d*-states and a small HOMO–LUMO gap, the resolution of the KS equations for a transition metal system is not straightforward (the energy minimization procedure often diverges). In CPMD, the density functional is minimized using direct methods for optimizations in many dimensions (in our case, the coefficients of the expansion of the KS orbitals in a PW basis set) ***(60)***. A useful trick for metal–porphyrin systems consists in performing several short-runs ($\approx$ 20–40 steps) using the conjugate gradient method (the number of steps should be decreased if performing a linear minimization along the searching direction). Once the electronic gradient is small (e.g., < 10E–03), the use of an acceleration scheme such as the DIIS (direct inversion in the iterative subspace) ***(93)*** quickly brings the system to the ground state surface.
11. Two different structure optimization approaches are available in CPMD. One of them uses the nuclear gradients and the Hessian matrix to extrapolate a new point in the potential energy surface. This method is very similar to those commonly implemented in quantum chemistry packages. The second approach is based on using the Car-Parrinello MD equations, applying a friction term to the nuclear degrees of freedom and/or the electronic ones. The resulting dynamics equations are a powerful method to simultaneously optimize the atomic structure and the KS orbitals. Within the CPMD program, this approach is referred as *annealing* of the nuclear and/or electronic velocities, because of its resemblance to the simulated annealing techniques used in classical MD calculations. Although the latter method is used here, the same results could be obtained using standard gradient/Hessian techniques.

12. In the CPMD program, this information is dumped out in the "ENERGIES" file. Columns 1–6 of this file correspond to the MD step, the electronic kinetic energy (E_{el}^{kin}), the nuclear temperature (T), the KS energy (E^{DFT}), the classical energy (i.e., $E_N^{kin} + E^{DFT}$) and the total energy of the CP Lagrangian ($E_{tot}^{CP} = E_{el}^{kin} + E_N^{kin} + E^{DFT}$). Each row in "ENERGIES" corresponds to a time interval of 0.14 fs (example I) or 0.12 fs (examples 2 and 3).
13. Because of using a non atom-centered basis set (PWs), the basis set superposition error (BSSE) is absent in CPMD. In the case of using another DFT code, a correction needs to be applied ***(94)***.
14. To obtain the spin density in the CPMD program, it is necessary to perform an additional calculation that (in the current implementation of the code, version 3.7.2) dumps out a binary file "SPINDEN". This file contains the spin density in reciprocal space and can be converted in real space (cube format) using the *cpmd2cube* utility (*see* **Subheading 2.2.**). Any software capable to read cube format files, such as VMD (http://www.ks.uiuc.edu/Research/vmd) or MOLEKEL (http://www.cscs.ch/molekel/), can be used to visualize the spin density.
15. In this case two different electronic states lie very close in energy: the closed-shell singlet, in which all electrons are paired, and the open-shell singlet, in which two unpaired electrons with opposite spin are localized on Fe atom and the O_2 molecule (**Fig. 6**). Using the LSD approximation both states are accessible and this might lead to instantaneous departures of the BO surface. In this situation, the electronic system needs to be brought to the exact BO surface very often during the simulation (i.e., reminimizing the energy functional and restarting the calculation with the previous nuclear velocities). Because the potential energy surface for both states is very similar, one way to avoid this problem is not to use LSD. This restricts the calculation to only one energy surface (the closed-shell singlet), thus minimizing BO departure.
16. When starting from a structure that is relaxed to the global minimum the temperature ends up fluctuating around a value that is approximately half of the initial temperature. This is due to equilibration between kinetic and potential energy. The temperature can also be controlled using thermostat techniques such as the well-established Nosé–Hoover method *(95)*. In this case, both approaches would lead to the same results.

Acknowledgments

The author thanks Jürg Hutter, Ignacio Fita, Enric Canadell, and Xevi Biarnés for a critical reading of this manuscript. This work was supported by Grants 2001SGR-00044 and BQU2001-04587-CO02. The computer resources were provided by the CEPBA-IBM Research Institute of Barcelona. The author also thanks the financial support from the Ramon y Cajal program of the MCYT and the ICREA foundation.

References

1. Stryer, L. (1997) Portrait of an allosteric protein, in: *Biochemistry*, Freeman, New York, NY, pp. 147–180.

2. Perutz, M. F., Fermi, G., Luisi, B., Shaanan, B., and Liddington, R. C. (1987) Stereochemistry of cooperative mechanisms in hemoglobin. *Acc. Chem. Res.* **20,** 309–321.
3. Choudhury, S. B., Lee, J.-W., Davidson, G., Yim, Y.-I., Bose, K., Sharma, M. L., Kang, S.-O., Cabelli, D. E., and Maroney, M. J. (1999) Examination of the nickel site structure and reaction mechanism in Streptomyces seoulensis superoxide dismutase. *Biochemistry* **38,** 3744–3752.
4. Koch, W. and Holthausen, M. C. (eds.) (2000) *A Chemist's Guide to Density Functional Theory*. Wiley-VCH, Weinheim, Germany.
5. Parr, R. G. and Yang, W. (eds.) (1989) *Density Functional Theory of Atoms and Molecules.* Oxford University Press, New York, NY.
6. Car, R. and Parrinello, M. (1985) Unified approach for molecular dynamics and density-functional theory. *Phys. Rev. Lett.* **55,** 2471–2474.
7. Ziegler, T. (1991) Approximate Density Functional Theory as a practical tool in molecular energetics and dynamics. *Chem. Rev.* **91,** 651–667.
8. Koch, W. and Holthausen, M. C. (2000) The Kohn-Sham approach, in: *A Chemist's Guide to Density Functional Theory.* Wiley-VCH, Weinheim, Germany, pp. 41–64.
9. Frank, I., Hutter, J., Marx, D., and Parrinello, M. (1998) Molecular dynamics in low-spin excited states. *J. Chem. Phys.* **108,** 4060–4069.
10. Filatov, M. and Shaik, S. (1998) Spin-restricted density functional approach to the open-shell problem. *Chem. Phys. Lett.* **288,** 689–697.
11. Casida, M. E. (1996) in: *Recent Developments and Applications of Modern Denisty Functional Theory.* (Seminario, J. M., ed.) Elsevier, Amsterdam, The Netherlands.
12. Kohn, W., Meir, Y., and Makarov, D. (1998) van der Waals energies in Density Functional Theory. *Phys. Rev. Lett.* **80,** 4153–4156.
13. Eltsner, M., Hobza, P., Frauenheim, T., Suhai, S., and Kaxiras, E. (2001) Hydrogen bonding and stacking interactions of nucleic acid base pairs: a density-functional-theory based treatment. *J. Chem. Phys.* **114,** 5149–5155.
14. Gervasio, F. L., Carloni, P., and Parrinello, M. (2002) Electronic structure of wet DNA. *Phys. Rev. Lett.* **89,** 108,102–108,105.
15. Aqvist, J. and Warshel, A. (1993) Simulation of enzyme reactions using valence bond force fields and other hybrid quantum/classical approaches. *Chem. Rev.* **93,** 2523–2544.
16. Laio, A., VandeVondele, J., and Rothlisberger, U. (2002) A Hamiltonian electrostatic coupling scheme for hybrid Car–Parrinello molecular dynamics simulations, *J. Chem. Phys.* **116,** 6941–6947.
17. Ordejon, P., Artacho, E., and Soler, J. M. (1996) Selfconsistent order-N density-functional calculations for very large systems. *Phys. Rev. B.* **53,** R10441.
18. Iannuzzi, M., Laio, A., and Parrinello, M. (2003) Efficient exploration of reactive potential energy surfaces using Car–Parrinello molecular dynamics, *Phys. Rev. Lett.* **90,** 238–302.
19. Siegbahn, P. E. and Blomberg, M. R. A. (1999) Density Functional Theory of biologically relevant metal centers, *Annu. Rev. Phys. Chem.* **50,** 221–249.

20. Friesner, R. A. and Beachy, M. D. (1998) Quantum mechanical calculations on biological systems. *Curr. Op. Str. Biol.* **8,** 257–262.
21. Carloni, P., Röthlisberger, U., and Parrinello, M. (2002) The role and perspective of ab initio molecular dynamics in the study of biological systems. *Acc. Chem. Res.* **35,** 45–464.
22. Himo, F. and Siegbahn, P. E. M. (2003) Quantum chemical studies of radical-containing enzymes. *Chem. Rev.* **103,** 242–2456.
23. Dudev, T. and Lim, C. (2003) Principles governing Mg, Ca, and Zn binding and selectivity in proteins. *Chem. Rev.* **103,** 773–787.
24. Dölker, N., Maseras, F., and Siegbahn, P. E. M. (2004) Stabilization of the adenosyl radical in coenzyme B_{12} – a theoretical study. *Chem. Phys. Lett.* **386,** 174–178.
25. Torrent, M., Musaev, D. G., and Morokuma, K. (2001) The flexibility of carboxylate ligands in methane monooxygenase and ribonucleotide reductase: a density functional study. *J. Phys. Chem.* **105,** 322–327.
26. Piana, S., Carloni, P., and Parrinello, M. (2002) Role of conformational fluctuations in the enzymatic reaction of HIV-1 protease. *J. Mol. Biol.* **319,** 567–583.
27. Guidoni, L. and Carloni, P. (2002) Potassium permeation through the KcsA channel: a density functional study. *Biochim. Biophys. Acta* **1563,** 1–6.
28. Sagnella D. E., Laasonen K., and Klein M. L. (1996) *Ab initio* molecular dynamics study of proton transfer in a polyglycine analog of the ion channel gramicidin A. *Biophys. J.* **71,** 1172–1178.
29. Rousseau, R., Kleinschmidt, V., Schmitt, U. W., and Marx, D. (2004) Modeling protonated water networks in bacteriorhodopsin. *Phys. Chem. Chem. Phys.* **6,** 1848–1859.
30. Stec, B. and Phillips, G. N. (2001) How the CO in myoglobin acquired its bend: lessons in interpretation of crystallographic data. *Acta Cryst.* **D57,** 751–754.
31. Collman, J. P. (1997). Functional analogs of heme protein active sites. *Inorg.Chem.* **36,** 5145–5155.
32. Ghosh, A. and Bocian, D. F. (1996) Carbonyl tilting and bending potential energy surface of carbon monoxyhemes. *J. Phys. Chem.* **100,** 6363–6367.
33. Rovira, C., Kunc, K., Hutter, J., Ballone, P., and Parrinello, M. (1997) Equilibrium geometries and electronic structure of iron-porphyrin complexes: a density functional study. *J. Phys. Chem. A.* **101,** 8914–8925.
34. Spiro, T. G. and Kozlowski, P. M. (1998) Discordant results on FeCO deformability in heme proteins reconciled by density functional theory. *J. Am. Chem. Soc.* **120,** 4524–4525.
35. Rovira, C. (2003) The structure and dynamics of the Fe–CO bond in myoglobin. *J. Phys. Cond. Mat.* **15,** 1809–1822.
36. Lim, M., Jackson, T. A., and Anfinrud, P. A. (1995) Binding of CO to myoglobin from a heme pocket docking site to form nearly linear Fe–C–O. *Science* **269,** 962–966.
37. Rovira, C. and M. Parrinello. 2000. Harmonic and anharmonic dynamics of Fe-CO and Fe–O_2 in heme models. *Biophys. J.* **78,** 93–100.
38. Kaupp, M., Rovira, C., and Parrinello, M. (2000) Density functional study of ^{17}O NMR chemical shift and nuclear quadrupole coupling tensors in oxyheme model complexes. *J. Phys. Chem. B.* **104,** 5200–5208.

39. Franzen, S. (2002) Spin-dependent mechanism for duatomic ligand binding to heme. *Proc. Nat. Acad. Sci.* **99,** 16,754–16,759.
40. De Visser, S. P., Shaik, S., Sharma, P. K., Kumar, D., and Thiel, W. (2003) Active species of horseradish peroxidase (HRP) and cytochrome P450: two electronic chameleons. *J. Am. Chem. Soc.* **125,** 15,779–15,788.
41. Martí, M. A., Scherlis, D. A., Doctorovich, F. A., Ordejón, P., and Estrin, D. A. (2003) Modulation of NO trans effect in heme proteins: implications for the activation of soluble guanylate cyclase. *J. Biol. Inorg. Chem.* **8,** 595–600.
42. Loewen, P. C., Carpena, X., Rovira, C., Ivancich, A., Perez-Luque, R., Haas, R., Odenbreit, S., Nichols, P., and Fita., I. (2004) Structure of Helicobacter pylori catalase, with and without formic acid bound, at 1.6 Å resolution. *Biochemistry* **43,** 3089–3103.
43. Mourant, J. R., Braunstein, D., Chu, K., Frauenfelder, H., Nienhaus, G. U., Ormos, P., and Young, R. D. (1993) Ligand binding to heme proteins. II. Transitions in the heme pocket of myoglobin. *Biophys. J.* **65,** 1496–1507.
44. Amara, P. and Field, M. (1998) Combined quantum mechanical and molecular mechanical potentials, in: *Encyclopedia of computational chemistry* (Schleyer, P. v. R., Allinger, N. L., Clark, T., Gasteiget, J., Kollman, P. A., Schaefer III, H. F., Schreiner, P. R., eds.) John Wiley & Sons, Chichester, UK.
45. Reuter, N., Dejaegere, A., Maigret, B., and Karplus, M. (2000) Frontier bonds in QM/MM methods: A comparison of different approaches. *J. Phys. Chem. A* **104,** 1720–1735.
46. Laio, A., VandeVondele, J., and Röthlisberger, U. (2002) A hamiltonian electrostatic coupling scheme for hybrid Car-Parrinello molecular dynamics simulations. *J. Chem. Phys.* **116,** 6941–6947.
47. Magistrato, A., DeGrado, W., Laio, A., Röthlisberger, U., VandeVondele, J., and Klein, M. L. (2003) Characterization of the dizinc analogue of the synthetic diiron DF1 using an initio hybrid quantum/classical molecular dynamics simulations. *J. Phys. Chem. B.* **107,** 4182–4188.
48. Rovira. C., Schulze, B., Eichinger, M., Evanseck, J. D., and Parrinello, M. (2001) Influence of the heme pocket conformation on the structure and vibrations of the Fe-CO bond in myoglobin. A QM/MM density functional study. *Biophys. J.* **81,** 435–445.
49. Schöneboom, J. C., Lin, H., Reuter, N., Thiel, W., Cohen, S., Ogliaro, F., and Shaik, S. (2002) The elusive oxidant species of cytochrome P450 enzymes: characterization by combined quantum mechanical/molecular mechanical (QM/MM) calculations. *J. Am. Chem. Soc.* **124,** 8142–8151.
50. CPMD program, Copyright IBM Corp. 1990–2003, Copyright MPI für Festkörperforschung, Stuttgart 1997-2001. URL: http://www.cpmd.org.
51. Hohenberg, P. and Kohn, W. (1964) Inhomogeneous electron gas. *Phys. Rev. B* **136,** 864–871.
52. Kohn, W. and Sham, L. J. (1965) Self-consistent equations including exchange and correlation effects. *Phys. Rev. A.* **140,** 1133–1138.
53. Koch, W. and Holthausen, M. C. (2000) The quest for approximate exchange-correlation functionals, in: *A Chemist's Guide to Density Functional Theory*, Wiley-VCH, Weinheim, Germany, pp. 65–91.

54. Kriegl, J. M., Nienhaus, K., Deng, P., Fuchs, J., and Nienhaus, G. U. (2003) Ligand dynamics in a protein internal cavity. *Proc. Natl. Acad. Sci.* **100,** 7069–7074.
55. Becke, A. D. (1986) Density functional calculations of molecular bond energies. *J. Chem. Phys.* **84,** 4524–4529.
56. Perdew, J. P. (1986) Density-functional approximation for the correlation energy of the inhomogeneous electron gas. *Phys. Rev. B.* **33,** 8822–8824.
57. Lee, C., Yang, W., and Parr, R. G. (1988) Development of the Colle–Salvetti correlation-energy formula into a functional of electron density. *Phys. Rev. B.* **37,** 785–789.
58. Perdew, J. P., Burke, K., and Ernzerhof, M. (1996) Generalized gradient approximation made simple. *Phys. Rev. Lett.* **77**, 3865–3868, Erratum: (1997) *Phys. Rev. Lett.* **78,** 1396.
59. Becke, A. D. (1993) A new mixing of Hartree–Fock and local density-functional theories. *J. Chem. Phys.* **98,** 5648–5652.
60. Payne, M. C., Teter, M. P., Allan, D. C., Arias, T. A., and Joannopoulos, J. D. (1992) Iterative minimization techniques for ab initio total-energy calculations: molecular dynamics and conjugate gradients. *Rev. Mod. Phys.* **64,** 1045–1097.
61. Kendall, R. A., Dunning, T. H., Jr., and Harrison, R. J. (1992) Electron affinities of the first-row atoms revisited. Systematic basis sets and wave functions. *J. Chem. Phys.* **96,** 6796–6806.
62. Pickett, W. E. (1989) Pseudopotential methods in condensed matter applications *Comput. Phys. Rep.* **9,** 115–197.
63. Marx, D. and Hutter, J. (2000) Ab initio molecular dynamics: theory and implementation, in: Modern *methods and algorithms of Quantum Chemistry* (Grotendorst, J., ed.), John von Neumann Institute for Computing, Julich, Germany, pp. 301–409.
64. Verlet, L. (1967) Computer "experiments" on classical fluids. I. Thermodynamical properties of Lennard-Jones molecules. *Phys. Rev.* **159,** 98–103.
65. Car, R., Parrinello, M., and Payne, M. (1991) Comment on "error cancelation in the molecular dynamics method for total energy calculations." *J. Phys. Cond. Matt.* **3,** 9539–9543.
66. Remler, D. K. and Madden, P. A. (1990) Molecular dynamics without effective potentials via the Car–Parrinello approach. *Mol. Phys.* **70,** 921–966.
67. Tse, J. S. (2002) *Ab initio* molecular dynamics with density functional theory. *Annu. Rev. Phys. Chem.* **53,** 24–290.
68. Scherlis, D. A. and Estrin, D. A. (2002) Structure and spin-state energetics of an iron–porphyrin model: An assessment of theoretical methods. *Int. J. Quant. Chem.* **87,** 158–166.
69. Jensen, K. and Ryde, U. (2003) Theoretical prediction of the Co–C bond strength in cobalamins. *J. Phys. Chem. A.* **107,** 7539–7545.
70. Troullier, M. and Martins, J. L. (1991) Efficient pseudopotentials for plane-wave calculations, *Phys. ReV. B* **43,** 1993–2006.
71. Louie, S. G., Froyen, S., and Cohen, M. L. (1982) Non-linear ionic pseudopotentials in spin-density-functional calculations, *Phys. Rev. B.* **26,** 1738–1742.

72. Vojtechovsky, J., Chu, K., Berendzen, J., Sweet, R. M., and Schlichting, I. (1999) Crystal structures of myoglobin-ligand complexes at near-atomic resolution. *Biophys. J.* **77,** 2153–2174.
73. Shaanan, B. (1982) The iron–oxygen bond in human oxyhaemoglobin. *Nature* **296,** 683–684.
74. Jameson, G. B., Rodley, G. A., Robinson, W. T., Gagne, R. R., Reed, C. A., and Collman, J. P. (1978). Structure of a dyoxygen adduct of (1-methylimidazole)-meso-tetrakis(α,α,α,α–o-pivalamidophenyl) porphinatoiron(II). An iron dioxygen model for the heme component of oxymyoglobin. *Inorg. Chem.* **17,** 850–857.
75. Jameson, G. B., Molinaro, F., Ibers, J. A., Collman, J. P., Brauman, J. I., Rose, E., and Suslick, K. S. (1980) Models for the active site of oxygen-binding hemoproteins. Dioxygen binding properties and the structures of (2-methylimidazole)-meso-tetra(α,α,α,α–o–pivalamidophenyl)poprhyrinatoiron(II)-ethanol and its dioxygen adduct. *J. Am. Chem. Soc.* **102,** 3224–3237.
76. Collman, J. P., Brauman, J. I., Iverson, B. L., Sessler, J. L., Morris, R. M., and Gibson, Q. H. (1983) O2 and CO binding to iron(II) porphyrins: a comparison of the "picket fence" and "pocket" porphyrins. *J. Am. Chem. Soc.* **105,** 3052–3064.
77. Wang, M. R., Hoffman, B. M., Shire, S. J., and Gurd, F. R. N. (1979) Oxygen binding to myoglobin and their cobalt analogs. *J. Am. Chem. Soc.* **101,** 7394–7397.
78. Weiss, J. J. (1964) Nature of the iron-oxygen bond in oxyhemoglobin. *Nature (London)* **202,** 83–84.
79. Pauling, L. (1964) Nature of the iron-oxygen bond in oxyhemoglobin. *Nature (London)* **203,** 182–183.
80. Bade, D., Parak, F. (1978) Mössbauer spectroscopy on oxygenated sperm whale myoglobin: Evidence for an Fe^{3+}–O_2^- coupling at the active center. *Z. Naturforsch* **33,** 488–494.
81. Momenteau, M. and Reed, C. A. (1994) Synthetic heme-dioxygen complexes *Chem. Rev.* **94,** 659–698.
82. Herman, Z. S. and Loew, G. H. (1980) A theoretical investigation of the magnetic and ground-state properties of model oxyhemoglobin complexes. *J. Am. Chem. Soc.* **102,** 1815–1821.
83. Yamamoto, S. and Kashiwagi, H. (1993) CASSCF calculation on dioxygen heme complex with extended basis set. *Chem. Phys. Lett.* **205,** 306–312.
84. Rovira, C. and Parrinello, M. (1998) Oxygen binding to iron-porphyrin: a Density functional study using both LSD and LSD+GC schemes. *Int. J. Quant. Chem.* **70,** 387–394.
85. Shaanan, B. (1983). Structure of human oxyhaemoglobin at 2.1 Å resolution. *J. Mol. Biol.* **171,** 31–59.
86. Spartalian, K., Lang, G., Collman, J. P., Gagne, R. R., and Reed, C. A. (1975) Mössbauer spectroscopy of hemoglobin model compounds: evidence for conformational excitation. *J. Chem. Phys.* **63,** 5375–5382.
87. Mispelter, J., Momenteau, M., Lavalette, D., and Lhoste, J.-M. (1983) Hydrogen-bond stabilization of oxygen in hemoprotein models. *J. Am.Chem. Soc.* **105,** 5165–5166.

88. Oldfield, E., Lee, H. C., Coretsopoulos, C., Adebodum, F., Park, K. D., Yang, S., Chung, J., and Phillips, B. (1991) Solid state ^{17}O nuclear-magnetic-resonance spectroscopic studies of [O_2-^{17}O] picket-fence porphyrin, myoglobin, and hemoglobin. *J. Am. Chem. Soc.* **113,** 8680–8685.
89. Bowen, J. H., Shokhirev, N. V., Raitsimring, A. M., Buttlaire, D. H., and Walker, F. A. (1997) EPR studies of the dynamics of rotation of dioxygen in model cobalt(II) hemes and cobalt-containing hybrid hemoglobins. *J. Am. Chem. Soc.* **101,** 8683–8691.
90. Nicholls, P., Fita, I., and Loewen, P. C. (2001) Enzymology and structure of catalases, *Adv. Inorg. Chem.* **51,** 51–106.
91. Xu, X. and Goddard III, W. A. (2004) The X3LYP extended density functional for accurate descriptions of nonbond interactions, spin states, and thermochemical properties. *Proc. Nat. Acad. Sci. USA* **101,** 2673–2677.
92. Kleinman, L. and Bylander, D. M. (1982) Efficacious form for model pseudopotentials *Phys. Rev. Lett.* **48,** 1425–1428.
93. Pulay, P. (1980) Convergence acceleration of iterative sequences. The case of scf iteration. *Chem. Phys. Lett.* **73,** 393–398.
94. Salvador, P., Paizs, B., Duran, M., and Suhai, S. (2001) On the effect of the BSSE on intermolecular potential energy surfaces. Comparison of *a priori* and *a posteriori* BSSE correction schemes. *J. Comp. Chem.* **22,** 765–786.
95. Nosé. S. (1984) A molecular dynamics method for simulations in the canonical ensemble. *Mol. Phys.* **52,** 255–268.

Index

D

FSC
www.fsc.org
MIX
Papier | Fördert
gute Waldnutzung
FSC® C083411

Kolibri 360 GmbH
Ferdinand-Jühlke-Straße 7
99095 Erfurt, Deutschland
produktsicherheit@kolibri360.de